22.00



PROBABILITY, STATISTICS, AND DATA ANALYSIS

# PROBABILITY, STATISTICS, AND DATA ANALYSIS

Oscar Kempthorne
Professor of Statistics
Distinguished Professor
in Sciences and Humanities
IOWA STATE UNIVERSITY

Leroy Folks
Professor of Mathematics and Statistics
OKLAHOMA STATE UNIVERSITY

THE IOWA STATE UNIVERSITY PRESS · AMES, IOWA, U.S.A.

Printed in the U.S.A.
First edition, 1971
International Standard Book Number: 0-8138-1285-2
Library of Congress Catalog Card Number: 74-114799

International Standard Book Number: 0-8138-2470-2



## About the Authors

Oscar Kempthorne is Distinguished Professor in Sciences and Humanities and Professor of Statistics at Iowa State University, where his field of specialization is statistical inference with special reference to experiments, design of experiments, and probability and statistics of genetic systems. He holds the Sc.D. degree from Cambridge University. He is a fellow of the American Statistical Association, the Institute of Mathematical Statistics, and the American Association for the Advancement of Science and a member of the International Statistical Institute. He has contributed to scientific journals for the past 25 years and is a frequent participant in professional statistical meetings. In 1965 he delivered the Fisher Memorial Lecture of the joint statistical societies. He has authored two books, *The Design and Analysis of Experiments* and *An Introduction to Genetic Statistics.*

John Leroy Folks is Professor of Mathematics and Statistics, Chairman of the Statistics Unit, and Director of the Statistical Laboratory at Oklahoma State University. He has served as operations research scientist and manager of operations research in industry. He holds the Ph.D. degree from Iowa State University and is a member of the American Society of Quality Control, the American Statistical Association, the Institute of Mathematical Statistics, the Biometric Society, and the International Association for Statistics in the Physical Sciences. His writings include numerous papers and journal articles, and he is associate editor for *Technometrics.*

To: Jill, Joan, and Peter (O.K.)

To: Sue (J.L.F.)

# Contents

# Preface

The authors have felt for many years that there is a need for an introductory description of the ideas of probability and statistics, which uses the elementary calculus and addresses itself not only to probability theory, of which distribution theory is a part, but to the collection and assessment of data. The problems in giving a reasonable account of the whole range of ideas involved are considerable, and we wish to discuss them briefly.

To give basic ideas on probability is not simple because of the existence of widely different usages in the whole literature. Thus probability has been used to express uncertainty of belief or knowledge, as a model for the behavior of physical systems, as a purely mathematical abstract idea, and as a basis for a theory of incomplete logical implication. Throughout, we have taken the view that probability is an intellectual construct based on the idea of relative frequency. This seems to be appropriate because our overall objective is to bring probability ideas to bear on the interpretation of observational data. We have not stressed axiomatic development of probability because the purpose of our development is to aid individuals on how to use probability ideas, how to handle data, and how to develop ideas on ways of examining data. It seems clear that axiomatization generally follows rather than precedes ideas in the development of a science, and a purely axiomatic development often loses the spirit of the subject. Probability and statistics are intrinsic to processes of the acquisition of knowledge, opinion, and belief; to suppose one can encompass this range of activities by careful choice of a few axioms which seem entirely innocent and totally compelling is naive and dangerous.

Our dominant aim is to present ideas of statistics which have direct bearing on what informative operations one can perform on data. Our treatment of probability is strongly tailored toward this aim, so that there is essentially no attention to the mathematically deep questions of the theory of probability. Furthermore, we wished to write a book on statistics and not one on the mathematics of statistics. So for our purposes, mathematics and probability theory are tools and occasionally our arguments are heuristic. The great bulk of our presentation of elemental ideas is made in terms of discrete probability spaces for which the necessary set arithmetic operations are part of present-day high school mathematics. A rigorous treatment of continuous probability requires the advanced mathematics of measure theory and is beyond our attempted level. Continuous distributions are essential, but the necessary tools in our development are basic calculus and Riemann integration, which are normally covered in freshman or sophomore university mathematics. This is possible because we are not concerned with infinite dimensional sample spaces, and the only convergence ideas used are those of convergence in distribution. In the ideas we present, the essential role of continuous distributions is either as convenient representation of situations or as approximation of sums by Riemann integrals.

Because our main target is statistics, which relates to numerical attributes of populations, real or conceptual, we have adopted the unusual, but not unique, idea of presenting ideas of the theory of distributions, averages, moments, and so on, as devices for the partial description of real populations rather than as attributes of probability structures. This has the real merit, it seems to us, of bringing to the fore an extremely useful mode of thinking about data situations, which is to visualize what would be appropriate if one had an indefinitely large amount of data. The ideas of relative frequency then lead naturally into ideas of probability.

We have devoted a large portion of the book to the subject of inference, a term which we use in a much broader sense than is traditional in statistics books. We have attempted to distinguish between the formation of opinion and decision making; for instance, we distinguish between tests of significance and tests of hypotheses. Although we cannot claim any great originality in doing so (Fisher certainly made the distinction), we felt it necessary to discuss the distinction. In fact, we are convinced that many practicing statisticians perform significance tests while continuing to give lip service to the theory of hypothesis testing. Of course, part of the difficulty lies in the connotations of the word test. A significance test is not really an accept-reject rule of inductive behavior, but an evaluation of significance, or strength of evidence. Perhaps we should have adopted a new word, but we chose not to do so. The word significance has a long history in statistics and gives the correct connotation. We did, however, choose to adopt the expression consonance interval to emphasize a distinction from confidence interval. To explain our preference for the use of the word consonance, we state emphatically that in many investigative situations we are not sampling at random from some population with a structure that is partly known. We choose to embed our data in a population of repetitions which appeals to us. We determine a model incorporating random elements which conforms with the data. To say that we have confidence in such a model seems to us a misleading use of the English language. To say, however, that the model is consonant to a stated extent with the data expresses nicely what we have done, no more or less compelling than it really should be.

Our title lumps together probability, statistics, and data analysis. This serves to indicate the general content of the book. We should emphasize, however, that much of applied statistics, e.g., baseball statistics to take a very prosaic example, is aimed solely at the building of a historical record and does not involve probability ideas at all deeply. Similarly, there are processes of data analysis that involve only summarization and presentation of data by statistics numerical and/or graphical without direct utilization of probability models. It is the case, however, that insofar as such processes are used to give ideas of unknowns in the future or past, they must incorporate probability ideas.

The level of presentation of this book is suitable for a senior or first-year graduate course in probability and statistics for students in mathematics, physical sciences, or engineering. We anticipate that it could be a good text for any undergraduate with a basic calculus prerequisite. It will be suitable as a first course for graduate students who have had no background in mathematical statistics. Also the text will lead reasonably to specialized courses and texts on sample survey, design of experiments, econometrics, psychometrics, and so on, as well as into those on mathematical theory. For those who have had considerable exposure to mathematical statistics, we hope that our presentation will cast some light on the aims and methods of individuals who have spent many years as practicing statisticians with exposure to inferential problems in many spheres of investigation.

If the teacher has to make a selection of material, he can omit Chapter 7 without loss of continuity. He can also omit Chapters 15 and 16. We believe however that such omissions should be avoided if at all possible. The ideas of stochastic processes in Chapter 7 are critical for the development of models for data. The developments in Chapters 15 and 16 discuss application of the ideas of all previous chapters to data situations that are of common occurrence, and they illustrate how the basic ideas are related to common statistical practice.

It is important to warn the reader, particularly the beginning student, that there are in probability and statistics, as in mathematics and all intellectual disciplines, deep elemental questions which are unresolved.

Perhaps we should have ignored these, but in some cases we felt that brief, even cryptic, mention should be made. To guide the reader we have indicated the degree of difficulty of cited material by a code, which we hope will save him the trouble of researching references which may be beyond his level.

It is impossible to write a book of the present type without some intrusion of the personal outlook of the authors. We have attempted to be undogmatic and to indicate carefully when our view is different from one commonly held. We have attempted to give good references to other views which are not presented at length. We have the belief that there is no single panacea, no single exactly appropriate approach to data interpretation. Most of the approaches to which we refer have some role, although we believe the one we present has the most relevance to real problems. We plead with the student to follow up some of the references given, particularly to the older literature, which discusses "modern" controversies in length. Furthermore, we suggest occasional rereading of Chapters 1, 8, and 17.

It is appropriate to mention here some debts of the authors.

Both authors acknowledge a most profound intellectual debt to R. A. Fisher. Hardly a page of the book does not carry an imprint of his ideas. To have known him personally was the fortune of one of us (O.K.) and was to have had contact with an intellectual giant.

Oscar Kempthorne is indebted to Iowa State University for a stimulating statistical environment and to the National Science Foundation and Wright-Patterson Aerospace Research Laboratories for support of research activities which have contributed substantially to his outlook on the whole area. He is also indebted to his colleagues T. A. Bancroft, H. T. David, W. Fuller, E. Pollak, G. Zyskind, and to former colleagues D. Jowett and R. J. Buehler, for ideas often contrary to his own. He has gained tremendously by association with former graduate students, particularly M. B. Wilk, R. F. White, and G. Zyskind. He has a deep debt to a large number of scientists who have shared their problems with him.

Leroy Folks is indebted to Carl E. Marshall and L. Wayne Johnson for their interest and active support at Oklahoma State University, to the Research Foundation of Oklahoma State University, and to the National Aeronautics and Space Administration for support during the intermediate stages of writing the book.

We are indebted to several senior graduate students, including G. L. Ghai, L. Jensen, P. C. Papaioannou, W. D. Warde, and E. N. West of Iowa State University and R. C. Littell and J. Murphy of Oklahoma State University for critical reading of parts of the manuscript.

The writing of this book has spanned a number of years and the authors are grateful to all those who helped with typing and retyping.

They wish to mention particularly Mrs. Alice Riggins and Mrs. Beverly Richardson who typed several versions of the manuscript and to Mrs. Dee Anna Clabaugh, Mrs. I. Zeliadt, and Mrs. N. Perry for preparation of the final manuscript. The authors thank the Iowa State University Press and particularly the editor, Nancy Schworm, for excellent understanding and cooperation.

Finally, we are indebted to the literary executor of the late Sir Ronald A. Fisher, F.R.S., Cambridge; to Dr. Frank Yates, F.R.S., Rothamsted; and to Messrs. Oliver and Boyd, Ltd., Edinburgh, for permission to reprint Table 8.6 from their book *Statistical Tables for Biological, Agricultural, and Medical Research*, to C. Hastings, Jr., for certain formulae, to the International Business Machines Company for information about pseudo-random number generation, and to several other sources noted in the tables.

PROBABILITY, STATISTICS, AND DATA ANALYSIS

# 1

# General Background

The purpose of this first chapter is to give a general setting for the material of our book. To do this we have to review several different lines of thought which converge on our overall objective—the development of some understanding of probability models, statistics, and data analysis. We suggest that an occasional rereading of this chapter may be desirable if the reader later feels that he is becoming lost in a maze of mathematical details and formulation. It does seem essential that if the present book is to make sense, the ideas and processes must be embedded understandably in a long chain of thought processes that has developed gradually for centuries.

The type of outlook on the collection and interpretation of data which is presented in this book and in the many books on probability and statistics now available is understandably a very recent development in intellectual activity. Indeed, we can say that almost all the ideas we present have come to the fore only in the past 100 years—even in the twentieth century alone, though discussions leading up to them were clearly evident in the late nineteenth century.

## 1.1 Ideas of Probability

The use of some idea of probability as an expression of uncertainty of outcome of a process and as an expression of a peculiar type of partial knowledge or uncertain opinion certainly extends back into antiquity. Aristotle, for instance, made considerable use of a word which is translated as "probable."

The initial impetus for the formulation of a definite idea of probability came, it appears, from the need of individuals to have guiding rules for gambling. The idea developed, in a very vague intuitive way, of perfect shuffling of a deck of playing cards or of a perfectly balanced die. The idea of perfect shuffling led to the idea of probability as relative frequency in a large number of trials. So, for instance, a tossing of two perfect pennies will give the results $HH$, $HT$, or $TT$ and if one makes a very large number of tosses, say $N$, then close to $N/4$ of the tosses will give $HH$, close to $N/2$ of the tosses will give $HT$ and close to $N/4$ of the tosses will give $TT$. This type of perfect gambling device led to a wide variety of gambling games and a related array of probabilities for each game. For example, in stud poker two cards are dealt initially, one down or hidden and one up or shown. A player sees his own cards and all the up cards of the other players and can compute the probability $P$ that an individual with an ace showing has another ace hidden. Any such probability is a relative frequency in the sense that if one dealt again and again with perfect shuffling, a proportion $P$ of the deals would have an ace hidden under an exposed ace. The computation of probabilities for such gambling devices has fascinated humanity, and the topic is by no means exhausted.

## 1.2 Ideas of Statistics

The original idea of statistics was the collection of information about the "state," and it is still true that by far the greatest statistical activity is the collection of facts about countries and such identifiable portions of them as the farms, the steel industry, the labor force, etc. Indeed a major problem at present is the collection of reliable statistics about food production and the population of the world. In all these areas statistics is a discipline directed toward the collection and interpretation of data about *existent* populations. Every nation has huge armies of data collectors who are quite appropriately called statisticians. Within our own country we have large agencies like the Bureau of the Census, and almost all our governmental divisions have statistical departments.

## 1.3 The Use of Probability Ideas in Existent Populations

The description of real existent populations has led to the use of probability ideas in a very simple way. A study of the sex ratio has led to our adopting such probability statements as "The probability that an unborn child will be male is around 1/2." We have probabilities of survival which enable insurance companies to determine reasonable rates for life insurance; we have statistics of road accidents which lead to rates for automobile insurance. We can continue this list indefinitely, but the basic idea is that a relative frequency observed in the past is used as a guide for risks in the future—a probability, if you like.

The ideas of probability as derived from gambling devices have an impact on our ways of representing real populations. A simple case is the observation, say, of 10 successful heart transplants in 40 trials, for which we use the idea of the mathematical binomial distribution. Another common example is the use of the normal distribution. It is doubtful that this distribution would have been discovered merely by looking at actual data. How this distribution can be derived from a simple probability model is part of the mathematical theory of probability and statistics and is covered later. The interesting and fascinating aspect of the matter in the present discussion is that this mathematical distribution has been found to be useful to a remarkable extent as a good approximate description of real populations, e.g., the distribution of human height. This illustrates how ideas of random devices and probability have an impact on modes of characterizing real populations. The distributions called normal or Gaussian dominate this usage, but a wide variety of mathematical distributions derived from simple probability models has found use.

## 1.4 Populations Made According to Specification

With the rise of experimental science we are continually in the act of forming opinions about populations that do not exist but which we can develop if we wish. We have to envisage the population of children taught by the "new math," the population of light bulbs which would result from a particular production process, and the natural population of biological organisms which would result from particular use policies for 2,4-D or DDT. In all these cases we are interested in what large populations, developed according to specified processes, would be like. The reality of such questions cannot be doubted; modern society and a modern technology is based on the answers. It appears also that modern science is based partially on this process.

## 1.5 The Development of the Idea of Random Sampling

We see from the above that the purposes of humanity require partial understanding of many populations—populations of mice, of light switches, rockets, etc., almost indefinitely. We must develop means of forming opinions about populations by examining only small portions of them. Furthermore, the process of forming opinions must be communicable and have force for all users.

A remarkable development of this, in the twentieth century, is the application of the idea of the perfect gambling device to the problem of learning about a population by examining a portion of it. This is the idea that there is a process called "random sampling." Consider a population of $N$ individuals. We can associate with each individual a number from 1 to $N$. We can make up $N$ chips, numbered from 1 to $N$, as similar as possible. We can place these chips in an urn, shake them, and draw out $n$ chips. We then look at the individuals whose numbers are on the chips drawn. The information on these $n$ individuals "tells us" something about the population; it does this as a consequence of the theory of probability. If, for example, we find that 1/4 of the sample have an attribute value less than 50, we are able by means of the techniques of statistics to express in some way our uncertainty on how many in the population have a value less than 50.

It is most important that the idea of random sampling be understood, and we give the following very simple example. Suppose we have a population of 3 individuals—$A$, $B$, and $C$. Then we say that a process of drawing a subset of 2 individuals is random sampling if in a large number of repetitions we obtain $A$ and $B$ with relative frequency 1/3, $A$ and $C$ with relative frequency 1/3, and $B$ and $C$ with relative frequency 1/3. Thus the process cannot be judged as random sampling merely on the basis of one trial. To determine if a sample of a definite finite population is a random sample, we have to know the process by which it was obtained. We have to consider the relative frequency with which different possible samples would be obtained. These relative frequencies we call *the* probabilities of the samples, and they will be determined by the sampling process. To show that a sampling process is random, we have to show that it leads to the relative frequencies, defined as the probabilities, of the various possible samples. So we are led to the necessity of constructing a physical process which mimics the mathematics of probability.

If we have a real process which we can regard as giving realizations of our conceptual random process, we can use it for a wide variety of sampling problems. We will have a logical structure by which we can calculate how the frequencies in the population affect the possible samples we can get. Finally we will have the logical problem that we have a random sample from a population and must have rational processes of forming opinions and making decisions about the nature of the population.

On this type of basis we have seen in the twentieth century a vast development of processes incorporating random devices for obtaining data, information, conclusions, and decisions about real existent populations by the examination of small portions of them. The processes have probability properties which are properties of relative frequency in populations of defined repetitions, the repetitions incorporating the use of "proven" random devices. The nub of the whole logical problem is the question of proving that a device is "random" or, to go to the elemental aspect, to decide what we mean by "proving" that a device is random.

## 1.6 Dynamic Populations

From many points of view the problems we discuss above are simple compared to the problem of developing a model for a system that is developing in time under some unknown complex of forces. The preeminent example of this is our own economic system. This is an ongoing entity changing in response to the multitude of economic forces we can readily imagine and which surely have relevance. Our problem here is that we have an abstract of the history of the process up to the present time, and we have to develop a model for its dynamic properties. The difficulties are tremendous since experiments are impossible. A model based only on associations derived from the historical record can be entirely erroneous for predicting outcomes of economic actions.

## 1.7 Data Analysis

There are many fields of investigation in which experiments can be performed; but there are also many in which we can merely take observations on existent populations, so that we are unable to incorporate probability ideas directly into the collection of data. Also, even in the case of planned experiments, we are limited in the extent to which we can use random devices. As a result any investigation—whether purely observational, as the collection of data from an existent population, or experimental, as the collection of data from experimentally induced populations—leads us to the problem of data interpretation.

It is here that we face a fundamental dilemma. Every possible observation can be explained by stating that there was a unique chain of causes which led to it. But this is of no aid in developing a mental construct of the totality of observations. It does not lead to predictions of what new observations will be, and the making of predictions of future observations is the cornerstone of scientific methodology. The simplest type of prediction occurs when an observational fact $A$, say, is always followed in time by observational fact $B$, as for instance, that the application of heat to a body always produces an increase in the temperature of the body, or a magnetic force always produces a deflection in a stream of electrons. But this type of elementary causation is from many points of view more often the exception than the rule. More commonly the imposition of a stimulus tends to produce a defined response with some relative frequency not equal to unity. We are aware that a medical treatment does not always produce the desired response. We find that in the population of patients only some show a favorable response. The reason for this uncertainty is that the population of patients is variable, and no two patients are really totally alike. We are forced to the type of conclusion that the treatment produces a cure in a high proportion of cases of a loosely defined population. We could well find that the treatment which produces a cure with high frequency for the population of the United States fails to do so for the population of Greenland, and the reason would be that the population of Greenland is different from the population of the United States in certain respects. When such an observational result is obtained, the question arises of why the two populations differ in their response; and subsequent description of the populations may suggest reasons which then have to be checked by further observations.

A consequence of this type of occurrence is that the analysis of observational data is of crucial importance. We may conduct an inquiry with preformulated questions and with extensive use of random devices to assure ourselves that we are looking at random members of a chosen population. This enables us to establish certain probabilistic measures of weight of evidence or certain decision rules with known operating characteristics.

The basic problem remains of examining the data with a view to developing a better picture of what is happening. We mention in this connection the remarkable data analysis of Kepler who, by examining records of the positions of planets, was able to formulate a good approximate description of the motion of planets around the sun. This example is informative from several points of view. In particular, Kepler did not obtain a perfect description, and we can now understand why. A complete model for this motion with present-day ideas requires the solution of the many-body problem, and even at the present time we do not have a complete answer to the consequences of the gravitational forces between bodies on their joint motion. A model obtained by data analysis is not the true model but is a step along the road toward the development of a perfect model, which in general is never achieved. The example is highly informative also because it exhibits the power of intelligent data analysis and because the model was obtainable only through data analysis. The "laws" of physical science are, in the last resort, generalizations of experience which have been shown to have wide validity, often incorporating elements of the randomness exhibited by simple gambling devices.

We see then that data analysis is a fundamental activity of scientific methodology. We can also surmise readily that data analysis unaccompanied by the formulation of models leads us into a vast sea of pure empiricism out of which there will be no path.

So an intrinsic part of data analysis is introspection leading to models for the observation. Furthermore, because we are faced with imperfect repeatability of phenomena, we have no alternative to the building of models which incorporate probability or randomness of elements. The history of science gives us two very remarkable examples of how the incorporation of randomness in the model, which is an explicit incorporation of ignorance, has led to great advances. The first of these examples is genetics, in which Mendel made the fantastic step of proposing that what happens in the mating of two genetic types is explicable by a model which incorporates equiprobable random assortment of hypothesized entities called genes—entities which were later found to exist in the sense that gene models give excellent predictions of the probability–relative-frequency type. The second example, perhaps of greater importance in the development of scientific models, is the incorporation of randomness and probability into the description of the atom and elementary particles. The current introductory texts on physics commonly use the analogy of coin tossing to motivate the student to the probability models which are deemed essential to the understanding of elemental phenomena.

The message that these and many other examples give us is that the growth of knowledge is often based in an essential way on two activities: (1) the development of probability models in which probability is relative frequency in a large set of similar trials and (2) the development of processes of data analysis which are meshed with the necessity of using probability models.

An important aspect of the whole problem of developing knowledge is that it is only in very specialized and usually very elemental cases, if ever, that we can isolate and control all the forces which produce variability. In the great bulk of investigative situations, we are unable to obtain experimental units (e.g., mice or physical specimens of an alloy) which are so nearly alike that we can ignore the differences.

It is critical, therefore, that we analyze our data from as many points of view as occur to us. If we are so fortunate as to find by data inspection a model which perfectly explains our results, we may reach a model in which the probability aspects enter only trivially with probabilities of unity or zero. But such an extreme case is to be regarded, it appears, only as one point in a spectrum of possibilities in which probabilities are intermediate between zero and 1. In general, we shall achieve such representation of data as a combination of forces which produce system and forces which produce unexplainable variability. The unexplainable variability we shall represent as arising from a specified random process. As our knowledge increases, we hope to be able to explain more and more of the hitherto unexplained variability by identifiable nonrandom forces. We may reach an impasse very early, as happened in the science of genetics, and decide to accept a very basic unexplained probability mechanism as intrinsic to the whole process. The fantastic progress made in this field is a testimony to the utility of making such a choice.

## 1.8 Our Main Objective

The purpose of this book is to introduce the reader to the ideas of describing populations by means of probability models in which probability is relative frequency in hypothetical populations of repetitions. We begin by consideration of ways of partially describing real existent populations. Then we go into the probability calculus as a means of developing models for populations which exhibit variability. This leads to a whole array of probabilistic models which have been found useful for the representation of data. We then turn to the main body of the book, which deals with how data can be examined from the viewpoint of being representable by probability models. The basic problem is the analysis of data to determine reasonable probability models. These will incorporate constants with unspecified values plus a probability structure for deviations of the observations from true values which are determined by the constants. There is no complete solution to this problem; we can develop many models for representing a fixed set of data. It follows that the problem involves the making of a compromise between mutually conflicting aims: (1) to obtain a simple model, with the difficulty that simplicity is an undefined concept; (2) to obtain a model with good predictive properties; (3) to obtain as good a representation of the data as possible, i.e., a communicable and understandable summary which is a fair representation of the data.

While we regard this type of problem as the main aim of statistics, there is a large class of problems in which the probability model can be taken as known apart from some unknown constants. Our problem, then, will be to develop rules for making conclusions, or rules for making such terminal decisions as "Reject the batch of 1000 switches if a random sample of 10 shows one defective switch." The development of such rules is an integral part of a modern technology with all its apparatus of mass production. Thus we first consider the basic problems of data analysis—determination of a model and specification of plausible values for unspecified constants. Following this is a short introduction to the class of situations in which the model can be assumed known and we have to formulate decision rules. We then present an introduction to data analysis in which the problem is to determine how some specified observations are explainable in terms of others. Finally, there is a short epilogue on fundamental ideas of data interpretation and probability which we hope will be understandable in the light of the development given in the whole work.

# 2
# The Nature of Real Populations

## 2.1 Introduction

The subject of statistics is aimed largely at the understanding of populations such as the people, business establishments, or labor force of a state or country; or of the many existent biological populations, plants, or animals; or of physical populations such as the automobiles on the roads of the United States, the steel output, or the vitamin output. In many cases we can imagine having an army of investigators who enumerate and observe every member of a population. In other cases we can enumerate the whole population but cannot examine it completely without destroying it if the attribute of interest is obtainable only by a destructive test. In still other cases we can envisage populations which we could produce by developmental and production processes, but we are confined for a variety of reasons to producing for each a few members which we hope will be representative of the very large potential population. Finally we have situations in which we have a few actual cases, and we can only imagine a large population of similar cases.

In almost all cases of interpretation of data we can envisage a population of examples like the one under examination, so a critically important first step is to consider the representation of populations that are completely available. For this reason our first chapter has nothing to do with probability or randomness.

## 2.2 Definition of Individual and Population

The concept of a population of individuals is easily recognized in the case of a human population, such as that of the United States, or a biological population, such as the animals or plants of a defined area. It is also easy to apply the concept to collections of such inanimate objects as automobiles or light bulbs. The individual is easily recognized as something with some degree of permanence, recognizable as a property of a collection of similar individuals which we call a population.

However, even in these obvious cases of identifiability of an individual there is an unlimited number of populations of which we may regard the individual as being a member. To take the simplest case of a human being, we may wish to regard him as a member of any one of many populations—e.g., (1) the population of males, (2) the residents of New York City, (3) the high school students of 1968, (4) the people in the United States on December 25, 1964, who wore spectacles, etc. The joint definition of individuals and populations is determined in any particular instance by the interests of the observer.

A less obvious but equally important class of individuals and populations is that in which the individual is an identifiable section of the whole space-time structure. For instance, an individual in a climatic study may be a calendar year, with the population as the collection of years. In a psychological experiment it may be appropriate to define the individual as the life of John Smith from 10:30 A.M. to 11:30 A.M. on May 27, 1975, and the population as the same portion of the lives of the student body of Iowa State University.

The sole criterion for the definition of individual and population is that a set of operational rules can be applied unambiguously; i.e., given an individual we must be able to state categorically whether or not he is a member of the population.

## 2.3 Types of Measurement of an Individual

There are three types of measurement of an individual; i.e., the result of observing an individual will be a combination of three primitive types of measurement:

1. The placing of an individual in a class according to an unordered classificatory scheme as, for instance, classifying biological organizations by a taxonomic system or humans as members of particular ethnic groups. In this case, there is a classificatory scheme with classes $C_1, C_2, \ldots, C_k$, say, and an individual is placed in one of the classes. There is, furthermore, no ordering of the classes in the sense that $C_1$ is "less than" $C_2$ in some respect. This is the most primitive type of observation.
2. The placing of an individual into a class according to an ordered classificatory system, i.e., into one of the classes $C_1, C_2, \ldots, C_k$, say, these classes having the property that in some defined sense $C_1$ is lower

than $C_2$, $C_2$ is lower than $C_3$, etc., and the ordering is transitive. This type of classification is widely used in the observation of social behavior, and measurement in psychology uses such scales very extensively.

3. The assignment of a real number to an individual, this having all the properties we associate with the ordinary concept of number. We call the collection of individuals with their numbers a number population. Conceptually there are two types of measurement by number, in that the number may be a count, such as number of legs, which is restricted to integers or to a quantification of magnitude not restricted to integers such as length or weight. It seems important, however, to note that any such measurement reduces to the representation of the magnitude by one of a discrete set of numbers because any such measurement involves the use of a grid of measurement.

Observation of a real population will frequently involve the making of measurements of all types. A more complex type of observation is the ranking of an individual within a set of individuals. In this case the result should be regarded as a joint property of the individual and the set of individuals, and not of the individual alone.

## 2.4 Number Populations

When we speak of variability in a population of physical items, we realize that there may be, in fact, several associated number populations, one for each possible way of measuring the items and there will be variability in each of these number populations. For example, most of us would agree there is variability in a shipload of iron ore. The weights of the individual rocks in the load is a number population which would exhibit variability. Another number population of interest might be the ore content of five-grain samples of ore, which would lead to several different number populations, depending upon the method of analysis used. Any physical population will have several attributes and a number population for each. There is also a population of ordered sets of numbers, a specified number of the ordered set referring to a particular attribute. Such populations are discussed at the end of this chapter. First we shall be dealing with a single-number population. In applications where it is clearly understood which number population is being studied, it is common to speak of the physical population and the number population as being one and the same.

A simple technique for describing a number population is to note the frequency with which each distinct number in the population occurs. To

**Table 2.1** Frequency Distribution—Number of bends required to break transistor leads in a population of twenty

| (1) Number of Bends | (2) Frequency of Occurrence | (3) Relative Frequency | (4) Cumulative Frequency | (5) Cumulative Relative Frequency |
|---|---|---|---|---|
| 0-4 | 0 | 0 | 0 | 0 |
| 5 | 1 | 1/20 | 1 | 1/20 |
| 6 | 1 | 1/20 | 2 | 2/20 |
| 7 | 3 | 3/20 | 5 | 5/20 |
| 8 | 5 | 5/20 | 10 | 10/20 |
| 9 | 6 | 6/20 | 16 | 16/20 |
| 10 | 3 | 3/20 | 19 | 19/20 |
| 11 | 1 | 1/20 | 20 | 20/20 |

illustrate this concept, the frequency distribution for a number population of interest in transistor technology is displayed in Table 2.1. A population of twenty transistors was submitted for testing. One lead on each transistor was bent successively until a break occurred, and the number of bends required to break the lead was recorded.

Column (2) simply gives the frequency of each number represented in the number population, it being understood that the frequency of all numbers greater than 11 is 0. Column (3) is obtained by dividing the

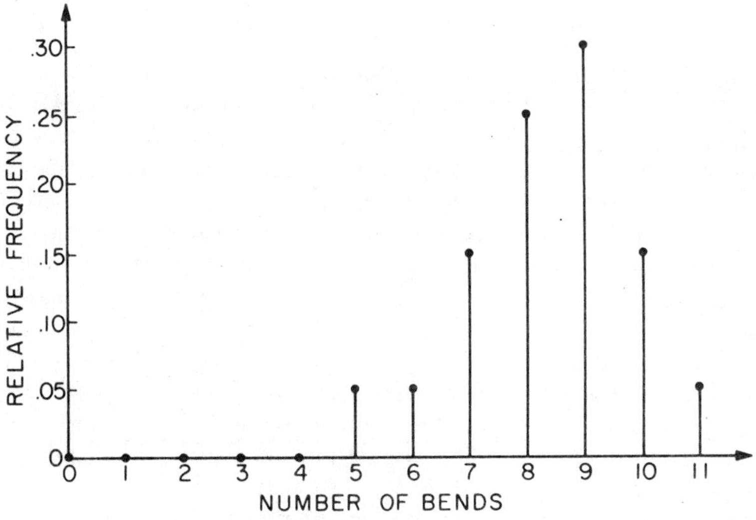


**Fig. 2.1** Relative Frequency of Occurrence of Number of Bends

entries of the second column by 20, thus giving the frequency *relative* to the total frequency of 20. Columns (4) and (5) are obtained from columns (2) and (3) respectively. The entire table is presented to exhibit several of the ways we can describe a number population. The particular way one chooses is a matter of preference. As a matter of fact, however, the relative frequency and cumulative relative frequency are most commonly used. It is often worthwhile to graph the relative frequencies and cumulative relative frequencies as in Figure 2.1 and Figure 2.2.

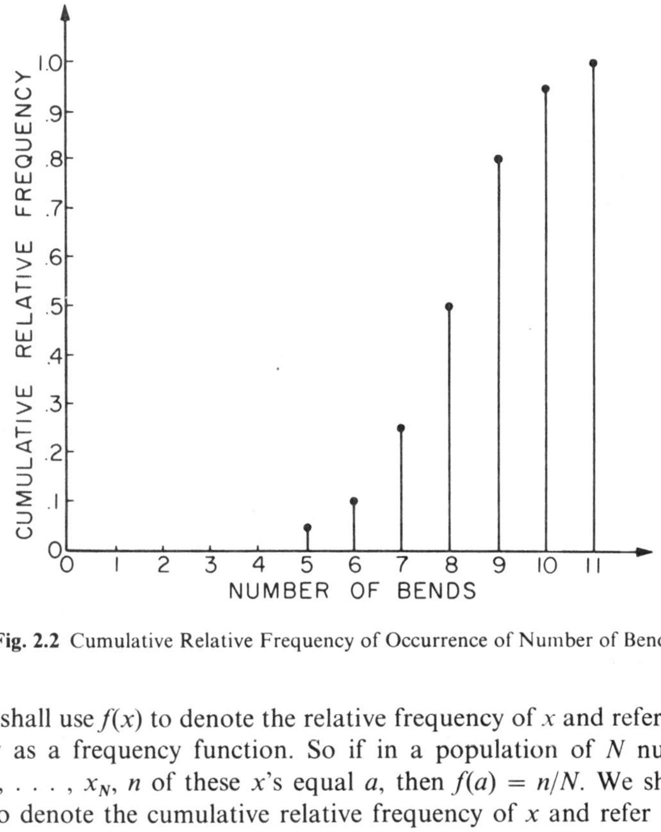


**Fig. 2.2** Cumulative Relative Frequency of Occurrence of Number of Bends

We shall use $f(x)$ to denote the relative frequency of $x$ and refer to $f(x)$ simply as a frequency function. So if in a population of $N$ numbers, $x_1, x_2, \ldots, x_N$, $n$ of these $x$'s equal $a$, then $f(a) = n/N$. We shall use $F(x)$ to denote the cumulative relative frequency of $x$ and refer to $F(x)$ simply as a distribution function. Both $f(x)$ and $F(x)$ will be defined for all $x$. Thus in the example under consideration the values of $f(x)$ for $x = 5, 6, \ldots, 11$ are as given in Table 2.1. For all other values of $x$, $f(x) = 0$. The distribution function $F(x)$ requires some other explanation if we are to regard it as being defined for all $x$. For the values of $x$ given in Table 2.1

$$F(x) = \text{cumulative relative frequency of values} \leq x \tag{2.1}$$

If we take this same statement as a definition for all other values of $x$, we obtain

$$\begin{aligned} F(x) &= 0 \quad , x < 5 \\ &= 1/20\ , 5 \leq x < 6 \\ &= 2/20\ , 6 \leq x < 7 \\ &\ \vdots \\ &= 1 \quad , x \geq 11 \end{aligned}$$

This is the function graphed in Figure 2.2 for integral $x$.

The example we have been considering is that of a population with a finite number of members and an associated population of numbers with a finite number of members also. We have the view that any measurement process necessarily gives one of a discrete set of results, but the number of possible results may be exceedingly large. In such cases the use of a frequency distribution for description of the population can become exceedingly tedious, and we may choose to group the data.

For example, suppose that an automobile manufacturer sets out to describe the length of time $x$ that purchasers keep a new automobile before it is sold, destroyed, or stolen. Data are obtained from 10,000 purchasers, who give the desired information in days. Obviously the frequency distribution for $x$ in days will be unnecessarily detailed for most purposes, and we may group the values by months—a month being defined as 30 days. Then frequencies are recorded for the groups: 1 month (0–30 days), 2 months (31–60 days), etc. The construction of the frequency distribution is now considerably simplified.

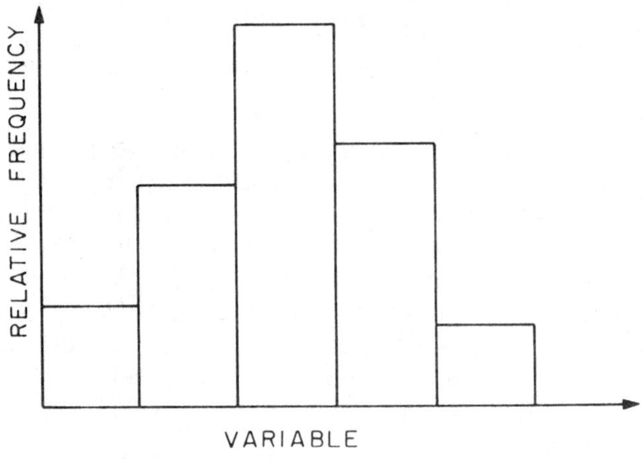


**Fig. 2.3** Example of a Histogram

With grouping of this sort it is more meaningful to give the frequency distribution in terms of a histogram, as in Figure 2.3. Here the drawing should be scaled so that the area of each bar is equal to the relative frequency of the values represented by that bar.

## 2.5 Approximate Description of Number Populations

While it is true that any number population is completely specified by the relative frequencies $f(x)$ or the cumulative relative frequencies $F(x)$ for all $x$, such a specification is too detailed for most purposes. We are led therefore to the consideration of approximate descriptions, or more precisely, of informative partial specifications. These fall into two classes, those based solely on ordering and those based on arithmetic computations.

### 2.5.1 Ordering Attributes of Populations

A complete specification of the population is given by the function $F(x)$, which has to be defined over the possible range of $x$. A useful partial specification is by measuring the value of $x$ at which $F(x)$ equals 1/2, this being the median. The intuitive idea of a median value is that 50% of the population have larger values and 50% have smaller values. It is easily seen that there is no number satisfying this property for the population 2, 3, 6, 7, 8.

This example might lead us to define the median as the value $m$ such that the fraction of values greater than $m$ is less than 50%. This definition satisfies the previous example in that it specifies the number 6 as the median. It is not adequate, however, for the example 2, 3, 4, 7, 8, 11. We are thus led to the following definition:

*Definition 2.1. A median of a population is a value such that the fraction of values greater is less than or equal to 50%, and the fraction of values less is less than or equal to 50%.*

This definition allows for the possibility of many medians. In the second example any number from 4 to 7 inclusive would be a median. It is conventional in this case to take 5 1/2 as the median. Note that the median is not required to be a member of the population.

The median can be regarded as a measure of "central tendency" or of location of the population on the number line. Knowledge of the median tells us little about the form of the population unless we know other attributes. If the population has a known shape, knowledge of the median is highly informative; but merely to know that a population of numbers has a median of 50 tells us little. For instance, we may be told as citizens that the median family income is \$5500 per annum, and this may appear to us to be an income adequate for a decent standard of living. But it is quite possible that a high proportion (in fact anything below 50%) could have a family income of less than \$1000.

Further information about the nature of the population is given by the quartile values of $x$—i.e., the values of $x$ for which $F(x) = 1/4$ and $F(x) = 3/4$. Also the difference between these two $x$ values tells us something about the variability in the population. More detailed information is given by the percentiles—i.e., the values of $x$ at which $F(x) = k/100$, $k = 1, 2, \ldots, 100$.

Another simple aspect of the population based on relative frequencies is the mode—i.e., the value $x$ at which $f(x)$ is maximum. This is the "most frequent value." In the case of a mathematically defined continuous population (Sect. 2.8) this mode will turn out to be based partly on arithmetic properties.

### 2.5.2 Arithmetic Properties of Populations

For many reasons, some obvious and elementary and others quite subtle, properties of populations based on arithmetical operations on the constituent members are very useful. The obvious and elementary aspect of a population of numbers is its mean, and this is the most commonly used measure of the "central tendency" or location of a population of numbers. So with a discrete population of individuals having values $x_i$, $i = 1, 2, \ldots, n$, the mean is $\sum_i x_i f(x_i)$. Here the summation is over the set of $x_i$'s. We shall abbreviate this frequently to $\sum xf(x)$ without indicating the range of summation.

Just as the median is a very limited description of the population, so is the mean. If we are told, for instance, that the mean family income in country $X$ is \$500 per annum, we may form an erroneous picture of the whole population. We know that income distributions are not symmetrical; i.e., there may be many low incomes and a few very high incomes, so we may guess that perhaps the bulk of incomes are below \$500. The measure of variability or dispersion that is most easily managed from all points of view is the variance, defined as the average of the squares of deviations around the mean. With a set of $N$ numbers $x_1, x_2, \ldots, x_N$, this is equal to $\sum (x_i - \mu)^2/N$, where $\mu = \bar{x} = \sum x_i/N$, and in general it is

$$\mu_2 = \sum (x - \mu)^2 f(x) \tag{2.2}$$

where $\mu$ is $\sum xf(x)$. If $f(x) = 1$ for one value of $x$, $\mu_2$ is zero. The symbol $\mu_2$ is always used to denote variance, and frequently the symbol

$\sigma^2$ is used. The square root of the variance is a measure of spread in the same units as the population numbers and is called the standard deviation. It is almost universally designated by $\sigma$. A much less commonly used measure of variability is the mean absolute deviation defined as $\sum|x - \mu|f(x)$. This had considerable appeal many decades ago, but this has diminished, primarily because it is difficult to study mathematically.

The first moment is the center of gravity of the population and the variance is its moment of inertia. It is natural to consider higher moments. It is useful to define these initially around zero, so that

$$\mu'_k = \sum x^k f(x) = \text{Ave}(x^k) \tag{2.3}$$

where "Ave" is used to denote average.

These moments around zero are less valuable than moments around the mean, one elementary reason being that they are all nonzero for a population with $f(x) = 1$ at some nonzero value of $x$. Moments about the population mean are more useful because they indicate the variability. These are denoted by $\mu_k$ with

$$\mu_k = \sum (x - \mu)^k f(x) = \text{Ave}[(x - \mu)^k] \tag{2.4}$$

The central moments $\mu_k$ are simple functions of the moments about zero, and the transformations for low moments are requested in the problems.

It seems clear intuitively that a population of $N$ numbers $x_1, x_2, \ldots, x_N$ is determined completely by its first $N$ moments, whether central or defined around zero, because the $x$ values are given by the solution to $N$ equations in $N$ unknowns:

$$\sum_{i=1}^{N} x_i^k = N\mu'_k; \quad k = 1, 2, \ldots, N$$

We may expect these to have a single solution.

If the population is a mathematically defined infinite one, the question of whether the moments define the population is not trivial but is in fact one of the deep mathematical problems of the past 100 years, the Moment Problem.

There is an indefinitely large number of ways of representing a population of $N$ numbers, say $x_i$, $i = 1, 2, \ldots, N$. We can perform any transformation from the $x$ scale to the $z$ scale by a function $z = h(x)$ and characterize the population by means of properties of the population of numbers $z_1, z_2, \ldots, z_N$. If for instance, the $x$ values with nonzero frequency are nonnegative, we can envisage the population of square roots, the population of logarithms, etc. What we should do depends on several matters. If, for instance, the population of square roots has a familiar shape commonly recognized in the area of discourse, it will be informative to make the square root transformation. This is important in mathematical approximation of real populations.

In general one can envisage positive, negative, and fractional moments with populations of positive numbers. Negative and fractional moments have been used little except in some mathematical investigations, but can have a real role. If, for instance, savings are thought to be related to the square root of income, the 1/2th moment would be informative.

Many number populations arise from a counting process, so that the numbers in the population are among the nonnegative integers. In these cases factorial moments are useful (see Chapter 4), defined as follows:

$$\mu_{(1)} = \sum x f(x)$$
$$\mu_{(2)} = \sum x(x - 1) f(x)$$
$$\cdots\cdots\cdots\cdots\cdots\cdots$$
$$\mu_{(k)} = \sum x(x - 1) \ldots (x - k + 1) f(x) \tag{2.5}$$

It is desirable to mention that there appears to be no limit to the possible types of observation and that the nature of the observation delimits the set of useful statistics. The determining aspect is the topology of the observation space. A particularly interesting case is that of directions in a plane or in ordinary 3-dimensional space. The former includes the type of situation in which an observation is the time of day of an event. The origin in this case, e.g., 12 midnight, has no special significance, and observations should be recognized as unit vectors with fixed but arbitrary axes of reference. Alternatively, the observation can be thought of as an angle ranging from 0 to $2\pi$, with the proviso that it is measured relative to a fixed but arbitrary direction. In this case a set of angles say $\{\theta_i, i = 1, 2, \ldots, n\}$ is equivalent to another set of observations $\{\theta_i + \alpha, i = 1, 2, \ldots, n\}$, where $\alpha$ is fixed but arbitrary. A frequently used distribution for this case has the density function, $c \exp[\kappa(\theta - \theta_0)]$. Ordinary moments are no longer of general value, but should be replaced by trigonometric moments, which are given by Ave$(e^{im\theta})$, $m = 1, 2, \ldots$. There is an array of statistical procedures associated with this type of distribution, and we refer the reader to the excellent expository monograph of Batschelet (1965), and references contained therein. This type of measurement is of great importance in some aspects of biology, geology, and crystallography. Watson (1970) reviews the field, including his own extensive work, from a useful scientific viewpoint.

### 2.5.3 Characterization of Shape

Many of the ideas in the preceding sections have been motivated by a consideration of the graph of $f(x)$. As we study graphs of frequency functions, we see that it may be meaningful to attempt a characterization of the intuitive concept of shape of the distribution. One of the simplest ideas is that of symmetry. If the graph of $f(x)$ for values greater than $\mu$ can be obtained by simply folding over the graph for values less than $\mu$, we say that the frequency distribution is symmetric about $\mu$. This means that if $x$ is a possible value in the population, $\mu - (x - \mu) = 2\mu - x$ is also a possible value and $f(x) = f(2\mu - x)$. As the reader can verify, the odd moments about the mean are zero for any symmetric finite population. Departure from symmetry is referred to as skewness, and a widely used skewness parameter is $\alpha_3$ defined by

$$\alpha_3 = \frac{\mu_3}{\sigma^3} \tag{2.6}$$

This parameter is useful because it does not depend on the origin of the $x$ scale and does not change with uniform stretching or shrinking of the scale, as the reader should verify. It is clearly zero if the distribution is symmetric, but it may be zero with nonsymmetric distributions. A widely used measure of the "peakedness" or kurtosis of the population of $x$ values is the quantity

$$\alpha_4 = \frac{\mu_4}{\sigma^4} \tag{2.7}$$

This has the same invariance properties as $\alpha_3$ and is small if the frequency of observations close to the mean is high and the frequency of observations far from $\mu$ is low. Both of these measures of shape are widely used in mathematical developments. In particular, they aid judgment of representing one population that may have inconvenient form by another which has simple properties.

### 2.5.4 Chebyshev's Inequality

If one adopts a description of a population by percentiles, one obtains the frequency of values in any range. More interesting is a statement of frequency of observations in terms of moments. A theorem (usually stated in terms of probability) due to Chebyshev is of utmost importance.

***Theorem 2.1*** Let the standard deviation of a population of numbers be $\sigma$. Then the proportion of numbers in the population which deviate from the population mean by $k\sigma$ or more is less than or equal to $1/k^2$.

*Proof:* Suppose that the population consists of $N$ numbers $x_1, \ldots, x_N$ and that $r$ of them are such that $|x_i - \mu| \geq k\sigma$. Then we may without loss of generality suppose that these are the last $r$. But

$$N\sigma^2 = \sum_{1}^{N-r} (x_i - \mu)^2 + \sum_{N-r+1}^{N} (x_i - \mu)^2$$

$$\geq \sum_{N-r+1}^{N} (x_i - \mu)^2 \geq rk^2\sigma^2 \tag{2.8}$$

Hence

$$rk^2 \leq N$$

or

$$\frac{r}{N} \leq \frac{1}{k^2} \tag{2.9}$$

So the proportion deviating from $\mu$ by at least $k\sigma \leq 1/k^2$.

If then we know from some source other than knowing the totality of numbers that, say, $\mu$ is 10 and $\sigma$ is 1, we can say that the proportion of numbers outside the interval $[10 - k, 10 + k] < 1/k^2$. So the proportion outside the interval $[5, 15] < 1/25$, or 4%. As another example, suppose a manufacturer is producing articles which are to have length within the interval $(100 - 2, 100 + 2)$, and he knows that the mean length and standard deviation of length under this process are 100 and 1 respectively. Then, at most, 1/4 of the articles he produces will fail to meet specifications.

An examination of the proof will indicate that Chebyshev's inequality can be extended to higher moments. Let $k$ of the population of $N$ numbers be such that

$$(x_i - \mu)^a \geq c\mu_a \tag{2.10}$$

where $a$ is even. Then

$$N\mu_a = \sum_{1}^{N} (x_i - \mu)^a$$

$$N\mu_a \geq \sum_{N-k+1}^{N} (x_i - \mu)^a$$

$$\geq kc\mu_a$$

That is

$$\frac{k}{N} \leq 1/c \tag{2.11}$$

We then have the more general inequality that the fraction of the population deviating from the mean by $c^{1/a}\mu_a^{1/a}$ or more is less than or equal to $1/c$.

### 2.5.5 Generating Functions

The subject of transforms of one kind or another plays a central role in much of applied mathematics and engineering. As with many topics which occur in applied disciplines, it is difficult to capture the essential ideas outside the particular area of application. The basic idea of a transform is simple enough; a transform simply maps one function into another function.

In the elementary theory of statistics transforms are also important. We shall consider transforms of the frequency function of the form $\text{Ave}[g(x, t)]$. Then the operation of averaging with respect to $x$ can be regarded as transforming $f(x)$ into a function of $t$. We shall discuss briefly the transforms specified by the following six choices of $g(x, t)$:

1. $g(x, t) = x^t$
2. $g(x, t) = |x|^t$
3. $g(x, t) = x(x - 1)(x - 2) \ldots (x - t + 1)$
4. $g(x, t) = e^{tx}$
5. $g(x, t) = e^{itx}$, where $i = \sqrt{-1}$, so that $E(e^{itx}) = E[\cos tx + i \sin tx]$
6. $g(x, t) = t^x$ (2.12)

For nonnegative integral values of $t$, transforms (1), (2), and (3) all give quantities which should be familiar to us. Transform (1) simply yields the moments about the origin. Transform (2) gives the absolute moments, while (3) gives the factorial moments. For nonintegral values of $t$ the transforms still exist, but they are not generally used.

In a preceding section we presented the moments as a method of summarizing or characterizing a population. In fact, for a population of $N$ numbers a knowledge of the first $N$ moments would make it possible, theoretically at least, to know the numbers in the population. Transform (4) gives a remarkably compact summarization of the population. It is known as the moment generating function (m.g.f.) and is denoted by $m(t)$. Since

$$e^{tx} = 1 + tx + t^2x^2/2! + t^3x^3/3! + \cdots$$

$$m(t) = \text{Ave}(e^{tx}) = 1 + t\mu + t^2\mu_2'/2! + t^3\mu_3'/3! + \cdots \quad (2.13)$$

Thus the moments about the origin may be obtained by finding the m.g.f. in series form because

$$\left.\frac{d}{dt} m(t)\right|_{t=0} = \mu$$

$$\left.\frac{d^2}{dt^2} m(t)\right|_{t=0} = \mu_2'$$

$$\cdots$$

$$\left.\frac{d^k}{dt^k} m(t)\right|_{t=0} = \mu'_k \tag{2.14}$$

*Example*

$$f(x) = \frac{1}{2}, \qquad x = -1, 1$$
$$= 0, \qquad \text{otherwise}$$
$$m(t) = \text{Ave}(e^{tx})$$
$$= \left(\frac{1}{2}\right)e^{-t} + \left(\frac{1}{2}\right)e^{t}$$
$$= \cosh t$$

We can find the moments by differentiating. For instance,

$$\mu = \left.\frac{d}{dt}\cosh t\right|_{t=0} = \sinh(0) = 0$$
$$\mu'_2 = \left.\frac{d}{dt}\sinh t\right|_{t=0} = \cosh(0) = 1$$

For a population of $N$ real numbers, not including the pseudo-number $\infty$, the m.g.f. always exists. For mathematically defined populations considered later with infinite range it may not exist. A very useful m.g.f. is that defined around the mean, and equal to the average of $\exp[t(x - \mu)]$, where $\mu$ is the mean assumed to exist. The m.g.f. is used to obtain another set of parameters of the population by expanding the natural logarithm of the m.g.f. $m(t)$ as

$$\log m(t) = \kappa_1 t + \kappa_2 t^2/2! + \cdots \kappa_r t^r/r! + \cdots \tag{2.15}$$

These coefficients of $t^r/r!$ on the right-hand side of Equation 2.15 are called the cumulants of the distribution. (See Problem 10.) Transform (5) yields the characteristic function of the distribution. It is very important in advanced work because it always exists. It may be surmised and is true that use of this transformation leads to Fourier analysis ideas but these are beyond our level. The advanced reader may refer to Lukacs (1960) and Lukacs and Laha (1964).

Transform (6) is applicable only to the case of a population with integral values of $x$. It is called the probability generating function of the distribution because the coefficient of $t^r$ in $\text{Ave}(t^x)$ is equal to the frequency with which the value $r$ occurs.

We shall not discuss any of these transforms until we are examining mathematically specified populations.

### 2.5.6 Other Representations of Variability

We have described what we mean by a number population. We have also described the use of a frequency distribution for characterization of the variability of a number population. However, it may not be clear what number populations are appropriate for a given physical population. In particular, we may have to deal with populations which we cannot easily count. The book by Herdan (1960) is largely devoted to this type of situation. Consider for example a load of gravel, a container of basic plastic, a carload of coal, or a geological core sample. We might like to know what the variability of size of the gravel particles is. It is not at all obvious how one should characterize the size of a gravel particle. Should we consider the size to be measured by volume, by cross-sectional area, or by the largest linear dimension? There are many ways, conceptually at least. A road builder has to be interested in the size of particles, and should know what measure of size is really relevant to his aims—i.e., which measure enables him to separate good gravel for his building purposes from poor gravel. However much he might wish to measure size in a particular way, he has to use a measure which is reasonably easy.

If in fact the particles all have the same shape and are uniform in density, the measurements in terms of weight, volume, cross-sectional area, etc., are easily transformed from one to the other. It follows that one type of measurement will supply information about the frequency function of another type of measurement in such cases.

For example, suppose we have a set of sieves with which we are screening particles. Suppose that the particles are perfect spheres with various radii and the density of the material is constant. We find the weight of the material which passed through the successive sieves and thus obtain the information.

| Radius | Weight |
|---|---|
| $0 \leq r \leq r_1$ | $w_1$ |
| $r_1 < r \leq r_2$ | $w_2$ |
| . | . |
| . | . |
| . | . |
| $r_{k-1} < r \leq r_k$ | $w_k$ |

What can we say about the distribution of particles by weight? We can easily obtain the sample frequency distribution of particles by weight since the weight of a particle of radius $r$ is $4\rho\pi r^3/3$, where $\rho$ is the density.

| Weight ($w$) | Sample Relative Frequency |
|---|---|
| $0 \le w \le 4\rho\pi r_1^3/3$ | $w_1/\sum_1^k w_i$ |
| $4\rho\pi r_1^3/3 < w \le 4\rho\pi r_2^3/3$ | $w_2/\sum_1^k w_i$ |
| .................... | ....... |
| $4\rho\pi r_{k-1}^3/3 < w \le 4\rho\pi r_k^3/3r_2$ | $w_k/\sum_1^k w_i$ |

The interesting question concerns what we can say about the distribution of particles by size. For instance, what can we say about the frequency function of radii $f(r)$? We know that the proportion $f(r)\,dr$ has a radius between $r$ and $r + dr$, when $dr$ is small, and that the weight in this interval is $4\rho\pi r^3/3$. The proportion of weight, then, with radius between $r_1$ and $r_2$ is

$$\frac{\int_{r_1}^{r_2} \rho \frac{4\pi}{3} r^3 f(r)\,dr}{\int_0^{\infty} \rho \frac{4\pi}{3} r^3 f(r)\,dr} = \frac{\int_{r_1}^{r_2} r^3 f(r)\,dr}{\int_0^{\infty} r^3 f(r)\,dr}$$

The problem of estimating the density function $f(r)$ from the weights $w_1, w_2, \ldots, w_k$ is a difficult one and will not be discussed in this book. An even more complex situation would be the determination of the size distribution of a population of particles by observing the sedimentation rate of a sample. Such a procedure would have to rely on some physical law relating rate of sedimentation to size.

## 2.6 Variability of a Categorical Population

As we have seen, there are many ways of partially characterizing the variability of a number population. There appears to be a particularly important way of characterizing the variability of a categorical population. This measure of variability is called the entropy, and it quantifies the extent to which members of the population have identical attributes. It is a measure of the degree of disorder in the population and has been widely used in physical science, engineering, and communication theory. In our order of development of statistics a brief discussion is essential.

*Definition 2.2. Let there be M categories in a population denoted by i, i = 1, 2, . . . M, with respective relative frequencies $p_i$. Then the entropy of the population is defined to be*

$$E = -\sum_{i=1}^{M} p_i \ln p_i \tag{2.16}$$

It is interesting to note that this quantity is maximized when each $p_i$ is equal to $1/M$. To see this, use a Lagrange multiplier $\lambda$ and differentiate $-\sum p_i \ln p_i + \lambda(\sum p_i - 1)$ with regard to each $p_i$. Setting the derivatives equal to zero gives

$$-\ln p_i - 1 + \lambda = 0$$

so that $-\ln p_i$, and hence $p_i$ is the same for all $i$. Hence $E$ is maximized with each $p_i$ equal to $1/M$.

A highly readable development of the use of this measure in the theory of communication is given by Khinchin (1957). The development is in terms of probability, which for all purposes of this work is equivalent to relative frequency.

## 2.7 Variability in More Than One Variable

In the previous sections we were interested in some particular characteristic or attribute of each member of a given population. It seemed natural to talk about the measurement of this attribute and to discuss the population of measurements of it. For instance, if we were discussing machined parts, we found it convenient to shift our attention to the associated number population—i.e., to the population of measurements associated with the population of machined parts. In the great majority of situations, however, we are interested in more than one attribute of each member of the population. For example, most machined parts have more than one dimension of interest; transistor manufacturers are interested in more than one electrical property of each transistor; pharmaceutical companies are interested in more than one effect induced by a given type of pill. A population of individuals, each of which has more than one attribute, is called a multivariate population. With two attributes the population is called bivariate.

### 2.7.1 Joint Frequency Functions

Consider a population consisting of 100 specimens of a certain crystalline material, where the number of faults and the number of etch pits on each specimen have been recorded as in Table 2.2.

**Table 2.2** Frequency Table of Specimens of Crystalline Material

| | | Number of Etch Pits | | | |
|---|---|---|---|---|---|
| | | 0 | 1 | 2 | 3 |
| Number of Faults | 0 | 2 | 6 | 4 | 3 |
| | 1 | 4 | 9 | 7 | 5 |
| | 2 | 7 | 10 | 5 | 3 |
| | 3 | 7 | 15 | 4 | 9 |

These results could also be represented by a graph such as Figure 2.4. For populations where the number of distinct values is so small, such characterizations would be adequate, although graphing in three dimensions may be somewhat tedious. However, it is often easier to look at the relative frequencies, as shown in Table 2.3. The marginal totals give the relative frequencies over all possible values of the other variable.

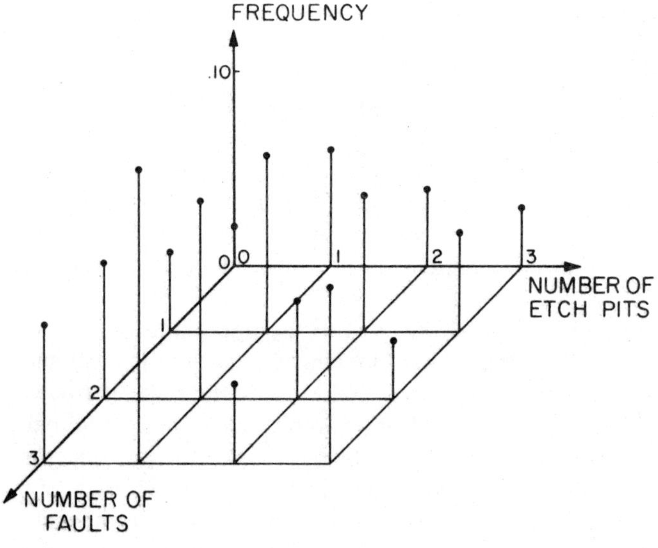


**Fig. 2.4** Graph of Frequency of Specimens of Crystalline Material

For discrete bivariate populations in general, let $f(x, y)$ give the relative frequency of $x$ and $y$. Then the marginal frequencies are given by

$$f_X(x) = \sum_y f(x, y)$$

$$f_Y(y) = \sum_x f(x, y) \tag{2.17}$$

**Table 2.3** Relative Frequencies

| | | Number of Etch Pits ($x$) 0 | 1 | 2 | 3 | Marginal Total |
|---|---|---|---|---|---|---|
| Number of Faults ($y$) | 0 | .02 | .06 | .04 | .03 | .15 |
| | 1 | .04 | .09 | .07 | .05 | .25 |
| | 2 | .07 | .10 | .05 | .03 | .25 |
| | 3 | .07 | .15 | .04 | .09 | .35 |
| Marginal Total | | .20 | .40 | .20 | .20 | 1.00 |

Note that $f(x, y)$ is not necessarily given by the product of the marginals. For example, in Table 2.3, $f(2, 3) = .04$, but $f_X(2)f_Y(3) = (.20)(.35) = .07$.

Quite often we examine subpopulations. Consider the subpopulation in Table 2.2 for individuals with one fault. Denote the relative frequency of etch pits by $f(x|1)$. This is referred to as the conditional frequency of $x$ given $y = 1$. Directly from Table 2.3 we can compute the relative frequencies to be as in Table 2.4. Upon examination of Table 2.3 we can see that $f(x|1) = f(x, 1)/f_Y(1)$.

**Table 2.4** Conditional Frequency

| $x$ | 0 | 1 | 2 | 3 |
|---|---|---|---|---|
| $f(x\|1)$ | .16 | .36 | .28 | .20 |

The complete set of conditional frequencies for this example may be seen in Table 2.5.

**Table 2.5** Complete Set of Conditional Frequencies

| $x$ | 0 | 1 | 2 | 3 |
|---|---|---|---|---|
| $f(x\|0)$ | 2/15 | 6/15 | 4/15 | 3/15 |
| $f(x\|1)$ | 4/25 | 9/25 | 7/25 | 5/25 |
| $f(x\|2)$ | 7/25 | 10/25 | 5/25 | 3/25 |
| $f(x\|3)$ | 7/35 | 15/35 | 4/35 | 9/35 |
| $y$ | 0 | 1 | 2 | 3 |
| $f(y\|0)$ | 2/20 | 4/20 | 7/20 | 7/20 |
| $f(y\|1)$ | 6/40 | 9/40 | 10/40 | 15/40 |
| $f(y\|2)$ | 4/20 | 7/20 | 5/20 | 4/20 |
| $f(y\|3)$ | 3/20 | 5/20 | 3/20 | 9/20 |

For the bivariate case in general the relative frequency of $x$ given $y$ is seen to be

$$f(x|y) = f(x, y)/f_Y(y) \tag{2.18}$$

and the relative frequency of $y$ given $x$ is seen to be

$$f(y|x) = f(x, y)/f_X(x) \tag{2.19}$$

It sometimes happens that the conditional frequency function of $x$ is the same for all $y$ as is the case in Table 2.6.

**Table 2.6** Joint Frequency Function

| | | $x$ | | | |
|---|---|---|---|---|---|
| | | 0 | 1 | 2 | Total |
| | 0 | .02 | .04 | .04 | .10 |
| | 1 | .03 | .06 | .06 | .15 |
| $y$ | 2 | .09 | .18 | .18 | .45 |
| | 3 | .06 | .12 | .12 | .30 |
| Total | | .20 | .40 | .40 | 1.00 |

The marginal frequencies are given in Table 2.7.

**Table 2.7** Marginal Frequencies

| $x$ | 0 | 1 | 2 | |
|---|---|---|---|---|
| $f_X(x)$ | .20 | .40 | .40 | |
| $y$ | 0 | 1 | 2 | 3 |
| $f_Y(y)$ | .10 | .15 | .45 | .30 |

The conditional frequencies are given in Table 2.8.

**Table 2.8** Conditional Frequencies

| $x$ | 0 | 1 | 2 |
|---|---|---|---|
| $f(x\|y = 0)$ | .20 | .40 | .40 |
| $f(x\|y = 1)$ | .20 | .40 | .40 |
| $f(x\|y = 2)$ | .20 | .40 | .40 |
| $f(x\|y = 3)$ | .20 | .40 | .40 |

**Table 2.8** Conditional Frequencies

| $y$ | 0 | 1 | 2 | 3 |
|---|---|---|---|---|
| $f(y\|x = 0)$ | .10 | .15 | .45 | .30 |
| $f(y\|x = 1)$ | .10 | .15 | .45 | .30 |
| $f(y\|x = 2)$ | .10 | .15 | .45 | .30 |

Thus $f(x|y)$ does not depend upon $y$ and further $f(x|y) = f_X(x)$ for all $y$. That is, the marginal frequency of $x$ is the same as the conditional of $x$ given $y$. Corresponding statements are seen to hold for the marginal of $y$ and the conditional densities of $y$. When this situation exists, $x$ and $y$ are said to be independent. In other words, independence is equivalent to either of the following statements:

1. $f(x|y) = f_X(x)$ for all $y$ and
   $$f(y|x) = f_Y(y) \text{ for all } x \tag{2.20}$$

2. $$f(x, y) = f_X(x)f_Y(y) \tag{2.21}$$

That is, if $x$ and $y$ are independent, the joint relative frequency is given by the product of the marginal relative frequencies.

To summarize the preceding discussion, any function $f(x, y)$ may serve as a frequency function for bivariate populations if

$$f(x, y) \geq 0$$

and

$$\sum_x \sum_y f(x, y) = 1 \tag{2.22}$$

where summation is over $x$ and $y$ values represented in the population. The marginal frequency functions are given by

$$f_X(x) = \sum_y f(x, y)$$

$$f_Y(y) = \sum_x f(x, y) \tag{2.23}$$

The conditional frequency functions are given by

$$f(x|y) = f(x, y)/f_Y(y)$$

$$f(y|x) = f(x, y)/f_X(x) \tag{2.24}$$

The concepts are easily generalized to include any number of attributes.

The conditional frequency functions are often useful for the process of obtaining a population average. Consider the average of a function

$z(x, y)$. Then

$$\begin{aligned}\text{Ave } z(x, y) &= \sum_{x,y} f(x, y)\, z(x, y) \\ &= \sum_{x,y} z(x, y)\, f_X(x)\, f(y|x) \\ &= \sum_{x} f_X(x) \sum_{y} f(y|x)\, z(x, y)\end{aligned}$$

By symmetry, this also equals

$$\sum_{y} f_Y(y) \sum_{x} f(x|y)\, z(x, y)$$

Note that

$$\sum_{y} f(y|x)\, z(x, y) \text{ and } \sum_{x} f(x|y)\, z(x, y)$$

are both averages in conditional arrays so we can write

$$\text{Ave } z(x, y) = \underset{x}{\text{Ave}}\ \underset{|x}{\text{Ave}}\ z(x, y) = \underset{y}{\text{Ave}}\ \underset{|y}{\text{Ave}}\ z(x, y)$$

where the subscript "$|x$" to "Ave" indicates averaging with fixed $x$, the subscript "$|y$" indicates averaging with fixed $y$, and the subscripts "$x$" and "$y$" indicate averaging over $x$ and $y$, respectively.

### 2.7.2. Partial Descriptions of Bivariate Populations

If we are given a population of $N$ ordered pairs $(x, y)$ in which each of $x$ and $y$ are one of the types of measurement, obviously we can give a partial description of the population by describing its marginal frequencies by any of the procedures available for a univariate population.

The commonly used case is that when $x$, $y$ are measurement numbers and we define the marginal components:

$$\begin{aligned}&r\text{th moment about the origin of } x = \text{Ave}(x^r) = \mu'_{r0} \\ &r\text{th moment about the mean of } x = \text{Ave}[x - (\text{Ave } x)]^r = \mu_{r0} \\ &s\text{th moment about the origin of } y = \text{Ave}(y^s) = \mu'_{0s} \\ &s\text{th moment about the mean of } y = \text{Ave}[y - (\text{Ave } y)]^s = \mu_{0s}\end{aligned} \tag{2.25}$$

These are power moments, and other types of moments may be possible and useful. One may also use ideas of medians, quartiles, and in general, fractiles to characterize partially the populations. However, these measures tell us only about the variability of each attribute separately. It is necessary to characterize the covariability of the attributes given by $x$ and $y$.

### 2.7.3 Covariability of Qualitative Attributes in a Population

The simplest case from a logical viewpoint is that arising by simple dichotomous classification of individuals of the population by two methods. This case goes back to antiquity, and an example which crops up in writings on logic follows. We are given a population of individuals which we can classify in two ways: either swans $S$ or nonswans $\bar{S}$ and either white $W$ or nonwhite $\bar{W}$. Complete enumeration of a population of $N$ individuals would give the $2 \times 2$ table

| | $W$ | $\bar{W}$ | Total |
|---|---|---|---|
| $S$ | $N_{11}$ | $N_{12}$ | $N_{1.}$ |
| $\bar{S}$ | $N_{21}$ | $N_{22}$ | $N_{2.}$ |
| Total | $N_{.1}$ | $N_{.2}$ | $N_{..} = N$ |

The interesting question underlying this example is consideration of the proposition "All swans are white." We shall not enter into this matter except to raise the question: "Does the observation that a nonwhite individual $\bar{W}$ is non-$S$, $\bar{S}$, lend support to the proposition that all $S$ are $W$?"

If the attributes in the table above are independent, then

$$N_{11}/N_{21} = N_{12}/N_{22}$$

and a natural measure of association is

$$Q_1 = \frac{N_{11}N_{22}}{N_{12}N_{21}} - 1 \tag{2.26}$$

or

$$Q_2 = \ln N_{11} + \ln N_{22} - \ln N_{12} - \ln N_{21} \tag{2.27}$$

Yule (1900) at the beginning of this century gave a measure of association

$$Q_3 = \frac{N_{11}N_{22} - N_{12}N_{21}}{N_{11}N_{22} + N_{12}N_{21}} \tag{2.28}$$

If the attributes are independent, $Q_3$ is zero; and $Q_3$ attains its maximum possible value of $+1$ only if $N_{12}N_{21}$ is zero, and its minimum value of $-1$ only if $N_{11}N_{22}$ is zero. Another measure of association is

$$Q_4 = \frac{N_{11}N_{22} - N_{12}N_{21}}{[N_{1.}N_{2.}N_{.1}N_{.2}]^{1/2}} \tag{2.29}$$

It is shown by Kendall and Stuart (1966, vol. 2, p. 539) that $Q_4$ has to be in the range $-1$ to $+1$.

These measures of association are not symmetric and cannot be extended readily to larger tables. We suggest that a notion of association can be based on the concept of entropy. We illustrate first with the $2 \times 2$ table in which we enter the proportions rather than the actual frequencies and use $A$, $\bar{A}$ and $B$, $\bar{B}$ for the attributes:

| | $B$ | $\bar{B}$ | Total |
|---|---|---|---|
| $A$ | $p_{11}$ | $p_{12}$ | $p_{1.}$ |
| $\bar{A}$ | $p_{21}$ | $p_{22}$ | $p_{2.}$ |
| Total | $p_{.1}$ | $p_{.2}$ | 1 |

The entropy of the whole system is

$$-p_{11}\ln p_{11} - p_{12}\ln p_{12} - p_{21}\ln p_{21} - p_{22}\ln p_{22}$$

The entropy of the classification by $A$, $\bar{A}$ is

$$-p_{1.}\ln p_{1.} - p_{2.}\ln p_{2.}$$

and of the classification by $B$, $\bar{B}$ is

$$-p_{.1}\ln p_{.1} - p_{.2}\ln p_{.2}$$

It can be shown (Khinchin, 1957, pp. 6, 7) that the entropy of the joint system is less than or equal to the sum of the entropies of the separate ones. So we can take

$$\begin{aligned} Q_5 = {} & -p_{1.}\ln p_{1.} - p_{2.}\ln p_{2.} - p_{.1}\ln p_{.1} - p_{.2}\ln p_{.2} \\ & +p_{11}\ln p_{11} + p_{12}\ln p_{12} + p_{21}\ln p_{21} + p_{22}\ln p_{22} \end{aligned} \tag{2.30}$$

This quantity $Q_5 \geq$ zero. It is zero if the classifications are independent. To show partially the relation of $Q_5$ to the previous measures, write

$$\begin{aligned} p_{11} &= p_{1.}p_{.1} + \Delta \\ p_{12} &= p_{1.}p_{.2} - \Delta \\ p_{21} &= p_{2.}p_{.1} - \Delta \\ p_{22} &= p_{2.}p_{.2} + \Delta \end{aligned}$$

with

$$\Delta = p_{11}p_{22} - p_{12}p_{21} \tag{2.31}$$

Rewriting $Q_5$ as

$$Q_5 = p_{11}\ln\left(\frac{p_{11}}{p_{1.}p_{.1}}\right) + p_{12}\ln\left(\frac{p_{12}}{p_{1.}p_{.2}}\right) + p_{21}\ln\left(\frac{p_{21}}{p_{2.}p_{.1}}\right) + p_{22}\ln\left(\frac{p_{22}}{p_{2.}p_{.2}}\right)$$

and using $\ln x \doteq x - 1$, we find that if $\Delta$ is small,

$$\begin{aligned}
Q_5 &\doteq p_{11}\left(\frac{p_{11} - p_{1.}p_{.1}}{p_{1.}p_{.1}}\right) + p_{12}\left(\frac{p_{12} - p_{1.}p_{.2}}{p_{1.}p_{.2}}\right) + p_{21}\left(\frac{p_{21} - p_{2.}p_{.1}}{p_{2.}p_{.1}}\right) \\
&\quad + p_{22}\left(\frac{p_{22} - p_{2.}p_{.2}}{p_{2.}p_{.2}}\right) \\
&= \frac{\Delta}{p_{1.}p_{.1}p_{2.}p_{.2}}(p_{11}p_{2.}p_{.2} - p_{12}p_{.1}p_{2.} - p_{21}p_{.2}p_{1.} + p_{22}p_{1.}p_{.1}) \\
&= \frac{\Delta^2}{p_{1.}p_{.1}p_{2.}p_{.2}}(p_{2.}p_{.2} + p_{.1}p_{2.} + p_{.2}p_{1.} + p_{1.}p_{.1}) \\
&= \Delta^2\left(\frac{1}{p_{1.}p_{.1}} + \frac{1}{p_{1.}p_{.2}} + \frac{1}{p_{2.}p_{.1}} + \frac{1}{p_{2.}p_{.2}}\right) = Q_4^2
\end{aligned} \tag{2.32}$$

In a general $r \times c$ table with one or both of $r$ and $c$ greater than 2, there may be no ordering of the categories by each mode of classification which can be regarded as positive association. In that case the entropy type of measure of association seems to have excellent properties.

With a general $r \times c$ table it will be

$$\begin{aligned}
Q_E = &\sum_{i=1}^{r}\sum_{j=1}^{c} p_{ij}\ln p_{ij} \\
&- \sum_{i=1}^{r}\left(\sum_{j=1}^{c} p_{ij}\right)\ln\left(\sum_{j=1}^{c} p_{ij}\right) \\
&- \sum_{j=1}^{c}\left(\sum_{i=1}^{r} p_{ij}\right)\ln\left(\sum_{i=1}^{r} p_{ij}\right)
\end{aligned} \tag{2.33}$$

The extension of this idea to the case of more than two modes of classification is fairly obvious but involves technicalities beyond the level of this text.

### 2.7.4 Covariability of Arithmetic Attributes in Populations

We give an example of the situation to be discussed. Suppose we take all adult males of the United States who achieved the 35th birthday in the year 1968, and that for each one we obtain two attributes $x_1$, $x_2$, where $x_1$ is the total income in 1967 and $x_2$ is the total amount paid in taxes. This will give us a finite but large population of ordered pairs of numbers $(x_1, x_2)$. Our job is to consider ways of characterizing the extent and the way in which the two attributes vary simultaneously.

Certainly the concept of fractiles can be applied. A simple process is to partition the population into a $2 \times 2$ table of frequencies by four classes: class $AB$ consists of the individuals above the median for both attributes, class $A\bar{B}$ above the median for $x_1$ and below the median for $x_2$, with classes $\bar{A}B$ and $\bar{A}\bar{B}$ defined correspondingly. We can then use any of the measures associated with the $2 \times 2$ table discussed in the previous section. A more general procedure would be to construct a $3 \times 3$ table or a $4 \times 4$ table or, in general, an $m \times m$ table by partitioning on the basis of the marginal populations according to tertiles, quartiles, or $m$-tiles. We would then construct the entropy measure of degree of deviation from lack of association.

Any of the above measures of association do not use fully the arithmetical property of the variables $x_1$, $x_2$. In fact they use only the ordinal properties that two values for $x_1$, say, equal to $x_1^{(1)}$, $x_1^{(2)}$, have the property of one being "below" the other. To use fully the arithmetical properties of $x_1$ and $x_2$, we can evaluate joint moments defined as follows:

$$\text{noncentral: } \mu'_{rs} = \text{Ave}(x_1^r x_2^s)$$
$$\text{central: } \mu_{rs} = \text{Ave}\{[x_1 - \text{Ave}(x_1)]^r[x_2 - \text{Ave}(x_2)]^s\} \quad (2.34)$$

Here, $\text{Ave}(x_1^r, x_2^s) = \sum_{x_1, x_2} f(x_1 x_2) x_1^r x_2^s$, with similar definitions throughout. Any Ave($z$) is the *average* of $z$ in a *real* population.

For a real population of ordered pairs $(x_1, x_2)$ in which values of infinity are not permitted, these functions of the assemblage can be calculated. It is the case, presumably (though we know of no proof) that a finite set of the pure power moments and of the mixed power moments will specify the population totally.

It is probably not unexpected that the most useful member of this whole array of measures of covariability is the one of lowest order:

$$\mu_{11} = \text{Ave}\{[x_1 - \text{Ave}(x_1)][x_2 - \text{Ave}(x_2)]\} \quad (2.35)$$

Suppose there is positive association of $x_1$ and $x_2$. Then we may expect that if $x_1$ is larger than the average value for an individual so also will $x_2$ be larger than its average value, and similarly for individuals below the average. Thus the product of $x_1 - \text{Ave}(x_1)$ and $x_2 - \text{Ave}(x_2)$ will tend to be positive.

The quantity $\mu_{11}$ in Equation (2.35) is defined to be the covariance of $x_1$ and $x_2$ in the population. In algebraic terms, we define

$$\mu_{10} = \sum_{x_1, x_2} x_1 f(x_1, x_2)$$

$$\mu_{01} = \sum_{x_1, x_2} x_2 f(x_1, x_2)$$

$$\mu_{11} = \sum_{x_1, x_2} (x_1 - \mu_{10})(x_2 - \mu_{01}) f(x_1, x_2) \tag{2.36}$$

The quantity $\mu_{11}$ is denoted very frequently by $\sigma_{12}$. It is clear that the magnitude of $\mu_{11}$ depends in an essential way on the units of measurement for $x_1$ and $x_2$. The removal of this dependence is easy in the following way. Recall the definitions of the pure moments of the second degree:

$$\sigma_1^2 = \mu_{20} = \sum_{x_1, x_2} (x_1 - \mu_{10})^2 f(x_1, x_2)$$

$$\sigma_2^2 = \mu_{02} = \sum_{x_1, x_2} (x_2 - \mu_{01})^2 f(x_1, x_2)$$

Define as a measure of association

$$\rho = \frac{\mu_{11}}{\sqrt{\mu_{20}\mu_{02}}} = \frac{\sigma_{12}}{\sigma_1 \sigma_2} . \tag{2.37}$$

This is called the correlation of $x_1$ and $x_2$ or, more explicitly, the product-moment correlation of $x_1$ and $x_2$.

A very basic inequality of mathematics is the *Cauchy-Schwarz inequality*.

***Theorem* 2.2** Consider two sequences of real numbers $a_1, a_2, \ldots, a_m$ and $b_1, b_2, \ldots, b_m$ with a sequence of weights which are greater than or equal to zero, denoted by $w_1, w_2, \ldots, w_m$. Then

$$\left(\sum w_i a_i b_i\right)^2 \leq \left(\sum w_i a_i^2\right)\left(\sum w_i b_i^2\right) \tag{2.38}$$

*Proof:* Consider $F = \sum w_i(a_i - \lambda b_i)^2$, where not all $w_i$ are zero. This is certainly greater than or equal to zero for any value of $\lambda$. Its minimum value over all $\lambda$ is certainly also greater than or equal to zero. $F$ is quadratic and has a minimum value when $\partial F/\partial \lambda = 0$ or at $\lambda_0 = (\sum w_i a_i b_i)/(\sum w_i b_i^2)$.

The value at $\lambda = \lambda_0$ is

$$\sum w_i a_i^2 - \left(\sum w_i a_i b_i\right)^2 / \sum w_i b_i^2$$

and this is equal to or greater than zero. Hence

$$(\sum w_i a_i b_i)^2 \leq (\sum w_i a_i^2)(\sum w_i b_i^2)$$

To apply this inequality, let

$$a_i = x_{1i} - \text{Ave}(x_1),\ b_i = x_{2i} - \text{Ave}(x_2)$$

and

$$w_i = f(x_{1i}, x_{2i})$$

We then have

$$\mu_{11}^2 \leq \mu_{20}\mu_{02}$$

or using the $\sigma$ symbol

$$\sigma_{12}^2 \leq \sigma_1^2 \sigma_2^2$$

Hence

$$\rho^2 = \frac{\sigma_{12}^2}{\sigma_1^2 \sigma_2^2} \leq 1$$

or

$$\rho = \frac{\sigma_{12}}{\sigma_1 \sigma_2} \text{ lies in the interval } [-1, 1]$$

The properties of the correlation coefficient $\rho$, which the student should verify, are:

1. $-1 \leq \rho \leq 1$
2. $\rho$ is scale invariant; i.e., if

$$x_1^* = ax_1 + c$$
$$x_2^* = bx_2 + d$$

where $a(>0)$, $b(>0)$, $c$, $d$ are constants, then

$$\rho(x_1^*, x_2^*) = \rho(x_1, x_2)$$

3. $\rho$ equals $+1$ or $-1$ if and only if $x_2 = \alpha x_1 + \beta$ for all individuals for some $\alpha$, $\beta$.

While the great bulk of past and current thinking about association of arithmetical attributes is based on the moments of the first and second degree and the derived correlation coefficient defined above, it is surely the case that other measures of association will be found useful as descriptive devices for particular cases. There is no limit to the possible measures of association, and the individual thinker should exercise his mental ingenuity to discover measures which are useful in his particular circumstances. The prime object of all summary partial descriptions of populations is to enable comparisons to be made which are meaningful and significant to the questions of the investigator. A critical aspect of the whole matter is that our interest in populations is almost never in a single population but in the comparison of populations or in the comparison of a single population with a conceptual population that seems desirable to us.

### 2.7.5 Concept of a Population Regression

In addition to measuring the degree of association of two attributes $(x_1, x_2)$ in a population, we will often wish to express in approximate form the nature of the average relationship of $x_1$ to $x_2$. For this purpose the concept of population regression is valuable. Suppose that we wish to determine constants $\beta_{10}$ and $\beta_{12}$ such that $(x_1 - \beta_{10} - \beta_{12}x_2)$ is on the average small in absolute magnitude. A natural requirement is to choose $\beta_{10}$ and $\beta_{12}$ so that

$$\text{Ave}(x_1 - \beta_{10} - \beta_{12}x_2)^2 \tag{2.39}$$

is as small as possible. It is clear that we can write

$$\begin{aligned} x_1 &= \mu_{10} + X_1 \\ x_2 &= \mu_{01} + X_2 \end{aligned} \tag{2.40}$$

and work in terms of $X_1$ and $X_2$ rather than $x_1$ and $x_2$, and we can then write

$$x_1 - \beta_{10} - \beta_{12}x_2 = (\mu_{10} - \beta_{10} - \beta_{12}\mu_{01}) + (X_1 - \beta_{12}X_2) \tag{2.41}$$

A consequence is that $\text{Ave}(X_1) = \text{Ave}(X_2) = 0$. We then have to find $\beta_{10}, \beta_{12}$ so that

$$(\mu_{10} - \beta_{10} - \beta_{12}\mu_{01})^2 + \text{Ave}(X_1 - \beta_{12}X_2)^2 \tag{2.42}$$

is a minimum. Clearly we can choose $\beta_{12}$ to minimize the second term and then take

$$\beta_{10} = \mu_{10} - \beta_{12}\mu_{01} \tag{2.43}$$

making the first term equal to zero. The second term is

$$\mu_{20} - 2\beta_{12}\mu_{11} + \beta_{12}^2\mu_{02}$$

and this is minimized when

$$\beta_{12} = \frac{\mu_{11}}{\mu_{02}} \tag{2.44}$$

So we say that the relationship

$$x_1 = \left(\mu_{10} - \frac{\mu_{11}}{\mu_{02}}\mu_{01}\right) + \frac{\mu_{11}}{\mu_{02}}x_2 \tag{2.45}$$

best expresses the linear relationship of $x_1$ to $x_2$ in the sense of mean square deviation. Equation (2.45) is called the population mean square linear regression of $x_1$ on $x_2$. Note that the slope, often called the regression coefficient, is $\mu_{11}/\mu_{02}$ which is equal to $\rho(\sigma_1/\sigma_2)$. Similarly the population mean square linear regression of $x_2$ on $x_1$ is given by the relationship

$$x_2 = \left(\mu_{01} - \frac{\mu_{11}}{\mu_{20}}\mu_{01}\right) + \frac{\mu_{11}}{\mu_{20}}x_1 \tag{2.46}$$

### 2.7.6 Bivariate Chebyshev Inequality

We saw in the univariate case that knowledge of the mean and the variance permitted a powerful partial characterization of the distribution by means of the Chebyshev inequality. We give one simple generalization to a bivariate population. We denote the pairs by $x_i$, $y_i$; the means by $\mu_x$, $\mu_y$; and second moments by $\sigma_x^2$, $\sigma_y^2$, and $\sigma_{xy}$.

***Theorem* 2.3** The fraction of number pairs $(x_i, y_i)$ on or outside the ellipse

$$a(x-\mu_x)^2 + b(x-\mu_x)(y-\mu_y) + c(y-\mu_y)^2 = d(a\sigma_x^2 + b\sigma_{xy} + c\sigma_y^2)$$

is not greater than $1/d$, where $a, c > 0$, $4ac > b^2$.

*Proof:* Suppose that $k$ of the pairs $(x_i, y_i)$ lie on or outside the ellipse, i.e., are such that

$$a(x-\mu_x)^2 + b(x-\mu_x)(y-\mu_y) + c(y-\mu_y)^2 \geq d(a\sigma_x^2 + b\sigma_{xy} + c\sigma_y^2) \tag{2.47}$$

Without loss of generality take these to be the last $k$ pairs. Now by definition

$$\sum_1^N [a(x_i-\mu_x)^2 + b(x_i-\mu_x)(y_i-\mu_y) + c(y_i-\mu_y)^2] = N(a\sigma_x^2 + b\sigma_{xy} + c\sigma_y^2)$$

Then

$$\sum_{N-k+1}^{N} [a(x_i-\mu_x)^2 + b(x_i-\mu_x)(y_i-\mu_y) + c(y_i-\mu_y)^2] \leq N(a\sigma_x^2 + b\sigma_{xy} + c\sigma_y^2) \tag{2.48}$$

It follows that

$$kd(a\sigma_x^2 + b\sigma_{xy} + c\sigma_y^2) \leq N(a\sigma_x^2 + b\sigma_{xy} + c\sigma_y^2)$$

because of the hypothesis. Then $k/N \leq 1/d$. As a special case suppose

$\sigma_x^2 = \sigma_y^2 = \sigma^2$, and the correlation is $\rho$. Then the fraction of pairs on or outside the ellipse

$$(x - \mu_x)^2 - 2\rho(x - \mu_x)(y - \mu_y) + (y - \mu_y)^2 = 2d\sigma^2(1 - \rho)$$

is not greater than $1/d$.

### 2.7.7 More Difficult Case of Two Variables

We wish merely to mention that a much less easy case is that of $N$ individuals with two attributes $(x_1, x_2)$ such that one attribute is categorical or merely ordered, while the other is arithmetical. It is not at all clear what are good measures of association and we invite the reader to give thought to the matter. Clearly one can categorize the arithmetical variable and then use a measure of association appropriate to two categorical variables.

This situation is exceedingly common, and the approach most frequently used is not to think in terms of measures of association, but to think in terms of partial descriptions of the several arithmetical populations determined by the nonarithmetical attribute. One may hope, for instance, that these several populations differ essentially in mean value only or in variance only.

### 2.7.8 Covariability in Three or More Attributes

The obvious extension of the preceding is to suppose that we have a population of $N$ individuals, each of which possesses attributes which we represent by the ordered $m$-tuple $x_1, x_2, \ldots, x_m$.

We can certainly envisage ideas of marginal frequencies and of conditional frequencies such as $f(x_1 | x_{20}, x_{30})$ which is the frequency of $x_1$ in the sub-population for which $x_2$ equals $x_{20}$ and $x_3$ equals $x_{30}$. Each of the variables $x_1, x_2, \ldots, x_m$ may be of one of the three types described in Sect. 2.3.

The simplest case is that in which each $x_i$ is arithmetical. We can then obtain all first moments, all second moments, and hence all correlations of pairs of attributes. We can also develop measures of association of pairs of attributes with restrictions on other attributes. This matter will be covered in the discussion of inferential statistics. We can also develop population mean square linear regressions of any one variable on any set of the other variables. So, for instance, we can find $\beta_{10}, \beta_{12}, \beta_{13}, \ldots, \beta_{1r}$ so that

$$\text{Ave}(x_1 - \beta_{10} - \beta_{12}x_2 - \beta_{13}x_3 \ldots - \beta_{1r}x_r)^2 \tag{2.49}$$

is a minimum. This topic will be discussed in connection with portions (samples) of populations in later chapters.

The commonly used approach to cases in which some of the variables are arithmetical and others are categorical is to attempt a partial description of each of the populations defined by the contributions of nonarithmetical variables. We take up these cases in Chapter 16.

## 2.8 Approximation of Frequency Distributions by Mathematical Distributions

### 2.8.1 Discrete and Continuous Mathematical Distributions

It is hoped that the reader is persuaded at this point that frequency functions are useful for the characterization of populations. However, it is apparent that such a characterization will be very tedious and algebraic manipulation can be only minimal unless the frequency function possesses some structure. For this reason, we often approximate the frequency function of an existing population by a function which can be given in terms of some algebraic formula. Let us abstract the essential properties of a frequency function. They are: $f(x) \geq 0$ and $\sum f(x) = 1$. For a function to be useful as an approximation for some frequency function, it must incorporate these two concepts in some manner. There are two types of functions which have proved to be useful for the purpose of approximating frequency functions—discrete and continuous density functions.

A discrete probability density function $f(x)$ is positive only on a countable set of numbers—i.e., on a set of numbers which is finite or corresponds one-to-one with the positive integers. It has the properties of nonnegativity and summing to unity on the set where it is positive. Of course, if $f(x) > 0$ on an infinite set—the positive integers, for instance—we require that the sum of $f(x)$ over the infinite set converge to unity.

A continuous probability density function $f(x)$ is a continuous function which is nonnegative and integrates to unity over the range where it is positive. Without loss of generality, we can say that

$$\int_{-\infty}^{\infty} f(x)\,dx = 1 \tag{2.50}$$

since we do not affect the value of the integral by integrating over any region where $f(x) = 0$. We shall use p.d.f. to denote probability density function.

The cumulative density function or mathematical distribution function $F(x)$ is defined to be

$$F(x) = \sum_{t \leq x} f(t) \quad \text{for the discrete case}$$

and

$$F(x) = \int_{-\infty}^{x} f(t)\,dt \qquad \text{for the continuous case} \qquad (2.51)$$

We shall use c.d.f. to denote cumulative distribution function.

In using mathematical distributions to approximate frequency functions of existing populations, the situation is fairly simple if we choose a discrete density function. We simply use $f(x)$, the ordinate of the density function, as our approximation for the relative frequency of $x$. If we use a continuous density function, however, the situation is changed somewhat. We take the view that $\int_a^b f(x)\,dx$ may be a reasonable approximation for the relative frequency of numbers between $a$ and $b$. However, as we let $b$ approach $a$, this gives us zero rather than an approximation for the relative frequency of $a$. Thus areas under the curve are meaningful rather than ordinate values. If $x$ is a number in our population, $f(x)\,dx$ is an approximation for the relative frequency that the variable lies in the interval $(x, x + dx)$.

Similarly a mathematical continuous distribution in $k$ attributes is defined by a nonnegative function $f(x_1, x_2, \ldots, x_k)$ such that the relative frequency with which $x_1$ is in $[a_1, b_1]$, $x_2$ is in $[a_2, b_2]$, . . . , $x_k$ is in $[a_k, b_k]$, is equal to the $k$-fold integral

$$\int_{a_1}^{b_1} \int_{a_2}^{b_2} \cdots \int_{a_k}^{b_k} f(x_1, x_2, \ldots, x_k)\,dx_1\,dx_2 \ldots dx_k$$

this being the ordinary multiple Riemann integral over a $k$-dimensional rectangle. In order for this to be viable, it is necessary that the integral exists for all generalized rectangles.

Discrete density functions arise quite naturally as approximating functions when the number population is generated by a counting process. If, for example, we are counting the number of particles emitted per hour by a radioactive source, it seems natural to take a density function defined on $x = 0, 1, 2, \ldots$. Of course there will actually be an upper bound to the number of particles that can be observed. However, if the density function gives extremely small mass or density to all the large values of $x$, it can still serve as a satisfactory approximation.

Continuous density functions are suggested as natural candidates when the number population is generated by a measurement process. Consider the measurement of horizontal deviations from the bull's-eye of shots fired at a target. There is a philosophical question involved in whether we can define precisely what we mean by horizontal deviation. However, it seems plausible that every possible value of $x$ within some range $a \leq x \leq b$ can occur; of course, we can never realize all possible values, but within some such range we would be reluctant to exclude any number as being impossible. We do not want to suggest that populations arising from measurement processes are continuous. On the contrary, we must acknowledge that there is a limit to the precision with which we take the measurements. We may be able to measure to the nearest millimeter with the gadgetry we use, but with a finer gadget we would measure to the nearest hundredth of a millimeter. Any measurement process, in fact, is intrinsically and basically a counting process in which we have a rod marked off as a ruler with numbered cells, and we conclude from taking the measurement that a value is in a certain cell, e.g., between 69 1/4 and 69 3/4 inches, which we label for brevity as 69 1/2 inches. From this point of view which seems quite incontrovertible, all observations are discrete.

### 2.8.2 Averaging with Mathematical Distributions

In the case of finite populations the arithmetic average of $g(x)$ is obtained by

$$\text{Ave}[g(x)] = \sum g(x)f(x) \tag{2.52}$$

where the summation is over distinct values in the population. For a discrete density function we define the average value of $g(x)$ in the same manner. A proviso is necessary with regard to all summations over infinite sets. It is then necessary that any summation be absolutely convergent. Otherwise we could get a variety of answers by different modes of summation. So we say that $\text{Ave}[g(x)]$ exists if and only if $\sum g(x)f(x)$ or $\int g(x)f(x)\,dx$ converges absolutely.

With these interpretations of $\text{Ave}[g(x)]$, moments, generating functions, entropy, etc., are defined for mathematical distributions. It should be emphasized that the use of mathematical distributions eliminates some problems by stripping away the messiness of dealing with somewhat irregular frequency functions, while creating other problems. As far as physical populations are concerned, there is no question of the existence of moments, moment generating functions, etc. However, for mathematical distributions the matter of convergence is of paramount concern.

## 2.9 The Main Substance of Probability and Statistics

A brief survey is included here of the overall processes covered in the ensuing chapters. The first basic concept is the development and study of mathematical populations based on the calculus of probability.

Rather than beginning with complete populations whose structure we wish to determine by analysis and dissection, we consider elementary probability processes which are defined mathematically and lead to mathematical populations. For almost all purposes, probability is relative frequency in repetitions. However, the concept of relative frequency is not adequate for continuous mathematical distributions. It is necessary to define a mathematical abstraction called probability for a mathematically defined broad class of sets of possible values of variables. This concept of probability leads to probabilities that are interpretable as relative frequencies.

From such probability calculations we can develop mathematical models for populations, and these mathematical models are then used to represent real populations. Chapters 3 to 6 are devoted to these matters. Chapter 7 is a short introduction to some aspects of complicated populations defined in terms of serially connected elementary probability models.

The second portion of the book deals with the general problem that most of the data sets which we meet are not usefully regarded as comprising complete populations, but should be regarded as portions, obtained by a process called random sampling, of infinite populations. This activity is called inferential statistics, and within it are many different complementary and competing aspects.

## PROBLEMS

1. Fifty 8-ohm resistors were handpicked from standard production for use in a special apparatus being assembled by a small electronics firm. Each resistor is individually tested and the resistances for this population of resistors are as follows:

| | | | | |
|---|---|---|---|---|
| 8.02 | 8.20 | 8.00 | 8.14 | 7.92 |
| 7.98 | 8.10 | 8.00 | 7.90 | 8.06 |
| 7.92 | 8.04 | 8.04 | 7.84 | 7.94 |
| 8.08 | 8.06 | 8.02 | 7.82 | 8.04 |
| 8.14 | 8.00 | 7.98 | 7.96 | 8.02 |
| 7.86 | 7.94 | 8.18 | 8.02 | 7.92 |
| 7.96 | 7.98 | 7.96 | 8.04 | 8.00 |
| 8.06 | 8.12 | 7.94 | 7.98 | 7.98 |
| 7.88 | 7.92 | 8.16 | 8.02 | 8.00 |
| 7.90 | 8.02 | 8.04 | 8.02 | 8.02 |

(a) Calculate the frequency function $f(x)$ and the distribution function $F(x)$. Sketch both.

(b) Find the mean and variance of the numbers in the population, using $f(x)$.

(c) Use Chebyshev's inequality to estimate the fraction of the population within $\pm .10$ of the mean and compare with the exact answer for this problem.
(d) Find the median, mode, and entropy.
(e) Evaluate the skewness and kurtosis parameters.

2. Group the data of the previous problem into five groups given by readings with values 7.82–7.88, 7.90–7.96, 7.98–8.04, 8.06–8.12, and 8.14–8.20.
(a) Using the frequencies and the midpoints of these intervals, calculate the frequency function $f(x)$ and the distribution function $F(x)$.
(b) Find the mean and variance and compare with the answers from the previous problem.

3. Archeologists studying Indian ruins in the southwestern United States have been able to make extensive use of tree-ring dating as well as carbon dating in estimating the age of artifacts. Suppose that in one collection the age (years) estimated by tree-ring dating $x$ and by carbon dating $y$ were as follows:

| $x$ | $y$ | $x$ | $y$ | $x$ | $y$ |
|---|---|---|---|---|---|
| 710 | 795 | 212 | 222 | 415 | 432 |
| 717 | 764 | 822 | 765 | 272 | 352 |
| 350 | 320 | 612 | 543 | 204 | 187 |
| 323 | 360 | 647 | 642 | 206 | 192 |
| 500 | 612 | 513 | 533 | 824 | 764 |
| 620 | 642 | 722 | 724 | 641 | 701 |
| 832 | 786 | 724 | 745 | 527 | 529 |
| 669 | 690 | 400 | 409 | 569 | 582 |
| 917 | 878 | 396 | 456 | 693 | 646 |
| 423 | 436 | 812 | 652 | 471 | 360 |

(a) Group the $x$ and $y$ measurements into five groups each and calculate the joint frequency function for the grouped data.
(b) Using the grouped frequency function, calculate the means, variances, and covariances and compare with the exact values for this population.
(c) Find the marginal and conditional frequency functions.

4. Show that the variance is given by $\sigma^2 = \text{Ave}(x^2) - \mu^2$.

5. Show that $\text{Ave}[(x - a)^2]$ is smallest when $a = \mu$.

6. Show that $\text{Ave}[|x - a|]$ is minimized by taking $a =$ median.

7. Show
(a) $\text{Ave}[g(x) + k] = \text{Ave}[g(x)] + k$
(b) $\text{Ave}[g_1(x) + g_2(x)] = \text{Ave}[g_1(x)] + \text{Ave}[g_2(x)]$

8. Express the first four factorial moments in terms of the moments about the origin.

9. Express the first four moments about the mean in terms of the moments about the origin.

10. The cumulant generating function is defined by $\kappa(t) = \ln m(t)$. The cumulants $\kappa$ are the coefficients of $t^i/i!$ in the power series expansion of $\kappa(t)$. By using the fact that $\ln(1 + x) = x - (x^2/2) + (x^3/3) \ldots$, verify that

$$\begin{aligned}\kappa_1 &= \mu_1\\ \kappa_2 &= \mu_2\\ \kappa_3 &= \mu_3\\ \kappa_4 &= \mu_4 - 3\mu_2^2\end{aligned}$$

11. Express the first four cumulants in terms of the factorial moments.
12. Show that the maximum entropy for a population with $N$ distinct values is $\ln N$.
13. Show that for any symmetric frequency function $\mu_3 = 0$.
14. Show that $\text{Ave}(x^2) \geq \mu^2$.
15. A population with the three distinct values 0, 1, 2 has mean 1 and variance 1. Find the frequency function.
16. Show that the covariance of $x_1$ and $x_2$ is given by $\sigma_{12} = \text{Ave}(x_1x_2) - \mu_1\mu_2$.
17. Show that $\text{Cov}(k_1x_1, k_2x_2) = k_1k_2\text{Cov}(x_1, x_2)$.
18. If $x$ and $y$ are independent, show that $\text{Ave}(xy) = \text{Ave}(x)\text{Ave}(y)$.
19. Prove the Cauchy-Schwarz inequality.
20. Show that a distribution function $F(x)$ has the properties:
    (a) $F(x)$ is nondecreasing as $x$ increases.
    (b) $\lim_{x\to\infty} F(x) = 1, \quad \lim_{x\to-\infty} F(x) = 0$
    (c) $F(x)$ is always continuous from the right but not necessarily from the left.
21. Prove that for the correlation coefficient $\rho, |\rho| \leq 1$..
22. Prove that if $\rho = \pm 1$, $x_1$ is a linear function of $x_2$. Also if $x_1$ is a linear function of $x_2$, $\rho = \pm 1$.
23. If $f_i(x)$, $i = 1, 2, \ldots$ is a denumerable family of discrete density functions, show that

$$\sum_{i=1}^{\infty} \alpha_i f_i(x), \qquad \alpha_i \geq 0, \ \sum_{i=1}^{\infty} \alpha_i = 1$$

    is a convergent series for all $x$ and is, therefore, a density function.
24. Repeat the previous exercise for a denumerable family of continuous density functions.
25. Suppose that we have a population in which the number of distinct values is so large that we are willing to approximate the frequency function by the density function:

$$\begin{aligned} f(x) &= 3x^2, \qquad 0 \leq x \leq 1\\ &= 0, \text{ otherwise}\end{aligned}$$

    However, our measuring technique is such that
    0 to 1/4 is recorded as 1/8
    1/4 to 1/2 is recorded as 3/8
    1/2 to 3/4 is recorded as 5/8
    3/4 to 1 is recorded as 7/8

(a) Obtain the frequency function of the recorded measurements.
(b) Find the mean and variance of the measurement data and compare with the mean and variance of the approximating continuous distribution.
(c) Sketch the two frequency functions on the same graph.
(d) Find the distribution functions for the two cases and sketch on the same graph.

26. Prove Chebyshev's inequality for discrete and continuous density functions.

27. Given the mathematical distribution

$$f(x) = (1/2)^x, \quad x = 1, 2, 3, \ldots$$
$$= 0, \text{ otherwise}$$

(a) Sketch $f(x)$.
(b) Find $F(x)$ and sketch.
(c) Find Ave($x$) using the definition.
(d) Find the moment generating function and differentiate to find the mean and variance.

28. Given the mathematical distribution with density function:

| $x$ | 0 | 1 | 2 | 3 | 4 | Other values |
|---|---|---|---|---|---|---|
| $f(x)$ | $k$ | $3k$ | $3k$ | $2k$ | $k$ | 0 |

(a) Find $k$.
(b) Sketch $f(x)$.
(c) Find and sketch $F(x)$.
(d) Find the mean and variance.
(e) Use Chebyshev's inequality to estimate the total density assigned to values such that $|x - \mu| \geq 2$ and compare with the exact answer for this problem.

29. Given the mathematical distribution

$$f(x) = k(x - 1)^2, \quad 0 \leq x \leq 1$$
$$= 0, \text{ otherwise}$$

repeat (a) through (e) of the preceding exercise.

30. Find all possible medians of the mathematical distribution

$$f(x) = (1/2)^x, \quad x = 1, 2, 3, \ldots$$
$$= 0, \text{ otherwise}$$

31. Find all possible medians and modes of the mathematical distribution

$$f(x) = \binom{5}{x}(1/3)^x(2/3)^{5-x}, \quad x = 0, 1, \ldots, 5$$
$$= 0, \text{ otherwise}$$

32. A given continuous density function is positive only for nonnegative values of $x$. Show that

$$\text{Ave}(x) = \int_0^\infty [1 - F(x)]\, dx$$

That is, show that the mean of the distribution is given by the area bounded by $x = 0$, $y = F(x)$, and $y = 1$.

33. Given the bivariate density function

$$f(x, y) = 4xy, \qquad 0 \le x \le 1, 0 \le y \le 1$$
$$= 0, \text{ otherwise}$$

(a) Sketch $f(x, y)$ and $F(x, y)$.
(b) Find the marginal and conditional density functions.
(c) Show that $x$ and $y$ are independent.
(d) Find $\sigma_1^2$, $\sigma_2^2$, $\sigma_{12}$, and $\rho$.
(e) Sketch the contours of the density function and distribution function.

34. If the density function $f(x_1, x_2) > 0$ within an elliptical region and is zero elsewhere, show that $x_1$ and $x_2$ cannot be independent.

35. Given the sequence of density functions

$$f(x) = kx^{k-1}, \qquad 0 \le x \le 1$$
$$= 0, \text{ otherwise}$$
$$k = 1, 2, 3, \ldots$$

evaluate the skewness and kurtosis parameters $\alpha_3$ and $\alpha_4$ for $k = 1, 2, \ldots, 10$ and graph $\alpha_3$ and $\alpha_4$ versus $k$.

36. Consider the sequence of density functions

$$f(x) = kx^2, \qquad -c \le x \le c$$
$$= 0, \text{ otherwise}$$

where $c$ and $k$ are positive and $k = 1, 2, 3, \ldots$.
(a) Express $c$ in terms of $k$ so that the functions integrate to unity.
(b) Evaluate the kurtosis parameter for various values of $k$ and graph $\alpha_4$ versus $k$.

37. Show that transform (5) mentioned in Sect. 2.5 can be used to generate moments. That is, show that

$$\mu_{(r)} = \frac{d^r}{dt^r}\left[\text{Ave}(t^x)\right]\Bigg|_{t=1}$$

38. Given the mathematical distributions:

$$f(y|x) = 3y^2/x^3, \qquad 0 \le y < x$$
$$f_X(x) = 2 - 2x, \qquad 0 \le x < 1$$

(a) Find $f(x, y)$.
(b) Find the joint moment generating function.

39. In this chapter we have described all existing populations as being finite and have considered the approximation of their frequency distributions by mathematical distributions, either finite or continuous. However some populations may be better approximated by a mixture of discrete and continuous density functions. Consider a population of 100 numbers consisting of 25 one's, 25 two's, and the numbers 3.02, 3.04, . . . , 4.00. Approximate the distribution function by a combination of

$$f(x) = 1/2, \qquad x = 1, 2$$
$$= 1, \qquad 3 \le x \le 4$$
$$= 0, \text{ otherwise}$$

Sketch the approximating distribution function.

40. Show that if $x_1, x_2, \ldots, x_n$ are independent with identical frequency functions, the m.g.f. of $\sum_{i=1}^{n} x_i$ is $[m(t)]^n$.

41. If $x_1$ and $x_2$ are independent, show that $\sigma_{12} = 0$.

42. Given the mathematical distribution

$$\begin{aligned} f(x) &= 1/5, \qquad x = -2, -1, 0, 1, 2 \\ &= 0, \text{ otherwise} \end{aligned}$$

show that $x$ and $x^2$ have zero covariance, although they are correlated.

# 3
# Calculus of Probability

## 3.1 Introduction

A surprising fact is that while there is no general agreement on the interpretation of probability, there is universal agreement about the rules—i.e., about how given or assumed probabilities are to be combined. This agreement has allowed advancement in spite of the lack of unanimity in interpretation. The basic idea is that one is given basic, also called elementary or atomic, events with their probabilities and then tries to obtain the probability of a particular complex event, this probability being a consequence of the probabilities of the basic or elementary events.

The mathematical idea of probability is an abstraction, and the theory an axiomatic theory without any necessary association with real-world phenomena. However, all the ideas in this mathematical theory can be understood by interpreting probability as relative frequency. Therefore, when we take the probability of heads in a toss of a coin to be one-half, we envisage an indefinitely large population of coin tosses, half of which give heads as the result. This is a conceptual idea, and the correspondence to a real coin is no more or less real than the correspondence of lines of geometry to "real" lines that can be drawn.

In Sections 3.1 through 3.12, we shall be concerned with development of some of the ideas of probability from the relative frequency point of view. Then in Section 3.13, we shall show how these same rules can be developed from an axiomatic basis.

## 3.2 Description of Events

Consider the result of tossing a die on a tabletop. The possible results obviously appear to be the showing of 1, 2, 3, 4, 5, or 6. We say "appear to be" because the number of dots on the top face of the die might not be an obvious way at all to describe the possible results for a person who has never seen a die before. He might, for instance, report whether or not the die rolled off the table, the distance of its final position from the nearest edge of the tabletop, etc. Upon a little reflection, it should be obvious that the set of possible results, often called the *sample space*, consists of a vast number of possibilities. When we report the occurrence of a 3, for example, we are choosing to ignore all the other properties associated with a particular toss. Equivalently, we are saying that the result belongs to the set of results which have the common characteristic of showing three dots on the top face of the die.

We refer to a set of possible results as an *event*, and when we speak of the probability of an event, we have in mind the probability of a result belonging to a particular set of results. For economy of language, however, we use expressions which are not quite so precise. In the throwing of a die we speak of the probability of a 2 rather than the probability that the result is a member of the set of all results which show a 2. In drawing a card from a bridge deck, we speak of the probability of a king rather than the probability that the result belongs to the set of results where a king shows.

In the example of the die, in most cases, we would think of the sets of results described by the number of dots on the top face as exhausting the space of possibilities. Also, in most cases we would not be interested in subdividing these sets into smaller sets. Thus these basic sets may be thought of as describing basic *events*, and all other events are combinations of these.

As another example, consider drawing a member of the population of the United States. The one we obtain may be (1) male or female, (2) less than 21 years old or greater than 21 years old, (3) blue-eyed or nonblue-eyed. If we designate the first of these alternatives by $A$, $B$, $C$ respectively and the second by $\bar{A}$, $\bar{B}$, $\bar{C}$, we can obtain the following possibilities with regard to these properties: $ABC$, $AB\bar{C}$, $A\bar{B}C$, $A\bar{B}\bar{C}$, $\bar{A}BC$, $\bar{A}B\bar{C}$, $\bar{A}\bar{B}C$, $\bar{A}\bar{B}\bar{C}$. The class $\bar{A}\bar{B}C$, e.g., contains females over twenty-one and blue-eyed. With each of the possibilities is an associated probability $P(AB\bar{C})$, which is the probability that the chosen individual is male, less than twenty-one years old, and nonblue-eyed.

In such problems we are usually given a set of basic events and their respective probabilities. In many problems every basic event has several attributes, and we can envisage classifying these events by the layout in Table 3.1, where the entries have no significance except to indicate what might be the case in a particular situation. In Table 3.1 $N$ is $2^3$, and in general with $r$ attributes $N$ is $2^r$.

Associated with each basic event will be a given probability and we can then envisage derived probabilities like $P(ABC)$ i.e., the probability that a result will have properties *A and B and C*—or $P(\bar{A}\bar{B}\bar{C})$—i.e., the probability that a result will have properties $\bar{A}$ *and* $\bar{B}$ *and* $\bar{C}$.

**Table 3.1** Classification of Basic Events

| Primitive Event | Attributes | | |
|---|---|---|---|
| | $A$ or $\bar{A}$ | $B$ or $\bar{B}$ | $C$ or $\bar{C}$ |
| 1 | $A$ | $\bar{B}$ | $C$ |
| 2 | $A$ | $B$ | $\bar{C}$ |
| . | . | . | . |
| . | . | . | . |
| . | . | . | . |
| $N$ | $\bar{A}$ | $\bar{B}$ | $\bar{C}$ |

The use of elementary set notation and Venn diagrams often helps in the description of events. In the Venn diagram in Figure 3.1, the region $\mathscr{X}$ represents the sample space, or the set of all possible results, the region $A_1$ represents the set of results which have property $A_1$, and the region $A_2$ represents the set of results with property $A_2$. If $\mathscr{X} - A_1$ denotes the results in $\mathscr{X}$ not in $A_1$, then $\mathscr{X} - A_1 = \bar{A}_1$, the complement of $A_1$. Similarly $\mathscr{X} - A_2 = \bar{A}_2$, the complement of $A_2$. The symbol $\cap$ denotes intersection and the intersection of sets $A_1$ and $A_2$, $A_1 \cap A_2$, is the set which lies in both $A_1$ and $A_2$. The symbol $\cup$ denotes union and the union of $A_1$ and $A_2$, $A_1 \cup A_2$ is the set of points which are in $A_1$, $A_2$, or both.

Some remarks must be made about the use of language. Although "$A$ union $B$" is a perfectly meaningful expression in set theory, some other choice of words seems necessary when a specific example is under consideration. Thus we never speak about the probability of an ace union king in a card problem, but rather about the probability of an ace or a king. In

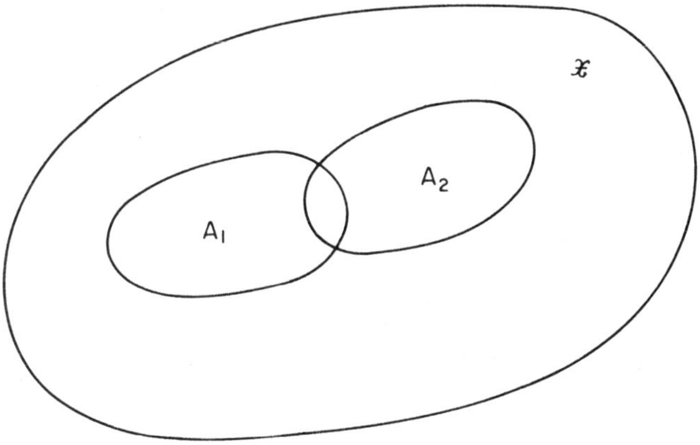


**Fig. 3.1** Description of Events

mathematics the word "or" usually means "and/or" so "$A$ or $B$" agrees with "$A$ and $B$", "$A$ and not-$B$", and "not-$A$ and $B$." The expression "$A$ intersection $B$" also becomes awkward in particular instances, and instead of talking about the probability of an ace intersection king, for example, we use the expression of an ace *and* a king. One final comment on notation seems in order. We shall often use the notation $AB$ instead of $A \cap B$ to mean "$A$ and $B$."

## 3.3 Mutually Exclusive Events

As should be obvious from the preceding section, the word event as used in probability theory has a double meaning. Sometimes it refers to a set of results having a particular property, and sometimes it refers to the property itself. Of course the difference in usage is not important as long as it is perceived and understood. In the first context mutually exclusive events are simply disjoint sets, i.e., sets with no points in common. In the second context $A$ and $B$ are mutually exclusive if properties $A$ and $B$ cannot occur together.

We have taken the point of view that probability is an abstraction of relative frequency. Consider a collection of $n$ objects, each of which has one and only one of the properties $A_1, A_2, A_3, \ldots, A_m$. Let the frequency of $A_i$ in the collection be $n_i$. Then the probability of $A_i$ is

$$P(A_i) = \frac{n_i}{n} \tag{3.1}$$

We may consider a classification of the objects by grouping together some of the classes $A_i$. Obviously if $B$ is equivalent to $A_1$ or $A_2$, then

$$P(B) = P(A_1 \text{ or } A_2) = P(A_1 \cup A_2) = P(A_1) + P(A_2) \tag{3.2}$$

In the above we assumed that the classification of objects is mutually exclusive and exhaustive. Obviously if $A_1, A_2, \ldots, A_m$ comprise mutually exclusive and exhaustive classes and if $\mathscr{X}$ is the class $A_1$ or $A_2$ or $A_3 \ldots$ or $A_m$, then

$$P(\mathscr{X}) = P(A_1) + P(A_2) + \cdots + P(A_m) \tag{3.3}$$

and because the relative frequency of an object being in class $\mathscr{X}$ is 1, we have

$$1 = P(A_1) + P(A_2) + \cdots + P(A_m) \tag{3.4}$$

It is also obvious that $A_1$ and $\bar{A}_1$ are mutually exclusive, so that

$$P(A_1) + P(\bar{A}_1) = 1 \tag{3.5}$$

If $\phi$ denotes the null set so that no objects fall in $\phi$, then $\phi$ and $\mathscr{X}$ are mutually exclusive and exhaustive, so that

$$P(\phi) = 0 \tag{3.6}$$

## 3.4 One or More Events

We are often interested in the probability that at least one of the events $A_1, A_2, \ldots, A_m$ will occur. For example, in an electronic system where parallel circuits are installed for the purpose of improving reliability, we would be interested in the probability that at least one of the circuits will operate successfully. The basic rule is

$$P(A_1 \cup A_2) = P(A_1) + P(A_2) - P(A_1 A_2) \tag{3.7}$$

This is easily motivated by the relative frequency concepts and by the use of the Venn diagram in Figure 3.1. The event $A_1 \cup A_2$ is equivalent to $(A_1 - A_1A_2) \cup (A_2 - A_1A_2) \cup A_1A_2$. That is, $A_1 \cup A_2$ is expressed in terms of mutually exclusive events. Let $n_{12}$ be the frequency of $A_1A_2$. Thus for a large number $n$ of objects

$$\begin{aligned} P(A_1 \cup A_2) &= \frac{n_1 - n_{12}}{n} + \frac{n_2 - n_{12}}{n} + \frac{n_{12}}{n} \\ &= \frac{n_1}{n} + \frac{n_2}{n} - \frac{n_{12}}{n} \\ &= P(A_1) + P(A_2) - P(A_1A_2) \end{aligned} \tag{3.8}$$

We can apply the same reasoning to obtain

$$\begin{aligned} P(A_1 \cup A_2 \cup A_3) = P(A_1) &+ P(A_2) + P(A_3) - P(A_1A_2) \\ &- P(A_1A_3) - P(A_2A_3) + P(A_1A_2A_3) \end{aligned} \tag{3.9}$$

In general we have the following rule:

$$\begin{aligned} P\left(\bigcup_1^n A_i\right) &= \sum_i P(A_i) - \sum_{i<j} P(A_iA_j) \\ &+ \sum_{i<j<k} P(A_iA_jA_k) - \cdots \pm P(A_1 \ldots A_n) \end{aligned} \tag{3.10}$$

A rule for remembering the algebraic sign is that probabilities involving even numbers of events occur with a negative sign, the others with positive signs.

## 3.5 Joint, Marginal, and Conditional Probabilities

Consider again the Venn diagram in Figure 3.1. For rather obvious reasons $P(A_1A_2)$ is referred to as a joint probability. For reasons which are somewhat less obvious but which will be stated shortly, $P(A_1)$ and $P(A_2)$ are called marginal probabilities.

We may also consider the results in $A_1$ and ask with what frequency they fall in $A_2$. This is the conditional probability of $A_2$ given $A_1$, $P(A_2|A_1)$ and is given by

$$P(A_2|A_1) = \frac{P(A_1A_2)}{P(A_1)} \tag{3.11}$$

From the relative frequency point of view we can envisage a large collection of $n$ objects with $n_1$ having attribute $A_1$, and $n_{12}$ having attributes $A_1$ and $A_2$. Then

$$\begin{aligned} P(A_1) &= n_1/n \\ P(A_1A_2) &= n_{12}/n \\ P(A_2|A_1) &= n_{12}/n_1 \\ &= \frac{P(A_1A_2)}{P(A_1)} \end{aligned}$$

Of course if $P(A_1) = 0$, the rule does not hold because we cannot divide by zero. This does not give us any trouble, however, because we shall simply not define the conditional probability $P(A_2|A_1)$ when $P(A_1) = 0$. In reading this and subsequent sections, the reader should keep this in mind because it will not always be explicitly mentioned.

In considering events which can be classified by two criteria, it is often helpful to portray the situation as in Table 3.2. At this point our use of the expression marginal probability should be obvious since $P(A)$, $P(\bar{A})$, $P(B)$, and $P(\bar{B})$ occur in the margins of Table 3.2. It should be emphasized

**Table 3.2** Classification by Two Criteria

| | $B$ | $\bar{B}$ | Marginal Total |
|---|---|---|---|
| $A$ | $P(AB)$ | $P(A\bar{B})$ | $P(A)$ |
| $\bar{A}$ | $P(\bar{A}B)$ | $P(\bar{A}\bar{B})$ | $P(\bar{A})$ |
| Marginal Total | $P(B)$ | $P(\bar{B})$ | 1 |

that the words "joint," "marginal," and "conditional" are not different types of probability but depend upon one's perspective.

In a certain sense all probabilities are conditional, because probabilities cannot be computed except on the basis of probabilities of basic events. We might, for instance, calculate that the probability of tossing a 1 or 6 with a die is 1/3, but we would be using a given datum such as that the probabilities of getting any one of 1, 2, 3, 4, 5, or 6 is 1/6. So strictly speaking we should say

$$P(1 \text{ or } 6 | \text{all numbers equally probable}) = 1/3 \qquad (3.12)$$

Frequently the conditioning statement in a probability statement is omitted either because it is obvious in the context of the problem or it is universally accepted. If we assert that the probability of obtaining a family of four sons is 1/16, we are implicitly including the basic probability of an offspring being male as 1/2, independently for each offspring. Of course this may not be the case and is not exactly so in the present instance.

We have already stated that probabilities are always conditional upon the evidence and knowledge available or assumed. In other words, relative frequencies of a type of individual depend totally on the population of individuals under consideration. Often, however, it is clear from the context of the problem what the background information or population $U$, say, is, and the probability $P(A|U)$ is simply written $P(A)$.

We are often interested in $P(A_1A_2A_3)$. Let the joint occurrence $A_2A_3$ be denoted by $B$. We know that

$$\begin{aligned} P(A_1B) &= P(A_1|B)P(B) \\ &= P(A_1|A_2A_3)P(A_2A_3) \end{aligned} \qquad (3.13)$$

We also know that

$$P(A_2A_3) = P(A_2|A_3)P(A_3)$$

Combining this equation with the preceding gives the interesting chain rule,

$$P(A_1A_2A_3) = P(A_1|A_2A_3)P(A_2|A_3)P(A_3) \qquad (3.14)$$

The student should verify that this can be written in several other ways by permuting the subscripts. For example,

$$P(A_1A_2A_3) = P(A_3|A_1A_2)P(A_2|A_1)P(A_1) \qquad (3.15)$$

The chain rule developed for this example can be developed in general for $n$ attributes as stated in the following rule:

$$P(A_1A_2 \ldots A_n) = P(A_1|A_2 \ldots A_n)P(A_2|A_3 \ldots A_n) \ldots P(A_{n-1}|A_n)P(A_n) \qquad (3.16)$$

## 3.6 Independence

The concept of statistical independence is a central one which must be distinguished from mathematical or functional independence. In fact, it seems to be a concept well established for most people in an intuitive sense, but not in a formal manner. Two events are said to be statistically independent if their joint probability is the product of their marginal probabilities. That is, $A_1$ and $A_2$ are independent if

$$P(A_1A_2) = P(A_1)P(A_2) \tag{3.17}$$

In general, $A_1, A_2, \ldots, A_n$ are independent if the joint probability of any subset is equal to the product of the marginal probabilities.

It is important to note that if we can regard an observation as arising from unrelated complexes of causes, we can achieve a considerable economy of representation. Let an observation be classified by a factor $A$ with mutually exclusive and exhaustive categories $A_1, A_2, \ldots, A_m$ *and* by a factor $B$ with mutually exclusive and exhaustive categories $B_1, B_2, \ldots, B_n$. Then any observation has the attribute $A_iB_j$ for some $i$ and $j$. If we have to regard the $mn$ classes $A_iB_j$ as fundamental and unrelated, we have to keep account of $mn$ probabilities or relative frequencies $P(A_iB_j)$. If on the other hand we can consider the $A$ and $B$ classifications to be made without connection, we would have $P(A_iB_j) = P(A_i)P(B_j)$. We would then have to account for only

$$\begin{aligned} &P(A_i), \qquad i = 1, \ldots, m \\ &P(B_j), \qquad j = 1, \ldots, n \end{aligned}$$

or $m + n$ in all. It is obvious then that the model of independence is useful. Indeed the aim of much data analysis is the development of a model using features of statistical independence.

We should briefly mention conditional probability in the case of independence. If $A_1, A_2, \ldots, A_n$ are independent, then all conditional probabilities which are defined are equal to the corresponding marginal probabilities. For example, if $A_1$ and $A_2$ are independent and $P(A_1) \neq 0$, then

$$\begin{aligned} P(A_2|A_1) &= \frac{P(A_1A_2)}{P(A_1)} \\ &= \frac{P(A_1)P(A_2)}{P(A_1)} \\ &= P(A_2) \end{aligned} \tag{3.18}$$

In general, we have the following theorem:

***Theorem 3.1*** If $A_1, A_2, \ldots, A_n$ are independent, the conditional probability of any one, say $A_k$, given any subset of the remaining, is equal to the marginal probability $P(A_k)$.

*Example 1*

This example uses several of the rules discussed in this and the preceding sections.

An electrical system consists of four components in circuit as shown in Figure 3.2.

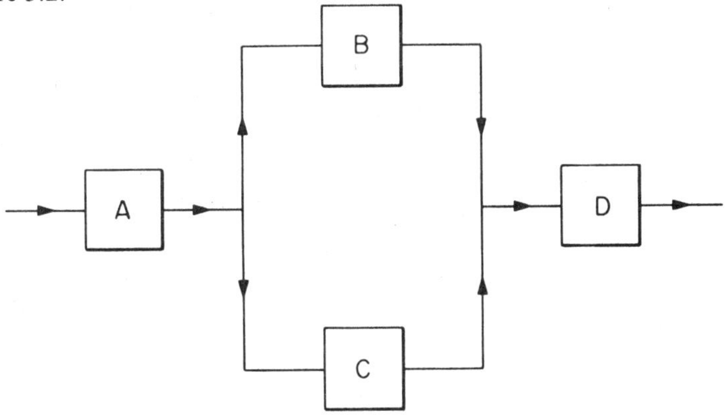


**Fig. 3.2** Hypothetical Electrical System

For the system to function, both $A$ and $D$ must operate and either $B$ or $C$ must operate. Probabilities of successful operation have been determined by extensive life test to be $P_A = .80$, $P_B = .90$, $P_C = .90$, and $P_D = .70$. If $A$ fails, the other components never get a chance to operate successfully. Assuming that the four components operate independently, what is the probability that the system will fail to operate?

There are three mutually exclusive ways in which the system can fail, namely:

$E_1$: failure of $A$
$E_2$: operation of $A$, failure of both $B$ and $C$
$E_3$: operation of $A$, operation of $B$ or $C$, failure of $D$

The probabilities of these three events are given by:

$$P(E_1) = 1 - P_A = .20$$

$$\begin{aligned} P(E_2) &= P_A(1 - P_B)(1 - P_C) \\ &= (.80)(.10)^2 = .008 \end{aligned}$$

$$\begin{aligned} P(E_3) &= P_A(P_B + P_C - P_{BC})(1 - P_D) \\ &= (.80)(.90 + .90 - .81)(.30) \\ &= .2376 \end{aligned}$$

Thus the probability of failure is given by $P(E_1) + P(E_2) + P(E_3) = .4456$.

An alternative way to compute the probability is to consider the way in which the system can operate successfully. $A$ must function, $B$ or $C$ must function, and $D$ must function. Thus the probability of successful operation is given by

$$P(A)P(B \cup C)P(D) = (.80)(.90 + .90 - .81)(.70)$$
$$= .5544$$

and the probability of failure is given by 1-Prob (success) which yields .4456, as before.

The idea of independence has two main uses. The simple one is that described above in which $P(A_1|A_2) = P(A_1)$, so that the relative frequency of $A_1$ in the subpopulation of individuals possessing $A_2$ is the same as in the overall population. The other use is critical in the development of probabilities. If we can assume that $A_1$ and $A_2$ are independent, then we can compute $P(A_1A_2)$ as $P(A_1)P(A_2)$. So if our population consists of elements $(x_1, x_2)$ in which $x_1$ is the result of a toss of a coin, $H$ or $T$, and $x_2$ is the result of a toss of a die, one of the integers 1 to 6, then we may assume that the two physical processes are unrelated in their results. We can then deduce $P(x_1, x_2) = P(x_1)P(x_2)$. Much of the theory of probability is concerned with a population of elements, the probabilities of which are obtained by this sort of process.

## 3.7 Combinatorics

While the concepts discussed thus far are elementary, the computational difficulties encountered sometimes become formidable. We have covered the fundamental concepts of probability, but the use of these concepts for the calculation of probabilities of generalized events can become very difficult. In the following section we begin by considering the probability that $n$ trials will give $r$ successes, given that the trials are independent and the probability of a success on any one trial is a quantity $p$. To take up such questions, we shall find permutations and combinations very useful.

Consider the number of ways in which $n$ objects may be arranged where the order of arrangement is important. That is, $(a, b, c)$ and $(c, b, a)$ are considered to be two different arrangements, called permutations. We are interested in the number of permutations possible. Consider the number of ways in which $n$ objects may be arranged. We may choose any one of $n$ objects for the first position in the arrangement, any one of $(n - 1)$ for the second position, etc. The total number of arrangements denoted by $P(n, n)$ is given by

$$P(n, n) = n(n - 1)(n - 2) \ldots (2)(1) = n! \tag{3.19}$$

We may be interested furthermore in the number of permutations possible when we choose $r$ objects from $n$. Suppose that we list all possible permutations of $n$ things taken $r$ at a time. The first position may be filled in $n$ ways, the second in $(n - 1)$ ways, etc., and the last position in $(n - r + 1)$ ways. Hence the number of arrangements of $n$ things taken $r$ at a time and denoted by $P(n, r)$ is

$$P(n, r) = n(n - 1)(n - 2) \dots (n - r + 1) = \frac{n!}{(n - r)!} \tag{3.20}$$

*Example 2*

A manufacturer uses a color code to identify lots of manufactured items. The code consists of stamping seven colored stripes on the container. The order of the colors is significant, and each identification uses all seven colors. Thus it will be possible to identify 5040 ($=7!$) lots before it is necessary to start repeating the code. Suppose that the manufacturer wishes to use seven stripes but has 15 shades or colors available. How many distinct markings can he get? This is merely the number of permutations of 15 things taken 7 at a time. That is,

$$P(15, 7) = \frac{15!}{8!} = 32{,}432{,}400$$

To illustrate the use of permutations in the calculation of probabilities, suppose that in the preceding example we want the probability that the first stripe is red. To use the frequency concept of probability we need

$$n = \text{number of cases where first stripe is red}$$
$$N = \text{total number of cases}$$

We already know $N$ to be $P(15, 7)$. Since there are 6 stripes left to be stamped from 14 remaining colors, $n = P(14, 6)$. Thus, the desired probability is given by

$$P(14, 6)/P(15, 7) = \frac{14!}{8!}\bigg/\frac{15!}{8!} = 1/15$$

Suppose that we have $n$ objects divided into groups with $r_1, r_2, \dots, r_k$ objects respectively. The groups are distinguishable from each other, but the members of any one group are indistinguishable. There are $r_1!$ times as many permutations when the elements of group 1 are distinguishable as when they are not. There are $r_2!$ times as many permutations when the elements of group 2 are distinguishable as when there are not, etc. Hence the number of permutations is

$$n!/r_1!r_2! \dots r_k! \tag{3.21}$$

For instance, consider the transmission of code involving two signals, a dot and a dash. How many five-letter words can be formed using 3 dots and 2 dashes? The number of possibilities is $5!/2!3! = 10$.

When arrangements are made without regard to order, they are called combinations. That is, $(a, b, c)$ and $(c, b, a)$ are considered to be the same combination of symbols. The number of combinations of $n$ things taken $r$ at a time will be denoted by $C(n, r)$, or

$$\binom{n}{r}$$

It follows that

$$C(n, r) = n!/[r!(n - r)!] = P(n, r)/r! \tag{3.22}$$

because any single combination arises from $r!$ permutations.

*Stirling's Approximation.* At one time the computational burden arising from formulas employing combinations and permutations was extremely heavy. This burden has been mitigated somewhat by the advent of modern computers and by the widespread distribution of combinatorial tables. However, for mathematical purposes and also for quick approximate calculation, we need Stirling's approximation for $n!$ This states that

$$\lim_{n \to \infty} \frac{n!}{\sqrt{2\pi n}\, n^n e^{-n}} = 1 \tag{3.23}$$

so Stirling's approximation to $n!$ is

$$n! = \sqrt{2\pi n}\, n^n e^{-n} \tag{3.24}$$

Some of the exercises are devoted to studying the behavior of this approximation. As $n$ increases, the absolute error becomes larger but the relative error decreases. A better approximation is given by

$$\ln n! \doteq (1/2)\ln(2\pi) + (1/2)\ln(n) + n\ln(n) - n - \frac{1}{12n} \tag{3.25}$$

## 3.8 Number of Successes in a Sample

We now consider the properties of subsets of a population, and to be specific suppose that we have a population of $N$ items which are classified into two categories that we shall call, purely for convenience, successes and failures. We consider furthermore subsets containing $n$ items of which there are

$$\binom{N}{n}$$

Each subset will contain a number of successes and a number of failures.

We now consider the population of all the

$$\binom{N}{n}$$

possible subsets and ask various questions about it. For instance, what is the relative frequency with which subsets contain, say, three successes, or the relative frequency with which such subsets contain less than two failures?

With this background, we can define random sampling. We say that random sampling from a finite population of $N$ items is a *process* of drawing a subset of $n$ from the population of $N$, such that the relative frequency of any particular subset in an indefinitely large number of repetitions is

$$1\bigg/\binom{N}{n}$$

A particular subset of $n$ that has been obtained by such a process is called a random sample. This leads then to statements of the form: The probability is 0.3 that the number of successes in a random sample is equal to 4.

The power of this idea is that we can discover physical processes which have with high confidence the property of selecting a subset with this probability or relative frequency. We can apply such a process to our population, obtaining a single sample which we call a random sample. We shall examine the random sample and then form opinions in a rational way about the nature of the population which has been sampled. In the probability calculus *per se* we are concerned only with the first aspect, the properties of random samples. It is worth noting that to ask for properties of a sample is meaningless unless the process by which the sample is obtained is specified. It is only when sampling incorporates some idea of randomness that we are able to develop useful ideas of the properties of a sample. Throughout this book the term "sample" will mean "random sample."

A problem may now be stated explicitly. Suppose there are $N$ items consisting of $N_s$ successes and $N_f$ failures. A sample of $n$ items is taken without replacement. That is, items are not replaced before drawing the next one. What is the probability that the sample will contain $k$ successes? We note that the number of samples containing $k$ successes (and consequently $n - k$ failures) is

$$\binom{N_s}{k}\binom{N_f}{n-k}$$

and that the total possible number of samples is

$$\binom{N}{n}$$

Therefore the answer is given by the following:

$$P(\text{number of successes} = k) = \frac{\binom{N_s}{k}\binom{N_f}{n-k}}{\binom{N}{n}} \tag{3.26}$$

This particular type of probability formula, or probability law (often called probability density function), is known as the hypergeometric probability law and will be studied further in Chapter 4.

## 3.9 Bernoulli Trials

If in the preceding section the sampling were done with replacement; i.e., if an item had been drawn at random, the result recorded, and then the item replaced, an item again drawn at random, etc., the probability of a success on any draw would have been $p = N_s/N$. The probability that any particular $k$ items in a sample of $n$ will be successes and the rest failures is then $p^k(1 - p)^{n-k}$. Since $k$ items may be chosen in

$$\binom{n}{k}$$

ways, the probability of exactly $k$ successes in a sample of $n$ is given by the following formula:

$$P(k \text{ successes}) = \binom{n}{k} p^k(1 - p)^{n-k}, \; k = 0, 1, \ldots, n \tag{3.27}$$

This formula is extremely well known. It is called the binomial probability law because the probabilities are the terms of the binomial expansion

$$(q + p)^n = \sum_{i=0}^{n} \binom{n}{i} p^i q^{n-i} \tag{3.28}$$

where $q = 1 - p$.

Independent trials such that the probability of a success on any trial is a constant $p$ are called Bernoulli trials; and whatever the origin of the problem, the probability law for the number of successes in $n$ Bernoulli trials is given by the binomial probability law, which will be studied further in Chapter 4.

## 3.10 Partitioning into More than Two Classes

In Sections 3.8 and 3.9 each item belongs to one of two possible classes. There are numerous examples where there are more than two possible classes. If a random sample of $n$ items is drawn without replacement from a set of $N$ items consisting of $N_1$ items in class 1, $N_2$ items in class 2, and ... $N_q$ in class $q$, the probability that the sample will contain $n_1$ in class 1, $n_2$ in class 2, ..., and $n_q$ in class $q$ is given by

$$\frac{\binom{N_1}{n_1}\binom{N_2}{n_2}\cdots\binom{N_q}{n_q}}{\binom{N}{n}} \tag{3.29}$$

This is called the generalized hypergeometric probability.

If sampling is done with replacement, the probability $P_i$ is given by $N_i/N$. Then the probability of $n_1$ items in class 1, $n_2$ items in class 2, etc., in any particular order is given by $P_1^{n_1}P_2^{n_2} \ldots P_q^{n_q}$. Since there are $n!/n_1!n_2! \ldots n_q!$ permutations with this probability, the desired probability is given by the multinomial law

$$\frac{n!}{n_1!n_2!\ldots n_q!} P_1^{n_1}P_2^{n_2}\ldots P_q^{n_q} \tag{3.30}$$

with the restrictions

$$\sum_{i=1}^{q} n_i = n, \sum_{i=1}^{q} P_i = 1$$

Often the multinomial law does not strictly hold, but $N$ is so large relative to $n$ that for practical purposes the multinomial suffices.

## 3.11 Occupancy Problems

In statistical mechanics, systems consisting of $p$ indistinguishable particles are considered in which each particle may be in any one of $n$ possible states (for example, $n$ different energy levels). To describe the state of the system, it becomes relevant to determine the number of different allocations of the $p$ particles to the $n$ states. The probability models constructed to describe such systems depend upon whether or not the Pauli exclusion principle is assumed to hold. This principle states that there can be at most one particle in a state. If the Pauli principle holds and all possible arrangements are given equal probability, the system is said to be described by Fermi-Dirac statistics. If the exclusion principle does not hold and all possible arrangements are given equal probability, the system is said to be described by Bose-Einstein statistics.

There arise numerous other problems similar in nature to those just mentioned. If we abstract the essential nature of these from their settings, we have the basic problem: In how many ways may $m$ balls be distributed among $n$ distinguishable urns? The answer to this problem depends upon how many balls are permitted to be in one urn and upon whether or not the balls are distinguishable one from another.

***Theorem** 3.2* The number of ways in which $m$ distinguishable balls can be assigned to $n(\geq m)$ distinguishable urns if at most one ball is permitted in an urn is $P(n, m) = n!/(n - m)!$.

*Proof:* Since each urn receives either 0 or 1 ball, the proof merely consists of seeing that the number of ways that we can choose the urns to be assigned balls is $P(n, m) = n!/(n - m)!$.

***Theorem** 3.3* The number of ways $m$ distinguishable balls can be assigned to $n$ distinguishable urns if no restriction is placed on the number of balls permitted in an urn is $n^m$.

*Proof:* Consider performing the assignment. The first ball can be assigned in $n$ ways. For each of these assignments the second ball can be assigned in $n$ ways, etc. Then the total number of possible assignments is $n^m$.

***Theorem** 3.4* The number of ways $m$ indistinguishable balls can be assigned to $n$ distinguishable urns when no restriction is placed on the number of balls permitted in any one urn is

$$\binom{n + m - 1}{m}$$

*Proof:* Represent the urns by the spaces between $n + 1$ bars and represent the objects by the letter $A$ as follows: $|A|AA|\,|AAA|||A|$ where there are seven objects and eight bars. There are then $n + m + 1$ positions filled with bars and with $A$'s. The first and last position will always be occupied by a bar. The problem is simply one of determining the number of ways of placing $m$ $A$'s in the remaining $n + m - 1$ positions. This is given by

$$\binom{n + m - 1}{m} = \binom{n + m - 1}{n - 1}$$

This theorem gives the formula which is the basis for the Bose-Einstein statistics.

***Theorem** 3.5* The number of ways $m$ indistinguishable balls can be assigned to $n$ distinguishable urns if only one ball is permitted per urn is

$$\binom{n}{m}$$

*Proof:* Since the balls are indistinguishable, the proof consists merely of observing that we want the number of ways of choosing $m$ urns from the $n$. This theorem gives the formula which is the basis of the Fermi-Dirac statistics.

## 3.12 Bayes's Theorem

Many of the important problems of science and engineering are concerned with the problem of causation. Since the philosophical difficulties connected with the problem of cause and effect are many, we admit that it is probably impossible to determine the cause and effect relationship. If $P(A|B) = 1$ and $B$ occurs, then of course $A$ occurs. Commonly it would be stated under these circumstances that $A$ is "because" of $B$ if $B$ precedes $A$ in time. Suppose that the occurrence of one of $N$ mutually exclusive events $A_1, A_2, \ldots, A_N$ is necessary for the occurrence of $B$. Given that $B$ has occurred, we wish to know which of the $A$'s preceded it, so we ask for the probability that $A_i$ has occurred, given that $B$ occurred. This is the problem which Bayes's theorem solves.

***Theorem 3.6***

$$P(A_i|B) = \frac{P(B|A_i)P(A_i)}{P(B)}, \ i = 1, 2, \ldots, N \tag{3.31}$$

*Proof:* The proof is really quite trivial. Since

$$P(A_i|B) = \frac{P(A_iB)}{P(B)}$$

and

$$P(A_iB) = P(B|A_i)P(A_i)$$

then

$$P(A_i|B) = \frac{P(B|A_i)P(A_i)}{P(B)}$$

But

$$P(B) = P(A_1B) + P(A_2B) + \cdots + P(A_NB) \tag{3.32}$$

So

$$P(A_i|B) = \frac{P(B|A_i)P(A_i)}{P(B|A_1)P(A_1) + P(B|A_2)P(A_2) + \cdots + P(B|A_N)P(A_N)}$$

$$= \frac{P(B|A_i)P(A_i)}{\sum_i P(B|A_i)P(A_i)} \tag{3.33}$$

This is of course nothing but an expanded form of the relation $P(A_i|B) = P(A_iB)/P(B)$ but is known as Bayes's theorem because Bayes (1763) was one of the first to consider this problem.

*Example 5*

Nearly all manufacturers now submit their finished product to lot acceptance sampling, where the probability of accepting the lot depends upon the quality of the lot. The following example, though artificial in its size and complexity, represents a real problem in industry. Suppose that lots of equal size are submitted to acceptance sampling. The percent defective in the lot is either 1%, 2%, 3%, or 4%, and the probabilities of acceptance in each case are given by:

| % Defective | Probability of Acceptance |
|---|---|
| 1 | .95 |
| 2 | .70 |
| 3 | .30 |
| 4 | .10 |

Given that a lot has been rejected, what is the probability that it is 1% defective?

We may use Bayes's theorem to solve this problem. Let

$$\begin{aligned} A_1 &= 1\% \text{ defective lot} \\ A_2 &= 2\% \text{ defective lot} \\ A_3 &= 3\% \text{ defective lot} \\ A_4 &= 4\% \text{ defective lot} \\ B &= \text{rejected lot} \end{aligned}$$

Then

$$\begin{aligned} P(B|A_1) &= .05 \\ P(B|A_2) &= .30 \\ P(B|A_3) &= .70 \\ P(B|A_4) &= .90 \end{aligned}$$

To use Bayes's theorem, we must also have $P(A_i)$. Suppose it is known from long experience that $P(A_1) = .20$, $P(A_2) = .30$, $P(A_3) = .30$, and $P(A_4) = .20$. Then

$$\begin{aligned} P(A_1|B) &= \frac{P(B|A_1)P(A_1)}{P(B|A_1)P(A_1)+P(B|A_2)P(A_2)+P(B|A_3)P(A_3)+P(B|A_4)P(A_4)} \\ &= \frac{(.05)(.20)}{(.05)(.20)+(.30)(.30)+(.70)(.30)+(.90)(.20)} \\ &= 10/490 \\ &= .020 \end{aligned}$$

Since the expression known as Bayes's theorem involves only the standard rules given earlier in the chapter, one may well ask why this elementary result is given special attention in a separate section. The answer lies in the use that has been made of this theorem. Suppose a person says that he is unwilling to assign probabilities to the events $A_i$ because he simply does not have a reference set in which to embed the problem. What should he do? Some say that he should then assign probabilities to the events $A_i$ according to certain rules. One such rule is called the principle of insufficient reason, which says that if not enough is known to arrive at a probability assignment of any kind then one should make equal probability assignments. This has been termed the Bayes postulate, though whether Bayes strongly advocated it is obscure. Our own view is that such rules amount to saying how a person must reason, and they cannot be defended on this basis. However, it is certainly worthwhile to compute the probability of $A_i$, given $B$, for various probability assignments to the $A_i$.

## 3.13 Axiomatic Theory: Probability as a Set Function

With the relative frequency concepts of the previous sections in mind, we now consider probability as a set function with the class of events $\mathscr{A}$ in the sample space $\mathscr{X}$ as its domain and the zero–one interval as its range. In other words, we consider probability as a mapping of each event into a number between zero and one. In the case of a finite sample space we take the class of events $\mathscr{A}$ to be the class of all subsets of $\mathscr{X}$, including the impossible event or empty set and sample space itself. It is then possible to take some of the relative frequency properties considered in this chapter as axioms and to show that the other relative frequency properties follow as natural consequences from these axioms. Let us take as axioms the following, where $\phi$ denotes the null set:

Axiom 1. $P(A) \geq 0$ for every event $A$
Axiom 2. $P(\mathscr{X}) = 1$
Axiom 3. $P(A \cup B) = P(A) + P(B)$, if $AB = \phi$

From these three axioms and the elementary rules of set theory we can obtain as logical consequences all the properties based upon relative frequency ideas. For example, since $\mathscr{X} \cup \phi = \mathscr{X}$, and $\mathscr{X} \cap \phi = \phi$, it follows from Axiom 3 that $P(\mathscr{X}) + P(\phi) = P(\mathscr{X})$, from which it follows that $P(\phi) = 0$. Similarly we can show that: $P(\bar{A}) = 1 - P(A)$, $P(A \cup B) = P(A) + P(B) - P(AB)$, $0 \leq P(A) \leq 1$, etc. It also follows that the additive property of Axiom 3 applies to any set of $k$ mutually exclusive events. That is,

$$P\left(\bigcup_1^k A_i\right) = \sum_1^k P(A_i) \text{ if } A_iA_j = \phi,\ i \neq j \tag{3.34}$$

As long as the space $\mathscr{X}$ is finite, the above is adequate because the class of all subsets contains all sets obtainable from others by forming a finite number of complements, unions, and intersections. When the space $\mathscr{X}$ is infinite, the definition of the class $\mathscr{A}$ of sets $A$ is much more difficult. It is natural to require that the class $\mathscr{A}$ be closed under finitely many complementations, unions, and intersections; but this is not enough. For complete development of the theory it is necessary that $\mathscr{A}$ be closed under denumerably many set operations of forming complements, unions, and intersections. Such a class of sets is called a $\sigma$-field of sets. For such a collection of sets it is necessary to require the axiom of countable additivity of probability:

$$\text{Axiom 4.} \quad P\left(\bigcup_{i=1}^{\infty} A_i\right) = \sum_{i=1}^{\infty} P(A_i), \text{ if } A_iA_j = \phi,\ i \neq j \tag{3.35}$$

Axiom 4 includes Axiom 3 as a special case.

A consequence is that the probability of a set formed by the intersection of an enumerable infinity of sets is obtainable.

Examples of $\sigma$-fields are: (1) $\{\phi, \mathscr{X}\}$, (2) $\{A, \bar{A}, \phi, \mathscr{X}\}$, and (3) Class of all subsets of $\mathscr{X}$.

For infinite sample spaces we cannot give a meaningful probability function for any $\sigma$-field. For example, if we wish to define a probability function to match the intuitive idea of choosing a point at random on the unit square, the probability of a set should be analogous to the area of a set. There is no meaningful way to generalize the concept of area to include the class of all subsets of the unit square. It is necessary to restrict consideration to some collection of sets.

A probability configuration consists of a space $\mathscr{X}$ of elements $\omega$, which may be regarded as outcomes of an experiment, with a $\sigma$-field $\mathscr{A}$ of sets $A$, having a probability defined for each set $A$, $P(A)$, which satisfies Axioms 1 to 4. This whole $\mathscr{X}$ configuration is the triple $(\mathscr{X}, \mathscr{A}, P)$. Real understanding of the role of $\sigma$-fields and Axiom 4 requires advanced mathematical knowledge beyond the level of this book. It is hoped, however, that the above presentation will enable the student to have an understanding of the triple $(\mathscr{X}, \mathscr{A}, P)$ which occurs in most of the mathematical writings on statistics.

## 3.14 Random Variables and Probability Distributions

We now take up the elementary cases for the space $\mathscr{X}$ with elements $\omega$. The simplest case is certainly the case in which $\mathscr{X}$ is the real line and the elements $\omega$ are real numbers $X$. The question that arises immediately is an appropriate nature for the class $\mathscr{A}$ of sets $A$. It is natural that there should be a probability for $X$ lying in any interval, and for $X$ taking any value. The development of ideas here is very subtle, so we abbreviate the picture sharply. The standard usage is to take the class $\mathscr{A}$ of sets to be the collection of all intervals $(a, b]$, open on the left and closed on the right, and all sets obtainable from this collection by an enumerable or countable number of operations of taking complements, unions, and intersections. A set so formed is called a Borel set and the class of all such sets is a $\sigma$-field called the Borel class of sets and is denoted by $\mathscr{B}$. We can then give:

*Definition 1: An entity $X$ is a univariate random variable if it takes on values on the real line and if $P(a < X \leq b)$ is defined for all $a$, $b$ consistently with the axioms.*

In terms of the previous ideas, the underlying $\mathscr{X}$ is the real line with elements $X$, and $\mathscr{A}$ is the class of Borel sets.

*Definition 2: In the case of an arbitrary probability space $(\mathscr{X}, \mathscr{A}, P)$ a real-valued function $X(\cdot)$ of the elements $\omega$ of $\mathscr{X}$ is a derived univariate or real random variable, if the probability defined in $\mathscr{X}$ results in the variable $X(\omega)$ being a univariate random variable. That is, for every $a$, $b$ there is a number which we denote by $P(a < X(\omega) \leq b)$ which is equal to $P[\{\omega : a < X(\omega) \leq b\}]$. This requires that the set of elements, $\omega$, such that $a < X(\omega) \leq b$ be a set in the class of sets defined over $\mathscr{X}$. Strictly speaking, the function $X(\cdot)$ is a random variable relative to $\mathscr{A}$.*

The simplest method of specifying a real random variable $X$ is to use the distribution function $F(x) = P(X \leq x)$, because we then have

$$P(a < X \leq b) = F(b) - F(a) \tag{3.36}$$

All the sets to be considered are obtainable from the collection of intervals $(a, b]$ by an enumerable number of set operations, so the axioms then lead to a probability for any such set. This probability of a set will be the same for any way of obtaining it. The function $F(x)$ is like that given in Section 2.8.1 and has the properties (1) $F(-\infty) = 0$, $F(\infty) = 1$, (2) $F$ is continuous from the right, and (3) $F$ is nondecreasing. There are many functions which satisfy these three conditions. In this book, however, we shall be concerned primarily with two types of distribution functions, discrete and continuous.

The discrete distribution function is a step function and the height of the step at $x = b$ gives the probability that $X = b$. That is,

$$P(X = b) = F(b) - \lim_{\varepsilon \to 0} F(b - \varepsilon) \tag{3.37}$$

$F(x)$ is obviously given by summing the step values over all values less than or equal to $x$. That is,

$$F(x) = \sum_{t \leq x} f(t) \tag{3.38}$$

where $t$ ranges over the points of positive probability. The function $f(t)$ is called, alternatively, the frequency function, probability mass function, probability function, probability density function, or simply the density function.

A continuous distribution function is of the form:

$$F(x) = \int_{-\infty}^{x} f(x)\,dx \tag{3.39}$$

where the integral is the ordinary Riemann integral. The function $f(x)$ is called the probability density function (p.d.f.), or simply the density function. It must be of such nature that the Riemann integral exists. It is required that

(1) $f(x) \geq 0$ for all $x$

and

(2) $\int_{-\infty}^{\infty} f(x)\,dx = 1.$

A class of distribution functions general enough for almost all purposes can be obtained by forming

$$\lambda F_d + (1 - \lambda)F_c \tag{3.40}$$

where (1) $0 \leq \lambda \leq 1$, (2) $F_d$ is a discrete distribution function, and (3) $F_c$ is a continuous distribution function.

A final note is useful. A univariate p.d.f. must be specified as a function defined over the totality of points of the real line. A point function can be written typically in one of two ways exemplified as follows: Suppose we define a function of a scalar variable $x$ as $f(x) = 1/\theta$ if $x$ is one of the integers $1, 2, \ldots, \theta$ where $\theta$ is a positive integer and zero, otherwise. For manipulative purposes, we often need a single specification of $f(x)$, which says what $f(x)$ is for every $x$. We can accomplish this by the alternative specification

$$f(x) = \frac{N(x)\delta(\theta, x)\delta(x, 1)}{\theta} \tag{3.41}$$

where $N(x) = 1$ if $x$ is a positive integer and zero, otherwise, and $\delta(u, v)$ is 1 if $u \geq v$ and is zero, otherwise. Equation (3.41) expresses the probability for all $x$. Similarly a continuous variable with $f(x) = 1/\theta$ for $0 \leq x \leq \theta$ and zero, otherwise, can be represented as having the density

$$f(x) = \frac{\delta(\theta, x)\delta(x, 0)}{\theta} \tag{3.42}$$

Again this form of $f(x)$ specifies in a single mathematical form the density for all $x$ between $-\infty$ and $+\infty$. The main role of representations such as

Equations (3.41) and (3.42) is that they state explicitly the dependence of probability or probability density on the value of the parameter $\theta$ for all $x$. See also Problem 35.

## 3.15 Multivariate Random Variables

When we turn to bivariate distributions on $X_1$, $X_2$, where each $X$ takes a value on the real line, the situation becomes much more complex. The two simple cases are: (1) the discrete case, in which probability mass is present only at a denumerable number of points of the plane and (2) the continuous case, in which

$$P(a_1 < X_1 < b_1; a_2 < X_2 < b_2) = \int_{a_1}^{b_1} \int_{a_2}^{b_2} f(t_1, t_2)\, dt_1 dt_2$$

where $f(t_1, t_2)$ is a function greater than or equal to zero, such that the integral exists as a Riemann integral for all $a_1$, $b_1$, $a_2$, $b_2$. Furthermore $P(X_1 = a_1, X_2 = a_2) = 0$ for all $a_1$, $a_2$, so that the $<$ signs can be replaced by $\leq$ signs wherever they occur. We can certainly envisage a mixture with positive weights adding to unity of two such distributions.

These two cases are, however, insufficient for general purposes. Consider the simple case of a distribution of $X_1$ and $X_2$ uniform over the square with corners (0, 0), (0, 1), (1, 0) and (1, 1). Now consider a transform of this distribution:

$$\begin{aligned} Z_1 &= X_1 \\ Z_2 &= 0 \text{ if } X_1 - X_2 \leq \tfrac{1}{2} \\ &= X_1 - X_2 \text{ if } X_1 - X_2 > \tfrac{1}{2} \end{aligned}$$

The joint distribution of $Z_1$ and $Z_2$ is a peculiar mixture in which the marginal distribution of $Z_1$ is continuous, and the marginal distribution of $Z_2$ is a mixture of a discrete and a continuous distribution. Certainly this type of distribution arises in probability calculations.

A general specification of a bivariate distribution requires specification of a probability function on the product class $\mathscr{B}_1 \times \mathscr{B}_2$, where $\mathscr{B}_1$ and $\mathscr{B}_2$ are the Borel fields for $X_1$ and $X_2$ respectively. That is, we have to specify $P(X_1 \in B_1, X_2 \in B_2)$ for all pairs of $B_1$ taken from $\mathscr{B}_1$ and $B_2$ taken from $\mathscr{B}_2$.

Alternatively, analogous to the univariate case, a bivariate distribution is specified by the cumulative distribution function

$$F(x_1, x_2) = \text{Prob}(X_1 \leq x_1, X_2 \leq x_2)$$

defined for all $(x_1, x_2)$ of the real plane.

This same type of analysis extends to a distribution defined over a space of $k$ dimensions. The simple cases in general are the two already given for one or two dimensions; namely, (1) the discrete case, in which probability mass is present only at a denumerable number of points in $k$ dimensions, or (2) the continuous case, in which the probability of any point in $k$ dimensions is zero, and the probability in a $k$-dimensional rectangle $a_i \leq x_i \leq b_i$ is given by the Riemann integral of a nonnegative function $f(t_1, t_2, \ldots, t_k)$ over the rectangle, i.e.,

$$\int_{a_1}^{b_1} \int_{a_2}^{b_2} \cdots \int_{a_k}^{b_k} f(t_1, t_2, \ldots, t_k)\, dt_1 dt_2 \ldots dt_k$$

For this case to arise it is necessary of course that the integral exists for all possible values of the $a_i$ and $b_i$. All integrals that arise are simply limits of Darboux sums over systems of generalized rectangles.

It is the case that the basic probability models considered in this book are either of one or the other of these types. It is also the case that the continuous type is the one that is much easier to work with because one can bring to bear all the processes of Riemann integration.

## 3.16 Independent Random Variables

*Definition 3. Random variables $X_1$ and $X_2$ will be said to be independent if*

$$P(X_1 \in B_1, X_2 \in B_2) = P(X_1 \in B_1)P(X_2 \in B_2) \tag{3.43}$$

*for all $B_1 \in \mathscr{B}_1$, $B_2 \in \mathscr{B}_2$.*

In the case that both $X_1$ and $X_2$ are discrete, it follows immediately, because a single point is a Borel set, that

$$P(X_1 = x_1, X_2 = x_2) = P(X_1 = x_1)P(X_2 = x_2)$$

for all $x_1$ and $x_2$. In case both variables are continuous random variables, it also follows that a necessary and sufficient condition for independence is

$$f(x_1, x_2) = f_1(x_1)f_2(x_2) \text{ for all } x_1 \text{ and } x_2$$

The definition of independence of random variables extends immediately to $k$ random variables by requiring for independence that

$$P(X_1 \in B_1, X_2 \in B_2, \ldots, X_k \in B_k) = \prod_1^k P(X_i \in B_i) \tag{3.44}$$

for all $B_i \in \mathscr{B}_i$, $i = 1, 2, \ldots, k$.

## 3.17 Conditional Distributions

We have given earlier in the chapter all that need be said about conditional distributions for discrete sample spaces—the summary statement being that if $X_1$ is a random variable with possible values represented by $x_1$, and $X_2$ is a random variable with possible values represented by $x_2$, then the conditional probability that $X_1$ equals $x_1$ given that $X_2$ equals $x_2$ is equal to $P(X_1 = x_1, X_2 = x_2)/P(X_2 = x_2)$. If $x_2$ is kept fixed and $x_1$ varies, this probability represents the conditional distribution of $X_1$, given that $X_2$ equals $x_2$. It is clear that $X_1, x_1, X_2, x_2$ can be vectors rather than scalars with no change in formula.

The definition of conditional distributions for continuous random variables presents considerable logical difficulties because if $X_2$ is a continuous random variable, $P(X_2 = x_2) = 0$ for all $x_2$.

The resolution of this problem can be made in a tight logical way only with advanced mathematics based on the theory of measure and Lebesgue integration. This is a very complex subject. We can, however, give a descriptive definition that is related to the sophisticated measure-theoretic definition.

This descriptive definition is motivated by a simple result for the case when both variables are discrete. Note that for this case

$$P(X_1 \leq x_1) = \sum_{x_2} P(X_1 \leq x_1 | x_2) P(X_2 = x_2)$$

In general, we give the following:

*Definition 4. Given a bivariate random variable* $(X_1, X_2)$, *let*

$$F_1(x_1) = P(X_1 \leq x_1) \text{ and } F_2(x_2) = P(X_2 \leq x_2)$$

*Then the conditional distribution function of* $X_1$ *given that* $X_2 = x_2$ *is a function of* $x_1$ *and* $x_2$, *say* $F_{1.2}(x_1|x_2)$, *such that*

$$F_1(x_1) = \int_{-\infty}^{\infty} F_{1.2}(x_1|x_2)\, dF_2(x_2) \tag{3.45}$$

*for almost all* $x_1$.

Here the integral is of a type called a Lebesgue-Steltjes integral discussed in advanced mathematical texts, and this general formulation covers both discrete and continuous cases. If $X_1$ and $X_2$ are jointly continuous, the integrals are ordinary Riemann integrals and we can utilize a conditional probability density function defined by

$$f_{1.2}(x_1|x_2) = f_{12}(x_1, x_2)/f_2(x_2)$$

for all $x_2$ such that $f_2(x_2)$ is greater than zero, where $f_{12}(x_1, x_2)$ is the joint probability density function of $X_1$ and $X_2$, and $f_2(x_2)$ is the probability density function of $X_2$.

The relation of this to Equation (3.45) is given by the following sequence:

$$F_1(x_1) = \int_{-\infty}^{\infty} \int_{-\infty}^{x_1} f_{12}(z_1, z_2)\, dz_1 dz_2$$

$$= \int_{-\infty}^{\infty} \left[ \int_{-\infty}^{x_1} \frac{f_{12}(z_1, z_2)}{f_2(z_2)}\, dz_1 \right] f_2(z_2)\, dz_2$$

## 3.18 Mathematical Expectation

A probability distribution is to all intents and purposes a relative frequency distribution for a population of infinite size. So one can envisage forming averages of functions just as in Section 2.8.2. In the case of probability distributions it is standard to use the symbol $E(\ )$ denoting expectation rather than Ave( ) denoting average. So if a random variable $X$ is discrete, the expression $E[g(X)]$ is an abbreviation for $\sum_x g(x)f(x)$, this existing if and only if the summation is absolutely convergent. In the case of a continuous distribution, $E[g(X)]$ is defined to be the Riemann integral

$$\int_{-\infty}^{\infty} g(x)f(x)\, dx$$

provided that the integral is absolutely convergent. If the integral is not absolutely convergent, $E[g(X)]$ is said not to exist.

The various transforms being averages of functions of a parameter $t$, described in Section 2.5, are applicable to probability distributions also. We describe in detail only one such transform, the moment generating function (sometimes abbreviated as m.g.f.), which is used extensively in the next chapters. For a univariate random variable $X$, this is defined to be $E[\exp(tX)]$. For a multivariate random variable $X_1, X_2, \ldots, X_k$, it is defined to be $E[\exp(t_1X_1 + t_2X_2 + \cdots + t_kX_k)]$. This is a function of $k$ mathematical variables $t_1, t_2, \ldots, t_k$. These moment generating functions are used widely to derive approximations to distributions. In advanced work and cases for which moment generating functions do not exist, recourse is taken to characteristic functions defined as $E[\exp(itX)]$ and $E\{\exp[i(t_1X_1 + t_2X_2 + \cdots + t_kX_k)]\}$ which always exist.

It is impossible in a book of the present level to present completely "tight" proof of every result we give. In the case of a finite sample space the results are essentially obvious. For instance, an m.g.f.

$$E[\exp(tX)] = \textstyle\sum_x P(X = x)e^{tx}$$

and is just a finite sum of functions if $x$ is bounded. For the case of sample space that has an enumerable infinity of points, convergence of infinite series must be verified in strict theory. In some cases this is easy, as in the m.g.f. for the Poisson

$$\sum_{x=0}^{\infty} e^{-\lambda} \frac{\lambda^x}{x!} e^{tx} = \sum_{x=0}^{\infty} e^{-\lambda} \frac{(\lambda e^t)^x}{x!}$$

The convergence of which for fixed $t$ results from the convergence of the exponential series. In general, however, proof of the convergence of infinite sums or of infinite integrals is nontrivial.

In the case of a multivariate probability distribution it is often useful to employ ideas of conditional expectation. Parenthetically, it is perhaps worth stating that in advanced treatments of conditional probability, the idea of conditional probability is derived from the idea of conditional expectation. For the simplest case of a bivariate discrete distribution it is easy to see for a function of a bivariate random variable, say $z(X_1, X_2)$, that

$$E[z(X_1, X_2)] = \sum_{x_1, x_2} P(X_1 = x_1, X_2 = x_2) z(x_1, x_2)$$

$$= \sum_{x_1} P(X_1 = x_1) \sum_{x_2} P(X_2 = x_2 | X_1 = x_1) z(x_1, x_2) \quad (3.46)$$

$$= \sum_{x_2} P(X_2 = x_2) \sum_{x_1} P(X_1 = x_1 | X_2 = x_2) z(x_1, x_2) \quad (3.47)$$

These two equations are written briefly as

$$\underset{X_1, X_2}{E} [z(X_1, X_2)] = \underset{X_1}{E} \underset{X_2|X_1}{E} [z(X_1, X_2)] \quad (3.48)$$

$$\underset{X_1, X_2}{E} [z(X_1, X_2)] = \underset{X_2}{E} \underset{X_1|X_2}{E} [z(X_1, X_2)] \quad (3.49)$$

For the continuous case, with probability density $f(x_1, x_2)$ and conditional probability densities $f(x_1|x_2)$, $f(x_2|x_1)$, we have

$$\int_{x_1} \int_{x_2} f(x_1, x_2) z(x_1, x_2)\, dx_1 dx_2$$

$$= \int_{x_1} f(x_1) \int_{x_2} f(x_2|x_1) z(x_1, x_2)\, dx_2 dx_1 \quad (3.50)$$

$$= \int_{x_2} f(x_2) \int_{x_1} f(x_1|x_2) z(x_1, x_2)\, dx_1 dx_2 \quad (3.51)$$

These equations we write briefly as

$$\underset{X_1, X_2}{E} [z(X_1, X_2)] = \underset{X_1}{E} \underset{X_2|X_1}{E} [z(X_1, X_2)] \quad (3.52)$$

$$= \underset{X_2}{E} \underset{X_1|X_2}{E} [z(X_1, X_2)] \quad (3.53)$$

The extensions are clear at a heuristic level, so that, for instance,

$$\underset{X_1, X_2, X_3}{E} [z(X_1, X_2, X_3)] = \underset{X_1}{E} \underset{X_2|X_1}{E} \underset{X_3|X_1, X_2}{E} [z(X_1, X_2, X_3)] \quad (3.54)$$

with all the possible permutations in the order of summation or integration. This may be written, for ease of typography, as

$$E[z(X_1, X_2, X_3)] = E(E\{E[z(X_1, X_2, X_3)|X_3]|X_2\}) \quad (3.55)$$

A further property holds for variances, which is very useful for bivariate random variables. Denote the overall expectation of $X_2$ by $E(X_2)$, the conditional expectation of $X_2$ given $X_1$ by $E(X_2|X_1)$, and the conditional variance of $X_2$ given $X_1$ by $\mathrm{Var}(X_2|X_1)$. Then

$$X_2 - E(X_2) = [X_2 - E(X_2|X_1)] + [E(X_2|X_1) - E(X_2)] \quad (3.56)$$

We square and take expectation with regard to $X_2$ given $X_1$:

$$\begin{aligned} \underset{X_2}{E}\{[X_2 - E(X_2)]^2|X_1\} &= \underset{X_2}{E}\{[X_2 - E(X_2|X_1)]^2|X_1\} \\ &+ \underset{X_2}{E}\{[E(X_2|X_1) - E(X_2)]^2|X_1\} \\ &+ 2\underset{X_2}{E}\{[X_2 - E(X_2|X_1)][E(X_2|X_1) - E(X_2)]|X_1\} \end{aligned}$$

The second term is $[E(X_2|X_1) - E(X_2)]^2$ because it does not depend on $X_2$. The third term is zero because it equals

$$2[E(X_2|X_1) - E(X_2)]\ \underset{X_2}{E}\{[X_2 - E(X_2|X_1)]|X_1\}$$

the second factor of which is identically zero. Therefore,

$$\underset{X_2}{E}\{[X_2 - E(X_2)]^2|X_1\} = \mathrm{Var}(X_2|X_1) + [E(X_2|X_1) - E(X_2)]^2 \quad (3.57)$$

Now taking expectations with regard to $X_1$, using

$$\underset{X_1}{E}\,\underset{X_2}{E}\{[X_2 - E(X_2)]^2|X_1\} = \underset{X_1, X_2}{E}[X_2 - E(X_2)]^2 = \mathrm{Var}(X_2)$$

and

$$\underset{X_1}{E}\,\underset{X_2}{E}(X_2|X_1) = E(X_2)$$

we have

$$\mathrm{Var}(X_2) = \underset{X_1}{E}\,[\underset{X_2}{\mathrm{Var}}(X_2|X_1)] + \underset{X_1}{\mathrm{Var}}[E(X_2|X_1)] \quad (3.58)$$

or the variance is the average of the conditional variance plus the variance of the conditional mean.

## 3.19 A Brief Review of Interpretations of Probability

In this chapter we have introduced probability as a calculus of relative frequencies in infinite populations, and we have presented the usual axiomatization of this. We think it is essential to present with references a sketch of interpretations of probability which have received appreciable attention in the history of human thought.

There are several accounts of the early history of probability, e.g., Todhunter (1949, originally 1865) and Cajori (1926). An excellent brief history of statistical method is given by Walker (1931), and we refer the interested student especially to this. It is certain that ideas of probability were used at least 2000 years ago. The theory of probability received increasing attention from around A.D. 1500 with regard to games of chance and gambling. From around 1800 very powerful mathematical minds, e.g., Laplace (1814), were attracted to the subject because of its relevance to all aspects of human life and thought. Gauss applied probability to the motion of celestial bodies. Uses of probability and statistics like those of the present time were initiated by Quetelet (1796–1814). This was the beginning of the "biometric" school of which Galton (1822–1911) and Karl Pearson were great innovators. Certainly R. A. Fisher (1890–1962) was a huge contributor to the use and development of statistical methods and his two basic books *Statistical Methods for Research Workers* (1958), first published in 1925, and *The Design of Experiments* (1935a) revised edition, 1960), are still regarded by many as mandatory reading.

Apart from some obscurities (which are not trivial) the outlook in all the development sketched above was that probability is a relative frequency in some conceptual infinite population of repetitions. It is important that the student be aware that other interpretations have been made.

The frequency idea of probability was initiated by the early workers on games of chance and was given a strong presentation by Venn (1866). It has dominated statistical method. It has been supported strongly by the philosopher Reichenbach (1949, p. 367) who says, "The criterion for the justification of an interpretation lies in its adequacy for purposes of prediction." The frequency idea was given also by von Mises (1964 and earlier) in his conception of an irregular collective.

A second class of interpretations of probability is that it is a logic of uncertain propositions analogous to the logic of certain or sure propositions. This view was developed by J. M. Keynes (1921), though it should be noted that the second half of his book is strongly statistical (rather than logical). More recently, Jeffreys (1961 and earlier) has taken a view which falls essentially within this logical class.

The third class of interpretations is that probability is strictly a personal degree of belief. This viewpoint was given first by Ramsey (1926), it appears; it received great impetus from de Finetti, and is the basis of the books by Savage (1954) and Lindley (1965). Good (1950) seems to take a position somewhat intermediate between the frequency and personalistic ideas. In all these cases, the development is made axiomatically from a small set of axioms which were considered by the authors to be strongly compelling. A purely axiomatic development of personalistic (or subjective) probability was given by Koopman (1940). A useful compendium of papers in this area is given by Kyburg and Smokler (1964).

The reconciliation of interpretations of probability is, it appears, extremely difficult, and certainly has not yet been achieved. A critique of the points of view would take us far afield, but we would be derelict in our professional responsibilities if we did not refer the student to the viewpoints.

Because our interest throughout this book is in data evaluation, the concept of probability as frequency in a population of repetitions seems crucial. Once the student has absorbed our material, he will find other interpretations easy to read, particularly in regard to axiomatic development because the basic theorems in all systems are essentially the same and the problems lie in the linguistic interpretations and in the mode of application of the ideas to the real world.

## PROBLEMS

1. The outcome of an experiment is classified according to criteria $A$ or not $A$ and $B$ or not $B$. The four possible outcomes with their probabilities $a$, $b$, $c$, and $d$ are represented in the following table:

| | $B$ | $\bar{B}$ |
|---|---|---|
| $A$ | $a$ | $b$ |
| $\bar{A}$ | $c$ | $d$ |

   Express in terms of $a$, $b$, $c$, and $d$:
   (a) The marginal probabilities $P(A)$, $P(\bar{A})$, $P(B)$, and $P(\bar{B})$.
   (b) The conditional probabilities $P(A|B)$, $P(A|\bar{B})$, $P(A|A \cup B)$.

2. In the preceding problem show that if $A$ and $B$ are independent then $\bar{A}$ and $\bar{B}$ are independent.

3. Two "loaded" dice are thrown with the following probabilities:

| | 1 | 2 | 3 | 4 | 5 | 6 |
|---|---|---|---|---|---|---|
| Die 1 | 1/6 | 1/6 | 3/12 | 1/12 | 1/6 | 1/6 |
| Die 2 | 1/6 | 1/6 | 1/12 | 3/12 | 1/6 | 1/6 |

   (a) Construct the table of joint probabilities and marginal probabilities.
   (b) What is the probability of getting a total of seven? How does this compare with fair dice?

4. Three fair dice are thrown. Calculate the probability of one or more three's by:
   (a) $1 - P(\text{no three's})$.
   (b) Use of the formula:

$$P(A \cup B \cup C) = P(A) + P(B) + P(C) - P(AB) - P(AC) - P(BC) + P(ABC)$$

5. A well-shuffled deck of 52 cards is dealt. What is the probability that at least one ace will be immediately succeeded by another ace?

6. $A$ and $B$ are independent and

$$P(A \cup B) = 5/8$$
$$P(A|B) = 1/4$$

   Find $P(B)$.

7. Three dice are tossed. One is red, one is green, one is black. Find $P(\text{two or more three's}|\text{green three})$.

8. What is the probability that a bridge hand will contain two or more aces given that it contains the ace of hearts?

9. What is the probability that a bridge hand will contain two or more aces given that it contains one ace?

10. A quality assurance department is using an acceptance sampling plan to accept or reject lots of items. Let $F$ = fraction defective in a lot. It has been determined that the plan has the following properties:

| $p$ | 0 | .1 | .2 | .3 | .4 | .5 | .6 | .7 | .8 | .9 | 1.0 |
|---|---|---|---|---|---|---|---|---|---|---|---|
| $P(\text{accept}\|F = p)$ | .99 | .95 | .90 | .80 | .60 | .40 | .20 | .10 | .05 | .01 | 0 |

   Historical data show that the distribution of the proportion defective is as follows:

| $p$ | 0 | .1 | .2 | .3 | .4 | .5 | .6 | .7 | .8 | .9 | 1.0 |
|---|---|---|---|---|---|---|---|---|---|---|---|
| $P(F = p)$ | .01 | .05 | .10 | .20 | .20 | .20 | .10 | .05 | .05 | .03 | .01 |

   Find $P(F = p|\text{accept})$ for $p = 0, 0.1, 0.2, \ldots, 1.0$.

11. $A$ and $B$ are mutually exclusive. What can we say about the independence of $A$ and $B$?

12. Derive the chain rule for $P(A_1A_2 \ldots A_n)$.

13. Prove Theorem 3.1 given in Section 3.6.

14. Prove Equation 3.10, Section 3.4

$$P\left(\bigcup_{i=1}^{n} A_i\right) = \sum P(A_i) - \sum P(A_iA_j) + \ldots \pm P(A_1A_2 \ldots A_n)$$

15. Describe the restrictions which must be placed upon $N_s$, $N_f$, $k$, and $n$ in the hypergeometric formula.

16. Show that the binomial is a limiting form of the hypergeometric.

17. A system consists of components which act independently with the probability of operating successfully as shown below. What is the probability of success of the system? That is, what is the probability of getting from $A$ to $B$ by one or more paths?

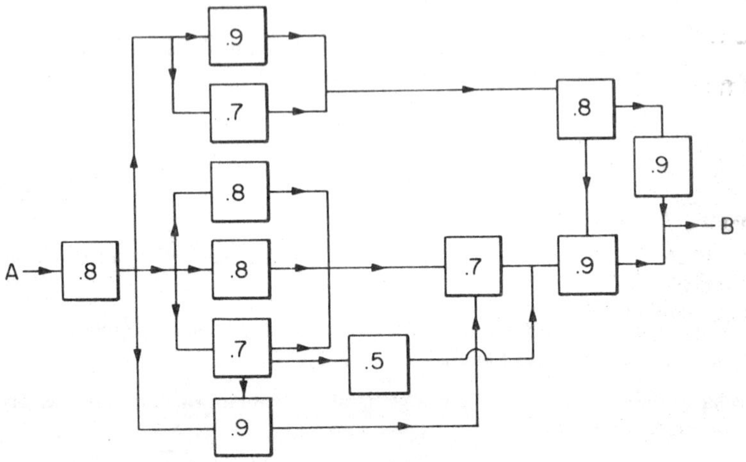


**Fig. for Ex. 3.17** Hypothetical System of Components

18. Give an example where $A$ and $B$ are independent but $P(AB|C) \neq P(A|C)P(B|C)$. That is, show by example that $A$ and $B$ can be independent, but not conditionally independent.

19. In a group of 50 people what is the probability that at least 2 have the same birthday given that each person is equally likely to have a birthday on any one of 365 different days?

20. A simple electrical system consists of $n_1$ components of type 1, $n_2$ of type 2, $n_3$ of type 3, all wired together in series. Components of type 1 each have an inherent probability of successful operation of $p_1 = .8$, components of type 2 have $p_2 = .7$, and components of type 3 have $p_3 = .9$. Type 1 components cost \$2.00 each, type 2 components cost \$1.75 each and type 3 components cost \$2.50 each. Assuming that the components operate independently so that the system probability of operating is $p_1^{n_1}p_2^{n_2}p_3^{n_3}$, maximize the system probability subject to the constraints that the cost must not exceed \$20.00 dollars and that at least two components of each type must be in the system.

21. Give the flow chart for a computer program to achieve the maximization in Problem 20 for arbitrary values of $p_1$, $p_2$, and $p_3$ and for component costs $c_1$, $c_2$, and $c_3$.

22. If we are tossing a coin for which the probability of heads $p_1$ is equally likely to be either 7/16, 1/2, or 9/16, what is the probability of getting heads on the $(n + 1)$st toss given that the first $n$ tosses give heads?

23. Verify numerically that the relative error in Stirling's approximation decreases as $n$ becomes larger.

24. A personnel survey for a company revealed that .380 of the employees were college graduates, .115 were female college graduates, .415 were females, .235 were females under twenty-five years of age, .500 were under twenty-five, and .200 of those under twenty-five were college graduates. If .035 were female college graduates under twenty-five, what fraction of males did not have college degrees?

25. The outcome of an experiment is classified according to three criteria with probabilities as shown in the following table:

| | $A$ | | $\bar{A}$ | |
|---|---|---|---|---|
| | $C$ | $\bar{C}$ | $C$ | $\bar{C}$ |
| $B$ | 1/4 | 0 | 0 | 1/4 |
| $\bar{B}$ | 0 | 1/4 | 1/4 | 0 |

Show that $A$ and $B$ are independent, $A$ and $C$ are independent, and $B$ and $C$ are independent but $A$, $B$, and $C$ are not independent. That is, show that pairwise independence does not imply joint, or mutual, independence.

26. Given that there are nine trumps between your hand and the board including the ace and king, most bridge players would advise you to lead your opponents out of trumps. What is the probability in this circumstance that there is, in fact, a two-two split in trumps?

27. Prove the identity

$$\binom{n+1}{j} = \binom{n}{j} + \binom{n}{j-1}$$

28. Give a flow chart for computing

$$\binom{n+1}{j}, j = 0, 1, 2, \ldots, n+1$$

for arbitrary $n$, making use of the identity proved in the previous exercise.

29. We have two urns $A$ and $B$. Urn $A$ has 3 white balls and 3 black balls. Urn $B$ has 2 white balls and 2 black balls. One of the two urns is chosen with probability 1/2, and 2 balls are selected from the chosen urn. Obtain probability (urn is $A$ given the result) for each of the three possible results.

30. One of two decks of cards, either an ordinary bridge deck or a pinochle deck, is chosen with probability 1/2. A card drawn at random from the chosen deck is a king. What is the probability that the deck chosen was the bridge deck?

31. Suppose the probability of urn $A$ is 1/3, of urn $B$ is 1/2, and of urn $C$ is 1/6. Suppose the probability of a white ball from urn $A$ is 1/4, from urn $B$ is 1/2, and from urn $C$ is 3/4. If a ball drawn at random from the selected urn was white, what are the conditional probabilities that the urn drawn was $A$, $B$, or $C$?

32. Thirteen cards are dealt without replacement from a well-shuffled deck. Let $X$ denote the number of aces and $Y$ the number of kings.
    (a) Find the probability density functions $f(x, y)$.
    (b) Find $F(x, y)$.
    (c) Find $E(X)$.
33. An observation $X$ is from distribution $F_1$ with probability $p$ and from distribution $F_2$ with probability $q$. Express the first and second moments of the distribution of $X$ in terms of the moments of $F_1$ and $F_2$.
34. Show that Axioms 1, 2, and 3 imply that if $A$ is contained in $B$ then $P(A) \leq P(B)$.
35. $f(x_1, x_2) = 2$, $0 < x_1 < x_2 < 1$. Show that $X_1$ and $X_2$ are not independent. Note that
$$f(x_1, x_2) = 2\delta(x_1, 0)\delta(1, x_1)\delta(x_2, 0)\delta(1, x_2)\delta(x_2, x_1)$$
and that $\delta(x_2, x_1)$ does not factor.
36. There are many generalizations of the Chebyshev inequality. Show that
$$E[(X - a)^2]/k^2 \geq P[|X - a| \geq k]$$
if $E[(X - a)^2]$ exists.
37. Show that Chebyshev's inequality discussed in Chapter 2 is a special case of the inequality obtained in the previous problem.
38. Show that if $E(|X - a|^r)$ exists then
$$P(|X - a| \geq k) \leq E(|X - a|^r)/k^r$$

# 4

# Some Commonly Occurring Mathematical Distributions

## 4.1 Introduction

In this chapter we present probability distributions which arise very frequently in simple data situations. Often, these serve as component parts of probability models for more complex data situations. We give four basic discrete distributions defined over the nonnegative integers, which are specified by the probability mass at each such integer. We then present the most important continuous distribution, the normal distribution. This distribution is important as an approximating distribution, so we discuss approximations of distributions briefly. We then give a few simple continuous univariate distributions. Turning to multivariate distributions, we discuss the multinomial distribution, the bivariate normal, and the multivariate normal, which we obtain as an approximation to the multinomial. The given distributions do not exhaust the useful ones. Their importance lies in their simplicity which makes them natural candidates for representing data situations, or as building blocks for complex probability models.

Our general procedure is to examine the probability or probability density function and simple properties given by low-order moments. We shall frequently use the moment generating function (m.g.f.). For a univariate random variable $X$, this is $E[\exp(tX)]$, and for a function $g(X)$ of $X$, it is $E\{\exp[tg(X)]\}$. For a $k$-variate random variable, say, $X_1, X_2, \ldots, X_k$, it is $E[\exp(t_1X_1 + t_2X_2 + \cdots + t_kX_k)]$. While these functions do not exist in general and recourse must be made to the characteristic function defined with each $t_j$ variable replaced by $iu_j$, say, with $i = \sqrt{-1}$, and $u_j$ real, which always exists, for our cases the functions exist and determine the distribution uniquely. We use two facts:

1. If $X_1, X_2, \ldots, X_k$ are independent and the m.g.f. exists, then

$$E[\exp(t_1X_1 + t_2X_2 + \cdots + t_kX_k)] = \prod_{i=1}^{k} E[\exp(t_iX_i)]$$

which is the product of the separate m.g.f.'s of $X_1, X_2, \ldots, X_k$ with variables $t_1, t_2, \ldots, t_k$ respectively.

2. If the joint m.g.f. exists and factors into a product $m_1(t_1)m_2(t_2)\ldots m_k(t_k)$ of m.g.f.'s, then $X_1, X_2, \ldots, X_k$ are independent.

Proof of (1) is simple at our level, involving either summation or Riemann integration in $k$-dimensional space. Proof of (2) is an advanced topic and will not be given.

## 4.2 The Binomial Distribution

The binomial distribution is useful in many situations where repeated trials are made and the result is judged as a success or failure, e.g., in sampling inspection for defective parts.

If the probability of a success on a single trial is $p$ and $n$ independent trials are made, the probability that the number of successes $X$ will equal $x$ is given by

$$P(X = x) = f(x) = \binom{n}{x} p^x q^{n-x}, \qquad x = 0, 1, 2, \ldots, n$$
$$= 0, \text{ otherwise} \tag{4.1}$$

where $q = 1 - p$.

We shall refer to this distribution by the expression $\text{Bi}(n, p)$. It is readily seen that $f(x)$ sums to unity because

$$\sum_{x=0}^{n} \frac{n!}{x!(n-x)!} p^x q^{n-x} \tag{4.2}$$

is the familiar binomial expansion of $(q + p)^n = 1^n = 1$. By using the expansion of $(q + p)^{n-1}$, we can find the mean as follows:

$$E(X) = \sum_{x=0}^{n} x f(x)$$

$$= \sum_{x=1}^{n} \frac{n!}{(n-x)!(x-1)!} p^x q^{n-x}$$

$$= np \sum_{x=1}^{n} \frac{(n-1)! p^{x-1} q^{n-x}}{(n-x)!(x-1)!}$$

$$= np \sum_{y=0}^{n-1} \frac{(n-1)!}{y!(n-1-y)!} p^y q^{n-1-y}$$

$$= np(q + p)^{n-1}$$

$$= np \tag{4.3}$$

To obtain the variance of $X$ we first find $E[X(X - 1)]$. The student can verify that this is easier than finding $E(X^2)$ directly.

$$E[X(X-1)] = \sum_{x=0}^{n} x(x-1)f(x) = \sum_{x=2}^{n} \frac{n!}{(x-2)!(n-x)!} p^x q^{n-x}$$

$$= n(n-1)p^2 \sum_{x=2}^{n} \frac{(n-2)!\,p^{x-2}q^{n-x}}{(x-2)!(n-x)!}$$

$$= n(n-1)p^2 \sum_{y=0}^{n-2} \frac{(n-2)!}{y!(n-2-y)!} p^y q^{n-2-y}$$

$$= n(n-1)p^2(q+p)^{n-2} = n(n-1)p^2 \quad (4.4)$$

Thus

$$E(X^2) = n^2p^2 - np^2 + np$$

and

$$\sigma^2 = E(X^2) - [E(X)]^2 = npq \quad (4.5)$$

The variance of $X/n$ is therefore $pq/n$.

The cumulative distribution function $F(x)$ is given by

$$F(x) = 0, \qquad x < 0$$

$$= \sum_{u=0}^{[x]} \binom{n}{u} p^u q^{n-u}, \qquad x \geq 0 \quad (4.6)$$

where $[x]$ is the largest integer less than or equal to $x$. A partial table is given in Table A.2 of the Appendix. More complete tables of the cumulative binomial are available.

The m.g.f. of the binomial distribution is $m(t) = (q + pe^t)^n$. This is easily found as follows:

$$m(t) = E(e^{tX}) = \sum_{x=0}^{n} e^{tx}f(x)$$

$$= \sum_{x=0}^{n} \frac{n!}{x!(n-x)!} (pe^t)^x q^{n-x}$$

$$= (q + pe^t)^n \quad (4.7)$$

The moments of the binomial distribution can be generated from $m(t)$.

For example,

$$\begin{aligned} E(X) &= \left.\frac{dm(t)}{dt}\right|_{t=0} \\ &= n(q + pe^t)^{n-1}pe^t|_{t=0} \\ &= np \end{aligned}$$

*Law of Large Numbers.* Now that we have obtained the mean and variance of the binomial distribution, we can give one simple form of the weak law of large numbers. Given a sequence of independent Bernoulli variables $X_1, X_2, \ldots$, the sum of the first $n$ variables $X = \sum_1^n X_i$ is a binomial variable with mean $np$ and variance $np(1 - p)$. Then from Chebyshev's inequality,

$$P\left(\left|\frac{X}{n} - p\right| \geq \varepsilon\right) \leq \frac{p(1 - p)}{n\varepsilon^2} \tag{4.8}$$

As $n \to \infty$ this probability converges to zero. That is,

$$\lim_{n\to\infty} P\left(\left|\frac{X}{n} - p\right| > \varepsilon\right) = 0 \tag{4.9}$$

This provides justification for the simple intuitive idea that as $n$ increases, the average will tend to be closer to the expected value.

## 4.3 The Negative Binomial Distribution

In the binomial situation we decide to make $n$ trials and observe the number of successes. An alternative would be to sample until we have obtained $x$ successes. The total sample size is then random and equals $n$ if the first $(n - 1)$ trials have given $(x - 1)$ successes in some order and the $n$th trial gives a success. The probability that the sample size is $n$ is therefore

$$P(N = n) = \frac{(n - 1)!}{(x - 1)!(n - x)!} p^x q^{n-x} \tag{4.10}$$

This distribution is clearly relevant to a sampling inspection procedure of the following type: Sample until 2 defectives are obtained; reject the lot if the total number sampled is less than 20. Note that we do not say that such a rule is a good one.

We shall, in fact, suppose that one samples until he obtains $r$ successes and observes $X$, the number of trials in excess of $r$. Thus

$$\begin{aligned} f(x) &= \binom{r+x-1}{x} p^r q^x, \quad x = 0, 1, 2, \ldots \\ &= 0, \text{ otherwise} \end{aligned} \tag{4.11}$$

The name negative binomial derives from the fact that $f(x)$ is one of the terms in the expansion of $p^r(1-q)^{-r}$. We can also make use of the fact that

$$\binom{r+x-1}{x} = \binom{-r}{x}(-1)^x \tag{4.12}$$

to write

$$\begin{aligned} f(x) &= \binom{-r}{x} p^r(-q)^x, \; x = 0, 1, 2, \ldots \\ &= 0, \text{ otherwise} \end{aligned} \tag{4.13}$$

To see that $f(x)$ sums to unity, we make use of the expansion of $(1-q)^{-r}$.

$$\begin{aligned} \sum_{x=0}^{\infty} f(x) &= p^r \sum_{x=0}^{\infty} \binom{r+x-1}{x} q^x \\ &= p^r(1-q)^{-r} \\ &= 1 \end{aligned} \tag{4.14}$$

The moments are obtained in a manner analogous to that used for the binomial. For example,

$$\begin{aligned} E(X) &= \sum_{x=0}^{\infty} x \binom{r+x-1}{x} p^r q^x \\ &= \sum_{x=1}^{\infty} \frac{(r+x-1)!}{(x-1)!(r-1)!} p^r q^x \\ &= \frac{rq}{p} \sum_{y=0}^{\infty} \frac{(r+1+y-1)!}{y!r!} p^{r+1} q^y \\ &= \frac{rq}{p} p^{r+1}(1-q)^{-r-1} \\ &= \frac{rq}{p} \end{aligned} \tag{4.15}$$

Similarly, we find the variance to be $rq/p^2$.

To obtain the m.g.f., we proceed as follows:

$$m(t) = \sum_{x=0}^{\infty} \binom{r+x-1}{x} e^{tx} p^r q^x$$

$$= \sum_{x=0}^{\infty} \binom{r+x-1}{x} p^r (qe^t)^x$$

$$= p^r(1 - qe^t)^{-r} \tag{4.16}$$

We can, of course, differentiate to find the moments.

## 4.4 The Hypergeometric Distribution

We encountered this distribution in Sect. 3.8 in discussing the probability of $x$ successes in a sample drawn without replacement. Let us now give the distribution in a slightly different form. Let the possible values for $p$ be 0, $1/N$, $2/N$, . . . , 1 and let $q = 1 - p$ as always. Then

$$f(x) = \frac{\binom{Np}{x}\binom{Nq}{n-x}}{\binom{N}{n}}, \qquad x = 0, 1, 2, \ldots, n$$

$$= 0, \text{ otherwise} \tag{4.17}$$

To see that this function sums to unity we make use of this identity

$$\sum_{x=0}^{k} \binom{r}{x}\binom{s}{k-x} = \binom{r+s}{k} \tag{4.18}$$

This identity is proved by equating the coefficient of $a^k$ in $(a + b)^r(a + b)^s$ to the coefficient of $a^k$ in $(a + b)^{r+s}$. Thus

$$\sum_{x=0}^{n} \binom{Np}{x}\binom{Nq}{n-x} = \binom{N(p+q)}{n} = \binom{N}{n}$$

Hence

$$\sum_{x=0}^{n} f(x) = 1 \tag{4.19}$$

The mean is found directly as follows:

$$
\begin{aligned}
E(X) &= \sum_{x=0}^{n} x \binom{Np}{x}\binom{Nq}{n-x} \Big/ \binom{N}{n} \\
&= Np \sum_{x=1}^{n} \binom{Np-1}{x-1}\binom{Nq}{n-x} \Big/ \binom{N}{n} \\
&= Np \binom{N-1}{n-1} \Big/ \binom{N}{n} \\
&= np \qquad (4.20)
\end{aligned}
$$

Similarly we find the variance to be $npq(N - n)/(N - 1)$. The m.g.f. is rather troublesome, and we shall not attempt to evaluate it.

## 4.5 The Poisson Distribution

The Poisson distribution has many applications in problems concerned with the rare occurrence of events in a fixed time interval. The classic example is the data of von Bortkiewicz on the number of men killed by horse kick in 10 army corps for 20 years. The density function is given by

$$
\begin{aligned}
f(x) &= \frac{e^{-\lambda}\lambda^x}{x!}, \qquad x = 0, 1, 2, \ldots \\
&= 0, \text{ otherwise} \qquad (4.21)
\end{aligned}
$$

with $\lambda > 0$.

The c.d.f. is given by

$$
\begin{aligned}
F(x) &= 0, \qquad x < 0 \\
&= \sum_{t=0}^{[x]} \frac{e^{-\lambda}\lambda^t}{t!}, \qquad x \geq 0 \qquad (4.22)
\end{aligned}
$$

where $[x]$ is the largest integer less than or equal to $x$. The function $F(x)$ is tabulated in Table A.3 of the Appendix. However, it can be shown that

$$
\begin{aligned}
F(x) &= 0, \qquad x < 0 \\
&= \frac{1}{[x]!} \int_{\lambda}^{\infty} t^{[x]} e^{-t}\, dt, \qquad x \geq 0 \qquad (4.23)
\end{aligned}
$$

It can be easily shown that $f(x)$ sums to unity because

$$\sum_{x=0}^{\infty} \frac{e^{-\lambda}\lambda^x}{x!} = e^{-\lambda} \sum_{x=0}^{\infty} \frac{\lambda^x}{x!} = e^{-\lambda}e^{\lambda} = 1 \tag{4.24}$$

The mean of the Poisson distribution is given by

$$\begin{aligned} E(X) &= \sum_{x=0}^{\infty} xf(x) = \sum_{x=1}^{\infty} xf(x) \\ &= \sum_{x=1}^{\infty} \frac{xe^{-\lambda}\lambda^x}{x!} = \sum_{x=1}^{\infty} \frac{e^{-\lambda}\lambda^x}{(x-1)!} \\ &= \lambda e^{-\lambda} \sum_{y=0}^{\infty} \frac{\lambda^y}{y!} = \lambda e^{-\lambda}e^{\lambda} \\ &= \lambda \end{aligned} \tag{4.25}$$

Thus, the parameter $\lambda$ of the Poisson distribution is the mean. The variance is found in a manner similar to that used for the binomial. That is,

$$\begin{aligned} E[X(X-1)] &= \sum_{x=0}^{\infty} x(x-1)f(x) \\ &= \sum_{x=2}^{\infty} \frac{x(x-1)e^{-\lambda}\lambda^x}{x!} = \sum_{x=2}^{\infty} \frac{e^{-\lambda}\lambda^x}{(x-2)!} \\ &= \lambda^2 \sum_{x=2}^{\infty} \frac{e^{-\lambda}\lambda^{x-2}}{(x-2)!} = \lambda^2 \sum_{y=0}^{\infty} \frac{e^{-\lambda}\lambda^y}{y!} \\ &= \lambda^2 \end{aligned} \tag{4.26}$$

Then

$$E(X^2) = \lambda^2 + \lambda$$

and

$$\sigma^2 = E(X^2) - \lambda^2 = \lambda \tag{4.27}$$

This is the well-known result that the variance equals the mean of the Poisson distribution.

Next let us consider the m.g.f.

$$E(e^{tX}) = \sum_{x=0}^{\infty} e^{tx} f(x) = \sum_{x=0}^{\infty} \frac{e^{-\lambda}\lambda^x e^{tx}}{x!}$$

$$= e^{-\lambda} \sum_{x=0}^{\infty} \frac{(\lambda e^t)^x}{x!} = e^{-\lambda} e^{\lambda e^t}$$

$$= e^{\lambda(e^t - 1)} \tag{4.28}$$

The moments can be generated by differentiating. For example

$$E(X^2) = \left.\frac{d^2 m(t)}{dt^2}\right|_{t=0}$$

$$= \left.\frac{d}{dt} \lambda e^t e^{\lambda(e^t - 1)}\right|_{t=0}$$

$$= \lambda^2 e^{2t} e^{\lambda(e^t - 1)} + \lambda e^t e^{\lambda(e^t - 1)} \Big|_{t=0}$$

$$= \lambda^2 + \lambda$$

The Poisson distribution is often used to approximate the binomial when $n$ is large ($n > 50$) and $p$ is small ($p < .01$) and $np$ is between 0 and 10. The justification for this approximation can be established upon the grounds that it has been found to be adequate and upon the basis of the following limiting argument.

Given the binomial distribution, let $n \to \infty$ in such a way that $np = c$, a constant. Now,

$$f(x) = \frac{n!}{x!(n-x)!} p^x (1-p)^{n-x}$$

$$= \frac{n(n-1)(n-2)\ldots(n-x+1)}{x!} \left(\frac{c}{n}\right)^x \left(1 - \frac{c}{n}\right)^{n-x}$$

$$= 1\left(1 - \frac{1}{n}\right)\left(1 - \frac{2}{n}\right)\cdots\left(1 - \frac{x-1}{n}\right)\frac{c^x}{x!}\left(1 - \frac{c}{n}\right)^{n-x} \tag{4.29}$$

As $n$ tends to $\infty$, $(1 - c/n)^{n-x}$ tends to $(1 - c/n)^n$, which tends to $e^{-c}$, and the initial factors all tend to unity, so that

$$\lim_{n \to \infty} f(x) = \frac{c^x e^{-c}}{x!} \tag{4.30}$$

Thus the limit of $f(x)$ is a Poisson probability with parameter $\lambda$ given by $np$.

## 4.6 The Normal Distribution

The examples of mathematical distributions we have considered so far have been discrete distributions in that the probability mass occurs at a finite or denumerable number of points on the real line. In such cases the probability associated with any interval on the real line is obtained by summing the probability function over the points in the interval where the probability mass is nonzero. We also wish to consider continuous mathematical distributions.

Of all continuous distributions there is none more familiar to the layman than the normal. While not applicable in all situations, its role in statistics is a central one because many existing populations are adequately described by the normal curve, and it has a central role as a limiting distribution.

The equation of the familiar bell-shaped curve shown in Figure 4.1 is given by

$$f(x) = \frac{1}{\sigma\sqrt{2\pi}} e^{-(x-\mu)^2/2\sigma^2},$$

$$-\infty < x < \infty, \sigma^2 > 0, \qquad -\infty < \mu < \infty \tag{4.31}$$

If this equation describes the density function of the random variable $X$, we shall say that $X$ is distributed normally with mean $\mu$ and variance $\sigma^2$. We shall denote this by $X \sim N(\mu, \sigma^2)$. Let us verify that $f(x)$ given by Equation (4.31) is a density function. Obviously $f(x) \geq 0$ for all $x$. It

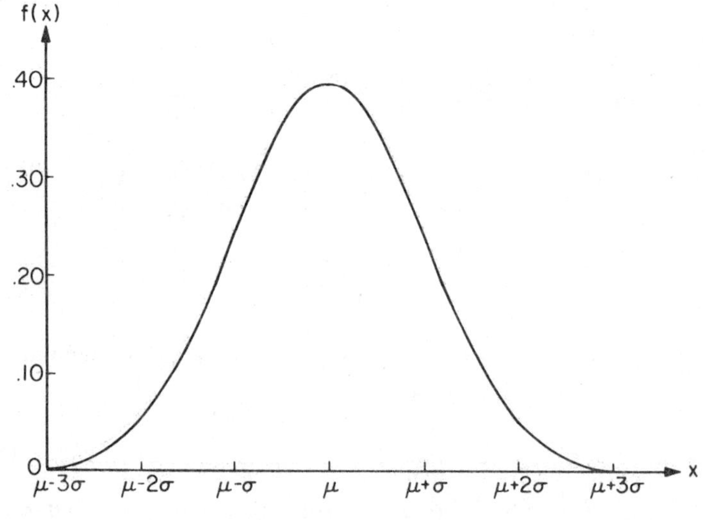


**Fig. 4.1** The Normal Curve

can be shown that $f(x)$ meets the additional requirement of integrating to unity as follows. Let

$$I = \int_{-\infty}^{\infty} f(x)\,dx$$

$$= \frac{1}{\sqrt{2\pi}} \int_{-\infty}^{\infty} e^{-t^2/2}\,dt \qquad (4.32)$$

Now

$$I^2 = \frac{1}{2\pi} \int_{-\infty}^{\infty} \int_{-\infty}^{\infty} e^{-(t^2+u^2)/2}\,dt\,du$$

$$= \frac{2}{\pi} \int_{0}^{\infty} \int_{0}^{\infty} e^{-(t^2+u^2)/2}\,dt\,du \qquad (4.33)$$

Let

$$u = r \sin \theta$$

$$t = r \cos \theta$$

Then the Jacobian is $r$, and

$$I^2 = \frac{2}{\pi} \int_{0}^{\pi/2} \int_{0}^{\infty} e^{-r^2/2} r\,dr\,d\theta$$

$$= \frac{2}{\pi} \int_{0}^{\pi/2} d\theta = 1 \qquad (4.34)$$

Since $I \geq 0$ and $I^2 = 1$, it follows that $I = 1$.

This result is related closely to the elementary theory of the gamma function which occurs widely in statistical theory and is defined by

$$\Gamma(\alpha) = \int_{0}^{\infty} t^{\alpha-1} e^{-t}\,dt \qquad (4.35)$$

which exists for all values of $\alpha$ other than zero and the negative integers. Our interest in statistics is in values of $\alpha > 0$. Integration by parts gives

$$\Gamma(\alpha + 1) = \alpha\Gamma(\alpha) \qquad (4.36)$$

Also

$$\Gamma(1) = 1 \qquad (4.37)$$

so

$$\Gamma(n + 1) = n!, \text{ for } n = 0, 1, 2, \ldots \qquad (4.38)$$

The integral of the previous paragraph may be used to show that

$$\Gamma(1/2) = \int_{0}^{\infty} t^{-1/2} e^{-t}\,dt = \sqrt{\pi} \qquad (4.39)$$

Putting $t = u^2/2$, we obtain

$$\Gamma(1/2) = \int_0^\infty \frac{\sqrt{2}}{u} e^{-u^2/2}\, u du = \sqrt{2} \int_0^\infty e^{-u^2/2}\, du$$

$$= \frac{1}{\sqrt{2}} \int_{-\infty}^\infty e^{-u^2/2}\, du = \frac{1}{\sqrt{2}} \sqrt{2\pi} = \sqrt{\pi} \tag{4.40}$$

The normal curve is symmetrical about $x = \mu$, and the distance from $\mu$ to the point of inflection on either side of $\mu$ is $\sigma$.

The reader may have inferred that $\mu$ and $\sigma$ are the mean and standard deviation of the normal density function. We have not yet shown that this choice of symbols is anything more than fortuitous. We must show that if $X$ has the density,

$$f(x) = \frac{1}{a\sqrt{2\pi}} e^{-(x-b)^2/2a^2}$$

then the mean and variance are $b$ and $a^2$ respectively. Now

$$E(X) = \frac{1}{a\sqrt{2\pi}} \int_{-\infty}^\infty x e^{-(x-b)^2/2a^2}\, dx$$

Let $t = (x - b)/a$. Then

$$E(X) = \frac{1}{\sqrt{2\pi}} \int_{-\infty}^\infty (at + b) e^{-t^2/2}\, dt$$

$$= \frac{1}{\sqrt{2\pi}} \int_{-\infty}^\infty at e^{-t^2/2}\, dt + \frac{1}{\sqrt{2\pi}} \int_{-\infty}^\infty b e^{-t^2/2}\, dt$$

$$= b \int_{-\infty}^\infty \frac{1}{\sqrt{2\pi}} e^{-t^2/2}\, dt = b \tag{4.41}$$

since the integral of an odd function is zero. Similarly, we can show that $E[(X - b)^2] = a^2$. Having once shown that the parameters $b$ and $a^2$ are the mean and variance, we use the symbols $\mu$ and $\sigma^2$ in the density function as in Equation (4.31). It is a remarkable feature of the normal density function that it is specified by the mean and variance.

$F(x)$ cannot be evaluated by the usual algebraic methods of integration, but must be evaluated by numerical means. The distribution function $F(x)$ has been extensively tabled, and one such table is given in Table A.4 of the Appendix for $\mu = 0$, $\sigma = 1$. The normal variable with mean zero and variance one is referred to as the standard normal variable and will be denoted by $Z$. Let $z_\alpha$ be such that $P(Z \geq z_\alpha) = \alpha$. From Table A.4 we see that $z_{.05} = 1.645$, $z_{.10} = 1.284$.

By subtracting the mean and dividing by the standard deviation, we can always obtain a new random variable with zero mean and variance unity. That is, if $X \sim N(\mu, \sigma^2)$, then $Z \sim N(0, 1)$ where $Z = (X - \mu)/\sigma$. This transformation enables us to use Table A.4 to evaluate $F(x)$ for a random normal variable with arbitrary mean $\mu$ and variance $\sigma^2$. This is done as follows;

$$\begin{aligned} F(x) &= P(X \leq x) \\ &= P[(X - \mu)/\sigma \leq (x - \mu)/\sigma] \\ &= P[Z \leq (x - \mu)/\sigma] \end{aligned} \tag{4.42}$$

Suppose $X \sim N(4, 25)$. Then

$$\begin{aligned} F(10) &= P(X \leq 10) \\ &= P[(X - 4)/5 \leq 6/5] \\ &= P(Z \leq 6/5) \end{aligned}$$

which from Table A.4 is found to be 0.8849.

In view of the very wide use of the standard normal distribution, we give two formulas associated with it, which are highly useful if one is performing calculations on a modern computer. These formulas are from Hastings (1955, pp. 167–91).*

The first formula is for computing $1 - F(x)$, i.e., $P(X > x)$. If $x$ is positive, we denote this by $\psi(x)$ for which we give a formula; if $x$ is negative,

$$1 - F(x) = F(-x) = 1 - \psi(-x) \tag{4.43}$$

so that we would calculate $1 - \psi(-x)$. The computing instructions for $\psi(x)$, $x > 0$ are:

Compute

$$\eta = 1/(1 + a_0 x/\sqrt{2}) \tag{4.44}$$

$$\psi(x) = (a_1\eta + a_2\eta^2 + a_3\eta^3)e^{-x^2/2}/\sqrt{\pi} \tag{4.45}$$

where

$$a_0 = .47047 \qquad a_2 = -.0849713$$
$$a_1 = .3084284 \qquad a_3 = .6627698$$

Equation (4.46) gives the abscissa $x_p$ such that a fraction $p$ is above $x_p$. That is, $x_p$ is such that $P(X > x_p) = p$. If $p \leq 0.5$, $x_p$ is given by

*Approximation for Digital Computers*, by Cecil Hastings, Jr. (© 1955 by the RAND Corporation). (Reprinted by permission of Princeton University Press.)

$\psi^*(p)$ below. If $p > 0.5$, the abscissa $x_p$ is given by $-\psi^*(1-p)$. The computing instructions for $\psi^*(p)$ are:

$$\eta = \sqrt{\ln(1/p^2)} \tag{4.46}$$
$$\psi^*(p) = \eta - (a_0 + a_1\eta)/(1 + b_1\eta + b_2\eta^2)$$

where

$$a_0 = 2.30753,\ b_1 = .99229$$
$$a_1 = .27061,\ b_2 = .04481$$

## 4.7 Approximating Distributions

In the previous section we considered approximation of binomial probabilities by the Poisson and showed that under a particular limiting process the binomial density approached a Poisson. Given a distribution dependent upon a positive integer $n$ (typically sample size), we often use the limiting distribution as $n \to \infty$ as an approximation. However, it may be the case that the limit of a probability distribution is not itself a probability distribution. Consider the sequence of random variables $Y_n$ with density functions

$$\begin{aligned} f_n(y) &= 1/n, \qquad y = 1, 2, \ldots, n \\ &= 0, \text{ otherwise} \end{aligned} \tag{4.47}$$

Now

$$\lim_{n\to\infty} f_n(y) = 0 \text{ for all } y \tag{4.48}$$

so the limit of $f_n(y)$ is not a p.d.f. We are interested in the case where the limit of a sequence of probability distributions is a probability distribution and give a definition in terms of distribution functions.

*Definition 4.1. Given the sequence of random variables $X_1, X_2, \ldots$, where $F_i$ is the distribution function of $X_i$ and the random variable $X$ with distribution function $F$, the sequence is said to approach $X$ in distribution if $\lim_{n\to\infty} F_n(x) = F(x)$ at all points where $F$ is continuous.*

This is often expressed as $X_n \xrightarrow{D} X$ or $X_n \xrightarrow{L} X$ where $D$ denotes "in distribution" and $L$ denotes "in law."

Of course it would be desirable if $F_n$ approached $F$ in the limit for all $x$. However it may happen that at the points of discontinuity of $F$, $F_n$ does not approach $F$. Suppose

$$\begin{aligned} F_n(x) &= 1/2 + nx/2, \qquad -1/n < x < 1/n \\ &= 0, \qquad x \le -1/n \\ &= 1, \qquad x \ge 1/n \end{aligned} \tag{4.49}$$

and

$$\begin{aligned} F(x) &= 0, & x &< 0 \\ &= 1, & x &\geq 0 \end{aligned} \tag{4.50}$$

Then

$$\begin{aligned} \lim_{n\to\infty} F_n(x) &= 0, & x &< 0 \\ &= 1/2, & x &= 0 \\ &= 1, & x &> 0 \end{aligned} \tag{4.51}$$

The limit of $F_n$ is not equal to $F$ at all points $x$. In fact, the limit of $F_n$ is not even a distribution function. However,

$$\lim_{n\to\infty} F_n(x) = F(x) \tag{4.52}$$

at all continuity points of $F$. The function $F$ thus seems to satisfy all our intuitive requirements for the limiting distribution.

We need ways of proving that a sequence of random variables approaches a given random variable in distribution. The most powerful method is given by the continuity theorem due to Levy (1937) and Cramér (1946).

***Theorem 4.1*** Given a sequence of random variables $X_1, X_2, \ldots$, with the corresponding sequence $\phi_1, \phi_2, \ldots$ of characteristic functions, a necessary and sufficient condition that the sequence approaches $X$ in distribution is that the sequence of characteristic functions converges to the characteristic function of $X$.

We have not developed characteristic functions in this book, but the theorem has an obvious analogue in terms of moment generating functions. We state without proof the following theorem due to Curtiss (1942).

***Theorem 4.2*** Given the sequence of random variables $X_1, X_2, \ldots$, with the corresponding sequence of m.g.f.'s $m_1, m_2, \ldots$, which exist for all $t$, $|t| < h$. If $X$ has an m.g.f. $m(t)$ defined for all $t$, $|t| \leq h_1 < h$ and

$$\lim_{n\to\infty} m_n(t) = m(t) \text{ for all } t,\ |t| \leq h_1 \tag{4.53}$$

then the sequence approaches $X$ in distribution.

For the case of vector random variables with $k$ components, the m.g.f. $M_n(t_1, t_2, \ldots, t_k)$ must tend as $n \to \infty$ for all $|t_i| < h$ to $M(t_1, t_2, \ldots, t_k)$ for the sequence of vector variables $X_1, X_2, \ldots$ to approach the vector variable $X$ in distribution.

To illustrate the utility of this theorem, let us consider again the Poisson approximation to the binomial. The binomial m.g.f. is

$[pe^t + (1 - p)]^n$. If we let $\lambda = np$ and take the limit as $n \to \infty$, we have

$$\lim_{n\to\infty}\left[1 + \frac{\lambda}{n}(e^t - 1)\right]^n = e^{\lambda(e^t - 1)} \tag{4.54}$$

which is the m.g.f. of a Poisson distribution with parameter $\lambda$.

We shall have several occasions to use the moment generating argument. The assessment of accuracy of such an approximation is very difficult and requires powerful mathematical analysis.

There are many other formulations of the idea of a sequence of random variables converging in some sense to another random variable. A brief discussion of a few of the underlying ideas follows.

*Definition 4.2. The sequence of random variables $X_1, X_2, \ldots$ approaches the random variable $X$ in quadratic mean if*

$$\lim_{n\to\infty} E[(X_n - X)^2] = 0 \tag{4.55}$$

The mean square difference is taken as a measure of the distance between $X_n$ and $X$; and if this distance converges to zero, the sequence converges to $X$, in a particular sense.

We have already encountered another type of convergence when we mentioned the weak law of large numbers in connection with Bernoulli trials. This is called convergence in probability, or stochastic convergence.

*Definition 4.3. The sequence of random variables $X_1, X_2, \ldots$ approaches $X$ in probability if*

$$\lim_{n\to\infty} P(|X_n - X| > \varepsilon) = 0 \text{ for every } \varepsilon > 0 \tag{4.56}$$

This is frequently expressed as $p \lim X_n = X$ or $X_n \xrightarrow{P} X$. It follows that these two types of convergence are related.

***Theorem 4.3*** If the sequence of random variables $X_1, X_2, \ldots$ converges in quadratic mean to $X$, the sequence converges in probability to $X$.

*Proof:*

$$\begin{aligned} E[(X_n - X)^2] &= \int_{|x_n - x| > c} (x_n - x)^2 f(x_n, x)\, dx_n\, dx \\ &\quad + \int_{|x_n - x| \le c} (x_n - x)^2 f(x_n, x)\, dx_n\, dx \\ &\ge \int_{|x_n - x| > c} (x_n - x)^2 f(x_n, x)\, dx_n\, dx \\ &\ge \int_{|x_n - x| > c} c^2 f(x_n, x)\, dx_n\, dx \\ &= c^2 P(|X_n - X| > c) \end{aligned} \tag{4.57}$$

So for any constant $\varepsilon > 0$,

$$P(|X_n - X| > \varepsilon) \leq E[(X_n - X)^2]/\varepsilon^2$$

and

$$\lim_{n\to\infty} E[(X_n - X)^2] = 0$$

so that

$$\lim_{n\to\infty} P(|X_n - X| > \varepsilon) = 0 \tag{4.58}$$

The same type of proof also shows that if $E(|X_n - X|) \to 0$, then $X_n \xrightarrow{P} X$.

We state the following without proof (see, e.g., Rao, 1965).

***Theorem 4.4*** $X_n \xrightarrow{P} X$ implies $X_n \xrightarrow{D} X$. If $X$ is equal to a constant $c$ with unit probability, then the distribution of $X_n$ tends to the distribution with unit mass at the point $c$.

We also state the following theorem.

***Theorem 4.5*** If $X_n \xrightarrow{P} X$ and if $g$ is a continuous function with probability unity, then $g(X_n) \xrightarrow{P} g(X)$ and hence $g(X_n) \xrightarrow{D} g(X)$.

This theorem is most useful because by Theorem 4.3 convergence in quadratic mean implies convergence in probability and hence by Theorem 4.4 convergence in distribution.

We have discussed three types of convergence of a sequence of random variables to another random variable. The question arises as to what can be done about convergence in the ordinary sense. Suppose that the result of an experiment gives a sequence of observed values $x_1, x_2, \ldots$ and an observed value $x$ for the random variable $X$. It is only natural to consider the convergence of the sequence of observed values $x_1, x_2, \ldots$ to $x$.

*Definition 4.4. The sequence of random variables $X_1, X_2, \ldots$ converges almost surely to the random variable $X$ if the probability of the observed sequence $x_1, x_2, \ldots$ converging to the observed value $x$ is unity.*

We state without proof the following:

***Theorem 4.6*** If the sequence $X_1, X_2, \ldots$ converges almost surely to $X$, the sequence converges to $X$ in probability.

Development of these ideas involves advanced mathematics, particularly measure theory. However, the basic ideas need presentation. For the purposes of this book the type of convergence that is relevant is in distribution, which is most easily based at our level on Theorem 4.2. It is also useful and interesting to derive approximations to discrete probabilities, and we show how the densities of the normal and multivariate normal distributions arise as approximations for the probabilities of the discrete binomial and multinomial distributions. In the case of the multivariate normal distribution the motivation for the density is otherwise a very difficult matter.

### 4.7.1 Normal Approximation to the Binomial

We wish to show that the binomial probability of $x$ successes in $n$ trials can be approximated by the ordinate of a normal density function with mean $np$ and variance $npq$. Denote the binomial probability by $P$. Using Stirling's approximation, we obtain

$$\begin{aligned}\ln P = &-\left(\frac{1}{2}\right)\ln(2\pi) + \left(n + \frac{1}{2}\right)\ln n \\ &-\left(x + \frac{1}{2}\right)\ln x - \left(n - x + \frac{1}{2}\right)\ln(n - x) \\ &+ x\ln p + (n - x)\ln q\end{aligned} \tag{4.59}$$

Now, if we let $x = np + \varepsilon$ where $\varepsilon$ is small, we obtain

$$\begin{aligned}&\left(x + \frac{1}{2}\right)\ln x + \left(n - x + \frac{1}{2}\right)\ln(n - x) \\ &\quad\doteq np\ln(np) + \varepsilon\ln(np) + \left(\frac{1}{2}\right)\ln(np) \\ &\quad + \varepsilon + \frac{\varepsilon^2}{np} + \frac{\varepsilon}{2np} - \frac{\varepsilon^2}{2np} - \frac{\varepsilon^2}{4n^2p^2} + nq\ln(nq) - \varepsilon\ln(nq) \\ &\quad + \left(\frac{1}{2}\right)\ln(nq) - \varepsilon + \frac{\varepsilon^2}{nq} - \frac{\varepsilon}{2nq} - \frac{\varepsilon^2}{2nq} - \frac{\varepsilon^2}{4n^2q^2}\end{aligned} \tag{4.60}$$

neglecting $\varepsilon^3, \varepsilon^4, \ldots$. Because of Chebyshev's inequality, $\varepsilon$ is of order $\sqrt{npq}$ with high probability, so we shall ignore $\varepsilon/np$, $\varepsilon^2/4n^2p^2$, $\varepsilon/nq$, and $\varepsilon^2/4n^2q^2$. Therefore $\ln P$ is given approximately by

$$\begin{aligned}\ln P = &-\left(\frac{1}{2}\right)\ln(2\pi) + \left(\frac{1}{2}\right)\ln n \\ &-\left(\frac{1}{2}\right)\ln(np) - \left(\frac{1}{2}\right)\ln(nq) \\ &-\varepsilon^2/2np - \varepsilon^2/2nq\end{aligned} \tag{4.61}$$

So $P$ is given approximately by

$$\frac{1}{\sqrt{2\pi}\sqrt{npq}}e^{-\varepsilon^2/2npq}$$

Thus the ordinate of a normal density function can be used to approximate the binomial probability of $x$ successes in $n$ trials. One surmises that the approximation is better for $p$ near 1/2 since the binomial distribution is in that case symmetric. The approximation also improves as $n$ increases.

It is much more common to use the cumulative normal to approximate the cumulative binomial. To justify this approximation, let us consider the m.g.f. of

$$Z = \frac{X - np}{\sqrt{npq}} \tag{4.62}$$

where $X$ is the number of successes in $n$ trials. Now $X$ can be considered as the sum of $n$ independent random variables $Y_i$, each with probability density

$$f(y_i) = p^{y_i}q^{1-y_i}, \qquad y_i = 0, 1$$

Then

$$m_{Y_i}(t) = q + pe^t$$

and

$$m_{Y_i - p}(t) = qe^{-pt} + pe^{qt} \tag{4.63}$$

Then

$$m_{(Y_i - p)/\sqrt{npq}}(t) = 1 + t^2/n2! + \text{terms which go to zero faster than } 1/n \tag{4.64}$$

Thus

$$\begin{aligned}\lim_{n\to\infty} m_{(X-np)/\sqrt{npq}}(t) &= \lim_{n\to\infty}\left[1 + \frac{t^2}{n2!}\right]^n \\ &= e^{t^2/2}\end{aligned} \tag{4.65}$$

We have therefore shown that the m.g.f. of $(X - np)/\sqrt{npq}$ approaches that of a normal distribution with mean zero and variance unity and hence, by Theorem 4.2, that the distribution of $Z$ approaches the normal distribution.

### 4.7.2 The Normal Approximation to the Poisson

If $X$ is a Poisson variable with parameter $\lambda$, consider the generating function of $Y = (X - \lambda)/\sqrt{\lambda}$.

$$\begin{aligned} m_X(t) &= e^{\lambda(e^t - 1)} \\ m_{X-\lambda}(t) &= e^{\lambda(e^t - 1 - t)} \\ &= e^{(\lambda t^2/2! + \lambda t^3/3! + \cdots)} \end{aligned} \quad (4.66)$$

Then $m_Y(t) \doteq \exp(t^2/2 + \text{terms which go to zero as } \lambda \text{ increases})$. So

$$\lim_{\lambda \to \infty} m_Y(t) = e^{t^2/2} \quad (4.67)$$

The conclusion of the matter is that for moderately large $\lambda$ the normal approximation to the Poisson is remarkably good.

## 4.8 The Rectangular Distribution

This class of distributions is used to describe populations which are uniformly distributed over some interval, such as rounding errors. The density function is given by

$$\begin{aligned} f(x) &= 1/(b - a), \qquad a < x < b \\ &= 0, \text{ otherwise} \end{aligned} \quad (4.68)$$

The cumulative distribution function is seen to be given by

$$\begin{aligned} F(x) &= 0, \qquad x \le a \\ &= (x - a)/(b - a), \qquad a < x < b \\ &= 1, \qquad x \ge b \end{aligned} \quad (4.69)$$

The mean is given by

$$\begin{aligned} E(X) &= \int_a^b x dx/(b - a) \\ &= (b^2 - a^2)/2(b - a) = (b + a)/2 \end{aligned} \quad (4.70)$$

In this case it can be seen that the expected value agrees with the intuitive concept of center of gravity. The variance can also be obtained directly and the reader should verify that $V(X) = (b - a)^2/12$.

## 4.9 The Negative Exponential Distribution

The negative exponential distribution has been used to a considerable extent in the theory of reliability and will be discussed at greater length in subsequent chapters. The density function is given by

$$\begin{aligned} f(x) &= (1/\theta)e^{-x/\theta}, \qquad 0 < x < \infty, \theta > 0 \qquad (4.71)\\ &= 0, \text{ otherwise} \end{aligned}$$

The distribution function is easily found.

$$\begin{aligned} F(x) &= \int_0^x (1/\theta)e^{-t/\theta}\,dt \\ &= (-e^{-t/\theta})\big|_0^x \\ &= 1 - e^{-x/\theta} \qquad (4.72) \end{aligned}$$

This can be evaluated for given $\theta$ from readily available exponential tables.

The moments can be found directly or by differentiating the m.g.f., which is easily found. Proceeding directly,

$$\begin{aligned} E(X) &= \int_0^\infty (x/\theta)e^{-x/\theta}\,dx \\ &= \theta \int_0^\infty te^{-t}\,dt \\ &= \theta\,\Gamma(2) \\ &= \theta \qquad (4.73) \end{aligned}$$

and

$$\begin{aligned} E(X^2) &= \int_0^\infty (x^2/\theta)e^{-x/\theta}\,dx \\ &= \theta^2 \int_0^\infty t^2e^{-t}\,dt \\ &= \theta^2\,\Gamma(3) \\ &= 2\theta^2 \qquad (4.74) \end{aligned}$$

Therefore

$$\begin{aligned} V(X) &= E(X^2) - [E(X)]^2 \\ &= \theta^2 \end{aligned}$$

The m.g.f. is given by

$$m(t) = E(e^{tX}) = \int_0^\infty (1/\theta)e^{-(1-\theta t)(x/\theta)}\,dx$$

$$= \frac{1}{1-\theta t}\int_0^\infty e^{-u}\,du$$

$$= 1/(1-\theta t) \tag{4.75}$$

In Chapter 7, a derivation of the Poisson distribution is given. If the number of occurrences $X$ of some event in a time interval of length $t$ has the Poisson distribution with parameter $\lambda t$, it can be shown that the time necessary to observe an occurrence is a random variable with the negative exponential distribution. This follows because

$$\begin{aligned} F(t) &= \text{Prob(time to occurrence} \le t) \\ &= 1 - \text{Prob(no occurrence before } t) \\ &= 1 - e^{-\lambda t}(\lambda t)^0/0! \\ &= 1 - e^{-\lambda t} \end{aligned} \tag{4.76}$$

Differentiating, we obtain

$$f(t) = \lambda e^{-\lambda t} \tag{4.77}$$

## 4.10 The Gamma Distribution

The gamma distribution is also encountered in reliability work and in other situations where time is the variable of interest. The random variable $X$ is said to have the gamma distribution with parameter $\alpha$ if the density function is given by

$$\begin{aligned} f(x) &= \frac{1}{\Gamma(\alpha)} x^{\alpha-1}e^{-x}, \qquad x > 0, \alpha > 0 \\ &= 0, \text{ otherwise} \end{aligned} \tag{4.78}$$

where $\Gamma(\alpha)$ is the gamma function. The distribution function

$$F(x) = \int_0^x \frac{1}{\Gamma(\alpha)} t^{\alpha-1}e^{-t}\,dt$$

cannot be evaluated by algebraic means, but is extensively tabled and is commonly referred to as the incomplete gamma function.

Two-parameter gamma distributions are rather common and are easily developed from the one-parameter case by introducing a scale factor.

Let $y = \beta x$. Then $dx = dy/\beta$ and

$$f(y) = \frac{y^{\alpha - 1} e^{-y/\beta}}{\beta^{\alpha}\Gamma(\alpha)} \tag{4.79}$$

We could obtain an alternate form by letting $x = \beta y$. In general, when we refer to the gamma distribution we will use the one-parameter distribution.

The moments of the one-parameter distribution can be obtained directly.

$$E(X^k) = \frac{1}{\Gamma(\alpha)} \int_0^{\infty} x^{k+\alpha-1} e^{-x}\, dx$$

$$= \frac{\Gamma(k + \alpha)}{\Gamma(\alpha)} \tag{4.80}$$

In particular

$$E(X) = \frac{\Gamma(\alpha + 1)}{\Gamma(\alpha)} = \alpha$$

$$E(X^2) = \frac{\Gamma(\alpha + 2)}{\Gamma(\alpha)} = \alpha(\alpha + 1)$$

The variance is given by

$$V(X) = E(X^2) - [E(X)]^2$$
$$= \alpha(\alpha + 1) - \alpha^2$$
$$= \alpha$$

The m.g.f. is easily obtained for both the one-parameter and two-parameter distributions.

The negative exponential distribution is a special case of the two-parameter gamma distribution given in Equation (4.79). This is seen by letting $\alpha = 1$.

## 4.11 The Beta Distribution

The beta distribution is used to describe variables which range from zero to one, such as the fraction of successes in a large number of trials. The density function is

$$f(x) = \frac{1}{B(\alpha, \beta)} x^{\alpha-1}(1 - x)^{\beta-1}, \qquad 0 \le x \le 1, \alpha, \beta, > 0$$
$$= 0, \text{ otherwise} \tag{4.81}$$

where

$$B(\alpha, \beta) = \int_0^1 y^{\alpha-1}(1 - y)^{\beta-1}\, dy = \frac{\Gamma(\alpha)\Gamma(\beta)}{\Gamma(\alpha + \beta)} \tag{4.82}$$

The distribution function $F(x)$ cannot be evaluated directly but has been extensively tabled and is commonly referred to as the incomplete beta function.

As in the case of the gamma distribution, the moments can be obtained in a direct fashion:

$$E(X^k) = \frac{1}{B(\alpha, \beta)} \int_0^1 x^{\alpha+k-1}(1-x)^{\beta-1}\,dx$$

$$= \frac{B(\alpha + k, \beta)}{B(\alpha, \beta)} \tag{4.83}$$

## 4.12 The Multinomial Distribution

This distribution is a generalization of the binomial distribution and was introduced in Chapter 3. Suppose that we perform $n$ independent trials and each trial can give a response in one of $k$ categories, $C_i$, $i = 1, 2, \ldots, k$. Suppose further that the probability of obtaining a response in category $C_i$ is $P_i$, and let $N_i$ denote the number of responses in category $C_i$. Then

$$P(N_1 = n_1, N_2 = n_2, \ldots, N_k = n_k) = \frac{n!}{n_1!n_2!\ldots n_k!} P_1^{n_1} P_2^{n_2} \ldots P_k^{n_k} \tag{4.84}$$

We are interested in presenting the multinomial distribution in more detail because all measurement data can be described by multinomial distributions. In a sample of $n$ observations the inevitable grouping error results in $N_1$ observations between $x_1$ and $x_2$, $N_2$ in observations between $x_2$ and $x_3$, etc.

We wish to consider the moments of the multinomial distribution, and we may ask for:

$$E(N_i), \qquad i = 1, 2, \ldots, k$$
$$V(N_i), \qquad i = 1, 2, \ldots, k$$
$$\mathrm{Cov}(N_i, N_j), \qquad i, j = 1, 2, \ldots, k, i \neq j$$

The quantities are easily obtained. To find $E(N_i)$ and $V(N_i)$, we note that we can collapse the situation into a binomial with two classes, $C_i$ and not-$C_i$ with probabilities $P_i$ and $1 - P_i$. Hence

$$E(N_i) = nP_i \tag{4.85}$$

and

$$V(N_i) = nP_i(1 - P_i) \tag{4.86}$$

To obtain the covariance of $N_i$ and $N_j$, we put together classes $C_i$ and $C_j$ to form class $D_1$, and the other classes together to form class $D_2$. Then $N_i + N_j$ is a binomial variable with probability $P_i + P_j$. So

$$E(N_i + N_j) = n(P_i + P_j)$$

as we already knew, and

$$V(N_i + N_j) = n(P_i + P_j)(1 - P_i - P_j)$$

We now use the fact that the variance of the sum of two random variables $X$ and $Y$ is

$$\begin{aligned} V(X + Y) &= E(X + Y - \mu_x - \mu_y)^2 \\ &= E[(X - \mu_x)^2 + (Y - \mu_y)^2 + 2(X - \mu_x)(Y - \mu_y)] \\ &= V(X) + V(Y) + 2\text{Cov}(X, Y) \end{aligned} \tag{4.87}$$

So $\text{Cov}(X, Y) = [V(X + Y) - V(X) - V(Y)]/2$.

In the case under discussion

$$\begin{aligned} \text{Cov}(N_i, N_j) &= (1/2)[n(P_i + P_j)(1 - P_i - P_j) - nP_i(1 - P_i) - nP_j(1 - P_j)] \\ &= (n/2)[(P_i + P_j) - (P_i + P_j)^2 - P_i + P_i^2 - P_j + P_j^2] \\ &= -nP_iP_j \end{aligned} \tag{4.88}$$

## 4.13 The Bivariate Normal Distribution

The normal distribution is one of the most frequently encountered distributions in the case of a single variable. In the case of two variables the bivariate normal distribution is frequently used. It will be seen it has the property that the marginal distributions are both normal and in addition the joint function is "smooth" and well behaved. The graph of the bivariate normal density function is a bell-shaped surface with elliptical contours, as shown in Figure 4.2.

The bivariate normal density function is given by

$$f(x, y) = \frac{1}{2\pi\sigma_x\sigma_y\sqrt{1 - \rho^2}} \exp\left\{-\frac{1}{2(1 - \rho^2)} \times \left[\frac{(x - \mu_x)^2}{\sigma_x^2} - \frac{2\rho(x - \mu_x)(y - \mu_y)}{\sigma_x\sigma_y} + \frac{(y - \mu_y)^2}{\sigma_y^2}\right]\right.$$

$$\begin{gathered} -\infty < x < \infty, \ -\infty < y < \infty \\ 0 < \sigma_x^2 < \infty, 0 < \sigma_y^2 < \infty \\ -\infty < \mu_x < \infty, \ -\infty < \mu_y < \infty, \ -1 < \rho < 1 \end{gathered} \tag{4.89}$$

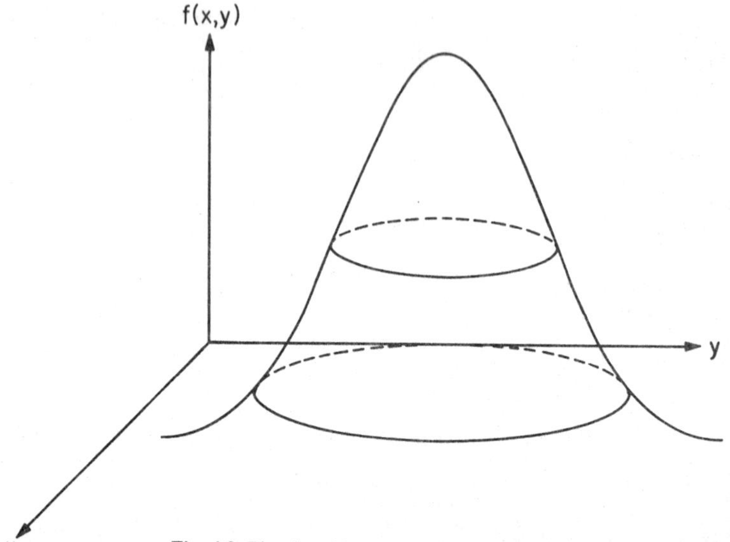


**Fig. 4.2** Bivariate Normal Density Function

The means, variances, and correlation coefficient are given by $\mu_x$, $\mu_y$, $\sigma_x^2$, $\sigma_y^2$, and $\rho$. We shall now derive this fact as well as show that this function is a bivariate density function, as we have asserted it to be. We shall also show that the marginal and conditional distributions are normal. Let us express the joint density as the product of two factors by using the elementary process of completing the square. Note that

$$a_{11}y_1^2 + 2a_{12}y_1y_2 + a_{22}y_2^2 = a_{11}\left(y_1 + \frac{a_{12}}{a_{11}}y_2\right)^2 + \left(a_{22} - \frac{a_{12}^2}{a_{11}}\right)y_2^2 \tag{4.90}$$

Then

$$f(x, y) = \frac{1}{\sqrt{2\pi}\sigma_x\sqrt{1-\rho^2}} \times$$
$$\exp\left\{-\frac{1}{2(1-\rho^2)\sigma_x^2}\left[x - \mu_x - \rho\frac{\sigma_x}{\sigma_y}(y - \mu_y)\right]^2\right\} \times$$
$$\frac{1}{\sqrt{2\pi}\sigma_y}\exp\left[-\frac{1}{2\sigma_y^2}(y - \mu_y)^2\right] \tag{4.91}$$

That is,

$$f(x, y) = g(x, y)h(y) \tag{4.92}$$

Making use of this factorization and using our knowledge of the univariate normal density function enables us to make the following statements:

1. $f(x, y)$ is a density function. Integrating first with respect to $x$ gives

$$\int_{-\infty}^{\infty} f(x, y)\, dx = h(y)$$

because $g(x, y)$ is, for a fixed $y$, a univariate normal density function and therefore integrates to unity. Since $h(y)$ is obviously a normal density function which integrates to unity,

$$\int_{-\infty}^{\infty}\int_{-\infty}^{\infty} f(x, y)\, dxdy = \int_{-\infty}^{\infty} h(y)\, dy = 1$$

2. The marginal distribution of $Y$ is normal with mean $\mu_y$ and variance $\sigma_y^2$. Integrating $f(x, y)$ with respect to $x$ gives the marginal density function of $Y$.
3. The conditional distribution of $X$ given $y$ is normal with mean $\mu_x + \rho(\sigma_x/\sigma_y)(y - \mu_y)$ and variance $\sigma_x^2(1 - \rho^2)$. This is obtained simply by dividing the joint density function by the marginal to obtain $f(x, y)/h(y) = g(x, y)$.
4. The marginal distribution of $X$ is normal with mean $\mu_x$ and variance $\sigma_x^2$. This can be seen by simply interchanging $x$ and $y$.
5. The conditional distribution of $Y$ given $x$ is normal with mean $\mu_y + \rho(\sigma_y/\sigma_x)(x - \mu_x)$ and variance $\sigma_y^2(1 - \rho^2)$. This is also seen by interchanging $x$ and $y$.

The correlation coefficient of $X$ and $Y$ is the parameter $\rho$. This is obtained by using the facts we have just derived and the properties of conditional expectation. From Section 3.18

$$\begin{aligned} E(XY) &= E[YE(X|Y)] \\ &= E\{Y[\mu_x + \rho(\sigma_x/\sigma_y)(Y - \mu_y)]\} \\ &= \mu_x\mu_y + \rho\sigma_x\sigma_y \end{aligned}$$

Therefore

$$\begin{aligned} \text{Cov}(X, Y) &= E(XY) - E(X)E(Y) \\ &= \rho\sigma_x\sigma_y \end{aligned} \tag{4.93}$$

and therefore the correlation coefficient is given by $\rho$.

The joint m.g.f. of $X$ and $Y$ is also obtained in much the same fashion:

$$\begin{aligned} &E[\exp(t_2 Y)\exp\{t_1[\mu_x + \rho(\sigma_x/\sigma_y)(Y - \mu_y)] + t_1^2\sigma_x^2(1 - \rho^2)/2\}] \\ &\quad = \exp\{t_1[\mu_x - \rho(\sigma_x/\sigma_y)\mu_y] + t_1^2\sigma_x^2(1 - \rho^2)/2\} \times \\ &\qquad E[\exp\{[t_2 + t_1\rho(\sigma_x/\sigma_y)]Y\}] \\ &\quad = \exp\{t_1[\mu_x - \rho(\sigma_x/\sigma_y)\mu_y] + t_1^2\sigma_x^2(1 - \rho^2)/2\} \times \\ &\qquad \exp\{[t_2 + t_1\rho(\sigma_x/\sigma_y)]\mu_y + [t_2 + t_1\rho(\sigma_x/\sigma_y)]^2\sigma_y^2/2\} \\ &\quad = \exp[t_1\mu_x + t_2\mu_y + (t_1^2\sigma_x^2 + 2t_1t_2\rho\sigma_x\sigma_y + t_2^2\sigma_y^2)/2] \end{aligned} \tag{4.94}$$

If $\rho = 0$, it is apparent from the m.g.f. that $X$ and $Y$ are independent, although it is not generally true that zero covariance implies independence.

A particular property of the bivariate normal distribution is that a linear function, say $aX + bY$ where $a$ and $b$ are constants, has a normal distribution with mean $a\mu_x + b\mu_y$ and variance $a^2\sigma_x^2 + 2ab\rho\sigma_x\sigma_y + b^2\sigma_y^2$. This is seen by considering the m.g.f. $m(t) = E[e^{t(aX+bY)}]$. By letting $t_1 = ta$ and $t_2 = tb$ in the joint m.g.f., we see that

$$m(t) = \exp[t(a\mu_x + b\mu_y) + t^2(a^2\sigma_x^2 + 2ab\rho\sigma_x\sigma_y + b^2\sigma_y^2)/2] \quad (4.95)$$

## 4.14 The Multivariate Normal Distribution

We have presented the normal distribution as a mathematical distribution so to speak, in its own right. We have indicated that this distribution often arises as an approximation to other distributions. We have seen that the probability of $r$ successes in $n$ trials can be approximated by an ordinate of a normal distribution and that the cumulative distribution of the binomial may under some circumstances be well approximated by the cumulative normal. Obviously, the density of a binomial is never approximated well for all values by a normal density because the binomial density is zero for nonintegers while the normal density is nonzero everywhere. The fact of the matter is that the c.d.f. for one variable may be approximated by another c.d.f. This is the usual sense in which one distribution is said to approximate another, though obviously one might wish to consider how one density approaches another.

Just as the normal distribution arises as a useful continuous approximation to many distributions of a single variable, the multivariate normal arises as a useful approximation in the case of several variables. One way of arriving at the multivariate normal distribution is to reason by analogy with the "derivation" of the normal distribution from the binomial distribution. The probability of $x$ successes in $n$ trials is closely approximated by

$$\frac{1}{\sqrt{2\pi npq}} \exp\left[-\frac{(x - np)^2}{2npq}\right]$$

A natural generalization of the binomial distribution is the multinomial distribution for which the probability of $n_1, n_2, \ldots, n_k$, $\sum_1^k n_i = n$ is, as we have seen,

$$P = \frac{n!}{\prod n_i!} \prod p_i^{n_i} \quad (4.96)$$

It is natural, therefore, to consider what happens to the probability (4.96) as $n, n_1, n_2, \ldots$ become large with $p_1, p_2, \ldots, p_k$ staying constant. Using Stirling's approximation, we obtain

$$\ln P = \left(\frac{1}{2}\right)\ln(2\pi) + \left(n + \frac{1}{2}\right)\ln n - n + \sum_{i=1}^{k} n_i \ln p_i$$

$$- \sum_{i=1}^{k}\left[\left(\frac{1}{2}\right)\ln(2\pi) + \left(n_i + \frac{1}{2}\right)\ln n_i - n_i\right]$$

$$= -\left(\frac{k-1}{2}\right)\ln(2\pi) + \left(n + \frac{1}{2}\right)\ln n + \sum_{i=1}^{k} n_i \ln p_i$$

$$- \sum_{i=1}^{k}\left(n_i + \frac{1}{2}\right)\ln n_i \tag{4.97}$$

Now if we let $n_i = np_i + \varepsilon_i$, where $\varepsilon_i$ is small, we obtain

$$\ln n_i = \ln(np_i + \varepsilon_i)$$

$$= \ln(np_i) + \ln\left(1 + \frac{\varepsilon_i}{np_i}\right)$$

$$= \ln(np_i) + \frac{\varepsilon_i}{np_i} - \frac{\varepsilon_i^2}{2n^2p_i^2}, \text{ neglecting } \varepsilon_i^3, \varepsilon_i^4, \ldots \tag{4.98}$$

and

$$(n_i + 1/2)\ln n_i = \left(np_i + \varepsilon_i + \frac{1}{2}\right)\ln(np_i + \varepsilon_i)$$

$$\doteq \left(np_i + \varepsilon_i + \frac{1}{2}\right)\left(\ln np_i + \frac{\varepsilon_i}{np_i} - \frac{\varepsilon_i^2}{2n^2p_i^2}\right)$$

$$= np_i \ln(np_i) + \varepsilon_i \ln(np_i) + \left(\frac{1}{2}\right)\ln(np_i)$$

$$+ \varepsilon_i + \frac{\varepsilon_i^2}{np_i} + \frac{\varepsilon_i}{2np_i} - \frac{\varepsilon_i^2}{2np_i} - \frac{\varepsilon_i^2}{4n^2p_i^2} \tag{4.99}$$

From Chebyshev's inequality, $\varepsilon_i$ is of the order of $\sqrt{np_i(1-p_i)}$ with high probability, so we shall ignore the terms $\varepsilon_i/np_i$ and $\varepsilon_i^2/4n_i^2p_i^2$. Hence we obtain

$$\ln P \doteq -\left(\frac{k-1}{2}\right)\ln(2\pi) + \left(\frac{1}{2}\right)\ln n - \left(\frac{1}{2}\right)\sum \ln(np_i) - \sum \frac{\varepsilon_i^2}{2np_i} \tag{4.100}$$

We have used the fact that $\sum \varepsilon_i = 0$. Let us put

$$\varepsilon_k = -\varepsilon_1 - \varepsilon_2 - \cdots - \varepsilon_{k-1}$$

Then we have

$$\ln P = -\left(\frac{k-1}{2}\right)\ln(2\pi) + \left(\frac{1}{2}\right)\ln n - \left(\frac{1}{2}\right)\ln\left(n^k \prod_1^k p_i\right)$$

$$-\left(\frac{1}{2}\right)\sum_{i=1}^{k-1} \frac{\varepsilon_i^2}{n}\left(\frac{1}{p_i} + \frac{1}{p_k}\right) - \frac{1}{2}\sum_{\substack{i,i'=1 \\ i\neq i'}}^{k-1} \frac{\varepsilon_i \varepsilon_{i'}}{np_k} \tag{4.101}$$

We see, therefore, that $P$ tends to the value

$$P^* = \frac{1}{(2\pi)^{(k-1)/2}(n^{k-1}p_1 p_2 \ldots p_k)^{1/2}} \exp\left(-\frac{1}{2}\sum_{i=1}^{k-1} a_{ii}\varepsilon_i^2 - \frac{1}{2}\sum_{\substack{i,i'=1 \\ i\neq i'}}^{k-1} a_{ii'}\varepsilon_i\varepsilon_{i'}\right) \tag{4.102}$$

where

$$a_{ii} = \frac{1}{n}\left(\frac{1}{p_i} + \frac{1}{p_k}\right)$$

and

$$a_{ii'} = \frac{1}{np_k}$$

We now take a step which may not be obvious to the reader, but is given in the elementary theory of determinants. The determinant $|(a_{ij})|$ is equal to

$$\frac{1}{np_1}\frac{1}{np_2}\cdots\frac{1}{np_{k-1}}\left[1 + \frac{1}{np_k}(np_1 + np_2 + \cdots + np_{k-1})\right]$$

$$= \frac{1}{np_1}\frac{1}{np_2}\cdots\frac{1}{np_k}(np_k + np_1 + np_2 + \cdots + np_{k-1})$$

$$= \frac{n}{\prod_{i=1}^{k} np_i} = \frac{1}{n^{k-1}\prod_{i=1}^{k} p_i} \tag{4.103}$$

We therefore have

$$P^* = \frac{1}{(2\pi)^{(k-1)/2}}[\det(a_{ij})]^{1/2}\exp\left[-\left(\frac{1}{2}\right)\sum_{i=1}^{k-1} a_{ii}\varepsilon_i^2 - \frac{1}{2}\sum_{\substack{i,j=1 \\ i\neq j}}^{k-1} a_{ij}\varepsilon_i\varepsilon_j\right] \tag{4.104}$$

This development suggests that we could consider as a density function for $m$ random variables the function

$$f(x_1, x_2, \ldots, x_m) =$$

$$\frac{1}{(2\pi)^{m/2}} \sqrt{\det(A)} \exp\left[-\left(\frac{1}{2}\right) \sum_{i,j=1}^{m} a_{ij}(x_i - \mu_i)(x_j - \mu_j)\right] \tag{4.105}$$

for $-\infty < x_i < \infty$, $i = 1, 2, \ldots, m$, where $A = (a_{ij})$.

Alternatively, we can develop the multivariate normal by giving the density function

$$f(x_1, x_2, \ldots, x_m) = K \exp\left\{-\left[\sum_{i,j=1}^{m} a_{ij}(x_i - b_i)(x_j - b_j)\right]/2\right\} \tag{4.106}$$

where $-\infty < x_i < \infty$, $i = 1, 2, \ldots, m$, and the matrix $(a_{ij})$ is symmetric. Then we find the value of $K$ so that the function integrates to unity and examine the nature of the constants $a_{ij}$ and $b_i$. This task is considerably simplified by the use of a few matrix results. Let

$$\begin{aligned} x' &= (x_1, x_2, \ldots, x_m) \\ b' &= (b_1, b_2, \ldots, b_m) \\ A &= [(a_{ij})] \end{aligned} \tag{4.107}$$

Then

$$\begin{aligned} f(x) &= f(x_1, x_2, \ldots, x_m) \\ &= K \exp[-(x - b)'A(x - b)/2] \end{aligned} \tag{4.108}$$

A fundamental theorem of matrix algebra states that for every real symmetric matrix $A$ there exists an orthogonal $P$ such that $P'AP = D$, a diagonal matrix with elements $\lambda_1, \lambda_2, \ldots, \lambda_m$. Let us make the change of variables

$$x_i - b_i = \sum_j p_{ij} y_j$$

or, in matrix notation,

$$x - b = Py \tag{4.109}$$

The Jacobian for this transformation is $\det(P) = \pm 1$, so

$$\int \cdots \int f(x_1, x_2, \ldots, x_m)\, dx_1 dx_2 \ldots dx_m$$

$$= K \int \ldots \int \exp\left[-\left(\sum_1^m \lambda_i y_i^2\right)/2\right] dy_1 \ldots dy_m \tag{4.110}$$

At this point we see that if any $\lambda$ is zero or negative, we would have a factor that is unbounded. This forces the requirement that all $\lambda$ are positive, or in other words, that $A$ is positive definite. If some $\lambda_i$ are zero, det($A$) would be zero, and we could not write the expression in Equation (4.105).

From Equation (4.110) we see that the integral is

$$K \prod_{i=1}^{m} (2\pi/\lambda_i)^{1/2} \left\{(\lambda_i/2\pi)^{1/2} \int \exp\left(-\frac{1}{2}\lambda_i y_i^2\right) dy_i\right\} \tag{4.111}$$

Each term in braces is the integral of a univariate normal with mean zero and variance $1/\lambda_i$, which we know to be unity. So we have

$$K(2\pi)^{m/2}|D|^{-1/2} = 1$$

But

$$P'AP = D$$

so

$$\begin{aligned} |A| &= |P|^2|D| \\ &= |D| \end{aligned} \tag{4.112}$$

Therefore, in order for $f(x_1, x_2, \ldots, x_m)$ to integrate to unity, we must have

$$K = |A|^{1/2} (2\pi)^{-m/2} \tag{4.113}$$

We can make use of the same algebra to obtain the properties of $b$ and $A$. From Equation (4.110) we see that the $Y_i$ are independent because the joint density factors into a product of univariate normal densities. Further, each $Y_i$ is normal with mean zero and variance $1/\lambda_i$. Therefore the expectation of $X_i$ is given by

$$\begin{aligned} E(X_i) &= E\left(\sum_j p_{ij} Y_j\right) + b_i \\ &= b_i \end{aligned} \tag{4.114}$$

So the constants $b_i$ appearing in the joint density are the mean values of the variables. The variance of $X_i$ is similarly seen to be

$$V(X_i) = \sum_j p_{ij}^2 (1/\lambda_j) \tag{4.115}$$

But

$$P'AP = D$$

so

$$A^{-1} = PD^{-1}P'$$

and

$$V(X_i) = a^{ii} \tag{4.116}$$

where $a^{ii}$ = $i$th diagonal element of $A^{-1}$.

Similarly,

$$\text{Cov}(X_i, X_j) = a^{ij}$$

where

$$a^{ij} = ij\text{th element of } A^{-1} \tag{4.117}$$

To emphasize the nature of $b$ and $A$, we commonly use the symbol $\mu$ for the vector of means and $\Sigma = A^{-1}$ for the matrix of variances and covariances. We then write the multivariate normal density as

$$|\Sigma|^{-1/2}(2\pi)^{-m/2} \exp[-(1/2)(x - \mu)'\Sigma^{-1}(x - \mu)] \tag{4.118}$$

To obtain the moment generating function, we again make use of the algebra already developed. Let $t' = (t_1, t_2, \ldots, t_m)$.

Then

$$E[e^{\Sigma t_i(X_i - \mu_i)}] = E[e^{t'(X-\mu)}]$$
$$= E(e^{t'PY}) = E(e^{\tau'Y}) \tag{4.119}$$

where $\tau' = t'P$.

So because the $Y_i$ are independent univariate normal variables, we have

$$E(e^{\tau'Y}) = e^{\Sigma\tau_i^2/2\lambda_i} = e^{\tau'D^{-1}\tau/2} \tag{4.120}$$

But

$$\tau'D^{-1}\tau = t'PD^{-1}P't = t'\Sigma t$$

So

$$E(e^{t'(X-\mu)}) = e^{t'\Sigma t/2}$$

and

$$E(e^{t'X}) = e^{t'\mu + t'\Sigma t/2} \tag{4.121}$$

A more general definition of a multivariate normal distribution is the distribution of a vector $X$ with an m.g.f. of the form in Equation (4.121). Note that here $\Sigma^{-1}$ may not exist, but in this case the density in the form in Equation (4.118) does not exist. Consider the vector of variables $Z$ given by $CX$. Now

$$m_Z(t) = E(e^{t'Z}) = E(e^{t'CX}) = E(e^{\tau'X}) \tag{4.122}$$

where $\tau' = t'C$. We know this to be

$$m_X(\tau) = e^{\tau'\mu + \tau'\Sigma\tau/2}$$

$$m_Z(t) = e^{t'C\mu + t'C\Sigma C't/2} \tag{4.123}$$

So the joint distribution of the variables $Z = CX$ is a multivariate normal with mean vector $C\mu$ and variance-covariance matrix $C\Sigma C'$.

Next we consider the question of marginal and conditional distributions. Partition the variables in the vector $X$ into two vectors, $X_1$ and $X_2$, and partition the mean vector and variance-covariance matrix accordingly. That is, let

$$\begin{pmatrix} X_1 \\ X_2 \end{pmatrix} \sim N\left[\begin{pmatrix} \mu_1 \\ \mu_2 \end{pmatrix}, \begin{pmatrix} \Sigma_{11} & \Sigma_{12} \\ \Sigma_{21} & \Sigma_{22} \end{pmatrix}\right] \tag{4.124}$$

where $X_1$ is a $(m - k) \times 1$ vector, and $X_2$ is a $k \times 1$ vector. It follows trivially from the result on linear combinations that $X_1$ is a multivariate normal vector with parameters $\mu_1$ and $\Sigma_{11}$. Let

$$C = (I|\phi)$$

so that

$$CX = X_1 \tag{4.125}$$

$$C\mu = \mu_1$$

and

$$C\Sigma C' = \Sigma_{11}$$

Similarly, we can show that $X_2$ is a multivariate normal with parameters $\mu_2$ and $\Sigma_{22}$ by letting $C = (\phi|I)$.

To obtain the conditional distribution of the variables of $X_1$ given the values of $X_2$, we shall proceed in the straightforward fashion of dividing the joint density function by the marginal density function. However, without loss of generality, let us take the means to be zero. Then we have

$$f(x_1|x_2) = \frac{|\Sigma|^{-1/2}(2\pi)^{-m/2}\exp(-x'\Sigma^{-1}x/2|)}{|\Sigma_{22}|^{-1/2}(2\pi)^{-k/2}\exp(-x_2'\Sigma_{22}^{-1}x_2/2)} \tag{4.126}$$

Before we attempt to simplify this expression, let us recall some facts about partitioned matrices. Let us denote the inverse of $\Sigma$ by $T$ and partition $T$ in conformity with partitioning of $\Sigma$. Then we have

$$\begin{pmatrix} \Sigma_{11} & \Sigma_{12} \\ \Sigma_{21} & \Sigma_{22} \end{pmatrix}\begin{pmatrix} T_{11} & T_{12} \\ T_{21} & T_{22} \end{pmatrix} = \begin{pmatrix} I & \phi \\ \phi & I \end{pmatrix} \tag{4.127}$$

This leads to the following equations:

$$\begin{aligned} \Sigma_{11}T_{11} + \Sigma_{12}T_{21} &= I \\ \Sigma_{11}T_{12} + \Sigma_{12}T_{22} &= \phi \\ \Sigma_{21}T_{11} + \Sigma_{22}T_{21} &= \phi \\ \Sigma_{21}T_{12} + \Sigma_{22}T_{22} &= I \end{aligned} \tag{4.128}$$

We can solve these equations for the elements of $T$ in terms of the elements of $\Sigma$. So

$$\begin{aligned}
T_{11} &= (\Sigma_{11} - \Sigma_{12}\Sigma_{22}^{-1}\Sigma_{21})^{-1} \\
T_{12} &= -\Sigma_{11}^{-1}\Sigma_{12}T_{22} = -T_{11}\Sigma_{12}\Sigma_{22}^{-1} \\
T_{21} &= -\Sigma_{22}^{-1}\Sigma_{21}T_{11} = -T_{22}\Sigma_{21}\Sigma_{11}^{-1} \\
T_{22} &= (\Sigma_{22} - \Sigma_{21}\Sigma_{11}^{-1}\Sigma_{12})^{-1}
\end{aligned} \tag{4.129}$$

The relations are symmetric in that we can exchange the symbols $T$ and $\Sigma$. For example $\Sigma_{22} = (T_{22} - T_{21}T_{11}^{-1}T_{12})^{-1}$. Recall also that we can obtain the determinant of a matrix by partitioning:

$$\begin{aligned}
|\Sigma| = \begin{vmatrix} \Sigma_{11} & \Sigma_{12} \\ \Sigma_{21} & \Sigma_{22} \end{vmatrix} &= |\Sigma_{22}|\,|\Sigma_{11} - \Sigma_{12}\Sigma_{22}^{-1}\Sigma_{21}| \\
&= |\Sigma_{11}|\,|\Sigma_{22} - \Sigma_{21}\Sigma_{11}^{-1}\Sigma_{12}|
\end{aligned} \tag{4.130}$$

This can be shown by noting that if $A$ and $C$ are nonsingular,

$$\begin{aligned}
&\begin{pmatrix} I & -BC^{-1} \\ \phi & I \end{pmatrix} \begin{pmatrix} A & B \\ B' & C \end{pmatrix} \begin{pmatrix} I & \phi \\ -C^{-1}B' & I \end{pmatrix} \\
&= \begin{pmatrix} A - BC^{-1}B' & \phi \\ \phi & C \end{pmatrix}
\end{aligned} \tag{4.131}$$

Taking determinants

$$\begin{vmatrix} A & B \\ B' & C \end{vmatrix} = |A - BC^{-1}B'|\,|C| \tag{4.132}$$

and, by reversing the roles of $A$ and $C$,

$$\begin{vmatrix} A & B \\ B' & C \end{vmatrix} = |C - B'A^{-1}B|\,|A|$$

Let us return to $f(x_1|x_2)$. The expression in the exponent will be

$$-(x_1'T_{11}x_1 + x_1'T_{12}x_2 + x_2'T_{21}x_1 + x_2'T_{22}x_2 - x_2'\Sigma_{22}^{-1}x_2)/2 \tag{4.133}$$

Let us "complete the square" on $x_1$ in the quantity in the brackets. We have

$$\begin{aligned}
&(x_1 + T_{11}^{-1}T_{12}x_2)'T_{11}(x_1 + T_{11}^{-1}T_{12}x_2) \\
&\quad + x_2'T_{22}x_2 - x_2'T_{21}T_{11}^{-1}T_{12}x_2 - x_2'\Sigma_{22}^{-1}x_2 \\
&= (x_1 - \Sigma_{12}\Sigma_{22}^{-1}x_2)'(\Sigma_{11} - \Sigma_{12}\Sigma_{22}^{-1}\Sigma_{21})^{-1}(x_1 - \Sigma_{12}\Sigma_{22}^{-1}x_2) \\
&\quad - x_2'\Sigma_{22}^{-1}x_2 + x_2'(T_{22} - T_{21}T_{11}^{-1}T_{12})x_2
\end{aligned} \tag{4.134}$$

But $\Sigma_{22}^{-1} = T_{22} - T_{21}T_{11}^{-1}T_{12}$. So we have in the exponent

$$(x_1 - \Sigma_{12}\Sigma_{22}^{-1}x_2)'\Sigma_{11.2}^{-1}(x_1 - \Sigma_{12}\Sigma_{22}^{-1}x_2) \tag{4.135}$$

where $\Sigma_{11.2} = \Sigma_{11} - \Sigma_{12}\Sigma_{22}^{-1}\Sigma_{21}$. The multiplicative constant in Equation (4.126) involves $|\Sigma|^{-1/2}/|\Sigma_{22}|^{-1/2}$. But this is equal to

$$|\Sigma_{11} - \Sigma_{12}\Sigma_{22}^{-1}\Sigma_{21}|^{-1/2} = |\Sigma_{11.2}|^{-1/2}$$

So finally we have

$$f(x_1|x_2) = |\Sigma_{11.2}|^{-1/2}(2\pi)^{(m-k)/2}$$
$$\exp[-(x_1 - \Sigma_{12}\Sigma_{22}^{-1}x_2)'\Sigma_{11.2}^{-1}(x_1 - \Sigma_{12}\Sigma_{22}^{-1}x_2)/2] \quad (4.136)$$

So, conditional on $X_2 = x_2$, $X_1$ has a multivariate normal distribution with mean vector $\Sigma_{12}\Sigma_{22}^{-1}x_2$ and variance-covariance matrix $\Sigma_{11.2}$. More generally, the mean vector with nonzero means $\mu_1$ and $\mu_2$ is

$$\mu_1 + \Sigma_{12}\Sigma_{22}^{-1}(x_2 - \mu_2) \quad (4.137)$$

Let us illustrate for the bivariate normal which we have already examined:

$$\begin{pmatrix} \Sigma_{11} & \Sigma_{12} \\ \Sigma_{21} & \Sigma_{22} \end{pmatrix} = \begin{pmatrix} \sigma_1^2 & \rho\sigma_1\sigma_2 \\ \rho\sigma_1\sigma_2 & \sigma_2^2 \end{pmatrix}$$
$$\Sigma_{11} - \Sigma_{12}\Sigma_{22}^{-1}\Sigma_{21} = \sigma_1^2 - \rho^2\sigma_1^2 = \sigma_1^2(1 - \rho^2)$$
$$\Sigma_{12}\Sigma_{22}^{-1} = \rho\sigma_1/\sigma_2$$

So the conditional mean of $X_1$ given $x_2$ is $\mu_1 + \rho(\sigma_1/\sigma_2)(x_2 - \mu_2)$, and the conditional variance is $\sigma_1^2(1 - \rho^2)$, as we obtained before.

## 4.15 The Limiting Distribution of the Multinomial

We have shown how the probabilities for the multinomial distribution tend with increasing sample size to the ordinates of the multivariate normal distribution. It is desirable to show that one distribution tends to the other. As usual, we use an m.g.f. argument. Consider

$$M(t_1, t_2, \ldots, t_k) = E\left\{\exp\left[\sum_{i=1}^{k} t_i(n_i - np_i)/\sqrt{n}\right]\right\}$$
$$= \exp\left(-\sqrt{n}\sum_{i=1}^{k} t_i p_i\right)\sum\left[\frac{n!}{\prod n_i!}\prod\left(p_i^{n_i}e^{t_i n_i/\sqrt{n}}\right)\right]$$
$$= \exp\left(-\sqrt{n}\sum_{i=1}^{k} t_i p_i\right)\left(\sum_{i=1}^{k} p_i e^{t_i/\sqrt{n}}\right)^n \quad (4.138)$$

Here, $\sum$ denotes summation over all $\{n_i\}$ such that $n_i \geq 0$, $\sum n_i = n$. Taking logarithms, we have

$$\ln M(t_1, t_2, \ldots, t_k) = -\sqrt{n}\sum_{i=1}^{k} t_i p_i + n\ln\left(\sum_{i=1}^{k} p_i e^{t_i/\sqrt{n}}\right) \quad (4.139)$$

But

$$\sum_{i=1}^{k} p_i e^{t_i/\sqrt{n}} = \sum_{i=1}^{k} p_i[1 + t_i/\sqrt{n} + t_i^2/2n + o(n^{-1})]$$

where $o(n^{-1})$ denotes a term which goes to zero faster than $n^{-1}$.

So

$$\sum_{i=1}^{k} p_i e^{t_i/\sqrt{n}} = 1 + \sum_{i=1}^{k} t_i p_i/\sqrt{n} + \sum_{i=1}^{k} p_i t_i^2/2n + o(n^{-1}) \tag{4.140}$$

Using $\ln(1 + x) = x - x^2/2 + o(x^2)$, we have

$$\ln M(t_1, t_2, \ldots, t_k) = -\sqrt{n} \sum_{i=1}^{k} t_i p_i + \sqrt{n} \sum_{i=1}^{k} t_i p_i + \sum_{i=1}^{k} p_i t_i^2/2 - \left(\sum_{i=1}^{k} t_i p_i\right)^2/2 + o(1)$$

Hence

$$\lim_{n\to\infty} \ln M(t_1, t_2, \ldots, t_k) = \frac{1}{2}\left[\sum_{i=1}^{k} p_i t_i^2 - \left(\sum_{i=1}^{k} p_i t_i\right)^2\right] \tag{4.141}$$

Then the limiting m.g.f. of $(n_i - np_i)/\sqrt{n}$, $i = 1, 2, \ldots, k - 1$ is given by $e^{Q/2}$, where

$$Q = \sum_{i=1}^{k-1} p_i(1 - p_i)t_i^2 - \sum_{i \neq i' = 1}^{k-1} p_i p_{i'} t_i t_{i'} \tag{4.142}$$

This is, however, the m.g.f. of the multivariate normal distribution with variances equal to $p_i(1 - p_i)$, and covariances equal to $-p_i p_{i'}$.

## PROBLEMS

1. Graph the binomial density function for $n = 2, 3, 4, 5$ and for $p = .1, .3, .5, .7, .9$.
2. Given the uniform density function

$$f(x) = 1, \quad 0 < x < 1$$
$$= 0, \text{ otherwise}$$

   (a) Find the mean and variance.
   (b) If five values are selected at random from this distribution, what is the probability that three out of the five will be greater than 1/4?
3. The m.g.f. of a discrete random variable was believed to be

$$m(t) = (1 + .9t + .5t^2 + .15t^3 + \cdots)^{10}$$

   Is this possible?

4. The number of particles emitted from a radioactive source in time $T$ is assumed to be a random variable with the Poisson density function

$$f(x) = e^{-\lambda T}(\lambda T)^x/x!$$
$$x = 0, 1, 2, 3, \ldots$$

The laboratory technicians prefer to talk in terms of particles per unit time. What is the density of $r = x/T$? What is the mean and variance of $r$?

5. The m.g.f. of a random variable is found to be $m(t) = (e^{e^t - 1})^2$. What is the density function of the random variable?

6. Graph the Poisson density function for $\lambda = 1, 5, 10, 15$.

7. A box contains 1000 fuses, 1% of which are defective. A random sample of 50 fuses is to be chosen. What is the probability that the sample contains exactly 2 defectives? Calculate the exact answer using the hypergeometric, and then calculate approximate answers using (a) the binomial approximation, (b) the Poisson approximation, and (c) the normal approximation.

8. The width of a slot on a forging is approximately normally distributed with mean 0.2 in. and standard deviation 0.01 in. If the specifications were given as 0.175 to 0.215, what percentage of forgings will be defective? If a random sample of 10 forgings is selected, what is the probability that the sample will contain fewer than 3 defective?

9. The voltage of a battery under specified conditions is required to be no less than 35 volts and no greater than 50 volts. If the voltage is approximately normally distributed with mean 40, what is the largest standard deviation that can be tolerated if the requirement is that at least 90% of batteries meet specifications?

10. If $X$ is a normal variable with mean zero and variance one, and $P(X \leq a) = .5$ and $P(X \leq b) = .7$, what is $P(-b \leq X \leq a)$?

11. If $X$ is a normal variable with mean zero and variance one, and $P(X \geq a) = .6$, find $P(X \geq 0 | X \leq a)$.

12. The m.g.f. of a random variable was determined to be

$$m(t) = e^{8t^2} \sum_{i=0}^{\infty} (2t)^i/i!$$

What is the density function of the random variable?

13. An engineer wishes to carry out a simulation study and requires a random number generator. He chooses a roulette wheel but decides to check it out before starting the experiment. The pointer is started from the same position each time, and the clockwise angular displacement is measured when the pointer comes to rest.
    (a) If the wheel is fair, what is the density function of the angular measurement?
    (b) If the pointer is spun 100 times, what is the probability that fewer than 25 of the measurements exceed $\pi$?
    (c) What is the probability that all 100 resting places will lie on a semicircle?
    (d) What is the mean and variance?

(e) What is the m.g.f.?
(f) How would you use this wheel to generate random normal deviates?

14. Graph the negative exponential for values of $\theta = 1/2, 1, 2$, and 4.
15. Given the negative exponential:
    (a) Find the median in terms of $\theta$.
    (b) Find the m.g.f.
    (c) Find $P(1000 \le X \le 1001 | X \ge 1000)$.
    (d) Find $P(1 \le X \le 2 | X \ge 1)$.
16. The time to failure of transistors often approximately follows the negative exponential distribution. What is the probability that a transistor chosen at random will survive longer than the average?
17. Graph the gamma density function given in Sect. 4.10 for $\alpha = 1, 2, 3, 4$.
18. Find the mean and variance of the gamma density function.
19. Graph the beta distribution for $\alpha = 1, 2, 3; \quad \beta = 1, 2, 3$.
20. Find the mean and variance of the beta density function.
21. Show by successively integrating by parts that

$$\sum_{x=0}^{r} \binom{n}{x} p^x q^{n-x} = \frac{1}{B(r+1, n-r)} \int_p^1 x^r (1-x)^{n-r-1}\, dx$$

22. If $X$ is a binomial random variable, find the mean of $\ln(1 + X)$.
23. Find the first three cumulants of the binomial distribution.
24. For a binomial distribution with $n = 4, p = 1/8$, find the mode and all medians of the distribution.
25. Construct an algorithm for finding the mode of a binomial distribution for any $n$ and $p$.
26. Find the mean of the negative binomial distribution.
27. Show that the m.g.f. of the negative binomial distribution is $[p/(1 - qe^t)]^r$.
28. Show that the cumulative Poisson distribution can be expressed in terms of the gamma, i.e., show that

$$\sum_{x=0}^{r} \frac{e^{-\lambda}\lambda^x}{x!} = \frac{1}{r!} \int_\lambda^\infty x^r e^{-x}\, dx$$

29. Find the first three cumulants of the Poisson distribution. Find the $r$th cumulant.
30. For the univariate distributions discussed in this chapter, derive and display in a table: (a) mean and variance, (b) mode(s) and median(s), (c) the m.g.f. and the first three cumulants, and (d) the skewness and kurtosis parameters $\alpha_3$ and $\alpha_4$.
31. Using the standard normal tables, express the interquartile range in terms of $\sigma$.
32. Show that the points of inflection for the normal curve are $\mu - \sigma$ and $\mu + \sigma$.

33. Show by successive integration by parts that $\Gamma(n+1) = n\Gamma(n)$.
34. If $n$ is a positive integer, show $\Gamma(n+1) = n!$.
35. Show that $\Gamma(1/2) = \sqrt{\pi}$.
36. Show that $B(a, b) = B(b, a)$.
37. Show that

$$B(a, b) = 2\int_0^{\pi/2} \sin^{2a-1}\theta \cos^{2b-1}\theta \, d\theta$$

38. Show that

$$\Gamma(a)\Gamma(b) = 4\int_0^{\pi/2} \sin^{2a-1} \cos^{2b-1} \, d\theta \int_0^{\infty} r^{2a+2b-1}e^{-r}\, dr$$

39. Show that

$$B(a, b) = \frac{\Gamma(a)\Gamma(b)}{\Gamma(a+b)}$$

40. Use Hasting's approximations for the standard normal distribution to calculate $P(X > 1.5)$ and to find $x_0$ such that $P(X > x_0) = .3$.
41. Give a flow chart for calculating $\Gamma(n)$ for any positive integral value of $n$.
42. Given the density function

$$\begin{aligned} f(x) &= p(1-p)^{x-1}, x = 1, 2, \ldots \\ &= 0, \text{otherwise} \end{aligned}$$

find the mean, variance, and moment generating function.

43. Given the density function,

| $x$ | 1 | 2 | 3 | other values |
|---|---|---|---|---|
| $f(x)$ | $f(1)$ | $f(2)$ | $f(3)$ | 0 |

find $f(1)$, $f(2)$, and $f(3)$, if $\mu = 14/6$, $\sigma^2 = 5/9$, $\mu_3' = 49/3$.

44. For the trinomial distribution show that the correlation coefficient $\rho(N_1, N_2)$ is given by

$$-\frac{p_1 p_2}{(p_1 + p_3)(p_2 + p_3)}$$

45. Show that the contours of the bivariate normal are transformed to the standard form for an ellipse by (a) letting $x_1' = (x_1 - \mu_1)/\sigma_1$ and $x_2' = (x_2 - \mu_2)/\sigma_2$, and (b) rotating the axis 45°. Discuss the major and minor axes of the ellipses in their standard form for various values of $\rho$.
46. Verify that the determinantal value given in Equation (4.103) is correct.
47. Show that for $p_1 + p_2 + p_3 = 1$,

$$\begin{pmatrix} np_1(1-p_1) & -np_1p_2 \\ -np_1p_2 & np_2(1-p_2) \end{pmatrix} \begin{pmatrix} \frac{1}{n}\left(\frac{1}{p_1} + \frac{1}{p_3}\right) & \frac{1}{np_3} \\ \frac{1}{np_3} & \frac{1}{n}\left(\frac{1}{p_2} + \frac{1}{p_3}\right) \end{pmatrix} = \begin{pmatrix} 1 & 0 \\ 0 & 1 \end{pmatrix}$$

This result shows that for the case of 3 multinomial classes the inverse of the variance-covariance matrix of $\varepsilon_1, \varepsilon_2$ is the matrix $a_{ij}$ in Equation (4.102).

48. Extend the proof given in the previous problem to include $k$ multinomial classes.
49. For the normal distribution show that $\mu_4 = 3\sigma^4$ and therefore that $\alpha_4 = 3$.
50. Use Chebyshev's inequality to show for the binomial distribution that

$$\lim_{n\to\infty} P(|X - np| > \varepsilon) = 0$$

51. Show that the m.g.f. of $X^2$ where $X \sim N(0, 1)$ is $(1 - 2t)^{-1/2}$.
52. Find the mean and mode of the beta distribution.
53. Find $E[(1/X)^r]$ for the one-parameter gamma distribution.
54. Find the mean and variance of the one-parameter gamma distribution.
55. The Laplace distribution has the density function

$$f(x) = \frac{1}{2\lambda}\exp(-|x - \mu|/\lambda), \quad -\infty < x < \infty$$

Find the mean, variance, skewness, and kurtosis.

56. If $n$ is distributed as a Poisson variable with parameter $\lambda$ and the conditional distribution of $X$ given $n$ is a binomial with parameters $n$ and $p$, find the conditional distribution of $n$ given $x$.
57. A random variable $X$ has the lognormal distribution if $\ln X$ has a normal distribution with mean $\mu$ and variance $\sigma^2$. Find the expectation of $X$.
58. Using the facts that

$$\frac{d}{du}\left(\frac{1}{u}e^{-u^2/2}\right) = -\left(1 + \frac{1}{u^2}\right)e^{-u^2/2}$$

$$\frac{d}{du}\left[\left(\frac{1}{u} - \frac{1}{u^3}\right)e^{-u^2/2}\right] = -\left(1 - \frac{3}{u^4}\right)e^{-u^2/2}$$

and

$$\left(1 - \frac{3}{u^4}\right)\frac{1}{\sqrt{2\pi}}e^{-u^2/2} < \frac{1}{\sqrt{2\pi}}e^{-u^2/2} < \left(1 + \frac{1}{u^2}\right)\frac{1}{\sqrt{2\pi}}e^{-u^2/2}$$

by integration show that

$$\left(1 - \frac{1}{t^2}\right)\frac{1}{\sqrt{2\pi}\,t}e^{-t^2/2} < \int_t^\infty \frac{1}{\sqrt{2\pi}}e^{-u^2/2}\,du < \frac{1}{\sqrt{2\pi}\,t}e^{-t^2/2}, \text{ for } t > 0$$

This is a very interesting inequality on an upper tail area of a normal distribution [from Feller (1957)].

59. Suppose $X$ given $\phi \sim N(\theta, \phi)$ where $\phi$ is distributed according to a distribution with m.g.f. $M_\phi(t)$. What is the m.g.f. of $X$? Consider further the case in which $\phi$ follows a gamma distribution.
60. Consider the sequence of random variables $X_n$, $n = 1, 2, \ldots$, with

$$P(X_n = 0) = 1 - \frac{1}{n}, \text{ and } P(X_n = n) = \frac{1}{n}$$

(a) Show that $X_n \xrightarrow{P} X$, where $P(X = 0) = 1$.
(b) Evaluate $E(X_n)$, and $E(X)$. Does $E(X_n) \to E(X)$?
(c) Evaluate $V(X_n)$ and $V(X)$. Does $\lim_{n\to\infty} V(X_n)$ exist? Comment on your results.

61. The Dirichlet distribution is a multivariate generalization of the univariate beta distribution.

The $k$-variate density function is

$$c\, x_1^{\alpha_1 - 1} x_2^{\alpha_2 - 1} \ldots x_k^{\alpha_k - 1}(1 - x_1 - x_2 - \cdots - x_k)^{\alpha_{k+1} - 1}$$

for $0 \le x_i, \sum_1^k x_i \le 1$, and zero, otherwise, and $\alpha_i > 0$, $i = 1, 2, \ldots, k + 1$. The normalizing factor is

$$\Gamma(\alpha_1 + \alpha_2 + \cdots \alpha_{k+1})/[\Gamma(\alpha_1)\Gamma(\alpha_2) \ldots \Gamma(\alpha_{k+1})]$$

Verify that this is a probability density for the case $k = 2$.

62. Suppose a random variable $X$ has the exponential distribution. Obtain, e.g., by differentiating an appropriate c.d.f., the distribution of $X^\alpha (\alpha > 0)$. Show also that the p.d.f. of $Z = X^{1/\beta} (\beta > 0)$ is

$$\frac{\beta}{\theta} z^{\beta - 1} \exp(-z^{\beta/\theta}), \qquad z > 0$$

63. Consider a simple random sample without replacement of size $n$ from $N$ balls: $N_1$ are red, $N_2$ are black, and the remainder are white. Show that the probability density function for the number $X$ of red balls and the number $Y$ of black balls is given by the bivariate hypergeometric density function

$$f(x, y) = \frac{\binom{N_1}{x}\binom{N_2}{y}\binom{N - N_1 - N_2}{n - x - y}}{\binom{N}{n}}$$

64. Show that as $N \to \infty$ with $N_1/N = P_1$ and $N_2/N = P_2$, the limit of the probability density function given in Problem 63 is the bivariate binomial density function

$$f(x, y) = \frac{N!}{x!y!(N - x - y)!} P_1^x P_2^y (1 - P_1 - P_2)^{N - x - y}$$

65. Show that as $N \to \infty$ with $NP_1 = \lambda_1$ and $NP_2 = \lambda_2$, we obtain as the limit of the bivariate binomial density function the bivariate Poisson density function

$$f(x, y) = \frac{e^{-\lambda_1 - \lambda_2} \lambda_1^x \lambda_2^y}{x!y!}$$

# 5

# Distributions of Functions of Random Variables

In this chapter random variables with either a known distribution or some known properties, such as the means and variances, are discussed as well as the distribution of a function or perhaps several functions of the original random variables. Considered first is the finding of approximate means and variances, then exact distributions for simple cases.

## 5.1 Statistical Differentials

### 5.1.1 Nonrandom Total Differential

Let $X$ be a variable and $Y = f(X)$ a differentiable function of $X$. Then from elementary calculus we know that

$$\begin{aligned} \Delta Y &= f(X + \Delta X) - f(X) \\ &= f'(X)\,\Delta X + \varepsilon \Delta X \end{aligned} \tag{5.1}$$

That is, the increment $\Delta Y$ is given approximately by the differential $f'(X)\,\Delta X$, the goodness of the approximation depending upon the linearity of $f(X)$ and the magnitude of $\Delta X$. This relationship can be used to approximate the error in $Y$ of $\Delta Y$ resulting from the error $\Delta X$ in $X$. For example, the diameter of a circle is found by measurement to be 5 inches. What effect would an error of 0.5 inches in measurement have on the calculated area? Since

$$A = \pi d^2/4, \; \Delta A \doteq \pi d \Delta d/2$$

or

$$\begin{aligned} \Delta A &\doteq \pi(5)(.5)/2 \\ &= 1.25\pi \end{aligned}$$

Note that the use of the exact formula in this case gives an upper limit of $7.5625\pi$ and a lower limit of $5.0625\pi$, or a maximum positive error of $1.3125\pi$ and a maximum negative error of $1.1875\pi$. This simple example illustrates a phenomenon of frequent occurrence—a small error in the input induces a sizable error in the output.

If $U$ is a function of several variables, say, $X$, $Y$, and $Z$, the total differential formula is

$$dU = \frac{\partial f}{\partial X}\,dX + \frac{\partial f}{\partial Y}\,dY + \frac{\partial f}{\partial Z}\,dZ$$

and the change $\Delta U$ is given approximately by

$$\Delta U \doteq \frac{\partial f}{\partial X}\,\Delta X + \frac{\partial f}{\partial Y}\,\Delta Y + \frac{\partial f}{\partial Z}\,\Delta Z \tag{5.2}$$

the "goodness" of the approximation again depending upon the linearity of $U = f(X, Y, Z)$.

*Example 1*

The preceding formulas are useful for determining the error induced in the variable of interest by specific errors in the measured quantities. If a good deal is known about the errors in question, it may be possible to determine the maximum and minimum error in $U$. For example, suppose that the cost per good item, $C$, in a manufacturing process is given by

$$C = \frac{C_1}{y_1} + \frac{C_2}{y_1 y_2}$$

where $y_1$ and $y_2$ are proportions of good items at two different stages of the process and $C_1$ and $C_2$ are costs. If it has been determined that $C_1 = \$5$, $C_2 = \$7$, $y_1 = .70$, and $y_2 = .80$, what is the maximum error in $C$ if the maximum errors in $y_1$ and $y_2$ are .10 and .05 respectively?

$$\Delta C \doteq -\left(\frac{C_1}{y_1^2} + \frac{C_2}{y_1^2 y_2}\right)\Delta y_1 - \frac{C_2}{y_1 y_2^2}\,\Delta y_2$$

We estimate the greatest error in this case to be given by

$$\begin{aligned}(\Delta C)\max &\doteq -\left(\frac{5}{.49} + \frac{7}{.392}\right)(.10) - \frac{7}{.448}(.05) \\ &= 3.587\end{aligned}$$

By simply evaluating $C$ we find that $C_{\max} = 23.89$, $C_{\min} = 16.54$, and that $C$ evaluated at the values actually measured is 19.64. Thus the maximum possible error is actually 4.25 rather than 3.587.

Sometimes it is interesting to express the error as a fraction of the quantity measured, e.g., $\Delta U/U$. This is referred to as the relative error. It may sometimes be found directly by logarithms since

$$d(\log U) = dU/U \tag{5.3}$$

For example, consider the relative error in computing the area of a circle when the radius is in error by $\Delta r$. Since $A = \pi r^2$, $\Delta A \doteq dA = 2\pi r dr$. Then

$$\Delta A/A = 2\pi r \Delta r/\pi r^2 = 2\Delta r/r$$

We could obtain this directly since

$$\log A = \log \pi + 2 \log r$$

and, as before

$$d \log A = 2\Delta r/r$$

### 5.1.2 Extension of Total Differential Concept to Random Errors

The material in the preceding section is useful in that one can specify particular errors in the measured quantities and determine approximately the resulting error in the computed quantity. However, a more common situation is the one in which the errors in the measured quantities are random and therefore the resulting error is a random error. The interest is then in characterizing the frequency distribution of the error in the resultant. We at least would like to have the mean and the variance.

Consider $U = f(X, Y, Z)$, where $X$, $Y$, and $Z$ are random variables. Suppose the means are given by $\mu_x$, $\mu_y$, and $\mu_z$ respectively; that the variances are given by $\sigma_x^2$, $\sigma_y^2$, and $\sigma_z^2$; and that the variables $X$, $Y$, and $Z$ have correlation coefficients $\rho_{xy}$, $\rho_{xz}$, and $\rho_{yz}$ respectively. Expanding $U$ in a Taylor series expansion about $\mu_x$, $\mu_y$, and $\mu_z$ gives

$$\begin{aligned} U = f(\mu_x, \mu_y, \mu_z) &+ \frac{\partial f}{\partial X}(X - \mu_x) \\ &+ \frac{\partial f}{\partial Y}(Y - \mu_y) + \frac{\partial f}{\partial Z}(Z - \mu_z) \\ &+ \text{higher order terms} \end{aligned} \tag{5.4}$$

where the derivatives are to be evaluated at $\mu_x, \mu_y, \mu_z$. If higher order terms are negligible, we have

$$\begin{aligned} E(U) &\doteq f(\mu_x, \mu_y, \mu_z) + \frac{\partial f}{\partial X} E(X - \mu_x) + \frac{\partial f}{\partial Y} E(Y - \mu_y) + \frac{\partial f}{\partial Z} E(Z - \mu_z) \\ &= f(\mu_x, \mu_y, \mu_z) \end{aligned} \tag{5.5}$$

and the variance of $U$ is found directly from the definition to be

$$E[U - E(U)]^2 \doteq E\left[\frac{\partial f}{\partial X}(X - \mu_x) + \frac{\partial f}{\partial Y}(Y - \mu_y) + \frac{\partial f}{\partial Z}(Z - \mu_z)\right]^2$$

$$\sigma_U^2 \doteq \left(\frac{\partial f}{\partial X}\right)^2 \sigma_x^2 + \left(\frac{\partial f}{\partial Y}\right)^2 \sigma_y^2 + \left(\frac{\partial f}{\partial Z}\right)^2 \sigma_z^2$$
$$+ 2\rho_{xy} \frac{\partial f}{\partial X}\frac{\partial f}{\partial Y}\sigma_x\sigma_y + 2\rho_{xz}\frac{\partial f}{\partial X}\frac{\partial f}{\partial Z}\sigma_x\sigma_z$$
$$+ 2\rho_{yz}\frac{\partial f}{\partial Y}\frac{\partial f}{\partial Z}\sigma_y\sigma_z \qquad (5.6)$$

where the derivatives are to be evaluated at $\mu_x$, $\mu_y$, and $\mu_z$. If the errors are independent, we have

$$\sigma_U^2 \doteq \left(\frac{\partial f}{\partial X}\right)^2 \sigma_x^2 + \left(\frac{\partial f}{\partial Y}\right)^2 \sigma_y^2 + \left(\frac{\partial f}{\partial Z}\right)^2 \sigma_z^2 \qquad (5.7)$$

This method of getting the means and variances of functions of random variables is called the method of statistical differentials. A more exact formula for $E(U)$ and $V(U)$ can be obtained by retaining additional terms of the Taylor expansion but it becomes a little complex.

We can apply the method of statistical differentials to two or more functions of random variables. If, for instance, $U = f(X, Y, Z)$, $V = g(X, Y, Z)$, then the approximate covariance of $U$ and $V$ is given by

$$\mathrm{Cov}(U, V) \doteq \left(\frac{\partial f}{\partial X}\frac{\partial g}{\partial X}\right)\sigma_x^2 + \left(\frac{\partial f}{\partial X}\frac{\partial g}{\partial Y} + \frac{\partial f}{\partial Y}\frac{\partial g}{\partial X}\right)\sigma_{xy}$$
$$+ \left(\frac{\partial f}{\partial X}\frac{\partial g}{\partial Z} + \frac{\partial f}{\partial Z}\frac{\partial g}{\partial X}\right)\sigma_{xz} + \left(\frac{\partial f}{\partial Y}\frac{\partial g}{\partial Y}\right)\sigma_y^2$$
$$+ \left(\frac{\partial f}{\partial Y}\frac{\partial g}{\partial Z} + \frac{\partial f}{\partial Z}\frac{\partial g}{\partial Y}\right)\sigma_{yz} + \left(\frac{\partial f}{\partial Z}\frac{\partial g}{\partial Z}\right)\sigma_z^2 \qquad (5.8)$$

### 5.1.3 Relative Error or Coefficient of Variation

If $U$ is in error by an amount $\Delta U$, it is often regarded as desirable to consider the relative error $\Delta U/U$. If, as discussed in the previous section, $\Delta U$ is a random variable, we often consider an analogous quantity, namely $\sigma_U/E(U)$, the standard deviation divided by the mean. This quantity is also called the coefficient of variation and is denoted by $C_U$. The coefficient of variation (frequently abbreviated C.V.) is of value because it is independent of the units of measurement. The motivation for considering this parameter arises from consideration of Chebyshev's inequality given in Chapter 2.

We have

$$P(|X - \mu| \geq C) \leq \frac{\sigma^2}{C^2}$$

So if we are interested in relative deviations of $X$ from $\mu$,

$$P\left(\left|\frac{X-\mu}{\mu}\right| \geq \delta\right) \leq \frac{\sigma^2}{\mu^2\delta^2} \tag{5.9}$$

Then if we desire a small relative deviation from the mean, we desire a small value for $\sigma/\mu$, the coefficient of variation. In some areas of application it is more customary to work with the reciprocal $\mu/\sigma$, which is called the signal-to-noise ratio.

### 5.1.4 Some Special Cases

If $U$ is given by a linear function of $X$, $Y$, and $Z$, for example,

$$U = aX + bY + cZ$$

the preceding formulas hold exactly because the Taylor series terminates with the linear terms and

$$E(U) = a\mu_x + b\mu_y + c\mu_z \tag{5.10}$$

$$\sigma_U^2 = a^2\sigma_x^2 + b^2\sigma_y^2 + c^2\sigma_z^2 + 2ab\rho_{xy}\sigma_x\sigma_y + 2ac\rho_{xz}\sigma_x\sigma_z + 2bc\rho_{yz}\sigma_y\sigma_z \tag{5.11}$$

If $X$, $Y$, and $Z$ are uncorrelated, then $\sigma_U^2 = a^2\sigma_x^2 + b^2\sigma_y^2 + c^2\sigma_z^2$. In general, if $U = \sum_1^k a_iX_i$,

$$E(U) = \sum_1^k a_i\mu_i \tag{5.12}$$

and

$$\sigma_U^2 = \sum_1^k a_i^2\sigma_i^2 + \sum_{i \neq j} a_ia_j\rho_{ij}\sigma_i\sigma_j \tag{5.13}$$

Suppose that $U$ is given by

$$U = X_1^{a_1}X_2^{a_2} \tag{5.14}$$

with $X_1$ and $X_2$ independent.
From previous sections we know that the mean is given approximately by

$$\mu_U \doteq \mu_1^{a_1}\mu_2^{a_2} \tag{5.15}$$

and the variance is given approximately by

$$\sigma_U^2 \doteq \left(\frac{\partial f}{\partial X_1}\right)^2\sigma_1^2 + \left(\frac{\partial f}{\partial X_2}\right)^2\sigma_2^2 \tag{5.16}$$

where the derivatives are evaluated at $\mu_1, \mu_2$. Thus

$$\sigma_U^2 \doteq (a_1\mu_1^{a_1-1}\mu_2^{a_2})^2\sigma_1^2 + (a_2\mu_1^{a_1}\mu_2^{a_2-1})^2\sigma_2^2 \tag{5.17}$$

and

$$\frac{\sigma_U^2}{\mu_U^2} \doteq a_1^2 \frac{\sigma_1^2}{\mu_1^2} + a_2^2 \frac{\sigma_2^2}{\mu_2^2} \tag{5.18}$$

Thus, we have the rather interesting result that the square of the coefficient of variation behaves with independence in the same way for products as the variance does for linear combinations. That is

$$C_U^2 \doteq a_1^2 C_1^2 + a_2^2 C_2^2 \tag{5.19}$$

In general, if

$$U = \prod_{i=1}^{k} X_i^{a_i}$$

the mean and coefficient of variation are given approximately by

$$E(U) \doteq \prod_{i=1}^{k} \mu_i^{a_i} \tag{5.20}$$

$$C_U^2 \doteq \sum_{i=1}^{k} a_i^2 C_i^2 \tag{5.21}$$

Suppose

$$U = \ln X_1^{a_1} X_2^{a_2} \tag{5.22}$$

with $X_1$ and $X_2$ independent, then

$$\sigma_U^2 \doteq \left( \frac{a_1 X_1^{a_1 - 1} X_2^{a_2}}{X_1^{a_1} X_2^{a_2}} \right)^2_{\mu_1, \mu_2} \sigma_1^2 + \left( \frac{a_2 X_1^{a_1} X_2^{a_2 - 1}}{X_1^{a_1} X_2^{a_2}} \right)^2_{\mu_1, \mu_2} \sigma_2^2$$

$$= a_1^2 \frac{\sigma_1^2}{\mu_1^2} + a_2^2 \frac{\sigma_2^2}{\mu_2^2}$$

Thus

$$\sigma_U^2 \doteq a_1^2 C_1^2 + a_2^2 C_2^2 \tag{5.23}$$

In general

$$\sigma_U^2 \doteq \sum_{i=1}^{k} a_i^2 C_i^2 \tag{5.24}$$

Suppose $U = e^{\Sigma a_i X_i}$. Then

$$\ln U = \sum a_i X_i \tag{5.25}$$

and

$$\text{Var}(\ln U) \doteq \sum a_i^2 \sigma_i^2 \tag{5.26}$$

From the preceding we know $\text{Var}(\ln U) \doteq C_U^2$; thus

$$C_U^2 \doteq \sum a_i^2 \sigma_i^2 \tag{5.27}$$

### 5.1.5 The Validity of Statistical Differentials

A logically complete discussion of the method of statistical differentials is beyond the level of this textbook. This must necessarily be based on ideas of approximating distributions of the type given in Sect. 4.7, and involves advanced ideas. We refer the interested reader to the paper of Mann and Wald (1943), of Chernoff (1956), and of Pratt (1959), in which is developed a whole probability theory of orders of magnitude, using symbols $O_p(\ )$ and $o_p(\ )$, which correspond to $O(\ )$ and $o(\ )$ of ordinary mathematics.

At our present level, however, we can give the following intuitive ideas. Consider the case of a random variable $X$ which is distributed as $\text{Bi}(n, p)$. We have seen that the distribution of $(X - np)/\sqrt{npq}$ approaches the normal distribution with mean zero and variance unity. Hence, if $n$ is large, we have

$$P[(X - np)/\sqrt{npq} \leq k] \doteq \int_{-\infty}^{k} \frac{1}{\sqrt{2\pi}} e^{-t^2/2}\, dt = \Phi(k)$$

though we have great difficulty in saying precisely what we mean by "large." Hence,

$$P(X \leq np + k\sqrt{npq}) = \Phi(k)$$

Consider now $\ln X$. We have

$$P[\ln X \leq \ln(np + k\sqrt{npq})] \doteq \Phi(k)$$

But

$$\ln(np + k\sqrt{npq} = \ln np + \ln\left(1 + k\sqrt{\frac{q}{np}}\right)$$
$$= \ln np + k\sqrt{\frac{q}{np}} + o\left(\frac{1}{\sqrt{n}}\right)$$

Hence, if $n$ is large,

$$P\left(\ln X \leq \ln np + k\sqrt{\frac{q}{np}}\right) \doteq \Phi(k)$$

or the distribution of $\ln X$ is given approximately by the normal distribution with mean $\ln np$ and variance $q/np$. The nature of the approximation must be clearly understood, and the following remarks may aid this. It is quite erroneous to say that $\ln X$ has a mean which is approximately $\ln np$, for instance, because $X$ equals zero with probability $q^n$, and hence with a useful but not exactly correct mode of expression, $\ln X$ takes the value

$-\infty$ with probability $q^n$ (which is of course greater than zero). Hence, the distribution of $\ln X$ has no moments of positive order. The approximation is in terms of the cumulative distribution function *only*. We could consider $\ln(X + 1)$, however, which will be close probability-wise to $\ln X$ if $n$ is large and this variate does have finite moments for any $n$ and its mean is close to $\ln(np + 1)$, provided $np$ is not "small." We suggest that the reader examine numerically the case of $\text{Bi}(2, p)$, for $0.1 \leq p \leq 0.9$ to convince himself of the utility of the approximation. Here $n$ is equal to 2 and is not "large" but the approximation is reasonable. This example for a particular original distribution and a particular function of the random variable possesses no special features and illustrates the basic idea behind the applicability of statistical differentials. The particular example shows in addition that the range of choice of mathematical distribution to represent real data may not be at all narrow.

## 5.2 Transformation of Mathematical Distributions

### 5.2.1 General Techniques

We have considered how a function of random variables $X_1, X_2, \ldots, X_n$ behaves. Actually we have given only approximations to elementary properties, the mean and variance. This is accurate enough for many purposes, and in many cases it is all we can do. We would like to solve the problem exactly, but it is only for special functions of random variables from special distributions that answers in usable form have been obtained.

The general problem of transforming mathematical distributions is that we have a $k$-variate random variable $X_1, X_2, \ldots, X_k$, and we wish to consider the distribution of an $m$-variate function of this random variable. Let

$$\begin{aligned} Y_1 &= g_1(X_1, X_2, \ldots, X_k) \\ Y_2 &= g_2(X_1, X_2, \ldots, X_k) \\ &\cdots\cdots\cdots\cdots \\ Y_m &= g_m(X_1, X_2, \ldots, X_k) \end{aligned} \tag{5.28}$$

The functions $Y_i$ are random variables, and the natural and obvious question is how these are distributed.

The simplest case is that of a discrete distribution, for in that case $P(X_1 = x_1, X_2 = x_2, \ldots, X_k = x_k)$ is some number $P(x_1, x_2, \ldots, x_k)$, and the probability that $g_1(X_1, X_2, \ldots, X_k) = b_1, g_2(X_1, X_2, \ldots, X_k) = b_2, \ldots, g_m(X_1, X_2, \ldots, X_k) = b_m$ is the sum of $P(x_1, x_2, \ldots, x_k)$ over all $k$-tuples $(x_1, x_2, \ldots, x_k)$ such that $g_1(x_1, x_2, \ldots, x_k) = b_1, g_2(x_1, x_2, \ldots, x_k) = b_2, \ldots, g_m(x_1, x_2, \ldots, x_k) = b_m$. In general this will be a very difficult problem. Apart from special tricks using combinatory arguments, which will be possible only if the functions $g$ have special forms, there is one general procedure which may be workable. This consists of determining the moment generating function of $Y_1, Y_2, \ldots, Y_m$. This method (or its more advanced counterpart, the use of the characteristic function) is the basic method of getting approximate distributions. The m.g.f. of $Y_1, Y_2, \ldots, Y_m$ is equal to $E[\exp(t_1 Y_1 + t_2 Y_2 + \cdots + t_m Y_m)]$ and is a function $M_{Y_1, \cdot, \ldots, Y_m}(t_1, t_2, \ldots, t_m)$ of $m$ variables. If we can recognize this is a joint m.g.f., and if this joint m.g.f. specifies uniquely a joint probability distribution which we can recognize, we have solved the problem. Additionally, if the function of $t_1, t_2, \ldots, t_m$ factors into a product of m.g.f.'s, say $m_1(t_1)m_2(t_2) \ldots m_m(t_m)$, we can conclude that $Y_1, Y_2, \ldots, Y_m$ are independent with distributions corresponding to $m_1(t_1), m_2(t_2), \ldots, m_m(t_m)$ respectively.

*Example 2*

Suppose $X_i$ are independent $\mathrm{Bi}(n_i, p)$, $i = 1, 2, \ldots, k$, and we want the distribution of $\sum X_i$. We have

$$E[\exp(t \textstyle\sum X_i)] = \prod_{i=1}^{k} E[\exp(tX_i)]$$
$$= \prod_{i=1}^{k} [pe^t + (1 - p)]^{n_i}$$
$$= [pe^t + (1 - p)]^N, \text{ where } N = \textstyle\sum n_i$$

But this is the m.g.f. of a random variable which is $\mathrm{Bi}(N, p)$. Hence $\sum_1^k X_i$ is $\mathrm{Bi}(N, p)$ where $N = \sum_1^k n_i$.

We now turn to the case of a continuous distribution. We suppose that we have a $k$-variate random variable $X = (X_1, X_2, \ldots, X_k)$ with a continuous distribution, and we wish to find the distribution of $g(X)$ or joint distribution of $g_1(X), g_2(X), \ldots, g_m(X)$. Such problems are, in general, very difficult. In this text we can give only an indication of methods that may be useful.

If we can obtain the m.g.f. and can recognize it as the m.g.f. of a distribution $D$, then we may be able to bring to bear the theorem which states that the m.g.f. determines uniquely the distribution under some conditions, and we would then have solved the problem.

*Example 3*

Suppose $X$ has the density function

$$f(x) = (1/\sqrt{2\pi})\, e^{-x^2/2}$$
$$-\infty < x < \infty$$

What is the density of $V = X^2$? The m.g.f. of $V$ is

$$E(e^{tV}) = \int_{-\infty}^{\infty} e^{tx^2} f(x)\, dx$$

$$= \frac{1}{\sqrt{2\pi}} \int_{-\infty}^{\infty} e^{-(1-2t)x^2/2}\, dx$$

$$= (1 - 2t)^{-1/2}$$

Thus the m.g.f. of $V$ is obtained in a straightforward fashion and is the m.g.f. of a variable $U$ with the density function

$$g(u) = (2\pi u)^{-1/2} e^{-u/2}, \qquad u \geq 0$$
$$= 0, \text{ otherwise}$$

*Example 4*

$$f(x_1, x_2) = \frac{1}{\sqrt{2\pi}} \exp\left(-\frac{x_1^2}{2}\right) \frac{1}{\sqrt{2\pi}} \exp\left(-\frac{x_2^2}{2}\right), \qquad \begin{array}{l} -\infty < x_1 < \infty \\ -\infty < x_2 < \infty \end{array}$$

Suppose we wish to find the joint distribution of

$$G_1 = X_1 + X_2$$
$$G_2 = X_1 - X_2$$

We have for the m.g.f. of $G_1$ and $G_2$,

$$\begin{aligned} &E\{\exp[t_1(X_1 + X_2) + t_2(X_1 - X_2)]\} \\ &= E\{\exp[(t_1 + t_2)X_1 + (t_1 - t_2)X_2]\} \\ &= E\{\exp[(t_1 + t_2)X_1]\}E[\exp(t_1 - t_2)X_2)] \\ &= \exp(t_1 + t_2)^2/2 \exp(t_1 - t_2)^2/2 \\ &= \exp(t_1^2 + t_2^2) \end{aligned}$$

But this is the m.g.f. of $U$, $V$ with $U$, and $V$ independently normal (0, 2). Hence $G_1$, $G_2$ has the bivariate normal density

$$f(g_1, g_2) = \frac{1}{\sqrt{4\pi}} \exp(-g_1^2/4) \frac{1}{\sqrt{4\pi}} \exp(-g_2^2/4)$$

The problems with the use of the m.g.f. as a mode of determining a derived distribution are: (1) The m.g.f. may not exist. This can be evaded by using the characteristic function, which always exists. (2) The passage from characteristic function to distribution is in general not easy. It requires some facility in the use of Fourier transforms and inverse Fourier transforms.

At the present level the moment generating method is useful if it leads to a recognizable m.g.f. for which the distribution is known and is uniquely determined.

We may sometimes be able to proceed directly from the definition to find the distribution function of $Y_1, Y_2, \ldots, Y_m$.

$$\begin{aligned} F(y_1, y_2, \ldots, y_m) = P[&g_1(X_1, X_2, \ldots, X_k) \le y_1 \\ &g_2(X_1, X_2, \ldots, X_k) \le y_2 \\ &\cdots\cdots\cdots\cdots\cdots\cdots \\ &g_m(X_1, X_2, \ldots, X_k) \le y_m] \end{aligned} \tag{5.29}$$

*Example 5*

Let us find the distribution function of

$$U = \max(X_1, X_2)$$

and

$$V = (X_1 + X_2)/2$$

where

$$\begin{aligned} f(x_1, x_2) &= 1, \qquad 0 < x_1 < 1, 0 < x_2 < 1 \\ &= 0, \text{ otherwise} \end{aligned}$$

The joint distribution function is readily obtained with the aid of Figure 5.1:

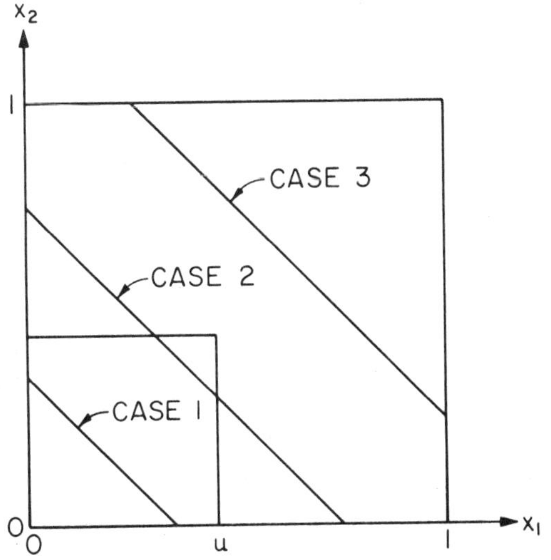


**Fig. 5.1** Determination of Cumulative Distribution Functions

*Case 1* $$2v \leq u$$

$$F(u, v) = P(X_{\max} \leq u, X_1 + X_2 \leq 2v) = (2v)^2/2$$

*Case 2* $$u \leq 2v \leq 2u$$

$$F(u, v) = u^2 - (2u - 2v)^2/2 = u^2 - 2(u - v)^2$$

*Case 3* $$2v \geq 2u$$

$$F(u, v) = u^2$$

Apart from the m.g.f. technique there are three general methods of obtaining the probability density of functions of continuous random variables. These are (1) differentiation of cumulative distribution function, (2) transformation of variables with integration if necessary, and (3) limit method applied directly.

### 5.2.2 Differentiation of Cumulative Distribution Function

Recall that $F(x)$ gives the cumulative probability that $X \leq x$. That is,

$$F(x) = P(X \leq x) \tag{5.30}$$

Further

$$\frac{d}{dx} F(x) = f(x) \tag{5.31}$$

*Example 6*

If $f(x) = e^{-x}, x > 0$ and $U = X^2$, let us find $g(u)$ by first finding the c.d.f. $G(u)$. Now $U \leq u$ holds if and only if $X \leq \sqrt{u}$,

so

$$G(u) = P(U \leq u) = P(X \leq \sqrt{u}) = \int_0^{\sqrt{u}} e^{-x}\, dx = 1 - e^{-\sqrt{u}}$$

Then

$$g(u) = u^{-1/2}e^{-u^{1/2}}/2$$

This procedure can be generalized, of course, to include many variables. To be explicit

$$f(x_1, x_2, \ldots, x_k) = \frac{\partial^k}{\partial x_1 \partial x_2 \ldots \partial x_k} F(x_1, x_2, \ldots, x_k) \tag{5.32}$$

*Example 7*

Let us illustrate this method with Example 5. Although the statement of the distribution function requires breaking the region into three cases, it is curious that when we differentiate with respect to $u$ and $v$, we obtain

$$f(u, v) = 4, \qquad u/2 \le v \le u$$
$$0 < u < 1$$
$$= 0, \text{ otherwise.}$$

*Example 8*

Suppose $X_1, X_2, \ldots, X_n$ are independent random variables from a distribution with c.d.f. $F(x)$. What is the probability density of $U = X_{\max}$, the maximum of $X_1, X_2, \ldots, X_n$?

We have (1) $P(X \le x) = F(x)$, (2) $U \le u$ if and only if every $X$ is $\le u$, so

$$P(U \le u) = [F(u)]^n,$$

and if $g(u)$ is the p.d.f. of $U$,

$$g(u) = \frac{d}{du}[F(u)]^n = n[F(u)]^{n-1} f(u)$$

where $f(x)$ is the p.d.f. of any $X$.

### 5.2.3 Transformation of Variables—Use of Jacobian

We first consider transformations of a univariate random variable. If $u = u(x)$ is such that there is a one-to-one correspondence between values of $u$ and $x$, the density function of $U$, $g(u)$ is given by

$$g(u) = f[x(u)]\left|\frac{dx}{du}\right|$$

This is seen by using the results of the previous section. For

$$\begin{aligned} g(u) &= \frac{d}{du} G(u) \\ &= \lim_{\Delta u \to 0} \frac{P(u \le U \le u + \Delta u)}{\Delta u} \\ &= \lim_{\Delta u \to 0} \frac{P[x(u) \le X \le x(u + \Delta u)]}{\Delta u} \\ &= \lim_{\Delta u \to 0} f(x')\left|\frac{\Delta x}{\Delta u}\right|, \text{ for } x' \text{ between } x(u) \text{ and } x(u + \Delta u) \end{aligned} \tag{5.33}$$

Then

$$g(u) = f[x(u)]\left|\frac{dx}{du}\right| \tag{5.34}$$

*Example 9*

If $u = 3x$ and $f(x)$ is given by $(x + 1)/8$, $-1 \leq x \leq 3$, $g(u)$ is given by

$$g(u) = (1/8)[x(u) + 1]\left|\frac{dx}{du}\right|$$

$$= (1/8)[(u/3) + 1](1/3)$$

$$= (u + 3)/72, \qquad -3 \leq u \leq 9$$

If for each value of $u$ there are several values of $x$, the procedure can be developed by use of Figure 5.2. The case is illustrated for which two values of $x$ correspond to one value of $u$.

$$P(u \leq U \leq u + \Delta u) = P(x_1 \leq X \leq x_1 + \Delta x_1) + P(x_2 \leq X \leq x_2 + \Delta x_2)$$
$$= f(x_1')\Delta x_1 + f(x_2')\Delta x_2 \qquad (5.35)$$

where

$$x_1 \leq x_1' \leq x_1 + \Delta x_1$$

and

$$x_2 \leq x_2' \leq x_2 + \Delta x_2$$

$$\frac{d}{du}G(u) = \lim_{\Delta u \to 0} \frac{P(u \leq U \leq u + \Delta u)}{\Delta u}$$

$$= -\lim_{-\Delta u \to 0} f(x_1')\frac{\Delta x_1(u)}{\Delta u} + \lim_{\Delta u \to 0} f(x_2')\frac{\Delta x_2(u)}{\Delta u} \qquad (5.36)$$

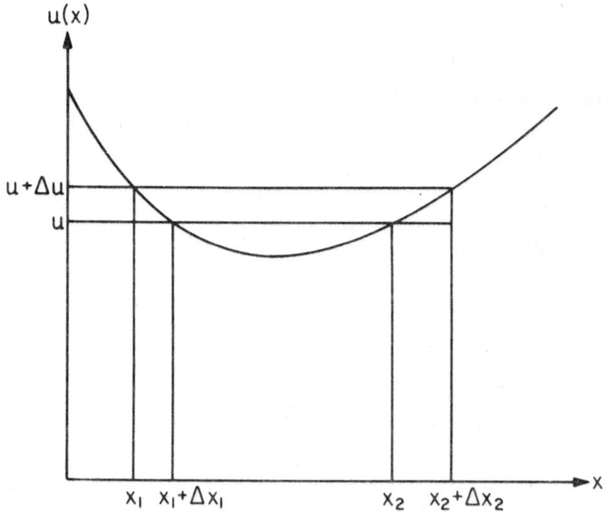


**Fig. 5.2** Use of Jacobian

That is,

$$g(u) = f(x_1)\left[-\frac{dx_1(u)}{du}\right] + f(x_2)\frac{dx_2(u)}{du} \tag{5.37}$$

Note that $-dx_1(u)/du$ is positive. In general, if values $x_1, x_2, \ldots, x_k$ give rise to the same value $u$,

$$g(u) = f(x_1)\left|\frac{dx_1(u)}{du}\right| + f(x_2)\left|\frac{dx_2(u)}{du}\right| + \cdots + f(x_k)\left|\frac{dx_k(u)}{du}\right| \tag{5.38}$$

*Example 10*

Let us find $g(u)$, where $u = x^2$, when $f(x)$ is given by $f(x) = (x + 1)/8$, $-1 \le x \le 3$. The range of $u$ will be from 0 to 9, and for values of $u \le 1$ there are two corresponding values of $x$. For values of $u > 1$ there is one corresponding value of $x$. If $u \le 1$

$$x_1(u) = -\sqrt{u}, \qquad x < 0$$

$$x_2(u) = \sqrt{u}, \qquad x > 0$$

For $u \le 1$,

$$g(u) = \frac{x_1(u) + 1}{8}\left|\frac{dx_1}{du}\right| + \frac{x_2(u) + 1}{8}\left|\frac{dx_2}{du}\right|$$

$$= \frac{-\sqrt{u} + 1}{8}\,\frac{1}{2\sqrt{u}} + \frac{\sqrt{u} + 1}{8}\,\frac{1}{2\sqrt{u}}$$

$$= \frac{1}{8\sqrt{u}}$$

For $u > 1$,

$$g(u) = \frac{\sqrt{u} + 1}{8}\,\frac{1}{2\sqrt{u}}$$

$$= \frac{\sqrt{u} + 1}{16\sqrt{u}}$$

This procedure generalizes to enable us to find the joint density of $U_1, U_2, \ldots, U_k$ where

$$\begin{aligned} u_1 &= u_1(x_1, \ldots, x_k) \\ u_2 &= u_2(x_1, \ldots, x_k) \\ &\cdots\cdots\cdots\cdots \\ u_k &= u_k(x_1, \ldots, x_k) \end{aligned} \tag{5.39}$$

The Jacobian is defined as the determinant

$$J = \begin{vmatrix} \dfrac{\partial x_1}{\partial u_1} & \dfrac{\partial x_1}{\partial u_2} & \cdots & \dfrac{\partial x_1}{\partial u_k} \\ \vdots & \vdots & \cdots & \vdots \\ \dfrac{\partial x_k}{\partial u_1} & \dfrac{\partial x_k}{\partial u_2} & \cdots & \dfrac{\partial x_k}{\partial u_k} \end{vmatrix} \tag{5.40}$$

If the transformation is such that only one point $(x_1, x_2, \ldots, x_k)$ corresponds to one point $(u_1, u_2, \ldots, u_k)$, the joint density function of the $U$'s is given by

$$f(u_1, u_2, \ldots, u_k) = f[x_1(u_1, \ldots, u_k), x_2(u_1, \ldots, u_k), \ldots, x_k(u_1, \ldots, u_k)]|J| \tag{5.41}$$

where $|J|$ is expressed in terms of $u_1, u_2, \ldots, u_k$, and $|J|$ is the absolute value of $J$. If more than one point $(x_1, x_2, \ldots, x_k)$ corresponds to a single point $(u_1, u_2, \ldots, u_k)$, the space of points $(x_1, x_2, \ldots, x_k)$ must be partitioned into regions such that within these regions the correspondence is 1 to 1.

*Example 11*

Let us continue with Example 5 discussed earlier. Let us find the joint density function of $U = \max(X_1, X_2)$ and $V = (X_1 + X_2)/2$, where

$$\begin{aligned} f(x_1, x_2) &= 1, \qquad 0 < x_1 < 1, 0 < x_2 < 1 \\ &= 0, \text{ otherwise} \end{aligned}$$

We must determine the regions within which the correspondence is 1 to 1:

Region 1

$U = X_1$

$V = (X_1 + X_2)/2$

$$\frac{1}{J_1} = \begin{vmatrix} 1 & 0 \\ \frac{1}{2} & \frac{1}{2} \end{vmatrix} = \frac{1}{2}$$

$|J_1| = 2$

Region 2

$U = X_2$

$V = (X_1 + X_2)/2$

$$\frac{1}{J_2} = \begin{vmatrix} 0 & 1 \\ \frac{1}{2} & \frac{1}{2} \end{vmatrix} = -\frac{1}{2}$$

$|J_2| = 2$

Then

$$\begin{aligned} f(u, v) &= f(x_1, x_2)|J_1| + f(x_1, x_2)|J_2| \\ &= 4 \end{aligned}$$

Hence we obtain the same result as we did previously.

The Jacobian is often written for brevity as

$$\left|\frac{\partial(x_1, x_2, \ldots, x_k)}{\partial(u_1, u_2, \ldots, u_k)}\right|$$

It is useful to note that

$$\left|\frac{\partial(x_1, x_2, \ldots, x_k)}{\partial(u_1, u_2, \ldots, u_k)}\right| \cdot \left|\frac{\partial(u_1, u_2, \ldots, u_k)}{\partial(x_1, x_2, \ldots, x_k)}\right| = 1 \tag{5.42}$$

so that one either works with $J$ or $1/J$, whichever is easier to handle.

The transformation procedure is applicable to the case of 1 to 1 transformations of a random variable $X_1, X_2, \ldots, X_k$, say. The common problem is that we have a random variable $X_1, X_2, \ldots, X_k$ and wish to find the distribution of $m$ variables ($m < k$) given by

$$U_i = u_i(X_1, X_2, \ldots, X_k),\ i = 1, 2, \ldots, m$$

For example, we have continuous random variables $X_1$, $X_2$, $X_3$ independently distributed with c.d.f. $F(x)$, and we wish to determine the distribution of $X_1 + X_2 + X_3$ and $X_1^2 + X_2^2 + X_3^2$. The transformation procedure is (1) to write $U_i = u_i(X_1, X_2, \ldots, X_k)$, $i = 1, 2, \ldots, m$, (2) to determine functions $U_i = u_i(X_1, X_2, \ldots, X_k)$, $i = m + 1, \ldots, k$ in such a way that the transformation $X_1, X_2, \ldots, X_k$ to $U_1, U_2, \ldots, U_k$ is 1 to 1, (3) to obtain the joint p.d.f. of $U_1, U_2, \ldots, U_k$, and finally (4) to integrate over $u_{m+1}, u_{m+2}, \ldots, u_k$ so as to obtain the p.d.f. of $U_1, U_2, \ldots, U_m$.

*Example 12*

Suppose $X_1$, $X_2$, $X_3$ has the p.d.f.

$$\left(\frac{1}{\sqrt{2\pi}}\right)^3 \exp[-(x_1^2 + x_2^2 + x_3^2)/2], \qquad -\infty < x_1, x_2, x_3 < \infty$$

What is the p.d.f. of $U_1 = X_1 + X_2 + X_3$ and $U_2 = X_1 + X_2 - 2X_3$? We write

$$\begin{aligned} U_1 &= X_1 + X_2 + X_3 \\ U_2 &= X_1 + X_2 - 2X_3 \\ U_3 &= X_1 - X_2 \end{aligned}$$

The absolute value of the Jacobian

$$\left|\frac{\partial(x_1, x_2, x_3)}{\partial(u_1, u_2, u_3)}\right|$$

is 1/6.

Also

$$X_1 = \frac{1}{6}(2U_1 + U_2 + 3U_3)$$

$$X_2 = \frac{1}{6}(2U_1 + U_2 - 3U_3), \text{ and}$$

$$X_3 = \frac{1}{3}(U_1 - U_2)$$

Hence for $-\infty < u_1, u_2, u_3 < \infty$,

$$g(u_1, u_2, u_3) = \frac{1}{6}\left(\frac{1}{\sqrt{2\pi}}\right)^3 \exp\left[-\left(\frac{u_1^2}{3} + \frac{u_2^2}{6} + \frac{u_3^2}{2}\right)\Big/2\right]$$

and for $-\infty < u_1, u_2 < \infty$,

$$g(u_1, u_2) = \frac{\sqrt{2}}{(\sqrt{2\pi})^2 6} \exp\left[-\left(\frac{u_1^2}{3} + \frac{u_2^2}{6}\right)\Big/2\right]$$

It is often the case that one will have to use a two-stage process of going from $X_1, X_2, \ldots, X_k$ to some $Z_1, Z_2, \ldots, Z_k$ and then to $U_1, U_2, \ldots, U_m$.

The method of transformation of variables can be very difficult to apply except in very simple problems. The problems arise because one will be involved in $k$-dimensional geometry, and it is rarely obvious how one completes a set of functions $U_1, U_2, \ldots, U_m$ to a set $U_1, U_2, \ldots, U_k$, so that the transformation is 1 to 1 throughout the space or is 1 to 1 in identifiable subspaces.

### 5.2.4 The Limit Method

Sometimes it is simpler to work with the defining equation of the derivative of the distribution function than it is to find an explicit function for the distribution function and then to take the derivative. Note that, in the case of a single random variable

$$f(x) = \lim_{\Delta x \to 0} \frac{P(x \le X \le x + \Delta x)}{\Delta x} \tag{5.43}$$

Similarly, for random variables $X$ and $Y$

$$f(x, y) = \lim_{\Delta x \to 0,\, \Delta y \to 0} \frac{P(x \le X \le x + \Delta x, y \le Y \le y + \Delta y)}{\Delta x\, \Delta y} \tag{5.44}$$

This result can be generalized in an obvious way to include any number of random variables.

*Example 13*

To illustrate the use of the method, consider the bivariate distribution:

$$\begin{aligned} f(x_1, x_2) &= 1, \qquad 0 < x_1 < 1, 0 < x_2 < 1 \\ &= 0, \text{ otherwise} \end{aligned}$$

An alternative way of writing this density function is

$$f(x_1, x_2) = \varepsilon(x_1, 1)\varepsilon(x_2, 1), \qquad -\infty < x_1, x_2 < \infty$$

where

$$\begin{aligned} \varepsilon(x, 1) &= 1, \qquad \text{if } 0 \le x \le 1 \\ &= 0, \text{ otherwise} \end{aligned}$$

Suppose we are asked to find the distribution of $U$ and $V$ where

$$\begin{aligned} U &= \max(X_1, X_2) \\ V &= (X_1 + X_2)/2 \end{aligned}$$

As a first step let us determine the range of $u$ and $v$ for positive values of $f(u, v)$. Since the average of two numbers cannot exceed the larger and must be at least one-half the larger, this region is described by $u/2 \le v \le u$, $0 < u < 1$, as shown in Figure 5.3.

For values of $u$ and $v$ within this region let us determine

$$P\left(u \le X_{\max} \le u + \Delta u, v \le \frac{X_1 + X_2}{2} \le v + \Delta v\right)$$

Suppose that $X_1$ is the larger. Then this probability is

$$P(u \le X_1 \le u + \Delta u, 2v - u' \le X_2 \le 2v - u' + 2\Delta v)$$

where $u \le u' \le u + \Delta u$.

From Figure 5.3 this is seen to be $2\Delta v \Delta u$. Since the other case is equally probable—i.e., $X_2 = X_{\max}$—the desired probability is $4\Delta u \Delta v$ and

$$f(u, v) = \lim_{\Delta u, \Delta v \to 0} \frac{4\Delta u \Delta v}{\Delta u \Delta v} = 4$$

Thus we have the result that

$$\begin{aligned} f(u, v) &= 4, \qquad u/2 \le v \le u, 0 < u < 1 \\ &= 0, \text{ otherwise} \end{aligned}$$

To obtain the marginals of $U$ or $V$, we can integrate out the other variable. However, it might be simpler to obtain the marginal directly. For instance

$$f(u) = \lim_{\Delta u \to 0} \frac{P(u \le X_{\max} \le u + \Delta u)}{\Delta u}$$

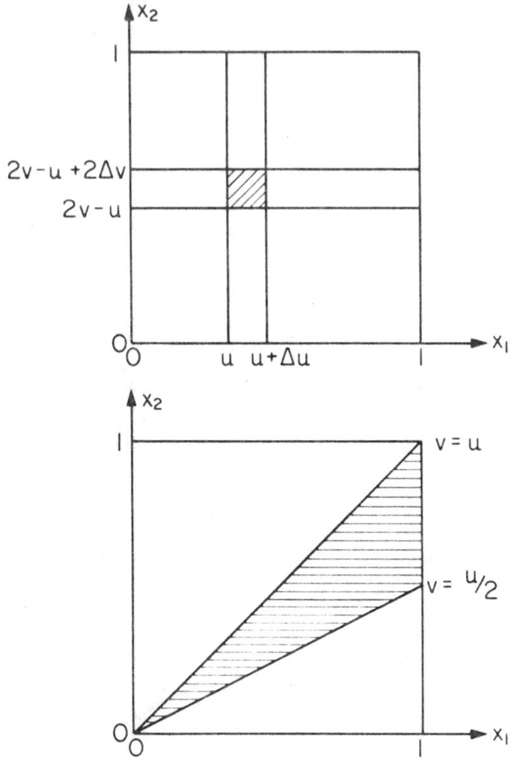


**Fig. 5.3** Transformation by Limits

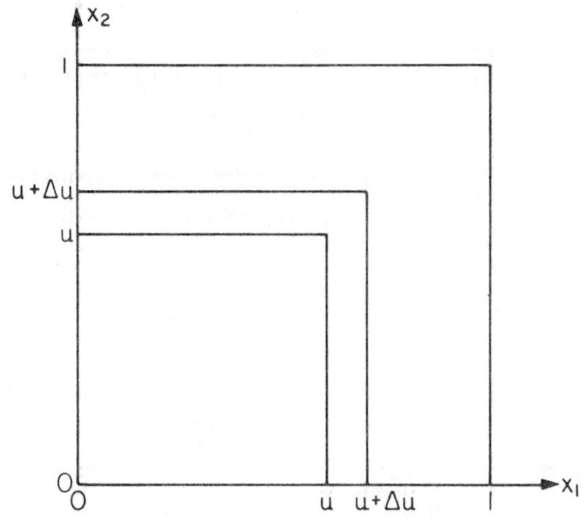


**Fig. 5.4** Transformations by Limits

From Figure 5.4 we can see that this is

$$\lim_{\Delta u \to 0} \frac{(u + \Delta u)^2 - u^2}{\Delta u} = \lim_{\Delta u \to 0} (2u + \Delta u) = 2u$$

A simple example of this process is the determination of the distribution of the median of a sample of size $n$, with $n$ odd and equal to $2m + 1$, from a continuous univariate distribution with density $f(x)$. The probability that $m$ values lie below $u$, one value lies in the interval $(u, u + \Delta u)$, and $m$ lies above $u + \Delta u$ is

$$\frac{(2m + 1)!}{m!m!} [F(u)]^m [F(u + \Delta u) - F(u)][1 - F(u + \Delta u)]^m$$

Dividing by $\Delta u$ and taking the limit as $\Delta u$ tends to zero, we obtain the density of $U$ equal to the median of $X_1, X_2, \ldots, X_{2m+1}$ to be

$$\frac{(2m + 1)!}{m!m!} [F(u)]^m [1 - F(u)]^m f(u)$$

with $u$ over the range of the original distribution.

## 5.3 The Use of Continuous Distributions

The preceding material on transformations of distributions leads naturally into a logical paradox. The following example, due to Barnard et al. (1962), illustrates the matter.

Suppose

$$\begin{aligned} f(x_1, x_2) &= 1/4, \qquad -1 < x_1 < 1, -1 < x_2 < 1 \\ &= 0, \text{ otherwise} \end{aligned} \tag{5.45}$$

Then let $Y_1 = X_1$, and $Y_2 = X_2/(1 + X_1^2)$.

This is a 1 to 1 transformation and the Jacobian is $1 + y_1^2$, so

$$\begin{aligned} f(y_1, y_2) &= \frac{(1 + y_1^2)}{4}, \qquad -1 < y_1 < 1, -\frac{1}{(1 + y_1^2)} < y_2 < \frac{1}{(1 + y_1^2)} \\ &= 0, \text{ otherwise} \end{aligned} \tag{5.46}$$

The marginal density of $Y_2$ can be found by integrating out $y_1$. It is

$$\begin{aligned} g(y_2) &= 2/3, \qquad |y_2| < 1/2 \\ &= \frac{1}{6}\left[\frac{2|y_2| + 1}{|y_2|}\right]\sqrt{\frac{1}{|y_2|} - 1}, \qquad 1/2 < |y_2| < 1 \\ &= 0, \text{ otherwise} \end{aligned} \tag{5.47}$$

Consider now the conditional density $f(y_1|y_2)$ for $|y_2| < 1/2$. This is simply

$$f(y_1|y_2) = \frac{3}{8}(1 + y_1^2), \qquad |y_2| < 1/2, \ -1 < y_1 < 1 \tag{5.48}$$

Then $f(y_1|0) = (3/8)(1 + y_1^2)$, but note that since $y_1 = x_1$, this can be written as $f(x_1|y_2 = 0)$. Note that $y_2 = 0$ if and only if $x_2 = 0$. Hence this should logically be the same as $f(x_1|x_2 = 0)$. But since $X_1$ and $X_2$ are independent, $f(x_1|x_2 = 0) = 1/2$, $-1 < x_1 < 1$. So we have reached a contradiction, or at least an apparent one. We assumed a property and by apparently logical development reached a conflicting property. We may generalize this by making the transformation

$$y_1 = x_1$$

$$y_2 = \frac{x_2}{1 + g(x_1)} \tag{5.49}$$

where $g(x_1) > 0$ and $g(0) = 1$. In this way we can get a wide variety of conditional distributions.

To see how the paradox arises from continuization of discrete probability, consider the discrete analogue of the above problem. Take $X_1$ and $X_2$ to be independently and identically distributed with probability $1/n$ assigned to each of the values

$$-1 + 1/n, \ -1 + 3/n, \ldots, 1 - 3/n, \ 1 - 1/n \tag{5.50}$$

where $n$ is odd.

Consider now what happens to $y_1$ and $y_2$. The possible values for $y_1$ are those for $x_1$, of course, and for $y_2$ the possible values are

$$\frac{-1 + \dfrac{2u - 1}{n}}{1 + \left(-1 + \dfrac{2v - 1}{n}\right)^2}, \qquad u, v = 1, 2, \ldots, n \tag{5.51}$$

The reader should make a sketch of the joint distribution. It is obvious that the conditional distribution of $Y_1$ given that $Y_2 = 0$ is, in fact, the conditional of $X_1$ given $X_2 = 0$, namely, the discrete uniform distribution between $-1$ and $1$. So in a real sense the continuous argument leads to the wrong answer if it is interpreted as an approximation to the actual discrete distribution.

The relationship in elementary cases between the mathematics of a continuous model and the mathematics of a discrete multinomial model obtained by incorporating a grouping error of measurement is simply the relationship between the Riemann integral and Darboux sums, the limit of which is the integral. This indicates that cumulative distributions as computed from the discretized continuous models will be closely approximated by Riemann integrals if the density functions are bounded. However, this does raise the question of the role of conditioning on the observed value of a continuous random variable, as a tool in data analysis and inference from *real* data.

## 5.4 Compound Distributions

The simple theory of random variables encompasses independent random variables from one mathematical distribution. The term "compound distribution" is used to refer to the case in which the final random variable is obtained as a result of a sequence of random variables. An elementary illustration is that of a random variable which follows a Poisson distribution with mean $m$, *and* the parameter $m$ itself follows a specified distribution, e.g., a gamma distribution. Or as another example, a nonnegative integer $r$ follows a binomial distribution with parameters $n$ and $p$ with the parameter $p$ itself following a distribution such as a beta-distribution (or the parameter $n$ following a specified Poisson distribution). Models of this type are important in the interpretation of data.

Consider the following situation. A random variable $X$ has conditionally on a parameter $\theta$, the distribution function $F(x, \theta)$, so that

$$P(X \leq x|\theta) = F(x; \theta) \tag{5.52}$$

Additionally, the parameter $\theta$ has the distribution function $G(y, \phi)$, so that

$$P(\theta \leq y|\phi) = G(u; \phi) \tag{5.53}$$

We can then seek to find the unconditional distribution of $X$. A partial answer is to use moment generating functions. It is clear that

$$E(e^{tX}) = \underset{\theta}{E}\, E(e^{tX}|\theta) \tag{5.54}$$

and if we are lucky we will be able to recognize the resulting form.

*Example 14*

$$P(X = x|N) = \binom{N}{x} p^x(1-p)^{N-x}, \qquad x = 0, 1, \ldots, N$$

$$P(N = n) = e^{-\lambda}\frac{\lambda^n}{n!}, \qquad n = 0, 1, 2, \ldots$$

Then

$$E(e^{tX}|N) = (pe^t + q)^N$$

$$E(e^{tX}) = \sum_{n=0}^{\infty} e^{-\lambda} \frac{\lambda^n}{n!} (pe^t + q)^n$$

$$= e^{-\lambda} e^{\lambda(pe^t + q)}$$

$$= e^{\lambda p(e^t - 1)}$$

This shows that $X$ has the Poisson distribution with mean $\lambda p$.

*Example 15*

$$P(X = x|\lambda) = e^{-\lambda} \frac{\lambda^x}{x!}, \qquad x = 0, 1, 2, \ldots$$

$$f(\lambda)\, d\lambda = \frac{1}{\Gamma(\alpha)} e^{-\lambda p}(\lambda p)^{\alpha - 1} p\, d\lambda, \qquad 0 < \lambda < \infty$$

Then

$$E(e^{tX}|\lambda) = \exp[\lambda(e^t - 1)]$$

and

$$E(e^{tX}) = \int_0^{\infty} \frac{1}{\Gamma(\alpha)} \exp[-\lambda p + \lambda(e^t - 1)](\lambda p)^{\alpha - 1} p\, d\lambda$$

$$= \int_0^{\infty} \frac{1}{\Gamma(\alpha)} \exp[-\lambda(1 + p - e^t)]\lambda^{\alpha - 1} p^{\alpha}\, d\lambda$$

$$= \frac{p^{\alpha}}{(1 + p - e^t)^{\alpha}}$$

This is the m.g.f. of a negative binomial random variable.

These examples are special cases of a general class of situations where the random variable $X$ is the sum of a random number of independent identically distributed random variables. Let the constituent random variables be $X_1, X_2, \ldots$. Then the c.d.f. of $X$ is

$$P(X \le x) = \sum_{n=0}^{\infty} P(N = n)\, P\left(\sum_{i=1}^{n} X_i \le x\right) \tag{5.55}$$

If the $X_i$ are discrete random variables,

$$P(X = x) = \sum_{n=0}^{\infty} P(N = n)\, P\left(\sum_{i=1}^{n} X_i = x\right) \tag{5.56}$$

Study of this class is straightforward with the moment generating function. Denoting the moment generating function of $X$ by $m_X(t)$, we have

$$m_X(t) = E(e^{tX}) = \sum_{n=0}^{\infty} P(N = n)\, E(e^{t\Sigma_1^n X_i}) \tag{5.57}$$

If now $E(e^{tX_i}) = m(t)$ for all $i$, we have

$$m_X(t) = \sum_{n=0}^{\infty} P(N = n)[m(t)]^n = E[m(t)^N] \tag{5.58}$$

It is hoped that the result of this development will be an m.g.f. we can recognize. In any case, we will be able to find moments of $X$.

The p.g.f. introduced in Chapter 2 is also useful for studying the sum of a random number of random variables. Recall that the p.g.f. for integer-valued random variables was defined by

$$\begin{aligned} g(t) &= E(t^X) \\ &= \sum_{x=0}^{\infty} t^x f(t) \end{aligned}$$

This sum obviously converges for $|t| < 1$. Probability generating functions are useful for calculating the probabilities of outcomes which are a result of simple probability outcomes. They are closely related to m.g.f.'s because

$$E(t^X) = E(e^{X \ln t})$$

For the distribution Bi($N$, $p$) the p.g.f. is easily seen to be $(pt + q)^N$. If now the number $N$ here is itself a random variable, $E(t^X) = E_1 E_2(t^X)$, where $E_2$ denotes expectation conditionally on $N$ and $E_1$ denotes expectation over the distribution of $N$. Hence $g_X(t) = E_1(pt + q)^N$.

Now if we have $N$ independent random variables with a common distribution and, necessarily, a common probability generating function $g(t)$, the probability generating function of $X$ is given by

$$\begin{aligned} g_X(t) = E(t^X) &= \sum_{n=0}^{\infty} P(N = n) E(t^{\Sigma_i X_i}) \\ &= \sum_{n=0}^{\infty} P(N = n)[g(t)]^n = E[g(t)^N] \end{aligned} \tag{5.59}$$

A common example of this compounding is the compound Poisson process in which the number $N$ has the Poisson distribution with mean $\lambda T$, say. This leads to

$$m_X(t) = \exp[-\lambda T + \lambda T m(t)] \tag{5.60}$$

where $m(t)$ is the m.g.f. of each $X_i$. By differentiating the m.g.f. we find that

$$E(X) = \lambda T E(X_i)$$
$$\mathrm{Var}(X) = \lambda T E(X_i^2) \tag{5.61}$$

It is interesting to note that if the $X_i$ are binomial variables with density $p^{x_i}(1-p)^{1-x_i}$, $x_i = 0, 1$

$$m(t) = e^{-\lambda T + \lambda T(q + pe^t)}$$
$$= e^{-\lambda T p(1 - e^t)} \tag{5.62}$$

Thus the sum $X = \sum_1^N X_i$ has a Poisson distribution with mean $\lambda Tp$. An example of the use of this distribution is the following. In the development of quantitative models for sales forecasting and inventory, it is common to assume that the number of orders $N$ received in time period $T$ is a Poisson random variable with some common distribution. Then the total volume of orders $S = \sum_1^N X_i$ received in the time interval $T$ has a compound Poisson distribution.

## PROBLEMS

1. For each of the following functions determine $\varepsilon$ and show that

$$\lim_{\Delta X \to 0} \varepsilon$$

is in fact zero, as mentioned in Section 5.1.:
(a) $Y = X^2$
(b) $Y = X$
(c) $Y = e^X$

2. Radiation thermometers make use of the Planck distribution law. If $E$ = intensity of radiation for the wavelength $\lambda$ and $T$ is the absolute temperature,

$$E = C_1 \lambda^{-5}/[e^{-C_2/(\lambda T)} - 1]$$

If $C_1$ and $C_2$ are fixed constants and $\lambda$ and $T$ are random variables with means $\mu_1$ and $\mu_2$ and variances $\sigma_1^2$ and $\sigma_2^2$, find the approximate mean and variance of $E$.

3. The following is an equation arising in the study of ballistics:

$$E = \frac{P(V - B)}{K - 1}$$

If $P$, $V$, $B$, and $K$ are random variables, approximate the mean and variance of $E$.

4. If $X$ and $Y$ are independent variables with variances $\sigma_x^2$ and $\sigma_y^2$, respectively, show

$$\mathrm{Cov}(aX + bY, cX + dY) = ac\sigma_x^2 + bd\sigma_y^2$$

5. Verify the variance formula given by Equation (5.6) of Section 5.1.

6. If $X_1, X_2, \ldots, X_n$ have variances $\sigma_i^2$ and if $\mathrm{Cov}(X_i, X_j) = \sigma_{ij}$, show

$$\mathrm{Cov}\left(\sum_i c_i X_i, \sum_i d_i X_i\right) = \sum_i c_i d_i \sigma_i^2 + \sum_{i \neq j} c_i d_j \sigma_{ij}$$

7. Show that

$$\begin{aligned}\text{Cov}(U + V, X + Y) &= \text{Cov}(U, X + Y) + \text{Cov}(V, X + Y)\\ &= \text{Cov}(U, X) + \text{Cov}(U, Y)\\ &\quad + \text{Cov}(V, X) + \text{Cov}(V, Y)\end{aligned}$$

8. A manufacturing process consists of processing items through the operations shown below. The fraction of items entering the $i$th operation which are acceptable after the operation is given by $P_i$. The cost of processing an item through operation $i$ is $C_i$. Set up a model giving the total cost per good unit in terms of the $C$'s and $P$'s. If the $C$'s and $P$'s are random variables with the means and variances given, approximate the mean and variance of the total cost per good unit. The product is divided equally after $O_1$.

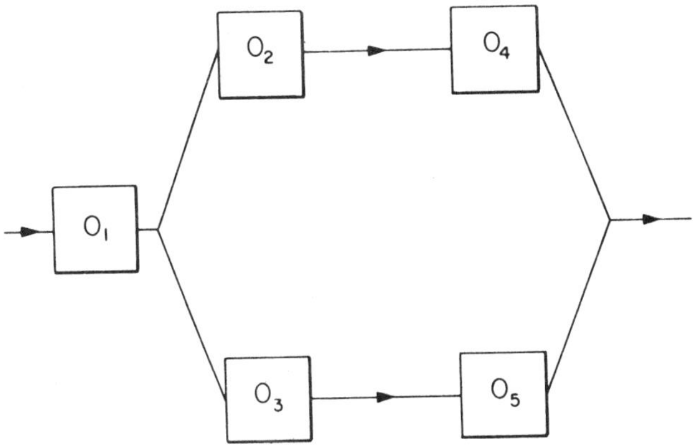


**Fig. Ex. 5.8** Hypothetical Manufacturing Process

| | $C_1$ | $C_2$ | $C_3$ | $C_4$ | $C_5$ | $P_1$ | $P_2$ | $P_3$ | $P_4$ | $P_5$ |
|---|---|---|---|---|---|---|---|---|---|---|
| Mean | 50 | 30 | 40 | 25 | 30 | .8 | .6 | .9 | .9 | .9 |
| Variance | 2 | 2 | 1 | 1 | 2 | .01 | .01 | .02 | .01 | .02 |

9. Type A resistors have a mean resistance of 4 ohms and a standard deviation of .05. Type B resistors have mean resistance of 10 ohms and standard deviation of 1. Type C resistors have mean resistance of 30 ohms and standard deviation of 5. We wish to connect resistors in series. Using normality in all cases, find the number of each type which should be used to minimize the total number of resistors used subject to the restriction $P(\text{total resistance} > 150) \geq .90$.

10. Two meshing gears are mounted on shafts which are 5 inches apart from center to center. If the distribution of outside diameter of the gears is approximately normal with mean of 4.95 inches and standard deviation of .05, what is the approximate probability of interference between two gears selected at random?

11. Suppose that the random variables $X_1$, $X_2$, and $X_3$ are independent with means and variances of (2, 3), (1, 5) and (−2, 12) respectively. Find the means and variances of the following linear combinations:
(a) $2X_1 - X_2 + X_3$
(b) $X_1 + 2X_2 - 3X_3$
(c) $X_1 - 3X_2 + X_3$

12. If $X$ is a random variable with density function $f(x)$, show that $U = F(X)$ is a uniform random variable.

13. If

$$f(x, y) = e^{-(x+y)}, \qquad x > 0, y > 0$$
$$= 0, \text{ otherwise}$$

and

$$U = X + Y$$
$$V = X - Y$$

find the density function $f(u, v)$ by each of the methods illustrated in Section 5.2.

14. If

$$f(x, y) = 1, \qquad 0 < x < 1, 0 < y < 1$$
$$= 0, \text{ otherwise}$$

and

$$U = X + Y$$
$$V = X - Y$$

find the means, variances, and the covariance of $U$ and $V$ without finding $f(u, v)$.

15. If

$$f(x, y) = 4xy, \qquad 0 \leq x < 1, 0 < y < 1$$
$$= 0, \text{ otherwise}$$

and

$$U = X + Y$$
$$V = X - Y$$

find the means, variances, and the covariance of $U$ and $V$ without finding $f(u, v)$.

16. For Equation (5.37) justify using

$$\lim_{\Delta u \to 0} \frac{\Delta x_1}{\Delta u} = -\frac{dx_1}{du}$$

and

$$\lim_{\Delta u \to 0} \frac{\Delta x_2}{\Delta u} = \frac{dx_2}{du}$$

17. If

$$f(x, y) = 1, \qquad 0 < x < 1, 0 < y < 1$$
$$= 0, \text{ otherwise}$$

find the density of $X$ and $Y$ given that $X + 2Y \leq 1/2$.

18. If

$$f(x, y) = 4xy, \qquad 0 < x < 1, 0 < y < 1$$
$$= 0, \text{ otherwise}$$

find the joint density of $U = X$, $V = Y^2$.

19. If

$$f(x, y) = kxy^2, \qquad 0 < y < x, 0 < x < 1$$
$$= 0, \text{ otherwise.}$$

(a) Find $k$.
(b) If $U = X + Y$, find $f(u)$.

20. If

$$f(x, y) = ye^{-xy}, \qquad 0 < x < \infty, 0 < y < 1$$
$$= 0, \text{ otherwise}$$

and

$$U = X^2$$
$$V = Y$$

find $f(u, v)$.

21. If

$$f(x, y) = ke^{-2x-y}, \qquad x \geq 0, y \geq 0, x^2 + y^2 < 1$$
$$= 0, \text{ otherwise}$$

and

$$U = 2X + Y$$
$$V = Y$$

find $f(u, v)$.

22. If

$$f(x, y) = e^{-x-y}, \qquad x > 0, y > 0$$
$$= 0, \text{ otherwise}$$

find $f(u)$, where $U = X - 2Y$.

23. If

$$f(x, y) = \frac{1}{\sqrt{\pi}} e^{-x^2}, \qquad -\infty < x < \infty, -1 < y < 0$$
$$= 0, \text{ otherwise}$$

and

$$U = X^2$$
$$V = X + Y$$

find $f(u, v)$.

24. If a random sample of $n$ observations is chosen from a population with mean $\mu$ and variance $\sigma^2$, use the material in Section 5.1.2 to show that the mean and variance of $\sum_1^n X_i$ are $n\mu$ and $n\sigma^2$. Show that the mean and variance of $\overline{X} = \sum_1^n X_i/n$ are $\mu$ and $\sigma^2/n$.

25. If $X$ is a random variable with mean $\mu$ and variance $\sigma^2$, use the material in Section 5.1.2 to find the mean and variance of $X - \mu$, $(X - \mu)/\sigma$, $\sqrt{n}(\bar{X} - \mu)/\sigma$.

26. Use the facts that the average of $n$ consecutive integers is (first plus last)/2, and that

$$\sum_{x=0}^{k-1} x^2 = k(k - 1)(2k - 1)/6$$

to show the mean and variance of the density function

$$\begin{aligned} f(x) &= 1/k, \qquad x = 0, 1, 2, \ldots, k - 1 \\ &= 0, \text{ otherwise} \end{aligned}$$

to be

$$\mu = (k - 1)/2$$

and

$$\sigma^2 = (k^2 - 1)/12$$

27. Given that $X_1, X_2, \ldots, X_k$ have the $k$-variate normal distribution as defined in Chapter 4, where $E(X_i) = \mu_i$, $V(X_i) = \sigma_{ii}$, and $\text{Cov}(X_i, X_j) = \sigma_{ij}$, show that the joint m.g.f. is

$$\exp\left[\sum_{i=1}^{k} t_i\mu_i + (1/2)\sum_{i,j=1}^{k} t_it_j\sigma_{ij}\right]$$

28. For the $k$-variate normal, show that $\sum_1^k c_iX_i$ has a univariate normal distribution with mean $\sum_1^k c_i\mu_i$ and variance $\sum_{ij} c_ic_j\sigma_{ij}$.

29. For the $k$-variate normal distribution, show that pairwise independence implies mutual independence.

30. Using the joint m.g.f., show that any subset of variables from a multivariate normal distribution is also multivariate normal.

31. If

$$\begin{aligned} f(x_1, x_2) &= 1, \qquad 0 < x_1 < 1, 0 < x_2 < 1 \\ &= 0, \text{ otherwise} \end{aligned}$$

use several methods to find $f(u, v)$, where
(a) $U = X_{\max}$, $V = X_{\min}$.
(b) $U = X_{\max} - X_{\min}$, $V = X_1 + X_2$.
(c) $U = X_{\max} - X_{\min}$, $V = X_{\min}$.

32. Determine the conditional distribution of $Y_1$ given that $Y_2 = 0$ for the discrete distribution in Section 5.3.

33. Suppose a random variable $Z$ is a function $f(\cdot)$ of a random variable $X$ with mean $\mu$, and variance $\text{Var}(X)$ which is small relative to $\mu^2$. Show that

$$\text{Var}(Z) \doteq [f'(\mu)]^2 \, \text{Var}(X)$$

34. With the ideas of Problem 33 suppose $\text{Var}(X) = c\mu$, where $c$ is a constant. Find a transformation $f(\cdot)$ so that the variance of $f(X)$ is approximately a constant independent of $\mu$.

35. Repeat Problem 34 with $\text{Var}(X) = c\mu^2$.

36. Repeat Problem 34 with $\text{Var}(X) = c\mu(1 - \mu)$, showing that $f(X) = \sin^{-1}\sqrt{X}$ gives the desired result. Relate this to transformation of percentage data.

37. The Cauchy distribution with density $f(x) = 1/[\pi(1 + x^2)]$, $-\infty < x < \infty$ occurs frequently in calculations of mathematical statistics.
(a) Verify that the density given is a probability density.
(b) Show that this distribution does not have a mean by showing that

$$\int_{-\infty}^{\infty} |x| f(x)\, dx = 2\int_0^{\infty} \frac{x}{\pi(1 + x^2)}\, dx$$

does not exist, i.e., the integral $I(A)$ from 0 to $A$, does not have a limit as $A \to \infty$.

38. Suppose $X$ has the Cauchy distribution. Obtain the distribution of $\tan^{-1} X$.

39. Let $X$, $Y$ be $IN(0, 1)$. Find the distribution of $U = X/Y$, by making the transformation $U = X/Y$, $V = Y$. Your result should be the Cauchy distribution of Problem 37.

40. Let $X \sim N(\mu, 1)$ and $Y \sim \gamma(v/2)$. Transform to obtain the density of $X\sqrt{v}/\sqrt{Y}$ as a series. A closed form does not exist.

41. The bivariate binomial distribution is obtained as follows. Let $X_1 \sim \text{Bi}(n_1, p_1)$, $X_2 \sim \text{Bi}(n_2, p_2)$, $X_{1.2} \sim \text{Bi}(n_{1.2}, p_{1.2})$, with $X_1, X_2, X_{1.2}$ independent. Show that the random variables $Z_1 = X_1 + X_{1.2}$, $Z_2 = X_2 + X_{1.2}$ have the m.g.f.

$$(p_1 e^{t_1} + q_1)^{n_1}(p_2 e^{t_2} + q_2)^{n_2}(p_{1.2} e^{t_1 + t_2} + q_{1.2})^{n_{1.2}}$$

where $q_1 = 1 - p_1$, $q_2 = 1 - p_2$, $q_{1.2} = 1 - p_{1.2}$. If we put $w_1 = e^{t_1}$, $w_2 = e^{t_2}$, then we obtain the p.g.f. $E(w_1^{Z_1} w_2^{Z_2})$ and we can then obtain $P(Z_1 = z_1, Z_2 = z_2)$ as the coefficient of $w_1^{z_1} w_2^{z_2}$ in this function.

42. The bivariate Poisson distribution is obtained as follows. Let $X_1 \sim Po(\lambda_1)$, $X_2 \sim Po(\lambda_2)$, $X_{1.2} \sim Po(\lambda_{1.2})$, where $Po(\lambda)$ means the Poisson distribution with parameter $\lambda$. Obtain the m.g.f. of $Z_1 = X_1 + X_{1.2}$, $Z_2 = X_2 + X_{1.2}$. Also obtain the p.g.f. of $Z_1, Z_2$ and hence $P(Z_1 = z_1, Z_2 = z_2)$. The general form is given by Teicher (1954).

# 6

# Distribution of Sample Statistics

## 6.1 Random Sampling

### 6.1.1 Random Sampling of a Real Population

Suppose we have a population of $N$ individuals, for example, the undergraduate body of college $X$ consisting of 1376 students. Suppose, furthermore, that we wish to obtain some ideas of the nature of this population with regard to some attributes that are difficult to measure. This restriction, we suppose, makes it impossible to examine every member of the population. The natural procedure, then, is to examine a subset which we hope to be "typical" or "representative" of the population. It is not possible, however, to give this suggested procedure any basis in logic because we cannot give a workable idea of typicality.

The way out of the dilemma which has been advocated generally by data gatherers of the twentieth century is based on the ideas of probability. We can manufacture physical devices, such as a coin and a tossing mechanism, which give results in repeated uses (or trials) which have as far as we can tell all the properties that we expect of our conceptual probability systems. We then take the view that the result of any use of the whole apparatus is a random variable with probability equal to relative frequency as observed in past trials. This enables us, conceptually at least, to obtain a realization of a random variable with probability 1/2, for example, and consequently with probability $1/M$, where $M$ is any positive integer. With such a device we can draw a random member from a finite population of $M$ elements. Use of the procedure has the consequence that each member of the population has equal probability of being drawn. The procedure is called *random sampling*.

It is necessary to distinguish between sampling with replacement and without replacement. In sampling with replacement the element drawn is replaced before another is drawn. In sampling without replacement the elements drawn are not replaced. We shall discuss the latter case. Let the members of the population be denoted by 1, 2, . . . , $N$, and let the respective attributes be denoted by $y_1, y_2, \ldots, y_N$. Let the attributes of the sample drawn without replacement be denoted by $x_1, x_2, \ldots, x_n$. Obviously, each element in the sample is one of the elements in the population, the correspondence being determined by the random sampling. It is possible to deal with this random correspondence algebraically. Let

$$\begin{aligned} \delta_i^j &= 1, \text{ if the } i\text{th element in the sample is the } j\text{th member of the population,} \\ &= 0, \text{ otherwise} \end{aligned} \tag{6.1}$$

We have defined a random variable with known distributional properties. For sampling without replacement,

$$\begin{aligned} \text{Prob}(\delta_i^j = 1) &= 1/N \\ \text{Prob}(\delta_i^j = 0) &= (N-1)/N \end{aligned} \tag{6.2}$$

The probabilities here are not really so obvious. Consider the possible ordered samples of size $k$, whose first member is a specified member of the population, say the $j$th. There are $(N-1)(N-2)\ldots(N-k+1)$ such ordered samples, but there are $N(N-1)(N-2)\ldots(N-k+1)$ ordered samples in all, so the probability that the first member in the sample is the $j$th in the population is $1/N$. Hence

$$E(\delta_i^j) = \text{Prob}(\delta_i^j = 1) = 1/N \tag{6.3}$$

Further

$$\begin{aligned} \text{Var}(\delta_i^j) &= E[(\delta_i^j)^2] - 1/N^2 \\ &= \text{Prob}(\delta_i^j = 1) - 1/N^2 \\ &= (N-1)/N^2 \end{aligned} \tag{6.4}$$

We can also find the covariances.

$$\begin{aligned} \text{Cov}(\delta_i^j, \delta_{i'}^j) &= -1/N^2, \quad i \neq i' \\ \text{Cov}(\delta_i^j, \delta_i^{j'}) &= -1/N^2, \quad j \neq j' \\ \text{Cov}(\delta_i^j, \delta_{i'}^{j'}) &= 1/N^2(N-1), \quad i \neq i', j \neq j' \end{aligned} \tag{6.5}$$

Since for fixed $i$, $\delta_i^j = 1$ for only one $j$, any observation $x_i$ is the realized value of the random variable $X_i = \sum_{j=1}^{n} \delta_i^j y_j$.

With this simple framework, it is possible to examine the probability distribution of the $x$'s. For example,

$$P(X_i = y_k) = 1/N, \text{ for } i = 1, 2, 3, \ldots, n$$

and

$$k = 1, 2, 3, \ldots, N$$

This is not to say that the conditional probability does not change as we proceed with the drawing. Rather,

$$P(X_2 = y_k | X_1 = y_m) = \frac{1}{(N-1)}$$

$$P(X_3 = y_k | X_1 = y_m, X_2 = y_n) = \frac{1}{(N-2)}$$

etc. Note that the $x$'s are not independent because

$$P(X_1 = y_k, X_2 = y_m, \ldots, X_n = y_u = \frac{1}{N(N-1)(N-2)\ldots(N-n+1)}$$

which is not equal to

$$P(X_1 = y_k)P(X_2 = y_m)\ldots P(X_n = y_n) = \frac{1}{N^n}$$

If, however, sampling is with replacement, the $X$'s are independent. Further, if $N$ is large relative to $n$ in sampling without replacement, the $X$'s are almost independent in that the joint density is approximately equal to the product of the marginal densities.

### 6.1.2 Random Sampling from Infinite Populations

We now describe briefly the concept of a random sample from mathematical infinite populations such as the binomial and the normal. We say that a set of $n$ random variables, $X_1, X_2, \ldots, X_n$ is a random sample from the discrete distribution if

$$P(X_1 = x_1, X_2 = x_2, \ldots, X_n = x_n) = P(X_1 = x_1)P(X_2 = x_2)\ldots P(X_n = x_n)$$

with the constituent factors arising from the discrete distribution under consideration. This case can be regarded as sampling from a finite population with replacement. In the case of a continuous distribution we say that $X_1, X_2, \ldots, X_n$ is a random sample if the joint probability density

$$f(x_1, x_2, \ldots, x_n) = f(x_1)f(x_2)\ldots f(x_n)$$

with the constituent factors arising from the continuous distribution under consideration. The random variables may be univariate or multivariate. The essential idea is that the $n$ random variables are identically and independently distributed.

## 6.2 Distribution of the Sample Mean

### 6.2.1 Properties for any Distribution

With the simple framework developed in Section 6.1.1 for random sampling without replacement from a population, consider the sample mean $\bar{x} = \sum_{i=1}^{n} x_i/n$. The mean and variance of $\bar{x}$ are found as follows:

$$\begin{aligned} E(\bar{x}) &= E\left(\sum_{i=1}^{n} x_i/n\right) = E\left(\frac{1}{n}\sum_{i=1}^{n}\sum_{j=1}^{N} \delta_i^j y_j\right) \\ &= \frac{1}{n}\sum_{i=1}^{n}\sum_{j=1}^{N} E(\delta_i^j)y_j = \frac{1}{n}(n)\sum_{j=1}^{N}(1/N)y_j \\ &= \frac{1}{N}\sum_{j=1}^{N} y_j \end{aligned} \tag{6.6}$$

Thus the expected value of the sample mean equals the population mean. We can also make use of the moments of the $\delta$'s to find the variance of $\bar{x}$.

Two very useful identities are the following:

$$\left(\sum_{i=1}^{N} a_i\right)^2 = \sum_{i=1}^{N} a_i^2 + \sum_{\substack{i,i'=1 \\ i\neq i'}}^{N} a_i a_{i'} \tag{6.7}$$

$$\begin{aligned} \left(\sum_{i=1}^{M}\sum_{j=1}^{N} a_{ij}\right)^2 &= \sum_{i=1}^{M}\sum_{j=1}^{N} a_{ij}^2 + \sum_{\substack{i,i'=1 \\ i\neq i'}}^{M}\sum_{j=1}^{N} a_{ij}a_{i'j} \\ &\quad + \sum_{i=1}^{M}\sum_{\substack{j,j'=1 \\ j\neq j'}}^{N} a_{ij}a_{ij'} + \sum_{\substack{i,i'=1 \\ i\neq i'}}^{M}\sum_{\substack{j,j'=1 \\ j\neq j'}}^{N} a_{ij}a_{i'j'} \end{aligned} \tag{6.8}$$

The computation of the variance of $\bar{x}$ follows:

$$\begin{aligned} \text{Var}(\bar{x}) &= \text{Var}\left(\frac{1}{n}\sum_{i=1}^{n}\sum_{j=1}^{N} \delta_i^j y_j\right) \\ &= (1/n^2)\Bigg[\sum_i\sum_j y_j^2 \,\text{Var}(\delta_i^j) + \sum_i\sum_{j\neq j'} y_j y_{j'} \,\text{Cov}(\delta_i^j, \delta_i^{j'}) \\ &\quad + \sum_{i\neq i'}\sum_j y_j^2 \,\text{Cov}(\delta_i^j, \delta_{i'}^j) + \sum_{i\neq i'}\sum_{j\neq j'} y_j y_{j'} \,\text{Cov}(\delta_i^j \delta_{i'}^{j'})\Bigg] \\ &= (1/n^2)\Bigg[\frac{n(N-1)}{N^2}\sum_j y_j^2 - \frac{n}{N^2}\sum_{j\neq j'} y_j y_{j'} \\ &\quad - \frac{n(n-1)}{N^2}\sum_j y_j^2 + \frac{n(n-1)}{N^2(N-1)}\sum_{j\neq j'} y_j y_{j'}\Bigg] \end{aligned}$$

$$= \frac{1}{n^2}\left[\sum_j \frac{n(N-n)}{N^2} y_j^2 - \sum_{j \neq j'} \frac{n(N-n)}{N^2(N-1)} y_j y_{j'}\right] \tag{6.9}$$

Since

$$(\sum y_j)^2 = \sum_j y_j^2 + \sum_{j \neq j'} y_j y_{j'}$$

$$\text{Var}(\bar{x}) = (1/n^2)\left[\sum_j \frac{n(N-n)}{N^2} \frac{N}{N-1} y_j^2 - \frac{n(N-n)}{N^2(N-1)}\left(\sum_j y_j\right)^2\right]$$

$$= \frac{N-n}{nN(N-1)}\left[\sum_j y_j^2 - \frac{\left(\sum_j y_j\right)^2}{N}\right]$$

$$= \frac{N-n}{n(N-1)} \sum_j \frac{(y_j - \bar{y})^2}{N} \tag{6.10}$$

A more commonly occurring form of this result is

$$\text{Var}(\bar{x}) = \left(\frac{N-n}{N}\right)\frac{S^2}{n} \tag{6.11}$$

where

$$S^2 = \sum (y_i - \bar{y})^2/(N-1) \tag{6.12}$$

Note that if $N$ is very large, the factor $(N - n)/N$ is essentially unity, and $S^2$ is essentially the population variance.

If sampling is done with replacement, so that an individual drawn is put back in the population and may be drawn again, we have

$$E(\delta_i^j) = 1/N \tag{6.13}$$

$$\text{Var}(\delta_i^j) = (N-1)/N^2 \tag{6.14}$$

and

$$\text{Cov}(\delta_i^j, \delta_{i'}^j) = 0, \qquad i \neq i'$$

$$\text{Cov}(\delta_i^j, \delta_i^{j'}) = -1/N^2, \qquad j \neq j'$$

$$\text{Cov}(\delta_i^j, \delta_{i'}^{j'}) = 0, \qquad i \neq i', j \neq j' \tag{6.15}$$

Thus

$$E(\bar{x}) = \bar{y}$$

and

$$\text{Var}(\bar{x}) = (1/n^2)\left[\sum_{i=1}^{n}\sum_{j=1}^{N} y_j^2 \,\text{Var}(\delta_i^j) + \sum_i \sum_{j \neq j'} y_j y_{j'} \,\text{Cov}(\delta_i^j, \delta_i^{j'})\right]$$

$$= (1/n^2)\left[\frac{n(N-1)}{N^2}\sum_j y_j^2 - \frac{n}{N^2}\sum_{j \neq j'} y_j y_{j'}\right]$$

$$= (1/n^2)\left[\sum_j \frac{n}{N} y_j^2 - \frac{n}{N^2}\left(\sum_j y_j\right)^2\right]$$

$$= \frac{\sum_j (y_j - \bar{y})^2}{nN} = \frac{\sigma^2}{n} \tag{6.16}$$

In this case the variance of the sample mean is exactly the population variance divided by $n$.

If we consider a random sample from a mathematical distribution with mean $\mu$ and variance $\sigma^2$, the mean and variance of the sample mean are obtained from the results of the previous chapter to be $\mu$ and $\sigma^2/n$. If $\sigma^2$ is infinite, then so is the variance of the sample mean.

### 6.2.2 Sampling from the Binomial Distribution

Consider the distribution of $\bar{X}$, where

$$f(x_i) = p^{x_i}(1-p)^{1-x_i}, \qquad x_i = 0, 1 \qquad i = 1, 2, \ldots, n \tag{6.17}$$

The distribution of $\bar{X}$ is easily determined by use of the m.g.f.:

$$m_{\bar{X}}(t) = m_{\Sigma X}\left(\frac{t}{n}\right) = \left[m_{X_i}\left(\frac{t}{n}\right)\right]^n$$

But

$$m_{X_i}\left(\frac{t}{n}\right) = E[\exp(tX_i/n)] = q + pe^{t/n}$$

So

$$m_{\bar{X}}(t) = (q + pe^{t/n})^n \tag{6.18}$$

This is the m.g.f. of a distribution with density function

$$f(y) = \binom{n}{ny} p^{ny}(1-p)^{n-ny} \tag{6.19}$$

$$y = 0, 1/n, 2/n, \ldots, 1$$

In other words, $n\bar{X}$ has the binomial distribution given in Chapter 4. The reader may verify that

$$E(\bar{X}) = p$$
$$\text{Var}(\bar{X}) = p(1-p)/n$$

### 6.2.3 Sampling from a Poisson Distribution

As was the case with the binomial, it is easier to find directly the distribution of $\sum X = n\bar{X}$ than that of $\bar{X}$. When a sample of size $n$ is taken at random from a distribution with density function

$$f(x) = \lambda^x e^{-\lambda}/x! \qquad x = 0, 1, 2, \ldots \tag{6.20}$$

the distribution of the sum of the observations $T = \sum X_i$ is another Poisson with parameter $n\lambda$. This is easily shown by finding the m.g.f. of $T$. Since the $X$'s are independent,

$$E(e^{tT}) = E(e^{t\Sigma X_i}) = \prod_1^n E(e^{tX_i}) = e^{n\lambda(e^t - 1)} \tag{6.21}$$

which is the m.g.f. of a Poisson variable with parameter $n\lambda$. Thus $E(n\bar{X}) = n\lambda$ and $\mathrm{Var}(n\bar{X}) = n\lambda$. Hence $E(\bar{X}) = \lambda$ and $\mathrm{Var}(\bar{X}) = \lambda/n$.

### 6.2.4 Sampling from the Uniform Discrete Distribution

A distribution assigning equal density to a set of numbers will be referred to as the uniform discrete distribution. Uniform assignment of density to the numbers 0 through 9 inclusive, denoted by $U[0\,(1)\,9]$, is of considerable interest. As we have mentioned, there are several tables of numbers printed in statistics books which simulate random sequences of numbers 0 through 9. Such tables furnish the basis for many simulation, or Monte Carlo, procedures which often consist of drawing random samples from a table of random numbers and examining the frequency distribution of some sample statistic. For small samples it is possible, in fact, to determine the distribution of sample statistics from $U[0\,(1)\,9]$.

Suppose we draw random samples of size two from $U[0\,(1)\,9]$. What is the distribution of the sample total? We find that

$$\mathrm{Prob}(X_1 + X_2 = 0) = \mathrm{Prob}(X_1 = 0, X_2 = 0) = 1/100$$

$$\mathrm{Prob}(X_1 + X_2 = 1) = P(X_1 = 1, X_2 = 0) + P(X_1 = 0, X_2 = 1) = 2/100$$

and in general

$$\mathrm{Prob}(X_1 = c, X_2 = k - c) = 1/100 \tag{6.22}$$

Therefore

$$P(X_1 + X_2 = k) = (k + 1)/100, \quad k = 0, 1, \ldots, 9$$
$$= (19 - k)/100, \quad k = 10, 11, \ldots, 18 \tag{6.23}$$

This is a discrete triangular distribution.

The distribution for samples of size 3 is also easily obtained. We see that

$$P(X_1 + X_2 + X_3 = k) = N(k)/1000 \qquad (6.24)$$

where $N(k)$ is the number of ways in which $X_1 + X_2 + X_3 = k$. $N(k)$ is easily determined by considering the product

$$(1 + t + \cdots + t^9)(1 + t + \cdots + t^9)(1 + t + \cdots + t^9)$$

The coefficient of $t^k$ in this product is $N(k)$. But this product is

$$\frac{(1 - t^{10})^3}{(1 - t)^3} = (1 - 3t^{10} + 3t^{20} - t^{30})\left(\frac{1.2}{2} + \frac{2.3t}{2} + \frac{3.4t^2}{2} + \cdots\right) \qquad (6.25)$$

We find after a little algebra that

$$\begin{aligned} N(k) &= \frac{(k+1)(k+2)}{2}, \qquad k = 0, 1, 2, \ldots, 9 \\ &= \frac{(k+1)(k+2)}{2} - \frac{3(k-9)(k-8)}{2}, \qquad k = 10, 11, \ldots, 19 \\ &= \frac{(k+1)(k+2)}{2} - \frac{3(k-9)(k-8)}{2} + \frac{3(k-19)(k-18)}{2}, \\ &\qquad k = 20, 21, \ldots, 27 \end{aligned} \qquad (6.26)$$

The probabilities for $n = 2$ are plotted in Figure 6.1, and it is noteworthy that the distribution looks a little like a normal distribution.

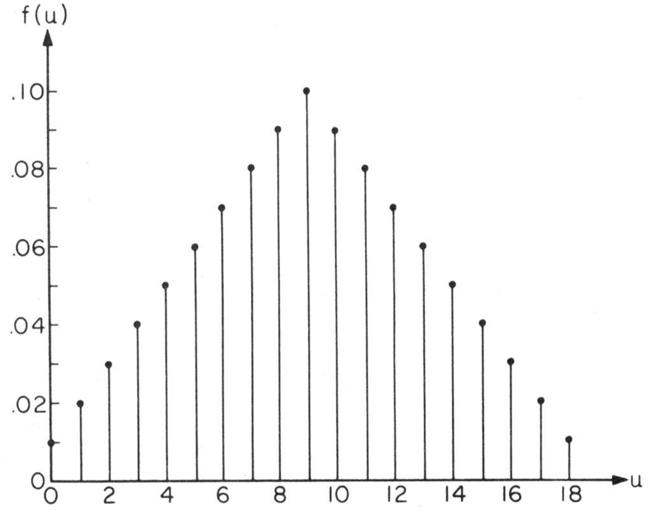


**Fig. 6.1** Distribution of Sample Total from $UD[0(1)9]$, $n = 2$

### 6.2.5 Sampling from a Normal Distribution

In the previous section we discussed the fact that $\bar{X}$ is distributed with mean $\mu$ and variance $\sigma^2/n$. When the parent population is normal, we can also say that $\bar{X}$ is normally distributed with mean $\mu$ and variance $\sigma^2/n$. More generally, it is true that $\sum a_i X_i$ is distributed normally with mean $\mu \sum a_i$ and variance $\sigma^2 \sum a_i^2$. This result was obtained in Chapter 4 for the multivariate normal distribution. For completeness we give the argument here for the univariate normal case.

Consider the m.g.f. for a normal distribution:

$$m_X(t) = E(e^{tX}) = \frac{1}{\sigma\sqrt{2\pi}} \int_{-\infty}^{\infty} e^{tx-(x-\mu)^2/2\sigma^2}\, dx$$

$$= \frac{1}{\sigma\sqrt{2\pi}} \int_{-\infty}^{\infty} e^{-[x^2-2x(\mu+\sigma^2 t)\,+\,\mu^2]/2\sigma^2}\, dx \qquad (6.27)$$

Completing the square in the exponent, we have

$$m_X(t) = \frac{1}{\sigma\sqrt{2\pi}} e^{t\mu+\sigma^2 t^2/2} \int_{-\infty}^{\infty} e^{-[x^2-2x(\mu+\sigma^2 t)+(\mu+\sigma^2 t)^2]/2\sigma^2}\, dx$$

$$= e^{t\mu+\sigma^2 t^2/2} \frac{1}{\sigma\sqrt{2\pi}} \int_{-\infty}^{\infty} e^{-[x-(\mu+\sigma^2 t)]^2/2\sigma^2}\, dx$$

$$= e^{t\mu+\sigma^2 t^2/2} \qquad (6.28)$$

since the remaining integral is the integral of a normal density with mean $\mu + \sigma^2 t$ and variance $\sigma^2$.

The uniqueness property of the m.g.f. means that any time we obtain an m.g.f. with this form, we know we have a normal distribution. For example, if we obtain

$$m_X(t) = e^{3t+8t^2}$$

then $X$ is a normal variable with mean 3 and variance 16.

Next, the m.g.f. of $\sum a_i X_i$ is given by $E(e^{t\Sigma a_i X_i})$. Since the $X$'s are independent

$$E(e^{t\Sigma a_i X_i}) = \prod_1^n E(e^{t a_i X_i})$$

Then

$$E(e^{t\Sigma a_i X_i}) = e^{t\mu\Sigma a_i + \Sigma a_i^2 \sigma^2 t^2/2} \qquad (6.29)$$

which is the m.g.f. of a normal variable with mean $\mu \sum a_i$ and variance $\sigma^2 \sum a_i^2$. Thus a linear function of independent normal variables is normally distributed. As we saw in Chapter 4, the property of independence is not necessary for this conclusion.

*Example 1*

With an elementary knowledge of the normal distribution it is possible to make several interesting probability statements. From previous sections we know that $(\bar{X} - \mu)/(\sigma/\sqrt{n})$ is normally distributed with mean 0 and variance 1 and that the standard normal tables may be used to obtain probabilities. Consider the following sample of yields from a manufacturing process: .70, .35, .40, .50, .45. From previous examination of the process, yields are believed to be distributed with mean yield .60 and $\sigma^2 = .0020$. Does the sample above indicate any change in the mean yield? The observed mean yield is .48. A natural way to proceed is to determine how likely it is that the sample mean yield would differ as much as .12 from the process mean yield if in fact there has been no change. Now

$$Z = \frac{\bar{X} - .60}{\sigma/\sqrt{n}} = \frac{.48 - .60}{.02} = -6$$

From Table A.4 in the Appendix we note the probability to be exceedingly small that $Z$ differs from zero in absolute value by more than 6. Thus we have the summary statement that either the process has changed or a highly deviate sample was obtained.

## 6.3 Distribution of the Sample Variance

If we regard the sample $X_1, X_2, \ldots, X_n$ as a population, it would be consistent with the definitions given previously to regard $\sum_1^n (X_i - \bar{X})^2/n$ as the sample variance. However, it is more customary to use a divisor of $n - 1$ rather than $n$. The consequences of this convention are very slight but do result in simpler expressions in many cases. We shall refer to $s^2 = \sum_1^n (X_i - \bar{X})^2/(n - 1)$ as the sample variance.

If random sampling without replacement is done from a population with $N$ members,

$$E(s^2) = S^2 = \sum_1^N (y_i - \bar{y})^2/(N - 1) \tag{6.30}$$

Thus for large $N$ the mean of $s^2$ is essentially the population variance. If sampling is done with replacement, $E(s^2) = \sigma^2$, as the following shows:

$$\begin{aligned}\sum_1^n (X_i - \mu)^2 &= \sum_1^n [X_i - \bar{X} + (\bar{X} - \mu)]^2 \\ &= \sum_1^n (X_i - \bar{X})^2 + n(\bar{X} - \mu)^2 \end{aligned} \tag{6.31}$$

$$E\left[\sum_1^n (X_i - \mu)^2\right] = E\left[\sum_1^n (X_i - \bar{X})^2\right] + nE[(\bar{X} - \mu)^2]$$

$$n\sigma^2 = E\left[\sum_1^n (X_i - \bar{X})^2\right] + n \operatorname{Var}(\bar{X})$$

$$= E\left[\sum_1^n (X_i - \bar{X})^2\right] + \sigma^2 \tag{6.32}$$

Thus

$$E\left[\sum_1^n (X_i - \bar{X})^2\right] = (n - 1)\sigma^2$$

and we have the useful result that

$$E(s^2) = E\left[\sum_1^n (X_i - \bar{X})^2/(n - 1)\right] = \sigma^2 \tag{6.33}$$

## 6.4 The Joint Distribution of $\bar{X}$ and $s^2$ from a Normal Distribution

We have shown that for random sampling from a normal distribution, $\bar{X}$ has a normal distribution with mean $\mu$ and variance $\sigma^2/n$. We have shown in the previous section that $s^2$ has mean $\sigma^2$ for random sampling from any distribution with finite variance. With sampling from a normal distribution, we can give not only the mean and variance of $s^2$ but also the density function. We introduce first the chi-square distribution and then indicate its relevance with respect to the distribution of $s^2$.

A random variable $U$ is said to have the $\chi^2$ distribution if its density function is given by:

$$f(u) = \frac{u^{(\alpha/2)-1}e^{-u/2}}{2^{\alpha/2}\Gamma(\alpha/2)}, \qquad u \geq 0$$

$$= 0, \text{ otherwise} \tag{6.34}$$

where the gamma function is the same as defined in Chapter 4. The parameter $\alpha$ is called the number of degrees of freedom. This is a skewed distribution which by a simple transformation becomes a one-parameter gamma distribution as is now shown. Let $Y = U/2$. Then

$$f(y) = \frac{2(2y)^{(\alpha/2)-1}e^{-y}}{2^{\alpha/2}\Gamma(\alpha/2)}$$

$$= \frac{y^{(\alpha/2)-1}e^{-y}}{\Gamma(\alpha/2)}$$

Thus the given function is a density function, since it is easily transformed to one we recognize.

The m.g.f. is given by

$$m_U(t) = E(e^{tU}) = \int_0^\infty \frac{e^{tu}e^{-u/2}u^{(\alpha/2)-1}}{2^{\alpha/2}\Gamma(\alpha/2)}\,du$$

$$= 2^{-\alpha/2}[\Gamma(\alpha/2)]^{-1}\int_0^\infty e^{-(u/2)(1-2t)}u^{(\alpha/2)-1}\,du \qquad (6.35)$$

Let $y = (1 - 2t)u$. Then $du = dy/(1 - 2t)$ and

$$m_U(t) = (1 - 2t)^{-\alpha/2}\int_0^\infty \frac{y^{(\alpha/2)-1}e^{-y/2}}{2^{\alpha/2}\Gamma(\alpha/2)}\,dy$$

$$= (1 - 2t)^{-\alpha/2} \qquad (6.36)$$

since the integral on the right is that of a $\chi^2$ density function which equals unity. Again, because of the uniqueness property any random variable with m.g.f. of the form $(1 - 2t)^{-\alpha/2}$ is a $\chi^2$ variable.

The moments of the $\chi^2$ distribution are readily obtained from the m.g.f. For example,

$$\text{Mean} = \left.\frac{d\,m_U(t)}{dt}\right|_{t=0} = \alpha[(1 - 2t)^{-(\alpha+2)/2}]\Big|_{t=0} = \alpha$$

$$\mu_2' = \left.\frac{d^2\,m_U(t)}{dt^2}\right|_{t=0} = \alpha(\alpha + 2)[(1 - 2t)^{-(\alpha+4)/2}]\Big|_{t=0} = \alpha(\alpha + 2)$$

Hence the mean is the number of degrees of freedom, and the variance is twice this.

We shall now show that $(n - 1)s^2/\sigma^2$ has the $\chi^2$ distribution with $n - 1$ degrees of freedom. This enables us to use the moments of a $\chi^2$ variable to find the moments of $s^2$. Since

$$E[(n - 1)s^2/\sigma^2] = n - 1$$

and

$$\text{Var}[(n - 1)s^2/\sigma^2] = 2(n - 1)$$

then

$$E(s^2) = \sigma^2$$

and

$$\text{Var}(s^2) = 2\sigma^4/(n - 1) \qquad (6.37)$$

To find the distribution of $s^2$, we use the procedure of transforming variables and obtain incidentally the joint distribution of $\bar{X}$ and $s^2$.

Let

$$Y_i = (X_i - \mu)/\sigma \tag{6.38}$$

Let

$$\begin{aligned} Z_1 &= (Y_1 + Y_2 + \cdots + Y_n)/\sqrt{n} \\ Z_2 &= (Y_1 - Y_2)/\sqrt{2} \\ Z_3 &= (Y_1 + Y_2 - 2Y_3)/\sqrt{6} \\ &\cdots\cdots\cdots\cdots\cdots\cdots \\ Z_n &= (Y_1 + Y_2 + \cdots + Y_{n-1} - (n-1)Y_n)/\sqrt{n(n-1)} \end{aligned} \tag{6.39}$$

Then

$$\begin{aligned} \sum_1^n Z_i^2 &= \sum_1^n Y_i^2 \\ \sum_2^n Z_i^2 &= \sum_1^n Y_i^2 - Z_1^2 \\ &= \sum_1^n (X_i - \mu)^2/\sigma^2 - n(\bar{X} - \mu)^2/\sigma^2 \\ &= \sum_1^n (X_i - \bar{X})^2/\sigma^2 \end{aligned} \tag{6.40}$$

Thus $\sum_1^n (X_i - \bar{X})^2/\sigma^2$ is distributed as $\sum_2^n Z_i^2$.

The $Z_i$'s are distributed as normal variables since each is a linear function of normal variables. Further, the $Z_i$'s, $i = 2, \ldots, n$, have zero means and variance unity, and are uncorrelated. It was shown in Chapter 4 that uncorrelated normal variables are independent. Thus $(n - 1)s^2/\sigma^2$ has the same distribution as the sum of squares of $n - 1$ independent normal variables with mean zero and variance one. Since $\bar{X}$ is a function of $Z_1$ which is independent of the other $Z$'s, we obtain the result that $\bar{X}$ and $s^2$ are independent.

To finish the derivation, we now show that a sum of squares of independent normal variables with means zero and variances unity is a $\chi^2$ variable. To do this, consider the m.g.f.:

$$\begin{aligned} E(e^{t\Sigma_1^k z_i^2}) &= \left(\frac{1}{\sqrt{2\pi}}\right)^k \int \cdots \int e^{-(1-2t)\Sigma_1^k z_i^2/2}\, dz_1 dz_2 \ldots dz_k \\ &= (1 - 2t)^{-k/2} \end{aligned} \tag{6.41}$$

Therefore, the sum of squares of $k$ independent standard normal variables is a $\chi^2$ variable with $k$ degrees of freedom.

We have the result, therefore, that $\bar{X}$ is normal with mean $\mu$ and variance $\sigma^2/n$, $(n - 1)s^2/\sigma^2$ is distributed as $\chi^2$ with $(n - 1)$ degrees of freedom, and $\bar{X}$ and $s^2$ are independent.

We may note, incidentally, that $(-2)$ times the exponent of an $m$-variate normal distribution is distributed as $\chi^2$ with $m$ degrees of freedom. Using the density in Equation (4.118) we consider the transformation $Y = C(X - \mu)$, where $C \Sigma C'$ is the identity matrix. Then the elements $Y_1, Y_2, \ldots, Y_m$ of $Y$ are independent normal with mean zero and variance unity. The variable $(X - \mu)' \Sigma^{-1} (X - \mu)$ becomes $\sum_1^m y_i^2$, which has the $\chi^2$ distribution with $m$ degrees of freedom.

## 6.5 The Noncentral Chi-Square Distribution

In the previous section, we showed that the sum of squares of $n$ independent $N(0, 1)$ variables is a $\chi^2$ variable with $n$ degrees of freedom. In the present section we wish to find the distribution of $\sum_1^n X_i^2$, when the $X_i$ are independent normal variables with means $\mu_i$ and variance unity. That is, $X_i \sim IN(\mu_i, 1)$. It is possible although quite tedious to obtain the distribution directly using the methods of Chapter 5. We shall instead use the m.g.f. argument.

Consider the m.g.f. of the square of a single normal variable with mean $\mu$ and variance unity:

$$E[e^{tX^2}] = \frac{1}{\sqrt{2\pi}} \int_{-\infty}^{\infty} e^{tx^2} e^{-(x-\mu)^2/2}\, dx \tag{6.42}$$

By completing the square in the exponent, we can see that this equals

$$e^{-[1 - 1/(1-2t)]\mu^2/2} \frac{1}{\sqrt{2\pi}} \int_{-\infty}^{\infty} e^{-[x^2(1-2t) - 2\mu x + \mu^2/(1-2t)]/2}\, dx$$

By making the change of variable $y = x(1 - 2t)^{1/2}$ and recognizing that a normal density integrates to unity, we finally obtain

$$e^{-\mu^2/2} e^{\mu^2/2(1-2t)} (1 - 2t)^{-1/2} \tag{6.43}$$

To obtain the m.g.f. of $\sum X^2$, we use the fact that the m.g.f. of a sum of independent random variables is the product of m.g.f.'s of the individual random variables. Thus

$$\begin{aligned} m_{\Sigma X^2}(t) &= \prod_1^n [e^{-\mu_i^2/2} e^{\mu_i^2/[2(1-2t)]} (1 - 2t)^{-1/2}] \\ &= e^{-\Sigma\mu_i^2/2} e^{\Sigma\mu_i^2/[2(1-2t)]} (1 - 2t)^{-n/2} \end{aligned} \tag{6.44}$$

By using the series representation of $e^y$ and letting $\sum \mu_i^2/2$ equal $\lambda$, we obtain the alternative form:

$$m_{\Sigma X^2}(t) = \sum_{j=0}^{\infty} \frac{\lambda^j e^{-\lambda}}{j!} (1 - 2t)^{-(n+2j)/2}$$

$$= \sum_{j=0}^{\infty} c_j (1 - 2t)^{-(n+2j)/2} \tag{6.45}$$

The m.g.f. of $\sum X^2$ is then seen to be a convex combination of $\chi^2$ m.g.f.'s with degrees of freedom $n$, $n + 2$, $n + 4$, $n + 6$, . . . . The coefficients appearing in the convex combination are, interestingly enough, merely Poisson probabilities. Formally, we can write the m.g.f. as

$$E[(1 - 2t)^{-n/2 - Y}] \tag{6.46}$$

where $Y$ is a Poisson variable with parameter $\lambda$.

It is now almost apparent that the distribution of $\sum X^2$ is the same Poisson convex combination of $\chi^2$ distributions:

$$f(u) = \sum_{j=0}^{\infty} \frac{e^{-\lambda}\lambda^j}{j!} \frac{u^{(n+2j)/2-1} e^{-u/2}}{2^{(n+2j)/2}\Gamma[(n + 2j)/2]} \tag{6.47}$$

Consider the m.g.f. of Equation (6.47). We obtain

$$m_U(t) = E(e^{tU})$$

$$= \sum_{j=0}^{\infty} \frac{e^{-\lambda}\lambda^j}{j!} \int_{-\infty}^{\infty} \frac{e^{-u(1-2t)/2}}{2^{(n+2j)/2}\Gamma[(n + 2j)/2]} du$$

$$= \sum_{j=0}^{\infty} \frac{e^{-\lambda}\lambda^j}{j!} (1 - 2t)^{-(n+2j)/2} \tag{6.48}$$

Therefore the density function of $\sum X^2$ is given by Equation (6.47).

The distribution specified by (6.47) is known as the noncentral $\chi^2$ distribution with $n$ degrees of freedom and $\lambda$ is called the noncentrality factor. Note that if $\lambda = 0$, $f(u)$ as given by (6.47) is simply the $\chi^2$ density function with $n$ degrees of freedom.

Some properties of the noncentral $\chi^2$ distribution are rather obvious but are worth mentioning. By differentiating the m.g.f., we obtain that the mean is $n + 2\lambda$ and the variance is $2n + 8\lambda$. It follows from the m.g.f. that the sum of two independent noncentral $\chi^2$ variables with parameters $(\lambda_1, n_1)$ and $(\lambda_2, n_2)$ is a noncentral $\chi^2$ variable with parameters $(\lambda_1 + \lambda_2, n_1 + n_2)$.

Finally, we note that if $X_i \sim IN(\mu_i, \sigma_i^2)$, then

$$\sum_{i=1}^{n} X_i^2/\sigma_i^2$$

is a noncentral $\chi^2$ variable with $n$ degrees of freedom and non-centrality parameter

$$\lambda = \sum_{i=1}^{n} \mu_i^2/2\sigma_i^2 \tag{6.49}$$

## 6.6 Distribution of Student's $t$

If we are sampling from a population with mean $\mu$ and variance $\sigma^2$, we can use Chebyshev's inequality to make statements about the proportion of values between $\mu + k\sigma$ and $\mu - k\sigma$. Further, when the population is normal, our statements can be made more exactly by using the standard normal tables. This is possible because we know the distribution of $Z = (\bar{X} - \mu)/(\sigma/\sqrt{n})$ to be a normal distribution with zero mean and variance unity. Suppose we replace $\sigma^2$ by $s^2$. How closely does the distribution of $t = (\bar{X} - \mu)/(s/\sqrt{n})$ approach that of $Z$? The distribution of $t$ is the subject of this section. We shall actually obtain the $t$ distribution for a more general formulation.

Note that $t$ as written above can be expressed as

$$t = \frac{\bar{X} - \mu}{\sigma/\sqrt{n}} \Bigg/ \sqrt{\frac{\sum (X - \bar{X})^2}{(n - 1)\sigma^2}} \tag{6.50}$$

Thus $t$ is the ratio of a standard normal variable and the square root of an independent $\chi^2$ variable divided by its degrees of freedom. In order to shorten the derivation, let us adopt a more compact notation.

Suppose $U \sim N(0, 1)$ independently of $V$ which is distributed as a $\chi^2$ with $b$ degrees of freedom. Then

$$f(u, v) = \frac{1}{\sqrt{2\pi}} e^{-u^2/2} \frac{v^{(b/2)-1} e^{-v/2}}{2^{b/2}\Gamma(b/2)} \tag{6.51}$$

for $-\infty < u < \infty$, $v > 0$, and 0, otherwise.

Make the transformation

$$t = \sqrt{b}u/\sqrt{v}, \qquad v = v$$

The Jacobian of this transformation is $\sqrt{v/b}$. So

$$f(t, v) = \frac{e^{(-v/2b)t^2} v^{(b/2)-1} e^{-v/2}}{\sqrt{2\pi} 2^{b/2}\Gamma(b/2)} \cdot \frac{\sqrt{v}}{\sqrt{b}} \tag{6.52}$$

We need to obtain the marginal density of $t$. Let $w = v(1 + t^2/b)$.
Then

$$f(t, w) = \frac{\Gamma[(b+1)/2]}{\sqrt{\pi b}\,\Gamma(b/2)}(1 + t^2/b)^{-(b+1)/2}\frac{w^{(b-1)/2}e^{-w/2}}{2^{(b+1)/2}\Gamma[(b+1)/2]}$$

for $-\infty < t < \infty$, $w > 0$, and 0, otherwise.

Since the term in $w$ is a $\chi^2$ density function which integrates to unity,

$$f(t) = \frac{\Gamma[(b+1)/2]}{\sqrt{\pi b}\,\Gamma(b/2)}(1 + t^2/b)^{-(b+1)/2} \tag{6.53}$$

for $-\infty < t < \infty$.

The density function is a symmetric function about zero and resembles the normal distribution in its general form. The distribution function is tabulated in Table A.6 in the Appendix. The tables give $t_\alpha$ such that $P(t \leq t_\alpha) = \alpha$. By convention the parameter $b$ is referred to as the degrees of freedom.

As a special case the distribution of $t = \sqrt{n}(\bar{X} - \mu)/s$ is given by

$$f(t) = \frac{\Gamma(n/2)}{\sqrt{\pi(n-1)}\,\Gamma[(n-1)/2][1 + t^2/(n-1)]^{n/2}} \tag{6.54}$$

It will be noticed that the sample size $n$ appears as a parameter in the density function. By common agreement $(n - 1)$ in this case is actually regarded as the parameter and is referred to as the degrees of freedom.

*Example 2*

To illustrate the use of this statistic, suppose that the thermal resistivity is measured on each of 10 samples of a certain test material. It is believed that the average thermal resistivity of that material is 25. Also, previous studies indicate that such resistivity readings are approximately normally distributed. The following quantities were computed from the resistivity readings: $\sum (X - \bar{X})^2 = 30$, $\bar{X} = 27$. Is this sample consistent with the idea that the overall resistivity average is 25?

Then

$$\frac{\bar{X} - 25}{s/\sqrt{n}} = \frac{27 - 25}{1/\sqrt{3}} = (1.732)(2) = 3.464$$

From Table A.6 of the Appendix we see that the probability of obtaining a value this large in absolute value by chance, if in fact 25 were the population average, is $<.01$. Thus we are inclined to believe that 25 is not the true average resistivity.

## 6.7 Distribution of Fisher's $F$

One commonly wishes to compare two processes, methods, or procedures. Of interest is the comparison of the variances in the two populations. Suppose that we have taken samples of size $n_1$ and $n_2$ from populations 1 and 2 respectively and have calculated $s_1^2$ and $s_2^2$. These statistics $s_1^2$ and $s_2^2$ are indicative of the population variances $\sigma_1^2$ and $\sigma_2^2$, and it is natural to try to form an opinion of the relative values of $\sigma_1^2$ and $\sigma_2^2$ from the sample values $s_1^2$ and $s_2^2$. We can do this easily only in the case of sampling from normal populations. Furthermore, the ratio $s_1^2/s_2^2$ is amenable to mathematical treatment and is both origin and scale independent.

In the section on the $\chi^2$ distribution we learned that

$$(n_1 - 1)s_1^2/\sigma_1^2 \sim \chi^2_{n_1-1} \text{ and } (n_2 - 1)s_2^2/\sigma_2^2 \sim \chi^2_{n_2-1} \qquad (6.55)$$

Then $(s_1^2/\sigma_1^2)/(s_2^2/\sigma_2^2)$ is actually the ratio of two independent $\chi^2$ variables after each has been divided by its degrees of freedom. That is,

$$\frac{s_1^2}{\sigma_1^2}\bigg/\frac{s_2^2}{\sigma_2^2} \sim \frac{\chi^2_{n_2-1}}{n_1 - 1}\bigg/\frac{\chi^2_{n_2-1}}{n_2 - 1} \qquad (6.56)$$

Let us find the distribution of $bU/aV = (U/a)/(V/b)$ when $U \sim \chi_a^2$ and $V \sim \chi_b^2$ independently of $U$. Now the joint density of $U$ and $V$ is

$$f(u, v) = K_1 u^{(a/2)-1} e^{-u/2} v^{(b/2)-1} e^{-v/2}, \; u, v > 0 \qquad (6.57)$$

The $K$ here and below are constants. Let $F = bu/av$, $v = v$.

The Jacobian for this transformation is $av/b$. Then

$$f(F, v) = K_2 (vF)^{a/2-1} e^{-avF/2b} v^{b/2} e^{-v/2} \qquad (6.58)$$

for $F$ and $v > 0$, and 0, otherwise.

Since $F$ is the variable of interest, we find the marginal density of $F$ by integrating out $v$. Thus

$$f(F) = K_2 F^{(a/2)-1} \int_0^\infty v^{(a+b-2)/2} e^{(-v/2)(1+aF/b)}\, dv \qquad (6.59)$$

Let $w = (v/2)(1 + aF/b)$. Then

$$f(F) = K_2 F^{a/2-1} \int_0^\infty [(2w)/(1 + aF/b)]^{(a+b-2)/2} e^{-w/2} dw/(1 + aF/b)$$

$$= K_3 F^{a/2-1}(b + aF)^{-(a+b)/2} \qquad (6.60)$$

By actually carrying the constants at each step the density is found to be

$$f(F) = \frac{a^{a/2} b^{b/2} F^{a/2-1}}{B(a/2, b/2)(b + aF)^{(a+b)/2}} \qquad (6.61)$$

The parameters $a$ and $b$ are referred to as the degrees of freedom for the numerator and denominator of $F$ respectively.

## 6.8 Distribution of an Ordered Sample

Let a sample of size $n$ be taken at random from a distribution with density function $f(x)$. Now order the values in the sample from low to high values as $Y_1 \leq Y_2 \leq Y_3 \leq \cdots \leq Y_n$. We want to consider several questions concerned with such an ordered sample. It is natural to consider the distribution of $Y_n$ the largest observation. We have seen in Sect. 5.2 that this has distribution function

$$G(y_n) = [F(y_n)]^n \tag{6.62}$$

If $F$ is continuous with density $f$, $Y_n$ has density

$$g(y_n) = n[F(y_n)]^{n-1}f(y_n) \tag{6.63}$$

For example, suppose $f(x) = 1/2$, $0 \leq x \leq 2$, and we take a sample of size $n$. Then

$$g(y_n) = (n/2)(y_n/2)^{n-1} \tag{6.64}$$

Next consider the matter of the joint distribution of the entire ordered sample. The event that $Y_1 \leq y_1, Y_2 \leq y_2, Y_3 \leq y_3, \ldots, Y_n \leq y_n$ where $y_1, y_2, \ldots, y_n$ is a specified set of numbers will occur when *one* of the $X$'s $\leq y_1$, *one* of the $X$'s $\leq y_2$, etc. Thus

$$P(Y_1 \leq y_1, Y_2 \leq y_2, Y_3 \leq y_3, \ldots, Y_n \leq y_n)$$

is given by the sum of the following probabilities:

$$P(X_1 \leq y_1, X_2 \leq y_2, X_3 \leq y_3, \ldots, X_n \leq y_n)$$
$$P(X_2 \leq y_1, X_5 \leq y_2, X_n \leq y_3, \ldots\ldots\ldots)$$
$$P(X_4 \leq y_1, X_2 \leq y_2, X_1 \leq y_3, \ldots\ldots\ldots), \text{ etc.}$$

Thus

$$\begin{aligned} G(y_1, y_2, \ldots, y_n) &= P(Y_1 \leq y_1, Y_2 \leq y_2, \ldots, Y_n \leq y_n) \\ &= n!F(y_1, y_2, \ldots, y_n), \end{aligned} \tag{6.65}$$

where $-\infty \leq y_1 \leq y_2 \leq \cdots \leq y_n < \infty$.

The joint density function is given by

$$g(y_1, y_2, \ldots, y_n) = n!f(y_1, y_2, \ldots, y_n) \tag{6.66}$$

If for example we take $n$ observations from the uniform density from zero to one,

$$g(y_1, y_2, \ldots, y_n) = n! \tag{6.67}$$

over the region $0 \le y_1 \le y_2 \le \cdots \le y_n \le 1$.

A not infrequent situation is the one in which the sample is taken in an ordered fashion. That is, the smallest observation is obtained first, the next larger one second, etc. This situation occurs in life testing, for example. The smallest lifetime is observed first, the second smallest lifetime is observed second, etc. In such a case it is of interest to know the distribution of the $r$ smallest observations in a sample of size $n$. We can obtain this result directly from $g(y_1, y_2, \ldots, y_n)$ by integrating out $y_{r+1}, y_{r+2}, \ldots, y_n$, or we can use an argument similar to the two preceding derivations. In either case we obtain

$$g(y_1, y_2, \ldots, y_r) = \frac{n!}{(n-r)!}[1 - F(y_r)]^{n-r} \prod_1^r f(y_i) \tag{6.68}$$

## 6.9 Distribution of the Sample Range

The difference between the largest and the smallest observations in a sample of size $n$ is called the sample range. It is one of the most commonly used statistics and is frequently calculated as an obvious measure of variation.

The joint density of the smallest and largest observations is found as follows:

$$g(u, v) = \lim_{\Delta u, \Delta v \to 0} \frac{P(u \le X_{\min} \le u + \Delta u, v \le X_{\max} \le v + \Delta v)}{\Delta u\, \Delta v} \tag{6.69}$$

$$= \lim_{\Delta u, \Delta v \to 0} \frac{n(n-1)f(u)\Delta u\, f(v)\Delta v\left[\int_u^v f(x)\,dx\right]^{n-2}}{\Delta u\, \Delta v} \tag{6.70}$$

$$= n(n-1)f(u)f(v)\left[\int_u^v f(x)\,dx\right]^{n-2} \tag{6.71}$$

We can obtain the density of $R$, the range, by making the transformation

$$R = x_{\max} - x_{\min} \tag{6.72}$$

$$w = x_{\min}$$

The Jacobian is unity so the joint density of $w$ and $R$ is

$$f(w, R) = n(n-1)f(w)f(w+R)\left[\int_w^{w+R} f(x)\,dx\right]^{n-2} \tag{6.73}$$

The marginal density of $R$ is then obtained by integrating with respect to $w$. We cannot get a general expression for all parent distributions. Rather this process must be carried through for each distribution. Some distributions will yield a closed expression for the density of $R$. Others will require that we resort to numerical evaluation. Suppose that a sample of size $n$ is taken from a normal distribution with mean $\mu$ and variance $\sigma^2$. The normal density function is such that we cannot obtain a closed expression for the density function of $R$ but we can evaluate its moments by resorting to the methods of numerical integration. It can be found that $E(R) = k\sigma$, where $k$ is given by the following table:

| $n$ | 2 | 3 | 4 | 5 |
|---|---|---|---|---|
| $k$ | 1.128 | 1.693 | 2.059 | 2.326 |

## 6.10 Monte Carlo Techniques

Suppose that one wished to evaluate the properties of a gambling strategy for roulette. A direct approach, if somewhat lacking in aesthetic appeal, would be to obtain a wheel and to determine the properties of a strategy by spinning the wheel many times while using that strategy. It would be necessary to try to simulate the actual playing conditions as nearly as possible. After spinning the wheel a number of times, it might become apparent that it is not really necessary to have a wheel. For instance we might simulate a wheel having 20 compartments by drawing 2-digit numbers from a table of random numbers where numbers 00 through 04 correspond to compartment 1, 05 through 09 to compartment 2, etc.

Not only can we simulate the process of drawing a random sample from a specified population but we can also simulate our algebraic manipulations, which we have described as random sampling from a mathematical distribution. It may be supposed that the most basic simulation of this type is that of simulating random sampling from the uniform discrete $U[0\,(1)\,9]$ distribution. Many techniques have been used to simulate random sampling from the $U[0\,(1)\,9]$ distribution, and many tabulations of such simulations have been made. These tabulations are referred to as random number tables. We now mention a few examples of such tables.

Hald (1952) presents a table of 15,000 digits which were compiled from drawings in the Danish State Lottery of 1948. After compilation the numbers were closely scrutinized, and certain sections of numbers were removed which did not conform with the idea of a random sequence. Certain statistical tests were used for scrutiny of the data. Fisher and Yates (1963) abstracted from Logarithmetica Britannica the 15th to 19th digits of the sequence of logarithms to the base 10 of numbers with a particular spacing and, after modifying them in a certain way, presented them as "random numbers." Of considerable interest is the Rand Corporation (1955) table of a million random digits. Certainly the largest random number table readily available, it is a notable example of artificially generated random numbers.

Many of the "random numbers" generating routines are refinements or modifications of the following simple procedure:

1. Use a constant multiplier.
2. Multiply by an arbitrary multiplicand.
3. Use the $k$ digits on the right for the second multiplicand and record the $p$ digits immediately preceding these $k$.

Table 6.1 was generated in this fashion.

**Table 6.1** Random Numbers

| | | | | | | | | | |
|---|---|---|---|---|---|---|---|---|---|
| 0 | 1 | 4 | 1 | 5 | 0 | 6 | 6 | 0 | 5 |
| 8 | 6 | 3 | 8 | 8 | 3 | 4 | 7 | 9 | 2 |
| 2 | 3 | 2 | 4 | 1 | 5 | 7 | 9 | 6 | 9 |
| 5 | 5 | 6 | 0 | 1 | 6 | 5 | 1 | 0 | 9 |
| 4 | 8 | 8 | 7 | 9 | 3 | 2 | 2 | 3 | 2 |
| 8 | 4 | 3 | 8 | 8 | 3 | 7 | 7 | 2 | 9 |
| 0 | 3 | 9 | 7 | 8 | 8 | 4 | 1 | 3 | 2 |
| 7 | 9 | 2 | 5 | 6 | 0 | 1 | 9 | 5 | 4 |
| 6 | 9 | 7 | 5 | 1 | 4 | 1 | 1 | 5 | 0 |
| 7 | 4 | 2 | 9 | 7 | 4 | 6 | 5 | 1 | 0 |

To generate these numbers, a multiplier of 273 was used to start the process and we used a multiplicand of 333. This gave a product of 90,909. The number 0 was recorded, and 909 was used as a new multiplicand. This gave 248,157. The number 8 was recorded, and 157 was used as the new multiplicand.

The frequency of digits is as follows:

| $x$ | 0 | 1 | 2 | 3 | 4 | 5 | 6 | 7 | 8 | 9 |
|---|---|---|---|---|---|---|---|---|---|---|
| Freq. | 9 | 11 | 10 | 9 | 10 | 11 | 9 | 10 | 10 | 11 |

If the reader will follow the sequence through, he will discover an interesting fact. The next 100 numbers generated will be the same sequence as this one. All such methods have a periodicity built into them. However, by using larger numbers, the period can be made quite large.

There are two conflicting ideas about the construction and use of random number tables. They are:

1. The table should give results "like" the results of a set of trials from a random device.
2. The table should be free from gross irregularities, even though such irregularities are bound to occur with any random device. Our own viewpoint is that even if we could get results from a purely random device, we probably would not want to record them as a random number table. This is because we wish to use the table to simulate a random device. Therefore, the table should be as free of all recognizable irregularities as possible.

As an illustration of this type of procedure the IBM company, in its reference manual on random number generation, suggests the following*:

1. Take as a starting number some odd integer $x_0$.
2. Choose a constant multiplier $u$ of the form $8t + 3$, which is near $2^{b/2}$, where $t$ is an integer, and $b$ is the word length.
3. Compute $x_0u$ and let $x_1$ be the first $b$ low order bits.
4. Compute $x_1u$ as in (2) to give $x_2$, etc., the successive numbers $x_1, x_2, \ldots, x_n, \ldots$ being "random" numbers of $b$ bits.

For a decimal computer with word size of $d$-digits:

1. Take any integer $x_0$ which is not divisible by 2 or 5.
2. Choose a constant multiplier $u$ which is of the form $200\,t + r$, where $t$ is an integer and $r$ is one of the numbers 3, 11, 13, 19, 21, 27, 29, 37, 53, 61, 67, 69, 77, 83, or 91, with $u$ close to $10^{d/2}$.
3. Form $x_0u$ and let the first $d$ digits give the number $x_1$.
4. Compute $x_1u$ as in (3) to give $x_2$, etc., the successive numbers $x_1, x_2, \ldots, x_n, \ldots$ being "random" numbers of $d$ digits and between 0 and $10d - 1$.

## 6.11 The Central Limit Theorem

In Section 6.2.4 it was noticed that for very small sample sizes the distribution of the sample total began to display properties similar to the normal distribution. This was not just a coincidence nor was it a circumstance peculiar to the uniform discrete distribution which was being sampled.

One of the most important theorems in mathematical statistics, the central limit theorem provides justification for the very common assumption that $\bar{X}$, or equivalently the sample total, is nearly normally distributed. If a random sample of size $n$ is taken from a population with mean $\mu$ and variance $\sigma^2$, the limiting distribution of $[\sqrt{n}(\bar{X} - \mu)]/\sigma$ as $n \to \infty$ is a

*Used by permission from publ. C20-8011, © by International Business Machine Corporation.

normal distribution with mean zero and variance one. Thus for large $n$ we feel justified in assuming that $\bar{X}$ is distributed normally with mean $\mu$ and variance $\sigma^2/n$.

To prove that the limiting distribution is the standard normal distribution, consider the m.g.f. of

$$V = \sqrt{n}(\bar{X} - \mu)/\sigma \tag{6.74}$$

Now $V$ can be expressed as the sum of independent random variables, i.e.,

$$V = \sum_{i=1}^{n} \frac{X_i - \mu}{\sigma\sqrt{n}} = \sum_{i=1}^{n} Z_i \tag{6.75}$$

From Chapter 2 we know that

$$m_V(t) = \prod_{i=1}^{n} m_{Z_i}(t)$$

Now $Y_i = (X_i - \mu)/\sigma$ are independent variables with mean zero and variance unity, so

$$m_{Y_i}(t) = e^{t^2/2} = 1 + \frac{t^2}{2} + \text{terms of higher order in } t \tag{6.76}$$

and the m.g.f. of $Z_i = (X_i - \mu)/\sigma\sqrt{n}$ becomes

$$m_{Z_i}(t) = 1 + \frac{t^2}{2n} + \text{terms of higher order in } \frac{t}{\sqrt{n}} \tag{6.77}$$

The higher order terms go to zero faster than $t^2/n$. Denote such terms by

$$o\left(\frac{t^2}{n}\right)$$

Then

$$\begin{aligned} m_V(t) &= \prod_{i=1}^{n}\left[1 + \frac{t^2}{2n} + o\left(\frac{t^2}{n}\right)\right] \\ &= \left[1 + \frac{t^2}{2n} + o\left(\frac{t^2}{n}\right)\right]^n \end{aligned} \tag{6.78}$$

Since

$$\lim_{n\to\infty}\left[o\left(\frac{t^2}{n}\right)\right] = \lim_{(t^2/n)\to 0}\left[o\left(\frac{t^2}{n}\right)\right] = 0 \tag{6.79}$$

and since

$$\lim_{n\to\infty}\left(1 + \frac{1}{n}\right)^n = e \tag{6.80}$$

$$\lim_{n\to\infty}\left[1 + \frac{t^2}{2n} + o\left(\frac{t^2}{n}\right)\right]^n = \lim_{n\to\infty}\left(1 + \frac{t^2}{2n}\right)^n = e^{t^2/2} \tag{6.81}$$

Thus the m.g.f. of $V$ approaches the m.g.f. of a normal variable with mean zero and variance unity. As pointed out in Chapter 4, this implies that the distribution of $V$ approaches a normal distribution. This theorem is unusually important because the rate of convergence can be incredibly high, as indicated by the examples in Section 6.2.4.

## PROBLEMS

1. Verify that the covariances of the $\delta$'s are as given by Equation (6.5).
2. In random sampling without replacement from a finite population, find the correlation between two member observations $x_i$ and $x_j$.
3. If $X_i$, $i = 1, 2, \ldots, n$ are independent Poisson variables with parameters $\lambda_i$, show $\sum_1^n X_i$ is a Poisson variable.
4. If the number of calls received at a switchboard during a given period of time is approximated by a Poisson random variable with $\lambda = 5$, what is the approximate probability that there will be no calls received in 10 such periods?
5. Prove the identity

$$\sum_i (X_i - \bar{X})^2 = \frac{1}{2n} \sum_{i \neq j} (X_i - X_j)^2$$

thereby showing that

$$s^2 = \frac{1}{2n(n-1)} \sum_{i \neq j} (X_i - X_j)^2 = \text{Ave}[(\text{difference})^2]/2$$

6. Smoothness readings are made on specimens of machined metal fittings. The readings are taken as approximately normally distributed. How large a sample must be taken so that the probability is about .95 that $\bar{X}$ differs from $\mu$ by less than $(.1)\sigma$?
7. If a random sample of size 10 is taken from a normal distribution, find the probability that at least two of the observations exceed $\mu + .5\sigma$.
8. Use the moments of the $\chi^2$ density function to find the mean and variance of $\sum_1^n (X_i - \mu)^2$ where the $X$'s constitute a random sample from a normal distribution with mean $\mu$ and variance $\sigma^2$.
9. Generate 50 samples of size 5 of random normal variables. For each sample compute $\bar{X}$ and $s^2$. Plot a scatter diagram of $\bar{X}$ and $s^2$. The plot should indicate that $\bar{X}$ and $s^2$ are nearly independent.
10. Given a random sample of size 10 from a normal population, find
    (a) $P[\sum_1^{10} (X_i - \mu)^2/\sigma^2 \leq 10]$.
    (b) $P[\sqrt{10}\,(\bar{X} - \mu)/\sigma \geq 2]$.
    (c) $P[\sum_1^{10} (X_i - \bar{X})^2/\sigma^2 \geq 5]$.
    (d) $P[\sqrt{10}\,(\bar{X} - \mu)/s \geq 2.2]$.
11. Graph Student's $t$ density function for degrees of freedom 1, 2, 3, 4, and 5 and compare with a graph of the $N(0, 1)$ density function.

12. Show that $t = \sqrt{n}\,(\bar{X} - \mu)/s$ and $s$ are uncorrelated. Does this mean that $t$ and $s$ are independent?
13. Graph Fisher's $F$ density function for numerator degrees of freedom 1, 2, and 3 and denominator degrees of freedom 1 and 2.
14. Given two random samples of size 10 and 15 respectively from a normal population, find
$$P\left(\frac{s_1^2}{s_2^2} \geq 2\right)$$
15. Graph the $\chi^2$ density function for degrees of freedom 1, 2, 3, 4, and 5.
16. For the transformation given in Sect. 6.4, verify that
 (a) $E(Z_i) = 0, i = 2, \ldots, n$.
 (b) $\mathrm{Var}(Z_i) = 1, i = 1, 2, \ldots, n$.
 (c) $\mathrm{Cov}(Z_i, Z_j) = 0, i \neq j$.
17. Given a random sample of size $n$ from a normal population, show that $\bar{X}$ and $X_i - \bar{X}$ are independent, $i = 1, 2, \ldots, n$. Can we conclude from this that $\bar{X}$ and $s^2$ are independent?
18. Given that $X_i \sim \mathrm{NID}(0, \sigma^2)$, where NID means normally and independently distributed and $i = 1, 2$, show that $X_1/X_2$ is a Student's $t$ variable with one degree of freedom.
19. Go through the derivation of Fisher's $F$ in Sect. 6.7, carrying the constants in explicit form, to verify that the density is as given in Equation (6.61).
20. Consider the finite population of 10 elements: 0, 1, 2, 3, 4, 5, 6, 7, 8, 9. Obtain the actual c.d.f. of the $t$-like statistic.
$$\frac{\bar{x} - \mu}{\sqrt{\dfrac{N - n}{N}\dfrac{s^2}{n}}}$$
for sample size 2.
21. Consider the finite population of 6 elements: 1, 1, 2, 2, 3, 3. Obtain the cumulative distribution of
$$\frac{\bar{x} - \mu}{\sqrt{\dfrac{N - n}{N}\dfrac{s^2}{n}}}$$
for sample size 2.
22. If $U$ is $\chi^2$ random variable with $a$ degrees of freedom, find the mean and variance of $1/U$.
23. Explore fractional and negative moments of a $\chi^2$ random variable.
24. Show that the m.g.f. of the noncentral $\chi^2$ can also be written as
$$(1 - 2t)^{-(\nu/2)} e^{2\lambda t/(1 - 2t)}$$
25. Find the mean and variance of the noncentral $\chi^2$ distribution.
26. If $X \sim N(\mu, \sigma^2)$, show that $X^2/\sigma^2$ is a noncentral $\chi^2$ variable with one degree of freedom, and $\lambda = \mu^2/2\sigma^2$.

27. Show that the sum of two independent noncentral $\chi^2$ variables is a noncentral $\chi^2$ variable.

28. Show that if $t$ is a Student's $t$ variable with $r$ degrees of freedom, $t^2$ is a Fisher's $F$ variable with one and $r$ degrees of freedom.

29. If $F$ is a Fisher's $F$ variable with $a$ and $b$ degrees of freedom, show that $aF/(b + aF)$ $aF/(b + aF)$ is a beta variable and find the density function.

30. If $X$ is a one-parameter gamma variable as described in Chapter 4 with parameter $\alpha$, let $Y = X/\beta$ and find the density function of $Y$. Call this a 2-parameter gamma distribution with parameters $\alpha$ and $\beta$.

31. What values of the parameters in the 2-parameter gamma density function derived in Problem 30 give a $\chi^2$ density function?

32. Show that the variance of $s^2$ is $2\sigma^4/(n - 1)$ where $s^2$ is calculated from a random sample of size $n$ from a normal distribution with mean $\mu$ and variance $\sigma^2$.

33. Find the mean, mode, and variance of the $F$ distribution, noting any restrictions that must be made on parameter values to insure existence.

34. Find the distribution of the $t$ statistic for a random sample of size 2 from $U[0\,(1)\,n]$.

35. Find the distribution of the $t$ statistic for a random sample of size 2 without replacement from the population of numbers: 0, 1, 2, . . . , $n$.

36. Compare the distributions obtained in the two previous exercises with Student's $t$ distribution with one degree of freedom with respect to
(a) Mean.
(b) Variance.
(c) Cumulative distribution function.

37. Derive the joint density of $X_{min}$, $X_{max}$ given in Equation (6.73) by
(a) Differentiating the c.d.f.
(b) Using a Jacobian.

38. Use the results of Sect. 6.8 and the standard normal tables to graph the density function of the largest observation in a random sample of 10 from a normal population with mean zero and variance one.

39. Use the results of Sect. 6.9 to find the density function of the sample range from a random sample of size 5 from the density function

$$f(x) = 4x^3, \quad 0 < x < 1$$
$$= 0, \text{ otherwise}$$

40. For the $U(0, \theta)$ distribution find the mean and variance of the sample range for samples of size $n$.

41. Consider an infinite population with the probability density function

| $x$ | 1 | 2 | 3 | other values |
|---|---|---|---|---|
| $f(x)$ | 1/2 | 1/3 | 1/6 | 0 |

(a) If a random sample of size 3 is drawn, find the probability density function of the sample total.

(b) Find the joint density function of the ordered observations.
(c) Find the marginal density functions of the ordered observations.

42. Use a random number table and draw 100 samples of size 4.
(a) Determine the frequency distribution of $\sum_1^4 X_i$.
(b) Determine the frequency distribution of the largest observation.

43. Extending the method used in Sects. 6.2.4 and 6.8, determine for random samples of size 4 from $U[0\ (1)\ 9]$:
(a) The density function of $T = \sum_1^4 X_i$.
(b) The density function of the largest observation in the sample. The results should be in general agreement with those of Problem 42.

44. Consider a random sample of size 4 from $U[0\ (1)\ 9]$.
(a) Find and graph the density function $T = \sum_1^4 X_i$.
(b) Find the joint density function of the ordered observations.
(c) Find the marginal density functions of the ordered observations.

45. Given a random sample of two observations $X_1$ and $X_2$ from the uniform discrete distribution $U[0\ (1)\ 9]$, find $f(u, v)$ where $U = X_{\max}$ and $V = X_1 + X_2$. Plot the marginals and compare with the results which would be obtained if the sample came from the continuous distribution with density

$$f(x_1, x_2) = 1/100, \qquad -0.5 < x_1 < 9.5,\ -0.5 < x_2 < 9.5$$
$$= 0, \text{ otherwise}$$

46. Devise a random number generator to generate digits 0 through 9. Generate 100 digits and check for randomness.

47. Write a computer program to generate numbers 0 through 9 using the IBM procedure described in Sect. 6.10. Generate 1000 numbers and check for randomness.

48. Devise if possible a computer program to generate random numbers which does not cycle, and give the flow chart for the program.

49. Do a little experimenting with the type of generator described in Sect. 6.10 and suggest some rules to aid one in generating a "random" sequence. For instance, certain multipliers must be ruled out because they will not yield all of the numbers 0, 1, 2, . . . , 9.

50. Let $S_n(x)$ = rel. freq. $(X \leq x)$, the observed c.d.f. of a random sample.
(a) What is the distribution of $S_n(x)$ for a fixed value of $x$?
(b) What are its mean and variance?
(c) What is the joint distribution of $S_n(x_1)$ and $S_n(x_2)$, and what are its means, variances and covariances?

We mention that the joint distribution of $S_n(x)$ for all values of $x$ is an advanced topic, requiring the apparatus of measure theory and a probability space of infinite dimensionality.

The random variable, the Kolmogorov-Smirnov statistic

$$\sup_x |S_n(x) - F(x)|$$

is used widely as a measure of the distance of an observed c.d.f. $S_n(x)$ from a hypothesized c.d.f. $F(x)$. Its distribution is described in most advanced texts.

51. Suppose $X_i$, $i = 1, 2, \ldots, n$ are independently normal with means $\mu_i$, $i = 1, 2, \ldots, n$, and common variance $\sigma^2$. Obtain a series expansion for the probability density of $\sqrt{n}\,\bar{X}/s$, where $(n - 1)s^2 = \sum (X_i - \bar{X})^2$. This is called the non-central $t$ distribution for obvious reasons.

# 7

# Stochastic Processes

## 7.1 Introduction

We take up in this chapter a short introduction to the study of random processes. Up to now we have been concerned with single random variables, or a set of random variables of specified size, as in a bivariate or $k$-variate normal distribution. A stochastic process is much more complicated in that it is concerned with the dynamics of a system, the progress of which is determined by elementary random variables. Examples are provided by the growth of populations, the theory of nuclear reactors, and the dynamics of a service system with variable times of arrival of users and/or variable times involved in service, a field known as queueing theory. The emphasis is on the dynamics of the overall system.

To motivate the development, consider two problems. In the first, a random walk, an object is moving along a line and takes a step to the right with probability $p$, say, and to the left with probability $q$. Suppose also that there are reflecting barriers. Then the position of the object after $T$ steps is a random variable. Indeed, the position of the object at times $T_1, T_2, \ldots, T_k$ is a $k$-variate random variable, and we can certainly imagine $k$ to be indefinitely large. In the second problem suppose we have a population of reproducing objects (e.g., cells) and suppose that in a unit time interval each cell produces offspring according to a Poisson distribution with mean $m$. After starting from a particular number of individuals, the population numbers at any set of future times $T_1, T_2, \ldots, T_k$ follow a $k$-variate distribution. Interest may be in the probability that the population will die out or explode in numbers (as in a chain reaction).

Stochastic process theory is concerned with an infinity of random variables, which can be labeled by a variable $t$ from some infinite set $T$. The infinite set may be denumerable as in the examples above or non-denumerable. If the infinite set is denumerable, a stochastic process is an infinite sequence of random variables, $X_1$, $X_2$, . . . . The interesting problems are of low dimensionality, such as the distribution of $X_\infty$ given $X_1$. The subject is important for two reasons: It enables the development of properties of probabilistic models, and it leads to probabilistic models for data, which can be subjected to the process of statistics. With regard to the second aspect the subject is not at all unique in that the processes of statistics are aimed generally at the interpretation of data with probability models. In this regard we refer the reader to the work of Billingsley (1961), Gani (1955), and Cox and Lewis (1966). The topic of stochastic processes is important in this book primarily from the second viewpoint.

The topics covered in this chapter form a very elementary introduction to the subject and do not comprise a general survey, but rather represent only a few situations. In the discussion on Markov chains we have attempted to use as few special words as possible, since the language in this area varies so much from writer to writer. The material on Poisson processes is oriented almost exclusively toward life testing. In the section on stationary time series we present some initial ideas on the characterization of the variability in infinite sequences.

## 7.2 Markov Chains

In presenting a few elementary topics from the theory of stochastic processes, we do not wish to give the impression that the subject consists merely of additional details about these specific topics. Rather we wish to give some general ideas about how these special topics fit into the general structure of the subject.

One way of classifying stochastic processes is to classify them as Markovian or non-Markovian. Markov processes can be described as being those which exhibit no carry-over effect. That is, the conditional distribution of random variables in the future given their values now is the same as their conditional distribution given their values now and at all times in the past. To state the matter in another way, the previous history of the random variables has no carry-over effect in determining the probability distribution in the future. Only their present values are of relevance.

Once we have classified a process as Markovian, it can still be classified in various other ways. For instance, we may ask whether the random variables are discrete or continuous and whether we are observing the process in time in a continuous fashion or in a discrete fashion. This leads to the four possibilities depicted in Table 7.1.

**Table 7.1** Classification of Markov Processes

| Random Variable | Time | |
|---|---|---|
| | Discrete | Continuous |
| Discrete | Case 1<br>Markov chains | Case 3<br>Markov processes |
| Continuous | Case 2<br>Markov chains | Case 4<br>Markov processes |

In this presentation we shall discuss only Case 1 and limit ourselves to the case of finite Markov chains, i.e., to the case where the random variable is not only discrete but can assume only a finite number of values.

The language used in the theory of Markov chains is heavily influenced by the areas of subject matter which have provided a great deal of stimulus for this particular subject. A great deal of statistical physics is concerned with predicting the state of a physical system in the future given the present state of the system. Although Markov chains have many applications other than in physics, we use the same language to describe Markov processes in general.

A well-known model was formulated by P. and T. Ehrenfest to describe some diffusion phenomena of statistical physics. Consider a chamber with a partition dividing the chamber into compartments $A$ and $B$. The chamber contains $n$ molecules. Consider now a sequence of trials in each of which a molecule is selected at random from one compartment and moved to the other compartment. Let the state of the system be the number of molecules in compartment $A$. There are then $n + 1$ possible states because the number can be any one of the integers 0 to $n$, and we desire the probability of each state given the state of the system at the preceding trial. Since the molecule is selected at random from the entire chamber, the probability of selecting a molecule from $A$ for transfer to $B$ is just the number of molecules in $A$ divided by $n$.

A little reflection indicates that the number of molecules in $A$ will either increase by 1 or decrease by 1. If there are $j$ molecules in $A$, the number will increase if the molecule selected is from $B$, for which the probability is $(n - j)/n$. The number of molecules will decrease if the molecule is selected from $A$, for which the probability is $j/n$.

The entire set of conditional probabilities $P(\text{no. in } A = j | \text{no. in } A = i$ at preceding trial) is indicated by Table 7.2.

**Table 7.2** Set of Conditional Probabilities

| | | j | | | | | | | |
|---|---|---|---|---|---|---|---|---|---|
| | | 0 | 1 | 2 | 3 | ... | $n-2$ | $n-1$ | $n$ |
| | 0 | 0 | 1 | 0 | 0 | ... | 0 | 0 | 0 |
| | 1 | $1/n$ | 0 | $(n-1)/n$ | 0 | ... | 0 | 0 | 0 |
| | 2 | 0 | $2/n$ | 0 | $(n-2)/n$ | ... | 0 | 0 | 0 |
| $i$ | 3 | 0 | 0 | $3/n$ | 0 | ... | 0 | 0 | 0 |
| | . | . | . | . | . | ... | . | . | . |
| | $n-1$ | 0 | 0 | 0 | 0 | ... | $(n-1)/n$ | 0 | $1/n$ |
| | $n$ | 0 | 0 | 0 | 0 | ... | 0 | 1 | 0 |

This diffusion model is an example of a finite Markov chain. In fact it is an example of a homogeneous finite Markov chain because the table of conditional probabilities remains constant from trial to trial, although this is not true for Markov chains in general. We shall discuss only the homogeneous case.

### 7.2.1 Transition Probabilities

Consider a system which has a finite number of possible states $s_1, s_2, s_3, \ldots, s_r$ and which changes states only at discrete steps or trials. We shall identify the times after successive trials by $t_0, t_1, t_2, \ldots$, with $t_0$ representing the starting point in time, $t_1$, the time of conclusion of the first trial, etc.

Let

$$p_i(0) = \text{Prob}(s_i \text{ at } t_0), \qquad i = 1, 2, \ldots, r \tag{7.1}$$

We can arrange the starting probabilities as a row vector

$$p'(0) = [p_1(0), p_2(0), p_3(0), \ldots, p_r(0)] \tag{7.2}$$

where, of course,

$$\sum_{i=1}^{r} p_i(0) = 1$$

Let

$$p_{jk} = \text{Prob}(s_k | s_j \text{ at preceding time}) \tag{7.3}$$

We can arrange the $p_{jk}$ in a rectangular array as follows:

$$P = \begin{pmatrix} p_{11} & p_{12} & p_{13} & \cdots & p_{1r} \\ p_{21} & p_{22} & p_{23} & \cdots & p_{2r} \\ \cdots & \cdots & \cdots & \cdots & \cdots \\ p_{r1} & p_{r2} & p_{r3} & \cdots & p_{rr} \end{pmatrix} \tag{7.4}$$

This matrix is called the transition matrix. The matrix displayed in the diffusion example is then the transition matrix for that chain. Since the $p_{jk}$'s are probabilities, it follows that $\sum_k p_{jk} = 1$ for all $j$; i.e., the probabilities in each row sum to unity. This is not true for columns. If the elements of $P$ do not depend on time, the transition probabilities are *stationary*.

### 7.2.2 Some Important Questions

Many probability models can be considered as Markov chains; however, it would not always be fruitful to do so. Because of the diversity of physical problems described by Markov chains and the many interesting questions peculiar to the various physical situations, a comprehensive treatment becomes voluminous. However, a certain appreciation of what the subject is about can be achieved by consideration of basic elementary questions. Given the initial probability vector $p(0)$ and the transition matrix $P$, let us consider the following:

1. What is the probability of $s_k$ at the $n$th time given that we start at $s_j$?
2. What is the unconditional probability of $s_k$ at the $n$th time?
3. What is the probability that the chain starting at state $j$ will eventually reach $k$? Note that this is not an empty question. The probability of reaching state $k$ from state $j$ in one step may be zero but nonzero for some finite number of steps.
4. What is the mean and variance of the number of steps required to reach state $k$ from state $j$?
5. What is the probability that the chain will reach a state and never leave that state? That is, what is the probability of the process terminating?
6. What is the mean number of steps required for the process to terminate?
7. What is the probability of returning to the starting state?
8. What is the mean number of steps required to return to the starting state?

General answers can be given to these questions in terms of $p(0)$ and $P$. More specific answers can be given for specific transition matrices since the answers depend upon the transition matrix. The study of the transition matrix has led to the classification of special types of Markov chains and the reader is encouraged to read further, e.g., Karlin (1966). We shall not give a detailed classification of Markov chains but shall consider briefly the questions raised.

### 7.2.3 Higher Transition Probabilities

In the preceding sections we introduced the vector $p(0)$ defined by

$$p'(0) = [p_1(0), p_2(0), \ldots, p_r(0)]$$

where $p_i(0)$ is the probability of the $i$th state at the beginning of the process. The transition matrix was defined by

$$P = \begin{pmatrix} p_{11} & p_{12} & \cdots & p_{1r} \\ p_{21} & p_{22} & \cdots & p_{2r} \\ \cdots & \cdots & \cdots & \cdots \\ p_{r1} & p_{r2} & \cdots & p_{rr} \end{pmatrix}$$

where $p_{jk}$ gives the probability of going from state $j$ to state $k$ in one step. We might well ask for the probability of going from state $j$ to state $k$, not in one step, but in $n$ steps. Denote such a probability by $p_{jk}(n)$.

Let us evaluate $p_{jk}(n)$ for various values of $n$ in terms of the elements $p_{jk}$. Now

$$p_{jk}(1) = p_{jk}, \text{ the } j,k\text{th element of } P \tag{7.5}$$

$$\begin{aligned} p_{jk}(2) &= p_{j1}p_{1k} + p_{j2}p_{2k} + \cdots + p_{jr}p_{rk} \\ &= \textstyle\sum_i p_{ji}p_{ik}, \text{ the } j,k\text{th element of } P^2 \end{aligned} \tag{7.6}$$

$$\begin{aligned} p_{jk}(3) &= \textstyle\sum_u p_{ju}(2)p_{uk} \\ &= \textstyle\sum_u \sum_i p_{ji}p_{iu}p_{uk}, \text{ the } j,k\text{th element of } P^3 \end{aligned} \tag{7.7}$$

These are special cases of the Chapman-Kolmogorov equations which play an important role in the theory of Markov processes. In general, the matrix $P(n)$ given by

$$P(n) = \begin{pmatrix} p_{11}(n) & p_{12}(n) & \cdots & p_{1r}(n) \\ p_{21}(n) & p_{22}(n) & \cdots & p_{2r}(n) \\ \cdots & \cdots & \cdots & \cdots \\ p_{r1}(n) & p_{r2}(n) & \cdots & p_{rr}(n) \end{pmatrix} \tag{7.8}$$

satisfies the matrix equation:

$$P(n) = P^n \tag{7.9}$$

Note that the elements of $P(n)$ supply answers to the first question raised in the previous section.

The second question raised in the previous section concerns the probability that the system will reach a given state, say state $k$, on the $n$th step. Denote such a probability by $p_k(n)$. Now

$$\begin{aligned} p_k(n) &= p_1(0)p_{1k}(n) + p_2(0)p_{2k}(n) + \cdots + p_r(0)p_{rk}(n) \\ &= \textstyle\sum_i p_i(0)p_{ik}(n) \end{aligned} \tag{7.10}$$

The vector

$$p'(n) = [p_1(n), p_2(n), \ldots, p_r(n)] \tag{7.11}$$

is then given by

$$p'(n) = p'(0)P(n) = p'(0)P^n$$

or

$$p(n) = P'(n)p(0) = (P')^n p(0) \tag{7.12}$$

We have thus far answered only the first two questions. In the remaining sections, we shall consider the others.

### 7.2.4 First Passage Times

We now examine questions 3, 4, and 5. We are interested in the time (number of steps) required to pass from state $j$ to state $k$ given that the system was initially in state $j$. Let $f_{jk}(n)$ be the probability of the first passage from $j$ to $k$ at step $n$. That is, let

$$f_{jk}(n) = \text{Prob}\left(\begin{array}{l}\text{system reaches state } k \text{ for}\\ \text{first time on step } n\end{array}\middle|\begin{array}{l}\text{system was initially}\\ \text{in state } j\end{array}\right) \tag{7.13}$$

To illustrate the meaning of $f_{jk}(n)$ without developing additional notation or theory, consider a Markov chain with only 2 states $s_0$ and $s_1$. What is $f_{00}(n)$? That is, what is the probability of returning to $s_0$ for the first time on the $n$th step? Clearly the first step must take us to $s_1$, the next $n-2$ steps must keep us in $s_1$, and the last step must return us to $s_0$. Then

$$f_{00}(n) = p_{01}p_{11}^{n-2}p_{10}, \text{ for } n \geq 2 \tag{7.14}$$

In a similar fashion the student should verify that

$$\begin{aligned}
f_{00}(1) &= p_{00}\\
f_{10}(n) &= p_{11}^{n-1}p_{10}\\
f_{01}(n) &= p_{00}^{n-1}p_{01}\\
f_{11}(1) &= p_{11}\\
f_{11}(n) &= p_{10}p_{00}^{n-2}p_{01},\ n \geq 2
\end{aligned} \tag{7.15}$$

Generally $f_{jk}(n)$ is referred to as the probability of a first passage from $j$ to $k$ on step $n$.

Let us consider the probability of a passage from $j$ to $k$. Let

$$f_{jk} = \text{Prob}\left(\begin{array}{l}\text{system reaches}\\ \text{state } k\end{array}\middle|\begin{array}{l}\text{system was initially}\\ \text{in state } j\end{array}\right) \tag{7.16}$$

The third question asked in the previous section concerned this quantity. The probability $f_{jk}$ is given by

$$f_{jk} = \sum_{n=1}^{\infty} f_{jk}(n) \tag{7.17}$$

However, it is obvious that for any particular transition matrix this may be a difficult number to find without additional theory and development.

The fourth question is easily answered if we have determined the first passage probabilities. The question is not meaningful, of course, unless $f_{jk} = 1$, i.e., unless the probability of ultimate passage is unity. In this event, the mean is given by

$$m_{jk} = \sum_{n=1}^{\infty} n f_{jk}(n) \tag{7.18}$$

and the variance by

$$\sigma_{jk}^2 = \sum_{n=1}^{\infty} (n - m_{jk})^2 f_{jk}(n) \tag{7.19}$$

The fifth question arises when there are one or more states such that the system will never leave them once it reaches them. Such states are called *absorbing states*. For instance, $k$ is an absorbing state if $p_{km} = 0$ for $m \neq k$. Then if $k$ is an absorbing state, $f_{jk}(n)$ is the probability of absorption into $k$ in $n$ steps, given that the system starts at $j$. The conditional probability of absorption into $k$ is $f_{jk}$ and the conditional probability of ultimate absorption is merely $\sum_k f_{jk}$, where the summation is over all the absorbing states.

The last three questions are merely special cases of preceding ones, and the answers have already been indicated.

### 7.2.5 Example

A small highly specialized industry builds one device each month. The devices are sold to three distributors and the total monthly demand is a random variable with the following distribution:

| Demand | 0 | 1 | 2 | 3 |
|---|---|---|---|---|
| $f(d)$ | 1/9 | 6/9 | 1/9 | 1/9 |

When the inventory level reaches 3, production is stopped until the inventory drops to 2. If the company is unable to meet the monthly demand, the distributors fill the remainder of the demand elsewhere. Let the states of the system be the inventory level. The transition matrix is found to be

$$P = \begin{array}{c} \\ 0 \\ 1 \\ 2 \\ 3 \end{array} \begin{array}{c} \begin{array}{cccc} 0 & 1 & 2 & 3 \end{array} \\ \begin{pmatrix} 8/9 & 1/9 & 0 & 0 \\ 2/9 & 6/9 & 1/9 & 0 \\ 1/9 & 1/9 & 6/9 & 1/9 \\ 1/9 & 1/9 & 6/9 & 1/9 \end{pmatrix} \end{array}$$

As we discussed earlier the elements of $P^n = P(n)$ give probabilities of proceeding from one state to another in exactly $n$ steps. Accordingly, let us evaluate $P^n$ for a few values of $n$.

$$P^2 = (1/81)\begin{pmatrix} 66 & 14 & 1 & 0 \\ 29 & 39 & 12 & 1 \\ 17 & 14 & 43 & 7 \\ 17 & 14 & 43 & 7 \end{pmatrix} \qquad P^3 = (1/729)\begin{pmatrix} 557 & 151 & 20 & 1 \\ 323 & 276 & 117 & 13 \\ 214 & 151 & 314 & 50 \\ 214 & 151 & 314 & 50 \end{pmatrix}$$

Suppose that we have started with 0 inventory. The probabilities of reaching the various states in $n$ steps are given by $p'(n) = p'(0)P(n)$. That is,

$$p'(1) = (1, 0, 0, 0)\begin{pmatrix} 8/9 & 1/9 & 0 & 0 \\ 2/9 & 6/9 & 1/9 & 0 \\ 1/9 & 1/9 & 6/9 & 1/9 \\ 1/9 & 1/9 & 6/9 & 1/9 \end{pmatrix} = (8/9, 1/9, 0, 0)$$

$$p'(2) = (66/81, 14/81, 1/81, 0/81)$$
$$p'(3) = (557/729, 151/729, 20/729, 1/729)$$

Inspection of these vectors reveals what one would expect. The probability of 0 inventory is declining at each stage and the probability is increasing for each of the other levels.

We can evaluate $P^n$ for any $n$ desired. It is also of interest to ask what happens as $n$ tends to infinity. The transition matrix here is an example of a positively regular Markov matrix, one in which all the elements of $P^n$ are positive ($>0$) for some $n$. In this example we see this condition is satisfied for $n = 3$. If $P$ is a Markov matrix, it can be shown that $P^n$ approaches a probability matrix $A$, where each row of $A$ is the same probability vector $\alpha'$. The vector $\alpha'$ is the unique probability vector such that $\alpha'P = \alpha'$, or $P'\alpha = \alpha$. To find $\alpha$, we must solve the equations

$$\begin{pmatrix} 8/9 & 2/9 & 1/9 & 1/9 \\ 1/9 & 6/9 & 1/9 & 1/9 \\ 0 & 1/9 & 6/9 & 6/9 \\ 0 & 0 & 1/9 & 1/9 \end{pmatrix}\begin{pmatrix} \alpha_1 \\ \alpha_2 \\ \alpha_3 \\ \alpha_4 \end{pmatrix} = \begin{pmatrix} \alpha_1 \\ \alpha_2 \\ \alpha_3 \\ \alpha_4 \end{pmatrix}$$

with the requirement $0 \le \alpha_i$, $i = 1, 2, 3, 4$, and $\sum \alpha_i = 1$. The unique solution to these equations is

$$\alpha' = (45/72, 18/72, 8/72, 1/72)$$

The limit matrix A is then

$$A = \begin{pmatrix} 45/72 & 18/72 & 8/72 & 1/72 \\ 45/72 & 18/72 & 8/72 & 1/72 \\ 45/72 & 18/72 & 8/72 & 1/72 \\ 45/72 & 18/72 & 8/72 & 1/72 \end{pmatrix}$$

and the limit of $p'(n)$ is (45/72, 18/72, 8/72, 1/72). Thus if the initial inventory level is zero, the system approaches the steady state where the probability of 0 inventory is 5/8. It may be of interest to compute the mean and variance of the steady state probability distribution.

The study of Markov chains can be developed along two lines: classification on the basis of properties of the transition matrix and classification on the basis of areas of application. Classification on the basis of properties of the matrix leads to a whole new vocabulary of special terms.

## 7.3 The Poisson Process

### 7.3.1 Introduction

The Poisson process is a model which describes a number of physical processes. Suppose that we are observing events of a particular type and are counting the events which have occurred since the start, $t = 0$. Let $N(t)$ be the number of events by time $t$. The process is said to be a Poisson process if the following conditions are satisfied for

$$t_0 < t_1 < t_2 < \ldots$$

1. $N(t_1) - N(t_0)$, $N(t_2) - N(t_1)$, $N(t_3) - N(t_2)$, etc., are independent; i.e., the number of events in any interval is independent of the number in any other interval.
2. $N(t_i) - N(t_j)$ is a Poisson random variable with mean $\lambda(t_i - t_j)$, i.e., the number of events occurring between $t_i$ and $t_j$ is a Poisson random variable with mean proportional to the difference $t_i - t_j$.

The Poisson process is the simplest Markov process (not Markov chain). It is Markovian because $P[N(t + \Delta t) = k_1 | N(t) = k_2]$ is the same regardless of the history prior to time $t$.

Note that if $N(t)$ and $M(t)$ arise from independent Poisson processes with parameters $\lambda$ and $\nu$, then $N(t) + M(t)$ is from a Poisson process with parameter $\lambda + \nu$.

One of the most common applications of a Poisson process is to the process of radioactive decay. It is assumed that all nuclei of a given element behave independently and have the same probability of decaying in a unit of time. The Poisson process is used widely in the theory of reliability of physical systems. A common assumption is that the number of failures of a system of $n$ components in an interval of time is distributed as a Poisson variable with mean proportional to the length of the interval. This assumption does not describe items with fatigue characteristics.

We have described the Poisson process in terms of counting the number of events observed in an interval of time. Actually, the reference system need not be time but can be, and often is, space. For instance, it is often assumed in statistical mechanics that the number of particles in a given region is a Poisson random variable with mean proportional to the volume of the region.

### 7.3.2 Derivation of Poisson Distribution

Probability density functions are often used to describe particular types of data because experience has so indicated. This is a valid reason. We have emphasized that probability density functions are only models to describe physical populations and at best are often only approximate. Thus the assumption that a Poisson density function will adequately describe a certain type of data is often based upon its use to describe similar data in the past. However, it should be recognized that the Poisson distribution can be derived from elementary "first principles"; i.e., it can be derived from elementary assumptions about the probability of certain events. One such derivation is given in this section.

Consider $X$, the number of times some event may occur in a time interval of length $\Delta t$. It is assumed that:

1. The probability that the event will occur exactly once is $\lambda\Delta t + o(\Delta t)$.
2. The probability that the event will occur more than once is $o(\Delta t)$. We use here the notation that $g(h)$ is $o(h)$ if the limit of $g(h)/h$ as $h \to 0$ is 0. For example, if

$$\text{Prob}(X > 1) = \lambda(\Delta t)^2$$

$$\lim_{\Delta t \to 0} \frac{\text{Prob}(X > 1)}{\Delta t} = \lim_{\Delta t \to 0} \lambda\Delta t = 0$$

3. The occurrence of the event in one interval is independent of the occurrence in another interval.

Let $P_u(t)$ denote the probability of $u$ events in the interval $(0, t)$. The event may occur $x$ times in the interval $(0, t + \Delta t)$ in the three mutually exclusive ways with probabilities as follows:

1. $x$ times in $(0, t)$ and 0 times in $(t, t + \Delta t)$: $P_x(t)(1 - \lambda\Delta t) + o(\Delta t)$.
2. $x - 1$ times in $(0, t)$ and once in $(t, t + \Delta t)$: $P_{x-1}(t)\lambda\Delta t + o(\Delta t)$.
3. $x - a$ times in $(0, t)$ and $a$ times in $(t, t + \Delta t)$: $o(\Delta t)$.

According to the addition rule, the probability of $x$ occurrences is given by

$$
\begin{aligned}
P_x(t + \Delta t) &= P_x(t)(1 - \lambda\Delta t) + P_{x-1}(t)\lambda\Delta t + o(\Delta t) \\
x &= 1, 2, \ldots \\
P_0(t + \Delta t) &= P_0(t)(1 - \lambda\Delta t) + o(\Delta t)
\end{aligned}
\tag{7.20}
$$

Then

$$\frac{P_x(t + \Delta t) - P_x(t)}{\Delta t} = \lambda[P_{x-1}(t) - P_x(t)] + \frac{o(\Delta t)}{\Delta t} \tag{7.21}$$

and

$$\frac{P_0(t + \Delta t) - P_0(t)}{\Delta t} = -\lambda P_0(t) + \frac{o(\Delta t)}{\Delta t} \tag{7.22}$$

Letting $\Delta t \to 0$, we obtain the linear differential equations

$$
\begin{aligned}
\frac{dP_x(t)}{dt} &= \lambda[P_{x-1}(t) - P_x(t)], \qquad x = 1, 2, \ldots \\
\frac{dP_0(t)}{dt} &= -\lambda P_0(t)
\end{aligned}
\tag{7.23}
$$

We must solve these differential equations with the boundary conditions that $P_0(0) = 1$ and $P_x(0) = 0$ for $x \geq 1$.

Since

$$
\begin{aligned}
\frac{dP_0(t)}{P_0(t)} &= -\lambda dt \\
\log P_0(t) &= -\lambda t
\end{aligned}
$$

and

$$P_0(t) = e^{-\lambda t} \tag{7.24}$$

Consider the linear equation in $P_1(t)$:

$$\frac{dP_1(t)}{dt} + \lambda P_1(t) = \lambda P_0(t)$$

The integrating factor is $e^{\lambda t}$. Multiplying by this, we have

$$e^{\lambda t}\frac{dP_1(t)}{dt} + \lambda e^{\lambda t} P_1(t) = \lambda$$

or

$$\frac{d}{dt}[P_1(t)e^{\lambda t}] = \lambda \tag{7.25}$$

Thus

$$P_1(t)e^{\lambda t} = \lambda t + C \tag{7.26}$$

Since $P_1(0) = 0$, $C = 0$. Therefore $P_1(t) = (\lambda t)e^{-\lambda t}$. Proceeding in this fashion, we can show that

$$P_x(t) = \frac{(\lambda t)^x e^{-\lambda t}}{x!} \tag{7.27}$$

### 7.3.3 Arrival Times

Suppose that we are observing a Poisson process with parameter $\lambda$ and note the times $T_1 < T_2 < T_3 < \ldots$, etc., at which the events occur. We call these times arrival times. The times between arrivals, $T_2 - T_1$, $T_3 - T_2$, $T_4 - T_3, \ldots$, we shall call interarrival times. That the distribution of $T_n$ is a two-parameter gamma distribution can be seen as follows. The cumulative density function of $T_n$ is given by

$$\begin{aligned} F_{T_n}(t) &= \text{Prob}(T_n \le t) \\ &= \text{Prob}[N(t) \ge n] \\ &= \sum_{j=n}^{\infty} \frac{e^{-\lambda t}(\lambda t)^j}{j!} \\ &= 1 - \sum_{j=0}^{n-1} \frac{e^{-\lambda t}(\lambda t)^j}{j!} \end{aligned} \tag{7.28}$$

Then the density function of $T_n$ is given by differentiating the c.d.f.:

$$\begin{aligned} \frac{dF(t)}{dt} &= -\sum_{j=1}^{n-1} \frac{e^{-\lambda t}(\lambda t)^{j-1} j\lambda}{j!} + \sum_{j=0}^{n-1} \frac{\lambda e^{-\lambda t}(\lambda t)^j}{j!} \\ &= \sum_{j=0}^{n-1} \frac{\lambda e^{-\lambda t}(\lambda t)^j}{j!} - \sum_{j=1}^{n-1} \frac{\lambda e^{-\lambda t}(\lambda t)^{j-1}}{(j-1)!} \\ &= \frac{\lambda e^{-\lambda t}(\lambda t)^{n-1}}{(n-1)!} \end{aligned} \tag{7.29}$$

Thus

$$f_{T_n}(t) = \frac{\lambda^n t^{n-1} e^{-\lambda t}}{(n-1)!}, \qquad t > 0$$
$$= 0, \text{ otherwise} \tag{7.30}$$

Of special interest is the density function of $T_1$, the time to the first arrival. Letting $n = 1$ in the above density function, we obtain:

$$f_{T_1}(t) = \lambda e^{-\lambda t}, \qquad t > 0$$
$$= 0, \text{ otherwise} \tag{7.31}$$

The average time to first arrival is the mean of this distribution, which is $1/\lambda$. This particular distribution is called the negative exponential, and since it follows from the Poisson process, the assumption of a Poisson process is occasionally referred to as the exponential assumption. Suppose that we have a large population of components described by the Poisson process. Then the distribution of times to failure in this population is described by the negative exponential distribution.

It can also be shown that the interarrival times $T_2 - T_1$, $T_3 - T_2$, etc., are independent random variables, each with the density function

$$f(t) = \lambda e^{-\lambda t}, \qquad t > 0$$
$$= 0, \text{ otherwise}$$

In summary, the Poisson process is such that the number of arrivals in a given time interval is a Poisson random variable, while the arrival times $T_1, T_2, \ldots, T_n$ are gamma random variables. Equivalently, the interarrival times are independently distributed negative exponential random variables.

### 7.3.4 Life Testing

Extensive life tests are conducted in many industries especially on electronic components such as transistors, capacitors, resistors, etc., to determine the failure pattern. Quite frequently the Poisson process is assumed to describe the failure of these components. It is desired to study the parameter $\lambda$ of the Poisson process by life testing a sample of $n$ components. The two most common procedures are:

1. Testing $n$ units until a number $r$, specified in advance, fail. This may be done with or without replacement of failures as they occur.
2. Testing $n$ units for a time $T^*$, specified in advance. Again this may be done with or without replacement of failures as they occur.

In most life testing situations, attention is focused on the parameters $\lambda$ and $1/\lambda$. The parameter $1/\lambda$ is the mean time to failure of the entire population of components. It is also the mean time between failures. The parameter $\lambda$ is called the failure rate. To understand this, consider the probability of a failure in the interval $\Delta t$ given that we have reached time $t$. This probability is given by

$$\frac{\int_t^{t+\Delta t} f(t)\,dt}{\int_t^{\infty} f(t)\,dt} = 1 - e^{-\lambda \Delta t} \tag{7.32}$$

But since

$$1 - e^{-\lambda\Delta t} = 1 - \sum_{j=0}^{\infty} \frac{(-\lambda\Delta t)^j}{j!}$$

$$= \lambda\Delta t - \frac{\lambda^2(\Delta t)^2}{2!} + \frac{\lambda^3(\Delta t)^3}{3!} - \cdots \tag{7.33}$$

for small $\Delta t$, $1 - e^{-\lambda\Delta t}$ is given approximately by $\lambda\Delta t$. This is, of course, not surprising since this was one of the assumptions in the derivation of the Poisson process. The name failure rate is thus motivated by the fact that the probability of failure is given approximately by a constant (the rate) times the length of the interval. A great deal of criticism has been made of the Poisson model for life testing because it leads to a constant failure rate.

We consider the first procedure with failures immediately replaced. This is equivalent to observing a Poisson process with parameter $n\lambda$ because there are always $n$ units on test. Denote by

$$X_1 < X_2 < X_3 < \ldots < X_r$$

the times at which the failures occurred and denote the total time logged by operative devices by $T$. Since there are $n$ operative devices at all times, $T = nX_r$. From Sect. 7.3.3 we know that $X_r$ is distributed as a gamma variable with density

$$f_{X_r}(x) = \frac{(n\lambda)^r x^{r-1} e^{-n\lambda x}}{(r-1)!}, \qquad x > 0 \tag{7.34}$$

$$= 0, \text{ otherwise}$$

Consider $U = 2\lambda T = 2\lambda n X_r$. The density of $U$ is given by

$$f(u) = \frac{(n\lambda)^r (u/2\lambda n)^{r-1} e^{-u/2}}{(r-1)!} \cdot \frac{1}{2\lambda n}$$

$$= \frac{u^{r-1} e^{-u/2}}{2^r \Gamma(r)}, \qquad u > 0 \tag{7.35}$$

Thus $2\lambda T$ is a $\chi^2$ variable with $2r$ degrees of freedom. It is this fact which is used for estimation of parameters, forming opinions about reasonable values of $\lambda$, etc.

Now suppose that failures are not replaced. The total time logged by the devices which fail is $\sum_1^r X_i$. The remaining $n - r$ devices are operative for the entire period, so they accumulate a time of $(n - r)X_r$. The total time accumulated by the units on test is then

$$T = \sum_1^r X_i + (n - r)X_r \tag{7.36}$$

In this case it can also be shown that $2\lambda T$ has the $\chi^2$ distribution with $2r$ degrees of freedom. Consider the joint distribution of $X_1, X_2, \ldots, X_r$. This is an ordered sample by virtue of the fact that the smallest observation occurs first, the second smallest second, etc. From the material in Chapter 6 the probability density is seen to be

$$f(x_1, x_2, \ldots, x_r; \lambda) = \frac{n!}{(n-r)!} \lambda^r e^{-\lambda[\Sigma_1^r x_i + (n-r)x_r]} \tag{7.37}$$

for $0 < x_1 < x_2 < \ldots x_n < \infty$, and 0, otherwise.

We can then show that $2\lambda T$ is a $\chi^2$ variable with $2r$ degrees of freedom by using the m.g.f., $m_{2\lambda T}(\theta)$. We have

$$\begin{aligned} m_{2\lambda T}(\theta) &= E(e^{2\lambda T\theta}) \\ &= \int \cdots \int \frac{n!}{(n-r)!} \lambda^r e^{-\lambda t} e^{2\lambda t\theta}\, dx_1 \ldots dx_r \end{aligned}$$

Putting $\lambda(1 - 2\theta) = v$, we get

$$\begin{aligned} m_{2\lambda T}(\theta) &= (1 - 2\theta)^{-r} \int \cdots \int f(x_1, x_2, \ldots, x_r; v)\, dx_1 \ldots dx_r \\ &= (1 - 2\theta)^{-r} \end{aligned} \tag{7.38}$$

Thus, $2\lambda T$ is distributed as $\chi^2$ with $2r$ degrees of freedom regardless of whether failures are replaced or not.

Let us briefly review a few facts about the $\chi^2$ distribution. The mean and variance of a $\chi^2$ variable with $v$ degrees of freedom are $v$ and $2v$. Then

$$E(2\lambda T) = 2r \text{ and } E(T/r) = 1/\lambda \tag{7.39}$$

The variance of $2\lambda T$ is $4r$, so that

$$\operatorname{Var}(T/r) = 1/r\lambda^2 \tag{7.40}$$

To summarize, the mean time to failure may be estimated by dividing the total time accumulated by the number of failures. The variance of this estimate is $1/r\lambda^2$.

The mean and variance of the reciprocal of a $\chi^2$ variable are

$$E(1/\chi^2) = 1/(v - 2), \qquad \mathrm{Var}(1/\chi^2) = 2/[(v - 2)^2(v - 4)] \tag{7.41}$$

Hence,

$$E(r/T) = r\lambda/(r - 1)$$
$$\mathrm{Var}(r/T) = \lambda^2 r^2/(r - 1)^2(r - 2)$$

and

$$E[(r - 1)/T] = \lambda$$
$$\mathrm{Var}[(r - 1)/T] = \lambda^2/(r - 2) \tag{7.42}$$

This suggests that we use $(r - 1)/T$ as a guess (or estimate) of the value of $\lambda$.

With the second procedure and replacement, we are observing a Poisson process with parameter $\lambda T^*$. Then the probability density function of number of failures observed is

$$f(r) = \frac{(\lambda T^*)^r e^{-\lambda T^*}}{r!} \tag{7.43}$$

Questions of inference about $\lambda$ are then merely questions about the parameter of a Poisson density function. In case the failures are not replaced as they occur, we have a simple binomial situation, $\mathrm{Bi}(n, p)$, with $p = 1 - e^{-\lambda T}$.

## 7.4 The Weibull Process

The Poisson process follows as a consequence of elementary assumptions about the probability of failure. There is a well-known probability density function used in the study of reliability known as the Weibull density function. We now show that it is possible to obtain the Weibull density from a simple set of assumptions. Instead of assuming that the probability of a failure in the interval $(t, t + \Delta t)$ is approximately proportional to $\Delta t$, let us assume that

$$\mathrm{Prob}[\text{failure in } (t, t + \Delta t)] = (\delta t^{\delta - 1}/\theta)\Delta t + o(\Delta t) \tag{7.44}$$

Neglecting terms of higher order than $\Delta t$, we use

$$\mathrm{Prob}[\text{one failure in } (t, t + \Delta t)] = (\delta t^{\delta - 1}/\theta)\Delta t$$
$$\mathrm{Prob}[\text{no failure in } (t, t + \Delta t)] = 1 - (\delta t^{\delta - 1}/\theta)\Delta t$$
$$\mathrm{Prob}[\text{two or more failures in } (t, t + \Delta t)] = o(\Delta t) \tag{7.45}$$

Proceeding as in the derivation of the Poisson distribution, we obtain

$$\begin{aligned}\text{Prob}[r \text{ failures in } (0, t + \Delta t)] &\\ &= P_r(t + \Delta t)\\ &= P_r(t)[1 - (\delta t^{\delta-1}/\theta)\Delta t] + P_{r-1}(t)[(\delta t^{\delta-1}/\theta)\Delta t] \qquad (7.46)\end{aligned}$$

Hence

$$\lim_{\Delta t \to 0} \frac{P_r(t + \Delta t) - P_r(t)}{\Delta t} = \frac{dP_r(t)}{dt}$$
$$= (\delta/\theta)t^{\delta-1}[P_{r-1}(t) - P_r(t)] \qquad (7.47)$$

Supplementing these differential equations with the initial conditions that

$$P_r(0) = 0,\ r \geq 1,\ P_0(0) = 1 \qquad (7.48)$$

we obtain

$$P_r(t) = \frac{(t^\delta/\theta)^r \exp(-t^\delta/\theta)}{r!} \qquad (7.49)$$

The distribution of times to failure is obtained by considering the cumulative distribution function $F(t)$:

$$\begin{aligned}F(t) &= 1 - \text{Prob(no failure before } t)\\ &= 1 - P_0(t) = 1 - \exp(-t^\delta/\theta)\end{aligned}$$

Then

$$f(t) = (\delta/\theta)t^{\delta-1}\exp(-t^\delta/\theta) \qquad (7.50)$$

This density function is known as the Weibull density function.

## 7.5 Characterization of the Variability of Infinite Sequences

As we have said, a stochastic process consists of an infinite collection of random variables. If the infinity is denumerable, the process is an infinite sequence of random variables, $X_1, X_2, \ldots$, and the underlying population is therefore a population of infinite sequences. In Chapter 2 we discussed ways of characterizing the variability of a population of scalars, of ordered pairs, and of ordered $n$-tuples. An obvious problem with stochastic processes is to develop useful ways of characterizing the variability of a population of infinite sequences. It is desirable to develop summary population statistics which will give a partial description of the whole population.

In earlier sections of this chapter the attention was primarily on properties of limiting features of the sequences, as for instance the probability that $X_\infty$ is equal to zero. Our problem now is to think about ways of describing the whole of the sequences. We use "time" to denote the indexing variable.

Certain summary statistics of the population are highly natural and come to mind immediately. We would certainly like to know $E(X_t) = \mu_t$ as a function of $t$. To characterize the variability, we would like to know $E[X_t - E(X_t)]^2 = \sigma_t^2$ or $\sigma_{tt}$. Clearly we would also like to know the covariance, which for times $t_1$ and $t_2$ is

$$E\{[X_{t_1} - E(X_{t_1})][X_{t_2} - E(X_{t_2})]\} = \sigma_{t_1 t_2} \tag{7.51}$$

This will be in general a function of two variables $t_1$ and $t_2$. If we had $N$ realizations of the sequence for times $t = 1, 2, \ldots, T$, we could certainly obtain an idea of the population attributes by computing the quantities:

$$\bar{X}_t = \text{Ave}(X_t)$$

$$s_{tt} = s_t^2 = \frac{1}{N-1} \sum (X_t - \bar{X}_t)^2$$

$$s_{t_1 t_2} = \frac{1}{N-1} \sum [(X_{t_1} - \bar{X}_{t_1})(X_{t_2} - \bar{X}_{t_2})] \tag{7.52}$$

We could also define the correlation

$$\rho(t_1, t_2) = \sigma_{t_1 t_2} / \sqrt{\sigma_{t_1 t_1} \sigma_{t_2 t_2}}$$

and get an idea of the population value by obtaining

$$r(t_1, t_2) = s_{t_1 t_2} / \sqrt{s_{t_1 t_1} s_{t_2 t_2}} \tag{7.53}$$

We can consider the joint variability for any three times $t_1$, $t_2$, and $t_3$ and indeed for all the observed times $1, 2, \ldots, T$.

If our stochastic process consists of the realizations of a physical process which can be repeated, we can obtain data to use for the type of computation indicated above. Consider the following elementary realizable example, which arises in a genetic model.

Suppose we have a supply of $4N$ black balls ($\theta$'s), $4N$ white balls ($\phi$'s), and two urns. We put $N$ $\theta$'s and $N$ $\phi$'s in urn I. Initially, urn II is empty. We shake and draw a ball from urn I. If it is a $\theta$, we put a $\theta$ in urn II; if it is a $\phi$, we put a $\phi$ in urn II. We replace the selected ball in urn I, shake, and draw again. We do this $2N$ times, using the same rule for putting balls in urn II. The result of this is that urn II contains, say, $X_1$ $\theta$'s and $(2N - X_1)$ $\phi$'s. Now use urn II as the original urn, empty urn I and sample from urn II to determine the contents of urn I. As a result, urn I will contain $X_2$ $\theta$'s and $(2N - X_2)$ $\phi$'s. Continue this for 100 times, say.

This will give a sequence $X_{11}, X_{12}, \ldots, X_{1,100}$. We can certainly start the process again and again and obtain $N$ sequences of length 100:

$$X_{21}, X_{22}, \ldots, X_{2,100}$$

$$X_{31}, X_{32}, \ldots, X_{3,100}$$

$$\cdots\cdots\cdots\cdots\cdots\cdots$$

$$X_{N1}, X_{N2}, \ldots, X_{N,100}$$

We can then obtain a partial idea of the nature of the population of sequences. The results of this process are in fact deducible if the sampling is random, such that the probability of obtaining a $\theta$ from the urn containing $N_1$ $\theta$'s and $(2N - N_1)$ $\phi$'s is $N_1/2N$.

Most of the real-world situations to which we wish to apply the above ideas are not repeatable physical processes. Our data may consist, for example, of the number of unemployed individuals in the United States for each week of the past 10 years (defining an unemployed individual for a week as one who had no employment in that week). The result is that we have one finite sequence of length 520, and we wish to regard this finite sequence as a section of a random infinite sequence. A more reasonable example, perhaps, is that we have for each of the past 100 years the number of days for which the maximum temperature was below freezing point. The problem is to develop a reasonable probabilistic model for the sequence of observations.

The simplest interesting model is to suppose that the underlying conceptual population of infinite sequences is stationary in some sense. There are various definitions of stationarity. If we have $E(X_t) = \mu$ independent of $t$, the process is mean stationary, speaking intuitively. If $\sigma_{tt}$ is equal to $\sigma^2$ independent of $t$, the process is variance stationary. If the covariance $C(t_1, t_2)$ is dependent only on the difference $t_2 - t_1$, the process is covariance stationary. If the joint distribution of $X_{t_1}, X_{t_2}, X_{t_3}, \ldots, t_1 < t_2 < t_3 < \ldots$ depends not upon the $t_i$, but only on $t_1, t_2 - t_1, t_3 - t_2, t_4 - t_3 \ldots$, the process is called stationary, and the resulting time series a stationary time series.

To apply the above ideas to a single segment of a time series, it is useful to make what is called an ergodic assumption. This is a highly technical matter. The intuitive idea is that an average over possible sequences at any point in time is equal to the average over time for any individual sequence. If we let $i$ index possible sequences so that a sequence is

$$X(i, t), \qquad t = \ldots, -2, -1, 0, 1, 2, \ldots$$

then the ergodic assumption gives

$$\underset{\substack{i \\ t \text{ fixed}}}{\text{Ave}} X(i, t) = \lim_{\substack{T \to \infty \\ i \text{ fixed}}} \frac{1}{2T} \sum_{t=-T}^{T} X(i, t) \qquad (7.54)$$

This is an extremely strong assumption, involving what may be termed no steady growth. In applications to meteorological observations it is equivalent to stating that over the long run nothing changes, which may be quite reasonable, even though climate has clearly changed over the very long run. In any application to economic time series, utility of the assumption seems most questionable in that the complex of economic forces has obviously changed markedly, even in a period of 25 years.

If one makes the assumptions of mean, variance, and covariance stationarity, the possible observations can be scaled to have mean zero and variance unity. The covariance function then becomes the autocorrelation function, say $R(t_1, t_2) = R(|t_2 - t_1|)$ with $R(0)$ equal to unity.

The elementary mathematical functions that are stationary in a natural sense are the cosine and sine functions. It is natural, therefore, to consider correlating $X(t)$ with $\cos \alpha t$ and $\sin \alpha t$ for various values of $\alpha$. This leads to the idea that a certain portion of the variance is associated with the frequency $\alpha$. Furthermore, the proportion of variance is found to depend only on the correlation function $R(|t_1 - t_2|)$. The decomposition of the variance of a stationary time series by means of a whole spectrum of frequencies is called spectrum analysis. This leads to the representation

$$\sigma^2 = \int_{-\infty}^{\infty} \Gamma(f)\, df \qquad (7.55)$$

in which $df$ is an element of frequency and $\Gamma(f)$ is the contribution to variance from this frequency. The mathematics and statistics of this matter are both very complicated, and the student may read with great advantage the relevant chapters of Kendall and Stuart (1966), the book of Jenkins and Watts (1968), and the early book of Blackman and Tukey (1959).

Jenkins and Watts describe spectrum analysis as nonparametric (or alternatively, as being the use of a model with an infinite number of parameters). The general topic is highly developed in physics and engineering, being used for development of understanding of a very wide variety of physical processes. The statistical problems are very complex due to three facts: Only a finite segment of a stationary series is available; the number of parameters is infinite, so complete disentanglement is impossible; and there is a limit to the accuracy of the record, whether it is treated by an analogue device or by a digital computer.

In contrast to the nonparametric nature of spectrum analysis for the treatment of stationary series is the use of autoregressive structures for interpretation of a discrete record. We give merely the barest ideas of this approach for an equally spaced record. The general form of such models is given by Jenkins and Watts to be

$$\begin{aligned} X(t) - \mu = {} & \alpha_1[X(t-1) - \mu] + \alpha_2[X(t-2) - \mu] \\ & + \cdots + \alpha_m[X(t-m) - \mu] + Z(t) + \beta_1 Z(t-1) \\ & + \cdots + \beta_r Z(t-r) \end{aligned} \tag{7.56}$$

in which the successive members of the sequence $X(t)$ are explained in terms of previous members, $X(t-1)$, $X(t-2)$, . . . , by a linear model with errors arising at each point of time $Z(t)$, $Z(t-1)$, . . . . For simplicity, these errors may be regarded as being normal and independent with mean zero and some variance $\sigma^2$. The above model involves $v = 1 + m + r$ parameters and leads to a joint distribution of the observations. This type of model is no different in *basic* nature from any of the models of previous chapters, though the distribution theory is very complex. This type of model leads also to a spectrum analysis, the true one being calculable. The model has several features which are advantageous for some purposes: the finiteness of number of parameters; a highly natural sequence of potential models by increasing $m$ or $r$ by unity, with the possibility of examining objectively the goodness of fit of successive models; and a nice prediction process. It is also possible to introduce nonstationarity of the mean by hypothesizing that the "true" mean at time $t$, $\mu(t)$ depends on $t$. An obvious candidate for this function is to suppose that it is also given by an "autoregressive" scheme, so that

$$\mu(t) = \gamma_1\mu(t-1) + \gamma_2\mu(t-2) + \cdots + \gamma_s\mu(t-s) \tag{7.57}$$

Those familiar with difference equations will recall that this has the consequence that $\mu(t)$ is a linear mixture of exponentials like $e^{\lambda t}$ and $t^j e^{\lambda t}$ with a finite set of $\lambda$'s. Such a model for the mean function is highly appealing in many situations. This type of approach is developed very extensively by Box and Jenkins (1970). Our reason for mentioning this autoregressive approach is that it is susceptible, though not necessarily at all easily, to the general inferential processes described in following chapters. This appears to be somewhat in contrast to spectrum analysis, which involves choices by the data analyzer that are much less transparent with regard to their consequences. The statistical interpretation of time series is a critical aspect of the development of understanding and explanation of any economic unit.

## PROBLEMS

1. For the transition matrices:

$$\begin{pmatrix} 0 & 1 \\ 1 & 0 \end{pmatrix}, \begin{pmatrix} 1/2 & 1/2 \\ 1/2 & 1/2 \end{pmatrix}$$

$$\begin{pmatrix} 1/3 & 2/3 & 0 \\ 0 & 1 & 0 \\ 1/2 & 0 & 1/2 \end{pmatrix}, \begin{pmatrix} 1 & 0 & 0 \\ 0 & 1 & 0 \\ 0 & 1/2 & 1/2 \end{pmatrix}$$

find $P(1)$, $P(2)$, and $P(3)$.

2. For the following transition matrix and vector of initial probabilities find $P(2)$, $P(3)$, $P(4)$ and the vectors $p(1)$, $p(2)$, $p(3)$, $p(4)$:

$$P = \begin{pmatrix} 1/6 & 2/3 & 1/6 \\ 0 & 1/3 & 2/3 \\ 0 & 0 & 1 \end{pmatrix}, p(0) = \begin{pmatrix} 1/3 \\ 2/3 \\ 0 \end{pmatrix}$$

3. In the preceding problem state 3 is an absorbing state in that once the system reaches state 3, it never leaves. What is the probability that the system will be absorbed by state 3 on the fourth step?

4. Given the following transition matrix and vector of initial probabilities, find $P(2)$, $P(3)$, $P(4)$, $p(2)$, $p(3)$, $p(4)$ and $F(2)$, $F(3)$, $F(4)$, where $F(n)$ is the matrix of probabilities $f_{jk}(n)$:

$$P = \begin{pmatrix} 1/2 & 1/2 & 0 \\ 0 & 1/2 & 1/2 \\ 1/2 & 0 & 1/2 \end{pmatrix}, p(0) = \begin{pmatrix} 1/6 \\ 2/3 \\ 1/6 \end{pmatrix}$$

5. For the transition matrix of the diffusion model let $n = 5$ and find $P(2)$, $P(3)$, $P(4)$, $p(2)$, $p(3)$, and $p(4)$ given that $p'(0) = (1/6, 1/6, 1/6, 1/6, 1/6, 1/6)$.

6. Using the differential equation in Sect. 7.3.2, verify that $P_2(t)$ is given by the Poisson probability with parameter $\lambda t$.

7. Use mathematical induction and the differential equations in Sect. 7.3.2 to show that $P_x(t)$ is a Poisson probability with parameter $\lambda t$ for $x = 0, 1, 2, \ldots$.

8. Show that the interarrival times $T_2 - T_1$, $T_3 - T_2$, etc., are independent random variables with the negative exponential distribution.

9. For the negative exponential distribution, verify that

$$\frac{\int_t^{t+\Delta t} f(x)\,dx}{\int_t^{\infty} f(x)\,dx} = 1 - e^{-(\Delta t)\lambda}$$

10. Express the median of the negative exponential distribution in terms of the mean $\lambda$.

11. Blood smears are taken from animals which have been exposed to radioactive material and are placed in a radioactive counter. Each smear is left in the counter until $r$ particles are counted, and the time required to count the particles $T$ is recorded. If we assume that particle emission is described by a Poisson process with mean time between emission of $\lambda$, what is the distribution of $2\lambda T$? What is the maximum probability of obtaining $T \geq 300$ for $\lambda < 2$ if $r = 50$?

12. Using the fact that an $F$ variable is the ratio of two independent $\chi^2$ variables divided by their degrees of freedom, what is the distribution of $T_1/T_2$, where $T_1$ and $T_2$ are two independent determinations of the time required to observe $r$ particles from the same source?

13. A researcher is worried about background radioactivity confusing the issue. He observes each slide twice until he observes 10 particles. If the background remains constant, the two times recorded should have the same $\lambda$. Using Problem 12, how reasonable is the idea that $\lambda_1 = \lambda_2$ if $r = 10$, $T_1 = 20$, $T_2 = 30$?

14. If radioactive counting is done in a different manner, the number of particles emitted may be a random variable. Suppose that the time to count is fixed at $T_0$, then the number of particles observed, $r$, is a Poisson random variable with parameter $\lambda T_0$. If several independent counts are made, the average count $\bar{r} = \sum_1^n r_i/n$ has mean $\lambda T_0$ and variance $\lambda T_0/n$. Assuming that $r$ is approximately normal, how large must $n$ be so that $\bar{r}$ will deviate from $\lambda T_0$ by less than 10% with probability .95?

15. An experimenter places 100 transistors on test until 6 fail without replacement of failures as they occur. The test results in the following data:

| Failure | Hours to Failure |
|---|---|
| 1 | 150 |
| 2 | 225 |
| 3 | 304 |
| 4 | 451 |
| 5 | 608 |
| 6 | 722 |

Assuming the failure mechanism to be a Poisson process, estimate $\lambda$ and $1/\lambda$ and state the variance of your estimates.

16. Rework Problem 15 assuming that the failures are replaced as they occur.

17. Suppose that the probability of an arrival in the interval $(t, t + \Delta t)$ is given by $\lambda_1 \Delta t + \lambda_2 t \Delta t + o(\Delta t)$. Derive the density function of arrival times and consequently of the number of arrivals in a fixed time interval.

18. Show that the joint density of the first $r$ ordered observations in a sample of size $n$ from the negative exponential is given by Equation (7.37).

19. Obtain the mean and variance of the reciprocal of a $\chi^2$ variable.

20. Use the differential equations in Sect. 7.4 to obtain $P_r(t)$.
21. Estimate the covariance function of the following data to detect the presence of any periodic components:

    1.6, 2.1, 0.4, 1.8, 2.9, 2.3, 1.7, −0.8, 2.9, 1.7, 2.3, 2.3, 1.5, 1.4, 1.8, 2.8, 1.2, 1.7, 2.2, 3.0, 3.3, 2.9, 1.9, 1.1, 2.9

    Ans. Signal: 1, −1, 1, −1, 1, −1, . . .
    Noise: Normal with mean 0, $\sigma = 2$
22. Estimate the covariance function from the following data to detect the presence of any periodic components:

    −1.4, 0.0, 0.7, −0.2, 3.8, 1.8, 1.0, −2.4, −2.7, −0.7, −2.1, 1.9, −0.2, −2.7, −0.2, 0.8, 0.9, −0.7, 2.6, −4.1

    Ans. Signal: 1, −1, 1, −1, 1, −1, . . .
    Noise: Normal with mean 0, $\sigma = 2$

# 8

# General Outline of Data Interpretation Problems

## 8.1 Introduction

The purpose of this short chapter is to give a general introduction to the wide variety of processes of interpreting data. The theory of probability is a mathematical discipline in its own right, with a wealth of conceptual ideas and axiomatic developments. Our prime interest in this book, however, is the use of ideas of probability in the assessment of data, what may be called loosely the making of sense of data and its use for many different purposes. It would be foolish to attempt to enumerate and classify *all* the analysis processes that are used. But we can make a rough partial classification, which will be of use in understanding the sequence of ideas of the following chapters.

## 8.2 General Ideas on Models

The whole of science and technology consists of the development of theories of observation—theories which relate observations, theories which lead to prediction of future observations, theories which suggest actions to produce desired consequences, etc.

The best way for the student to obtain an understanding of models and their use is to study the history of two or more disciplines of science. A book on physics such as the PSSC (Physical Science Study Committee) high school book exhibits with great clarity the sequence of models found useful in the development of understanding of physical systems, from the model of the ancient Greeks which attempted to explain phenomena in terms of four basic ideas—water, air, fire, and earth—through the beginnings of atomic theory by Dalton, to the present-day theories using protons, electrons, neutrons, mesons, etc. Each of the models received great favor for a period, sometimes extending over centuries. Each was displaced because it gave predictions which were not consonant with observations.

Each was an improvement on its predecessors in explaining both previously established observations *and* laws and some observations not explained before. Similarly, the field of biology exhibits a sequence of models from the essential uniqueness of classes of biological life to the idea of a continuous stream of origin of new types by evolution. Furthermore, the past century has seen the increasing use of ideas of physics and chemistry for improvement of the explanation of biological phenomena.

A striking feature of the development of models in both these sciences is the increasing use of ideas of probability to the point at which probability ideas are totally intrinsic. The basic ideas of structure of atoms and molecules are probabilistic, and insofar as any explanation is physical or chemical, the ultimate basis in our present understanding is probabilistic. The area of biology is interesting in that two quite distinct routes lead to probability—on the one hand via chemistry via physics and on the other hand via the Mendelian process which explains the inheritance of attributes of organisms by means of a probability mechanism, equivalent in the simplest case to a Bernoulli trial with probabilities of one-half for each of two alternative outcomes.

In the early days of any of the sciences, the models were totally deterministic in that a configuration of circumstances would produce a definite and certain outcome. The death of such models arose from the observation that configurations judged to be the same in all discernible characteristics did not give the same outcome. The simplest example is surely the phenomenon of biological reproduction in which the offspring produced by a male and a female have no basis for predictability, other than that the results of a mating are like independent tossings of a fair coin, i.e., like independent Bernoulli trials with probability (male) and probability (female) equal. It would be interesting to research the writings of the ancients to obtain their explanation of the irreproducibility of the phenomenon.

What lessons can we learn from such study of the history of the sciences? The dominant ones seem to be twofold. First, from a study of observations there came the development of laws, empirical laws which seemed to give essentially perfect relations between observables such as the universal gas law that $PV = RT$, where $P$, $V$, $T$ are respectively pressure, volume, and temperature and $R$ is a constant. Such laws are then taken to be truths which lead by mathematical deduction to the development of deterministic models and are accepted as reasonable bases for explaining phenomena. Explanation of this type consists of showing that observations are what would be expected by mathematical reasoning from accepted laws. Perhaps the best example is the whole of Newtonian mechanics based on Newton's three laws. Note, however, that these laws failed when pushed "to the wall" and were replaced by Einstein's theory of relativity. Second, it was found essential to modify deterministic laws by probability models which consist of likening processes to those of elementary chance mechanisms. The bridge between the types of model, deterministic and probabilistic, came by the use of laws of large numbers. While the result of a toss of a single fair coin is highly variable giving a head or a tail, a toss of 1000 fair coins will give a proportion of heads very close to 50%.

As examples of the pervasiveness of probability models we list the following:

1. The starting point of stellar dynamics, as presented by Chandrasekar (1960), is a probability model for what are called "residual velocities." This law is called Schwarzchild's law of ellipsoidal distribution of velocities and is in fact a trivariate normal distribution with independence.
2. A basis for statistical mechanics is a similar type of probability distribution.
3. A basis for quantum mechanics lies in the theory of stochastic processes, probability transition matrices, etc., and the solution of the Schrodinger wave equation gives probabilities.
4. We have already referred to the essential probabilistic model of Mendelian inheritance.
5. Modern theories of economics incorporate probabilistic elements, arising from physical happenings (e.g., weather) for which probability descriptions are used or from other types of force which are representable only by probability models.
6. In sociology an intrinsic aspect of the formation of opinion is the transmission of views, data, and opinions between members of the society. It is impossible to describe generally such a process except in probability terms.
7. In psychology the reaction of an individual to a stimulus (e.g., a test battery or a physical force) is not perfectly reproducible, and the variability must be incorporated in the overall picture by a probability model.

An essential aspect of the use of probability models is that they arise by data inspection and introspection. The models are used to represent partial ignorance. Therefore they obviously must not conflict with well-established observational regularities. In general the use of probability models is a "bootstrap" operation.

A probability model will be found not consonant with data, and a new probability model will be developed. In genetics, for instance, Mendel proposed an independence model, but this was rejected on the basis of observations. The occurrence of nonindependence led to the idea of linkage. This led to the idea of genes being arranged in linear order on chromosomes and thence, by a long series of investigations, to the recent Watson-Crick model for DNA.

It is important to note that the type of probability model used depends on the object of the investigation and on the available data. A simple example is the following. Suppose we wish to have a model which gives us an opinion about the temperature in Ames, Iowa, at noon on November 17, 1975. If we are assessing this question in 1970, perhaps for the purpose of deciding on the scheduling of a sports event, we shall probably rely on a historical record which seems to tell us that the noon temperatures on November 17 of different years are like a random sample from a certain mathematical distribution, with no trend over years. We shall use this model to make a probability prediction. If, however, the date is actually November 16, 1975, we shall use a different model, probably based on the historical record of days near November 17 of previous years and also on the known climatic configuration on a few days prior to November 17, 1975.

Another aspect of the matter is exhibited by the following example. Suppose we wish to predict the sex of a baby to be born sometime in the next few weeks. Suppose we know nothing else; i.e., we have no information which we judge to be relevant. Then we shall probably take the view that the probability of a male birth is about 0.51. Suppose, however, we have the additional information that the mating which has led to the new birth had already given five females in five births. The previous probability of 0.51 of a male birth is now somewhat suspect. The authors would not wish to wager a large sum on this probability.

A classic argument in science revolves around the role of what may be termed "pure observation" and the role of mathematical deductive argument.

Francis Bacon commonly is presented as a proponent of the idea that all science consists of observation and organization of observational facts. (Whether this is totally correct we do not say, because the matter is very complex.) The alternative is the idea, of which Eddington appears to have been a proponent, that all scientific knowledge is essentially derivable by introspection. This relates to the idea that "God is a mathematician." To resolve these opposing views is a deep matter appropriate for the historian of science. It would seem, however, that neither is totally compelling and the building of science is a result of interplay *between* the two types of approach.

## 8.3 Classes of Model Considered

The classes of model that will be considered in this book are mostly of the following form. We have "individuals"—humans, physical specimens, biological specimens, times of observation of a system, geographical places, or entities—which are like any of these in nature. Associated with each individual we have a set of observations which we represent as $y_1$, $y_2, \ldots, y_p; x_1, x_2, \ldots, x_q$. Any $y_i$ or $x_j$ is an observation, either a measurement number or an indicator of a class in which the individual lies; e.g., we could have $x_1 = 1$ to denote that the individual is male and $x_1 = 2$ to denote that the individual is female. In our listings of the observations on an individual we labeled some by $y$ with a subscript and some by $x$ with a subscript. The idea here is that the observations labeled by $y$ with a subscript are to be explained. In contrast, the observations labeled by $x$ are to be used for purposes of explanation (or prediction). If we contemplate using

$$\log P + \log V = \log R + \log T$$

as a means of explaining $\log P$ by means of $\log V$, $\log R$, and $\log T$, we would label $\log P$ as $y_1$, with $\log V$, $\log R$, and $\log T$ labeled as $x_1, x_2, x_3$ respectively. Normally, the explanatory variables are not regarded as random. Also a variable that is a $y$ for one individual may be an $x$ for another individual as in a time series $y_t$, $t = 1, 2, \ldots, T$ in which the individual is the point of time, and we might attempt to explain $y_t$ by means of $y_{t-1}$.

If there is only one variable to be explained, the problem is said to be univariate. If there is more than one variable of type $y$, we call the problem multivariate. The common way of "explaining" the variable is to use a univariate probability distribution, or family of such, for the univariate case and to use $k$-variate distributions for the case of $k$ variables to be explained.

# 9

# Goodness of Fit of a Completely Specified Model

## 9.1 Introduction

In this chapter we consider the very elemental situation that prior knowledge, whether of data which we deem to be like the data under consideration or of a probabilistic model based on hypothesized chance mechanisms, suggests that the data are adequately represented or describable as a random sample from some distribution or stochastic process.

A simple example is a sequence of results $S$ and $F$ which we think are representable as a realization of $n$ independent Bernoulli trials with the probability of success equal to 1/2. Our data is then a string, such as $S,S,F,S,F,S,S,F,F,S,S,F$. We can do many calculations on such a string or sequence. We can count the proportion of successes and should find it to be close to 1/2. We can plot the sequence and ask whether it appears to have any regularities or near regularities. It might appear, for instance, that $S$ is always followed by $F$, and $F$ by $S$. Occurrence of this would tend to negate the applicability of the model of independent Bernoulli trials. A second example is that prior knowledge and theory suggest that a set of numbers $x_1, x_2, \ldots, x_n$ is representable as a random sample from the normal distribution with zero mean and unit variance.

## 9.2 Plotting of Data

We now consider ways of plotting a set of data, $x_1, x_2, \ldots, x_n$, with a view to examining whether the data set is like a random sample from a completely specified distribution with cumulative distribution function $F(x)$. The most commonly used plotting method if $n$ is "large" is preparation of a histogram. By one rule or another we choose a set of $m$ values, say $a_1, a_2, \ldots, a_m$. We determine the observed proportion of observations in the intervals $-\infty < x \leq a_1$, $a_1 < x \leq a_2$, $\ldots$, $a_{m-1} < x \leq a_m$,

$a_m < x < \infty$. Let $a_0 = -\infty$ and $a_{m+1} = +\infty$. Let these proportions be denoted by $R_1, R_2, \ldots, R_{m+1}$. Then we construct the block diagram in which the height between $a_{i-1}$ and $a_i$ is equal to $R_i/(a_i - a_{i-1})$, for $i = 2, 3, \ldots, m$. This is a representation of our observations. If our hypothesized distribution is continuous, we can superimpose its probability density function on the graph. Alternatively, we can determine the expected proportions for each of the intervals as $F(a_i) - F(a_{i-1})$, and we can superimpose a histogram in which the height between $a_{i-1}$ and $a_i$ is $[F(a_i) - F(a_{i-1})]/(a_i - a_{i-1})$. In the case of a discrete distribution with probability mass at points $b_1 < b_2 < \ldots$, we can choose sets of points, $R_i$, $i = 1, 2, \ldots, m$, and make a spike diagram with $m$ pairs of ordinates: $S(R_i)$, which is the proportion of the observations in $R_i$, and $P(R_i)$, which is the proportion determined by the hypothesized distribution. If the hypothesized distribution has a single mode, it is common to make classes on the basis of contiguity. This mode of representing the relation of a set of data and a hypothesized distribution is widely used and we shall return to it later.

The above plotting procedure has the defect that the class intervals have to be chosen by some rule, and the resulting plot is a condensation of the whole data. A mode of plotting which represents the totality of information is to make a plot of the c.d.f. of the hypothesized distribution and of the sample c.d.f. $S_n(x)$ which is the proportion of the observations less than or equal to $x$.

The problem of determining if the set of data is like a random sample from $F$ is clearly that of determining if the two functions $S_n(x)$ and $F(x)$ are sufficiently close. One considers, therefore, ideas of distance between the two functions. Many are possible. The simplest is to choose a single $x$ value, e.g., $x_0$ such that $F(x_0) = p$, and to compare $S_n(x_0)$ and $F(x_0)$; $nS_n(x_0)$ is binomially distributed $(n, p)$ with $p = F(x_0)$. In the absence of any basis for picking an "$x_0$" or a "$p$," we wish to consider a distance which takes account of all possible values of $x$. A distance which has been studied extensively in the literature is the Kolmogorov-Smirnov distance, defined as the greatest difference over all $x$ values of $S_n(x)$ and $F(x)$, or if one likes, the greatest vertical distance between the two functions $S_n(x)$ and $F(x)$ when they are plotted simultaneously against $x$. This distance should not be too large for the model to be tenable. The picking out of the greatest vertical distance on this plot has defects. We know that the expected value of $S_n(x)$ for a particular value of $x$, say $x_0$, is $F(x_0)$ and its variance is $F(x_0)[1 - F(x_0)]/n$. So the vertical distances will have widely differing variances under random sampling from the hypothesized distribution. Suppose, for instance, that $n = 36$. Then the variance of $S_n(x_0)$ when $F(x_0) = 1/2$ is $(1/2)(1/2)/36$ and the standard deviation is $1/12$. So a deviation of $S_n(x_0)$ from 0.5 by as much as 0.08 is not at all unlikely. If on the other hand we consider $x_0$ such that $F(x_0) = 0.9$, the standard deviation of $S_n(x_0)$ is 0.05, and a deviation as large as 0.08 is moderately unlikely. So the vertical distances have unequal variances, particularly as between the center and the tails of the distribution $F$.

If the hypothesized distribution is continuous, it is possible to make a transformation and plot such that the hypothesized distribution appears as a straight line (in the case of a discrete distribution, as points on a straight line). This is obtained by a well-known transformation called the probability integral transformation.

Let the c.d.f. be $F(x)$. Then $F(x)$ ranges from 0 to 1 in a continuous and monotonically increasing way if $F$ is a continuous distribution. Hence, knowing $F(x)$, we can replace an observation $x_i$ by $z_i = F(x_i)$. Note that this transforms random variable $X$ with c.d.f. $F(x)$ to a random variable $Z$ which is uniformly distributed between zero and unity. Hence if $X_1$, $X_2, \ldots, X_n$ is a random sample from $F$, the transformed observations $z_i = F(x_i)$ are a random sample from $U(0, 1)$. We can then make a histogram of the observed distribution of the $z_i$ and compare it with the corresponding expected histogram. The transformation, $z = F(x)$, and the inverse transformation, $x = F^{-1}(z)$, can be used to make plots which should have the appearance of points scattered around a straight line. The two basic types of plots are described and illustrated by Wilk and Gnanadesikan (1968).

*Q-Q or Quantile Plots* Let the two continuous c.d.f.'s which we wish to compare be $F_1(x)$ and $F_2(x)$. Take a sequence of percentile values $0 < p_1 < p_2 \ldots < p_m < 1$. Then we find $x_1(p_i)$, $x_2(p_i)$ such that

$$\begin{aligned} F_1[x_1(p_i)] &= p_i \\ F_2[x_2(p_i)] &= p_i \end{aligned} \tag{9.1}$$

If $F_1(x) = F_2(x)$, we shall have $x_1(p_i) = x_2(p_i)$ for $i = 1, 2, \ldots, m$. If one of the $F_i(x)$ is in fact a sample c.d.f. and if the $p_i$ are not equal to a multiple of $1/n$, we can define $x(p_i)$ to be the $x$-value at which $F$ makes a jump from a value less than $p_i$ to a value greater than $p_i$. The commonly used values of $p_i$ for a sample of size $n$ are $p_i = (i - 1/2)/n$, $i = 1, 2, \ldots, n$. A plot of the population $x$-values against the sample $x$-values should be like points scattered around a straight line.

*P-P or Percentile Plots* In this case we take a sequence of values of $x$, say $a_1, a_2, \ldots, a_m$, and plot $F_1(a_i)$ against $F_2(a_i)$. If $F_1(x)$ and $F_2(x)$ are the same, we obtain points lying on a straight line from (0, 0) to (1, 1) on the unit square. If $F_2(x)$ is a sample c.d.f., we can take the set $\{a_1, a_2, \ldots, a_m\}$ to be the ordered observations $\{x_{(1)}, x_{(2)}, \ldots, x_{(n)}\}$ with some minor modification if there are ties.

It may be noted that *P-P* plots can be used for comparing a multivariate sample with a hypothesized multivariate distribution. But to do this, it will be necessary to choose an ordering of multivariate observations. Clearly there is a wide choice of possible orderings. One possibility is exemplified by the rule for a bivariate distribution "$x_1, x_2$ is before $x'_1, x'_2$ if $F(x_1, x_2) < F(x'_1, x'_2)$."

A very basic question arises in all cases of graphing data and seeing if the graph conforms to expectations under random sampling from some mathematically defined distribution. The general idea is that one should get a plot which has the appearance of points scattered haphazardly around a straight line. One can make a personal judgment in a particular case that points do not fall into this pattern, but this will not be regarded as totally acceptable. We will be satisfied only if we have some guidelines as to what constitutes "reasonable scatter" and what constitutes "reasonable closeness to a straight-line configuration." This is a complex problem which is approached only by the development of tests of significance to be considered later in this chapter.

## 9.3 Normal Probability Plotting

We discuss this case separately because it is very frequently used. Suppose we have a set of numbers $x_{(1)} < x_{(2)} < \ldots < x_{(n)}$ which we think are like or should be like an ordered random sample from a normal distribution, say $N(\mu, \sigma^2)$, where $\mu$ and $\sigma^2$ are known. Consider the quantiles $z_i$ which satisfy

$$\int_{-\infty}^{z_i} dN(0, 1) = (i - 1/2)/n \tag{9.2}$$

or more generally $z(p)$ such that

$$\int_{-\infty}^{z(p)} dN(0, 1) = p \tag{9.3}$$

We can construct paper with an arithmetic abscissa scale and with the ordinate scale labeled such that the value $z(p)$ is given the value $p$. Hence the zero ordinate is labeled 0.5, or 50%. The ordinate 1.96 is labeled 0.975, or 97 1/2%, etc. The ordinate value corresponding to 0% is $-\infty$ and to 100% is $+\infty$. It can be seen that the plotting of $z[(2i - 1)/2n]$ against $x_{(i)}$ will give a plot of points which has the appearance of haphazard scatter around a straight line, regardless of the values of $\mu$ and $\sigma$.

*Example 1*

It is desired to judge whether the following data could reasonably constitute a random sample from a normal distribution and, if so, to form an idea of the mean and variance: 4.68, 5.08, 5.13, 4.76, 4.12, 5.47, 5.04, 6.02, 3.97, 6.21, 4.42, 5.44, 4.15, 5.49, 5.67, 3.79, 5.13, 3.80, 5.89, 4.22.

$S_n(x)$ is calculated for this set of data and plotted on normal probability paper as shown in Figure 9.1. For the $i$th ranked observation, $(i - .5)/n$ is used to calculate $S_n(x)$. Note that a straight line appears to do a satisfactory job of describing the data. Since the normal distribution is symmetric about the mean, the 50% point can be taken as an indicator of the value of $\mu$. In this case it yields about 5.00. For $\sigma$ we have several options. Since the interval $(\mu - 2\sigma, \mu + 2\sigma)$ should contain about .95 of the area, we can use the 2.5 and 97.5 percentiles as values of $\mu - 2\sigma$ and $\mu + 2\sigma$. This gives us an indication of the value of $\sigma$.

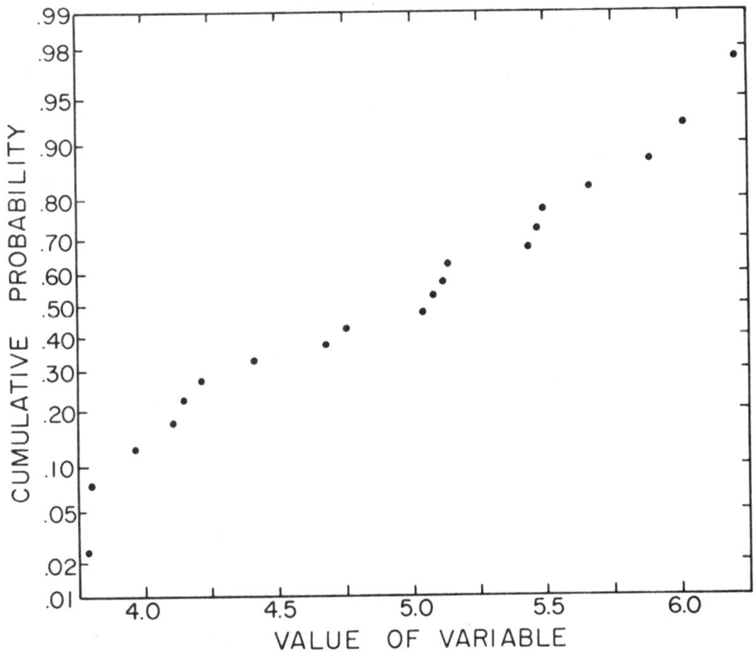


**Fig. 9.1** Normal Probability Plotting

## 9.4 Tests of Significance

We have seen that we can form an approximate idea of the tenability of a model by a graphical method, at least in very simple cases. We noted, however, that we have no idea as yet of what constitutes a big deviation from expectation based on the model. We have noted that $S_n(x)$ is distributed as Bi$[n, F(x)]$, and this suggests that a deviation $|S_n(x) - F(x)|$, say, should be judged in relation to what values we would get in repeated random sampling from a population specified by the model.

It might be thought that the probability of the actual result would be completely adequate. However, we do not know how to interpret this probability. To take a very simple case, suppose we have a model for which an observation $x$ comes from the binomial, Bi(1000, 1/2) and suppose that $x$ is 500, a result highly in conformity with the model from almost any point of view. However, the use of Stirling's approximation shows that the probability of the observation under the model is obviously very small. To circumvent this difficulty which occurs generally, tests of significance were developed. The basic idea is that we form a measure of deviation of the data from expectation under the model, and then compute the probability of a random sample being as deviant or more deviant than the actual data. In the above case of 500 successes in 1000 trials it is obvious that this is the least deviant of all possible random samples under the model from almost any point of view. We present the procedure from a slightly different (but intrinsically equivalent) viewpoint.

*Definition 9.1. Let the possible data sets under a probability model* $M_0$ *be denoted by* $\{D_i\}$. *A test of significance consists of: (1) the arrangement of the possible data sets,* $D_i$, *as a partially ordered set; if* $D_i$ *does not occur in the partial ordering after* $D_j$, *we write* $D_i \gg D_j$; *and (2) attaching to the observed data sets* $D_O$ *the number*

$$SL(D_O; M_0) = \sum_{D \gg D_O} P(D, M_0) \tag{9.4}$$

*which is called the significance level of the data sets* $D_O$ *with regard to the model* $M_0$ *for the partial ordering chosen.*

To illustrate, suppose there are seven different possible data sets under $M_0$, which we denote by $D_1$, $D_2$, $D_3$, $D_4$, $D_5$, $D_6$, and $D_7$. A partial ordering can be represented by the linear arrangement of sets of the $D_j$, e.g., $(D_1)(D_2, D_7)(D_3, D_6)(D_4)(D_5)$. In this arrangement $D_1$ is before all the other $D$'s; $D_2$ and $D_7$ occur after $D_1$ but not before or after one another. With this partial order, we write

$$\begin{gathered} D_1 \gg D_2,\ D_1 \gg D_3, \ldots, D_1 \gg D_7 \\ D_2 \gg D_3,\ D_2 \gg D_4, \ldots, D_2 \gg D_7 \\ \cdots\cdots\cdots\cdots\cdots\cdots \\ D_7 \gg D_2,\ D_7 \gg D_3, \ldots, D_7 \gg D_6 \end{gathered}$$

We say that with this ordering $D_i$ is at least as significant as $D_j$ if $D_i \gg D_j$.

It may be helpful also to note that the partial ordering is equivalent to a partition of the space of possible data sets into subsets with an order to the subsets; i.e., a partial ordering is an ordered partition.

The idea behind this formulation is that if the model $M_0$ is really true, we would get an infinite collection of data sets in repetitions. We wish to form a judgment of the distance, in some sense, of our actual data set from the infinite collection of possible samples. To illustrate, suppose that a data set is an ordered triplet of numbers $x_1, x_2, x_3$. We can define a distance between two points $D_1 = (x_1, x_2, x_3)$ and $D_2 = (x_1', x_2', x_3')$ in many ways. This type of thinking is very common in mathematics, and commonly used distances are:

$$d(D_1, D_2) = (x_1 - x_1')^2 + (x_2 - x_2')^2 + (x_3 - x_3')^2$$
$$d(D_1, D_2) = \max(|x_1 - x_1'|, |x_2 - x_2'|, |x_3 - x_3'|)$$
$$d(D_1, D_2) = \left[\sum (x_i - x_i')^p\right]^{1/p}, \qquad p > 0$$

It is clear that we can get a wide variety of answers, and even given a distance function, we could get different results by changing each $x$ to $h(x)$, where $h(\cdot)$ is a function. Furthermore, when we wish to consider the distance of an observed data set $D_O$ from a population of data sets, we are almost inexorably led to some sort of average distance, $\text{Ave}\, d(D_O, D)$, where "Ave" means average over the population of $D$'s.

Because the population is infinite and defined in probability terms, it is natural to construct the distance of an actual data set from the population on the probability scale, and this is done in a significance test. The distance is represented by the significance level, and whatever the problem we get a distance between 0 and 1, furthermore, a distance which does not depend on any choice of transformation of the observations. There is one slightly confusing aspect of the method. A significance level of 1 indicates that the distance is small, while a significance level of 0 indicates that the distance is very large, in fact "infinite."

We also note two very critical points. First, the choice of an ordering is very difficult and depends on our aims. We discuss this below and in Chapter 10. Second, we must be able to calculate the probabilities and are forced to use orderings whose properties are calculable given the model. We mention in this respect, for the benefit of readers of other texts and of the literature, that a test of significance must be based on an ordering which leads to calculable probabilities. For instance, we do not consider the Behrens-Fisher test to be a significance test in the strict sense.

We note an immediate consequence of the definition. If there is a finite number of possible data sets, the possible significance levels are denumerable. This happens because a significance level is equal to the sum of the probabilities of a subset of the possible data sets, and if there are $K$ possible data sets there are at most $2^K$ subsets. The set of possible significance levels will be called the achievable levels. If on the other hand there is a denumerable infinity, or with a continuous approximation a nondenumerable number of possible data sets, we may have a nondenumerable set of achievable significance levels, depending on the ordering that is used.

We state a fundamental theorem with regard to significance levels for a completely specified model $M_0$.

***Theorem 9.1*** $P[SL(D) \leq \alpha | M_0] = \alpha$ for all achievable $\alpha$, $0 \leq \alpha \leq 1$.

*Proof:* The significance level of a data set $D$ is less than or equal to $\alpha$ if and only if $D \gg D_1$, where $SL(D_1) = \alpha$. But

$$\begin{aligned} P(D \gg D_1) &= \sum_{D \gg D_1} P(D) \\ &= SL(D_1) \\ &= \alpha \end{aligned} \tag{9.5}$$

This theorem is the basis of tests of significance for the case of a completely specified model $M_0$. To illustrate its force, consider the case when all values between 0 and 1 are achievable significance levels, as in the continuous case. Then the distribution of a significance level under $M_0$ can be represented by the uniform distribution $U(0, 1)$ as depicted in Figure 9.2. In the case of a denumerable set of achievable levels the corresponding pictures are as shown in Figure 9.3.

The motivation for the use of a significance test is that we have replaced the data by a number between 0 and 1, the distribution of which is known if we are sampling from $M_0$. Our interest in calculating the number is to

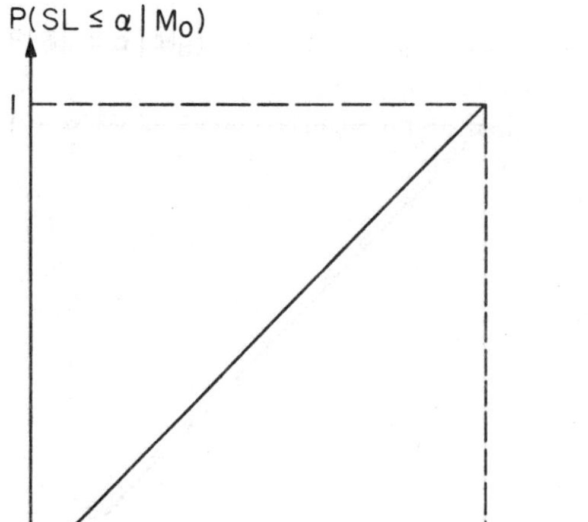


**Fig. 9.2** Distribution of $SL$ under $M_0$ for the Continuous Case

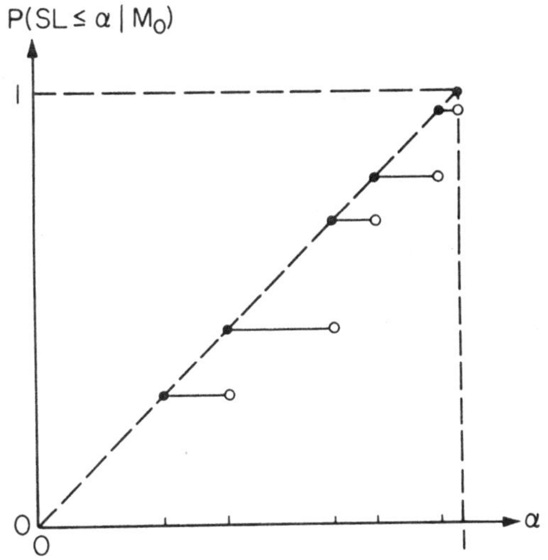


**Fig. 9.3** Distribution of $SL$ under $M_0$ for the Discrete Case

form an idea of the tenability of the model $M_0$. The question of tenability of a model $M_0$ does not arise *unless* we have in mind a class of alternative models which may be rather loosely specified. Suppose that there is an alternative model $M_1$. In random sampling from $M_1$ the $SL$ will have a distribution, and Figure 9.4 shows what may happen with a particular ordering of samples. We see that if this happened, the probability of low values for $SL$ is greater under $M_1$ than under $M_0$. It might happen for in-

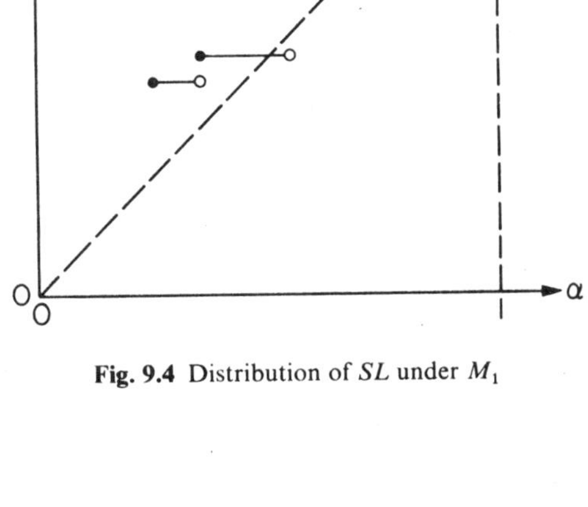

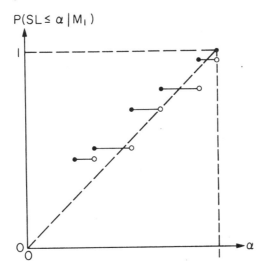

**Fig. 9.4** Distribution of $SL$ under $M_1$

stance, that we observe an $SL$ of .01 and that the probability of an $SL$ as low as this is 0.6 under $M_1$. We can then say that we have observed an event which has probability .01 under $M_0$ and .6 under $M_1$. This would generally be taken to be a fairly strong indication that $M_1$ is more tenable than $M_0$. The underlying idea is that a probability of zero for the data under a model $M_0$ is perfect evidence against $M_0$. An event of small probability is stronger evidence against $M_0$ than an event with large probability. If the situation in Figure 9.2 holds and we have adopted a rule of making conclusions that we will take $M_0$ to be untenable if the significance level is less than some value $\alpha_0$, we will reject $M_0$ when it is true a proportion $\alpha_0$ of times and will reject $M_0$ if $M_1$ is true a higher proportion of times.

It seems reasonable, therefore, to take $SL(D_O; M_0)$ as a measure of the tenability of the model $M_0$ with the ordering of data sets used in making up the test of significance. If $SL(D_1) = 0.2$ and $SL(D_2) = 0.03$, we say $D_2$ provides more evidence against $M_0$ than $D_1$, according to the ordering of possible data sets that has been used. We say that data $D_2$ is farther from $M_0$ than $D_1$ as measured by the ordering.

## 9.5 Ordering of Possible Data Sets

It is now necessary to consider how the possible data sets should be ordered. In general, the ordering should be based on consideration of possible alternative models.

In the case of goodness of fit tests with no specified alternative in mind, numerous orderings may be suggested. In general, with a given body of data, examination will often suggest the alternatives and hence an ordering. We shall be wise to examine the data in whatever respects they appear to deviate from expectations, and we will not use a model for data representation which gives a low significance level in any respect that occurs to us. For instance, on considering a set of data $x_1, x_2, \ldots, x_n$ as being from a Poisson distribution, we may be disturbed if the sample variance appears to be quite different from the mean; and to examine the possibility, we would order samples by $\sum (x_i - \bar{x})^2/\bar{x}$. Corresponding to any aspect of the data that we notice, we can develop an ordering and a test of significance. We can use any function of the data to map each possible set of data into the real line to give an ordering.

A natural idea is to base the ordering on $P(D|M_0)$, so that a data set is more significant than data sets of higher probability under $M_0$. We now explore this ordering to examine whether multinomial data are fitted adequately by a totally specified multinomial distribution. The situation is that we have $k$ classes with observed number $n_i$ in the $i$th class, $i = 1, 2, \ldots, k$, and the model is that the data are like a random sample from a known multinomial distribution, so that the probability of a sample $n_1, n_2, \ldots, n_k$ is

$$n! \prod_{i=1}^{k} \frac{p_i^{n_i}}{n_i!}, \qquad \sum n_i = n \tag{9.6}$$

in which the $p_i$ are known. Before proceeding, let us note that this case is much more general than it may appear at first glance. Any univariate distribution can be converted into a multinomial with a finite number of classes by grouping.

Let us suppose for the moment $n$ and the $n_i$ are large. We saw on page 114 that the joint distribution of the $n_i$ is close to the multivariate normal density

$$(2\pi)^{-(k-1)/2}\left(n^{k-1}\prod_{1}^{k} p_i\right)^{-1/2} \exp\left[-1/2\left(\sum_{i=1}^{k-1} a_{ii}\varepsilon_i^2 + \sum_{\substack{i,i'=1\\ i\neq i'}}^{k-1} a_{ii'}\varepsilon_i\varepsilon_i'\right)\right] \tag{9.7}$$

where

$$\varepsilon_i = n_i - np_i,\ a_{ii} = \frac{1}{n}\left(\frac{1}{p_i} + \frac{1}{p_k}\right),\ a_{ii'} = \frac{1}{np_k},\ i \neq i' \tag{9.8}$$

If we ordered possible data sets by their probabilities, we would order them according to the criterion $C$ where

$$C = \sum_{i=1}^{k-1} a_{ii}(n_i - np_i)^2 + \sum_{\substack{i,i'=1\\ i\neq i'}}^{k-1} a_{ii'}(n_i - np_i)(n_{i'} - np_{i'}) \tag{9.9}$$

We now try to put $C$ in a simpler form. We get

$$C = \sum_{i=1}^{k-1} \frac{1}{n}\left(\frac{1}{p_i} + \frac{1}{p_k}\right)(n_i - np_i)^2 + \sum_{\substack{i,i'=1\\ i\neq i'}}^{k-1} \frac{1}{np_k}(n_i - np_i)(n_{i'} - np_{i'})$$

$$= \sum_{i=1}^{k-1} \frac{(n_i - np_i)^2}{np_i} + \frac{1}{np_k}\left[\sum_{i=1}^{k-1}(n_i - np_i)^2 + \sum_{\substack{i,i'=1\\ i\neq i'}}^{k-1}(n_i - np_i)(n_{i'} - np_{i'})\right] \tag{9.10}$$

But

$$\sum_{i=1}^{v} u_i^2 + \sum_{\substack{i,i'=1\\ i\neq i'}}^{v} u_i u_{i'} = \left(\sum_{i=1}^{v} u_i\right)^2 \tag{9.11}$$

and

$$\sum_{i=1}^{k} n_i = n,\ \sum_{i=1}^{k} p_i = 1 \tag{9.12}$$

so

$$\sum_{i=1}^{k-1} (n_i - np_i) + (n_k - np_k) = 0$$

Hence

$$C = \sum_{i=1}^{k} \frac{(n_i - np_i)^2}{np_i} \tag{9.13}$$

Or to write this in a very general form,

$$C = \sum_{classes} \frac{(O_i - E_i)^2}{E_i} \tag{9.14}$$

where $O_i$ is the observed number in class $i$ and $E_i$ is the expected number in class $i$. Thus $C$ is a very simple criterion to compute, and in the final form above has intuitive appeal.

The development above has a very useful consequence. We have seen in Sect. 6.4 that $-2$(exponent of density of a $p$-variate normal distribution) is distributed according to the $\chi^2$ distribution with $p$ degrees of freedom. This suggests that $C$ be evaluated by reference to the mathematical $\chi^2$ distribution with $k - 1$ degrees of freedom.

To reinforce this suggestion somewhat, let us consider the mean of $C$ in random sampling from a multinomial $(p_i^*)$. This goes as follows:

$$C = \sum \frac{O_i^2}{E_i} - 2\sum O_i + \sum E_i$$

$$= \sum \frac{O_i^2}{E_i} - 2n + n = \sum_i \frac{O_i^2}{E_i} - n \tag{9.15}$$

If the $O_i$ are from the multinomial $p_1^*, p_2^*, \ldots, p_k^*$, we know

$$\begin{aligned} E(O_i) &= np_i^* \\ V(O_i) &= np_i^*(1 - p_i^*) \\ E(O_i^2) &= n^2 p_i^{*2} + np_i^*(1 - p_i^*) \\ &= np_i^* + n(n-1)p_i^{*2} \end{aligned} \tag{9.16}$$

So

$$E\left(\frac{O_i^2}{E_i}\right) = E\left(\frac{O_i^2}{np_i}\right) = \frac{p_i^*}{p_i} + (n-1)\frac{p_i^{*2}}{p_i} \tag{9.17}$$

If $p_i^* = p_i$, this equals $1 + (n - 1)p_i$ and

$$E(C) = \sum_{i=1}^{k} [1 + (n-1)p_i] - n = k - 1 \tag{9.18}$$

At least one $p_i^*$ must be greater than the corresponding $p_i$ if $\{p_i^*\} \neq \{p_i\}$ so as $n \to \infty$, $E(C) \to \infty$. Other moments of $C$ may be obtained. This is particularly easy if the $p_i$ are equal to $1/k$ (see Problem 9.12).

The significance level of an observed value $C_O$ is then approximately $P(C \geq C_O)$, which is obtainable from tables of the cumulative $\chi^2$ distribution. Because of the approximate distributional property of $C$, the procedure described is called the $\chi^2$ goodness of fit test of significance.

We should note that the distributional approximation may be somewhat poor if $n$ is not large and some $p_i$ are very small. Consider the case of $k = 2, n = 5, p_1 = 1/3, p_2 = 2/3$. The computations appear in Table 9.1. The reader should compare the actual c.d.f. with that given by the $\chi^2$ distribution with one degree of freedom.

**Table 9.1** Computation for Distributional Approximation

| Number of Class ($n_1$) | Prob | $C$ |
|---|---|---|
| 0 | 1/243 | 2.5 |
| 1 | 10/243 | 0.4 |
| 2 | 40/243 | 0.1 |
| 3 | 80/243 | 1.6 |
| 4 | 80/243 | 4.9 |
| 5 | 32/243 | 10.0 |

For testing that a set of numbers $x_1, x_2, \ldots, x_n$ is like a random sample from a distribution with specified c.d.f., we can use the $\chi^2$ test of significance but often have to convert the distribution into a multinomial with a finite number of classes. The suggestion based on empirical work by several individuals and theoretical work by Steck (1957) is that a reasonable procedure with $n$ observations is to partition the distribution into $n$ equal parts by proceeding from left to right of the distribution. It may not be possible to do this exactly. If $F$ is discrete, we will have to take lumps of probability as they occur and will be unable to achieve a probability of $1/n$ for each cell exactly. In the case of a continuous distribution our observations will be discrete as a result of inevitable grouping error of measurement, and again we will be unable to obtain classes which have exactly equal probabilities. However, when this is possible, the statistic is $\sum_j O_j^2 - n$ because $E_j$ for every cell is then one. It is recommended that this statistic be reduced by unity.

The above procedure is feasible only with very extensive tables of the cumulative distribution function of the distribution being fitted or with the use of a modern computer. It will be necessary to have convenient formulas for the computation of the abscissa corresponding to the fractiles $1/n, 2/n, \ldots, (n-1)/n$. A good formula for the case of the normal distribution is that given by Hastings (1955). If the partition of possible values into $n$ classes gives values which are observationally indistinguishable, like heights of 69.01 and 69.02 inches obtained with a standard tape, successive indistinguishable classes are to be pooled.

*Example 2*

To illustrate the mechanics of this procedure, suppose that data have been obtained from a physical population and we wish to examine the fit of a negative exponential with $\lambda = 100$. The data consist of the following ten rounded observations: 15, 18, 21, 22, 28, 34, 40, 67, 98, 186. Ten classes are then formed, each of which has probability of 1/10, and the statistic is calculated in Table 9.2. Then $\sum O_j^2 - n - 1 = 13$. From Table A.5 in the Appendix we see that the probability of a $\chi^2$ this large or larger is about .23. Thus there is no strong evidence against the negative exponential with $\lambda = 100$.

**Table 9.2** Calculation of the Statistic

| Class | Class Boundary | $O_j$ | $O_j^2$ |
|---|---|---|---|
| 1 | 0–10.5 | 0 | 0 |
| 2 | 10.5–22.3 | 4 | 16 |
| 3 | 22.3–35.7 | 2 | 4 |
| 4 | 35.7–51.1 | 1 | 1 |
| 5 | 51.1–69.3 | 1 | 1 |
| 6 | 69.3–91.6 | 0 | 0 |
| 7 | 91.6–120.4 | 1 | 1 |
| 8 | 120.4–161.0 | 0 | 0 |
| 9 | 161.0–230.2 | 1 | 1 |
| 10 | 230.2–∞ | 0 | 0 |
| | | | 24 |

This test is readily applied in the case of the bivariate normal distribution. The $n$ regions with probability content $1 - i/n$ are given by the $n - 1$ concentric ellipses given by the equation

$$\frac{1}{1-\rho^2}\left[\frac{(x-\mu_x)^2}{\sigma_x^2} - \frac{2\rho(x-\mu_x)(y-\mu_y)}{\sigma_x\sigma_y} + \frac{(y-\mu_y)^2}{\sigma_y^2}\right] = \chi^2_{2,\,i/n}$$

$$i = 1, 2, \ldots, n-1 \qquad (9.19)$$

where $\chi^2_{2,\,i/n}$ is the value of $\chi^2$ with 2 degrees of freedom above which a proportion $i/n$ occurs. Extensions to distributions of any dimensionality are obvious.

## 9.6 Other Goodness of Fit Tests of a Single Distribution

In the derivation of the $\chi^2$ test above we saw that the ordering of possible data sets is based on their probabilities under the model. There are many cases in which the probabilities of all the possible data sets are equal. The probability orderings would then give only two achievable levels, 1 and 0, and would therefore be essentially useless for testing goodness of fit. We give an example of this case in Sect. 9.7. In such cases other ways of ordering *must* be used. We now discuss in general terms a few other commonly used orderings.

In Sect. 9.2. we described several measures of distance between a sample c.d.f. and a population c.d.f. Associated with each measure of distance is a distribution of that distance under repeated random sampling from the hypothesized distribution, and we can envisage an evaluation of significance by taking $P(C \geq C_O)$, where $C_O$ is observed distance. The Kolmogorov-Smirnov distance has been studied extensively and is used fairly widely. A moderate amount of Monte Carlo study indicates, however, that it is only moderately sensitive as a general purpose test. The probability distribution of the Kolmogorov-Smirnov statistic is given in Table A.10 of the Appendix.

It is clear after reflection that there cannot be a single best goodness of fit test of significance. Every test of significance has the basic property that the distribution of significance level is uniform in the sense that

$$P(SL \leq \alpha; M_0) = \alpha$$

if $\alpha$ is an achievable level. But the distributions of the significance level under alternative models will vary considerably. There seems to be no recourse but for the data analyzer to contemplate what deviations from the hypothesized model are interesting to him and then to form tests of significance aimed in these directions.

If he is concerned with the possibility of there being an erroneous observation at the upper end of the scale, he will use what is called an outlier test. The reasoning process is as follows. The distribution of the largest of a random sample of size $n$ is known (Sect. 6.8). Let its c.d.f. be $G(x)$ and let the largest observation be $x_{(n)}$. Then the significance level is $1 - G[x_{(n)}]$. If the data analyzer is concerned with the possibility of there being too many observations in one or the other tail, he will have to think of an ordering of possible data sets which reflects this concern. There is an indefinitely large number of reasonable tests of goodness of fit. To develop any one requires the development of the random sampling distribution of the criterion, and the mathematical literature contains derivations for a wide variety of reasonable criteria with various distributions. It is impossible to give any survey of this work here. It is worthwhile, however, to mention one aspect. Some criteria do not use the mathematical nature of the particular hypothesized distribution and are called "distribution-free." Others such as outlier tests have to be based intimately on the particular mathematical form of the distribution function. The tests described above, the $\chi^2$ test and the Kolmogorov-Smirnov test, are distribution-free. Their sensitivity depends, however, on the particular form of the hypothesized distribution and the alternative distributions that are thought to be relevant.

There is a wide variety of tests for continuous univariate distributions based on the probability integral transformation. Let the actual distribution function be $F(x)$, and let the ordered values be $x_{(1)}, x_{(2)}, \ldots, x_{(n)}$. Form the variables $z_i = F[x_{(i)}]$, $i = 1, 2, \ldots, n$. These values should be like an ordered random sample from the uniform distribution $U(0, 1)$. Now construct $d_1 = z_1$, $d_i = z_i - z_{i-1}$, $i = 2, \ldots, n$, and $d_{n+1} = 1 - z_n$. These are called the spacings, and many functions of the set of $d$'s have been used as means of ordering data sets for tests of significance. The paper of Pyke (1965) contains an account of these.

## 9.7 Tests of Randomness of a Sequence

Consider the question of whether a sequence of $A$'s and $B$'s is like a sequence of independent Bernoulli trials, with $P(A) = P(B) = 1/2$. This question is a very important one. For instance, many probabilistic models are so complex that they cannot be examined mathematically in the present stage of statistics. Widely used in this case is the Monte Carlo method, which consists of developing on a computer sets of data which the model would give. To do this, however, one needs a "random number generator," which in the simplest case is a method of producing a sequence of $A$'s and $B$'s which has the property of being like a sequence of results of independent Bernoulli trials with $P(A) = P(B) = 1/2$.

Consider then a sequence $x_1, x_2, \ldots, x_n$. We note that the probability under the model of *any* sequence of $A$'s and $B$'s is $1/2^n$. So to make a test of goodness of fit, we cannot use the probability ordering of population sequences of length $n$. We must make a condensation of the data. Possible candidates which come to mind immediately are: (1) the number of $x_i$ equal to $A$, (2) the number of times $x_i$ equals $x_{i+1}$, and (3) the number of runs of the same digit. Candidates (1) and (2) are easy to examine by a $\chi^2$ test. For candidate (3) we have to obtain the distribution of runs.

A related question is whether a sequence of numbers $x_1, x_2, \ldots, x_n$ can be regarded as like a random sample from some distribution with regard to order. Are large values followed too frequently by small values, for instance? A procedure for examining this is based on the theory of runs

(Mood, 1940). We will suppose here that a median is well defined and there are $n$ numbers below this median and $n$ above. We consider only these $2n$ numbers and replace each by $A$ if it is above the median and by $B$ if below. This gives us a scrambled sequence of $n$ $A$'s and $n$ $B$'s, and if the sequence is a segment of a random sequence, there should be "insignificant" clumping. What is meant by "clumping" can be indicated by an example: Suppose $n = 6$ and the observed sequence is $A,A,A,B,A,A,B,B,A,B,B,B$. There appear to be too many $A$'s at the beginning of the sequence, and consequently too many $B$'s at the end. Define then a run of type 1 of length $i$ or an $A$ run of length $i$ as a sequence of letters $A$, beginning at the start of the sequence or immediately after a $B$ and terminating at the end or immediately before a $B$. Define similarly runs of type 2 or $B$ runs. Let $m_{ji}$ be the number of runs of type $j$ of length $i$. The limit on $i$ is $n$. For the observed sequence above, we have

$$m_{11} = 1, m_{12} = 1, m_{13} = 1, m_{1i} = 0, i > 3$$
$$m_{21} = 1, m_{22} = 1, m_{23} = 1, m_{2i} = 0, i > 3$$

The details of the ensuing partial derivation may be omitted by the reader. If the sequence is random, its structure would be "like" what we would get if we put $n$ chips labeled $A$ and $n$ chips labeled $B$ into an urn, then drew out chips one at a time, with repeated shaking, and recorded the sequence of letters that we obtained. The total number of ways of ordering $2n$ different objects is $(2n)!$. The total number of different ways of ordering $2n$ objects of which $n$ are $A$ and $n$ are $B$ is $(2n)!/n!n!$.

Consider now a sequence with the property that it consists of $m_{11}$ $A$ runs of length 1, $m_{12}$ $A$ runs of length 2, . . . , and $m_{21}$ $B$ runs of length 1, $m_{22}$ $B$ runs of length 2, . . . , such that $M_1$ is the total number of $A$ runs and $M_2$ is the total number of $B$ runs. $M_1$ and $M_2$ differ by zero or unity. If $M_1$ is equal to $M_2$, a sequence can start in two ways, with an $A$ run or with a $B$ run; but if $M_1$ and $M_2$ differ by unity, the sequence must start with the more frequent type. Note that $M_1 = \sum_i m_{1i}$ and $M_2 = \sum_i m_{2i}$. A sequence of this structure has $M_1 + M_2$ blocks. The blocks of $A$ runs can be permuted among themselves, and the blocks of $B$ runs can be permuted among themselves without altering the numbers $m_{ji}, j = 1, 2, i = 1, 2, \ldots, n$. The number of ways this can be done is

$$\frac{M_1!}{m_{11}!m_{12}! \ldots} \cdot \frac{M_2!}{m_{21}!m_{22}! \ldots} \tag{9.20}$$

The number of sequences having a particular set of values for all the $m_{ji}$ is

$$\frac{M_1!}{m_{11}!m_{12}! \ldots} \cdot \frac{M_2!}{m_{21}!m_{22}! \ldots} \alpha(M_1, M_2) \tag{9.21}$$

where $\alpha(M_1, M_2) = 2$ if $M_1 = M_2$, and equals 1 if $M_1 \neq M_2$. Then the probability of a sequence having these values for all the $m_{ji}$ is

$$\frac{M_1!}{m_{11}!m_{12}! \ldots} \cdot \frac{M_2!}{m_{21}!m_{22}! \ldots} \cdot \frac{n!n!}{(2n)!} \alpha(M_1, M_2) \tag{9.22}$$

The possible values of $m_{ji}$ are limited because $m_{ji} > 0$ and

$$\sum_{i=1}^{n} im_{ji} = n \tag{9.23}$$

For any particular value of $n$ the probabilities of all the different configurations could be calculated by a suitable computer program. The distribution is, however, quite complex.

To form a judgment as to whether a particular sequence has "random structure" (which is to say, has structure like that to be expected from a mathematically defined random sequence), we consider the quantification of degree of departure from what would be expected in the population of random repetitions. What should be used in the present case? The answer is not obvious, and there are several candidates. A reasonable quantification to consider is the number of runs, that is, $M_1 + M_2$. The exact distribution of $M_1$ and $M_2$ is obtained as follows by summing over all $m_{ji}$ so that $\sum_i m_{ji} = M_j$ and $\sum_i im_{ji} = n$. This process yields

$$\frac{(n-1)!}{(M_1 - 1)!(n - M_1)!} \cdot \frac{(n-1)!}{(M_2 - 1)!(n - M_2)!} \cdot \frac{n!n!}{(2n)!} \alpha(M_1, M_2) \tag{9.24}$$

Note that in this probability $M_1$ and $M_2$ are equal or differ by unity. This probability can be evaluated for any choice of $n$, $M_1$, and $M_2$. Consider finally the distribution of $M = M_1 + M_2$. If $M$ is even, $M_1 = M_2 = M/2$ and the probability of $M$ is

$$2[C(n - 1, M/2 - 1)]^2/C(2n, n) \tag{9.25}$$

where

$$C(a, b) = \binom{a}{b}$$

If $M$ is odd, $M_1 = (M - 1)/2$, $M_2 = (M - 1)/2 + 1$, or vice versa, and the probability is

$$2C(n - 1, M/2 - 3/2)C(n - 1, M/2 - 1/2)/C(2n, n) \tag{9.26}$$

For any particular $n$ and $M$, this probability can be calculated. It is relevant to our aims to consider the elementary properties of this distribution, in particular the mean and variance. The algebra for doing this is complicated and is not included. It turns out that

$$E(M) = n + 1$$

and

$$\text{Var}(M) = n(n - 1)/(2n - 1) \tag{9.27}$$

We can conduct our significance test for randomness by calculating the actual probabilities, or we may base a normal approximation upon the mean and variance of $M$. A value for $M$ which is too small is an indication of clumping, and a large value of $M$ is an indication of too much regularity of change. One is usually concerned about the former, so that given the observed value of $M$, say $M_O$, one finds the probability of a value less than or equal to $M_O$ if the sequence arose by a purely random process.

*Example 3*
In a development program the head engineer has asked for a sample of 20 batteries from typical production on a pilot line. He suspects that the pilot line people have learned about his objectives and have not given him typical batteries. He is interested in whether he can reasonably consider the sample of 20 readings as a random sample from a continuous distribution. The data are: 34, 33, 48, 45, 18, 32, 20, 21, 24, 38, 42, 52, 40, 51, 60, 56, 41, 46, 50, 26. Upon ranking the observations, it is found that the middle two are 40 and 41, so the sample median is taken to be 40.5. All observations greater than the median are replaced by $A$ and all those less are replaced by $B$. This results in the following: $B, B, A, A, B, B, B, B, B, B, A, A, B, A, A, A, A, A, A, B$. There are 4 runs of $B$'s and 3 runs of $A$'s. Note that there are exactly 10 $A$'s and 10 $B$'s. The total number of runs is $M = 7$. The mean and variance are computed by

$$\begin{aligned} E(M) &= n + 1 = 11 \\ \text{Var}(M) &= n(n - 1)/(2n - 1) \\ &= 4.74 \end{aligned}$$

The approximate standard normal deviate is $(7 - 11)/2.17 = -1.84$. From Table A.4 in the Appendix we see that the probability of a value this small or smaller is about .03 and the probability of a value as large or larger in absolute value is about .06.

*Example 4*
To illustrate the calculation of exact probabilities suppose we have the sequence 7, 9, 11, 15, 17, 13, 18, 19, 20. The derived sequence is $B, B, B, A, B, A, A, A$ which has 4 runs. Let us calculate the probability of 4 or fewer runs. The probabilities are:

$$\begin{aligned} M &= 2, \quad 2[C(3, 0)]^2/C(8, 4) = 2/70 \\ M &= 3, \quad 2C(3, 0)C(3, 1)/C(8, 4) = 6/70 \\ M &= 4, \quad 2[C(3, 1)]^2/C(8, 4) = 18/70 \end{aligned}$$

So the probability of getting 4 or less is 26/70, which is rather large. It seems that this test has low sensitivity, since the sequence obviously has a nonrandom structure.

A more sensitive test for the problem under discussion is to replace the observations by their rank orders and work with these. If there is a strong trend, we would expect the rankings of the observations to be $1, 2, 3, \ldots, n$ in order. Let $r_i$ be the rank for observation $i$. If we plotted $r_i$ against $i$, we would expect a haphazard distribution of points; but if there is a trend, we would expect $\sum ir_i$ to be small or large. Let us consider the behavior of $\sum ir_i$ if the observations arise from a random process. The random variables are the numbers $r_i$. The expectation of an $r$ number is

$$\frac{1}{n}(1 + 2 + 3 + \cdots + n) \tag{9.28}$$

which is $(n + 1)/2$. The expectation of the square of an $r_i$ is

$$\begin{aligned} \frac{1}{n}(1^2 + 2^2 + 3^2 + \cdots + n^2) &= \frac{1}{n}\frac{n(n + 1)(2n + 1)}{6} \\ &= \frac{(n + 1)(2n + 1)}{6} \end{aligned} \tag{9.29}$$

so the variance of an $r_i$ number is

$$\frac{(n + 1)(2n + 1)}{6} - \frac{(n + 1)^2}{4} = \frac{n^2 - 1}{12} \tag{9.30}$$

The expected value of the product of two $r_i$ values is

$$\frac{1}{n(n - 1)} \sum^{\neq} ii' \tag{9.31}$$

where $\sum^{\neq} ii'$ is the sum of products of all pairs of different numbers from 1 to $n$. But a general identity involving numbers $a_1$ to $a_n$ is

$$\sum_{i, i'}^{\neq} a_i a_{i'} = \left(\sum_{i=1}^{n} a_i\right)^2 - \left(\sum_{i=1}^{n} a_i^2\right)$$

so

$$\begin{aligned} \frac{1}{n(n - 1)} \sum^{\neq} ii' &= \frac{1}{n(n - 1)}\left[\left(\sum_{i=1}^{n} i\right)^2 - \left(\sum_{i=1}^{n} i^2\right)\right] \\ &= \frac{1}{n(n - 1)}\left[\frac{n^2(n + 1)^2}{4} - \frac{n(n + 1)(2n + 1)}{6}\right] \\ &= \frac{(n + 1)(3n + 2)}{12} \end{aligned} \tag{9.32}$$

Hence the covariance of two $r_i$'s is

$$\frac{(n+1)(3n+2)}{12} - \frac{(n+1)^2}{4} = \frac{(n+1)}{12}\left[(3n+2) - 3(n+1)\right]$$
$$= -\frac{(n+1)}{12} \qquad (9.33)$$

Consider now $\sum ir_i$. Its expected value is

$$\sum_i i[(n+1)/2] = n(n+1)^2/4 \qquad (9.34)$$

Its variance is

$$\sum_{i=1}^{n} i^2 \frac{(n^2-1)}{12} - \sum_{i,i'}^{\neq} ii' \frac{(n+1)}{12}$$
$$= \frac{(n^2-1)}{12}\frac{n(n+1)(2n+1)}{6} - \frac{(n+1)}{12}\frac{n(n-1)(n+1)(3n+2)}{12}$$
$$= \frac{(n-1)n(n+1)^2(2n+1)}{72} - \frac{(n-1)n(n+1)^2(3n+2)}{144}$$
$$= \frac{(n-1)n^2(n+1)^2}{144} \qquad (9.35)$$

*Example 5*

Suppose we observe the sequence 7, 9, 11, 15, 17, 13, 18, 19, 20. The ranks are 1, 2, 3, 5, 6, 4, 7, 8, 9. The quantity $\sum ir_i$ is equal to 282. We expect this number to be $9 \times 10 \times 10/4 = 225$. The variance is $8 \times 9 \times 9 \times 10 \times 10/144$ so that the standard deviation is $\sqrt{450} = 21.2$. If as an approximation we assume that $\sum ir_i$ is normally distributed, then we have observed a value of 282 as a supposed random number from a normal distribution with mean 225 and standard deviation 21.2. So we have observed a value of $(282 - 225)/21.2 = 57/21.2 = 2.69$ from a normal distribution with mean zero and standard deviation one. Table A.4 in the Appendix indicates that the probability of a deviation as large as this or larger is .008, or 0.8%. So we would have a deviation as large as we observed or larger in only 0.8% of repetitions of a random process. There is considerable strength of evidence against the model that the observed sequence arose as a random sequence. A considerable variety of evaluation procedures is possible; we have given only two. Of these two tests one appears to have little sensitivity to interesting alternative models, but it is commonly used. Other even more sensitive tests are possible if alternative models can be specified. In general they require more advanced knowledge and computational ability.

## 9.8 Concluding Note

We have attempted by our choice of words to maintain a sharp distinction between a data set and a random sample. Unless we have obtained our data by a random sampling process, using a physical or artificial random number generator from a population that is completely identified, we are not entitled to claim that we have a random sample from a population. The representation of data by the statement that the data are like or similar to a random sample from a certain mathematical distribution is aimed at representing our data in a simple way. Our only alternative to this is to state that we have a unique set of data which we present in its totality. Every individual we examine of a real population can be uniquely identified with little trouble. We can make no progress in the accumulation of empirical knowledge unless we partly ignore this uniqueness. We do this by using the analogy of random sampling from a population. In the absence of actual random sampling the population that we construct in our minds is hypothetical. As data on a real population are accumulated, we shall change our hypothetical population because we shall observe aspects of the accumulated data that are not consonant with the random sampling model arising from earlier data examination. If the accumulation of data leads to a specification of several populations within each of which the probability of a particular type of result is either zero or unity, we say that we have reached an empirical law. This never happens in actuality because of errors of measurement and because of inability to sample at random populations of future data sets. Additionally, to establish absolutely that a probability is exactly zero or exactly unity (or any other specified value) is impossible, but we may reach situations in which our relevant data indicate a probability extremely close to zero or unity. When we achieve this, no further model modification can produce improvement.

**PROBLEMS**

1. Use the methods of this chapter, i.e., the $\chi^2$ goodness of fit test and probability plotting, to choose distributions which represent the following sets of data. Your choice need not be unique for a given set of data.

(a)

| | | |
|---|---|---|
| 16.95 | 62.74 | 55.62 |
| 8.27 | 9.48 | 86.34 |
| 88.39 | 9.21 | 48.12 |
| 68.21 | 50.32 | 70.50 |
| 58.65 | 37.17 | 49.62 |
| 13.74 | 76.59 | 41.60 |
| 85.64 | 63.67 | 13.72 |
| 46.28 | 41.11 | 27.65 |
| 6.21 | 43.33 | 17.01 |
| 26.36 | 83.38 | 31.40 |

(b)

| | | | | |
|---|---|---|---|---|
| 1.632 | 1.474 | 1.539 | 1.666 | 1.959 |
| 1.273 | 0.938 | 0.912 | 1.016 | 2.188 |
| 1.286 | 1.620 | 1.236 | 1.912 | 0.785 |
| 2.547 | 2.705 | 1.543 | 1.678 | 2.366 |
| 2.916 | 1.009 | 1.792 | 3.357 | 1.606 |

(c)

| | | | | | |
|---|---|---|---|---|---|
| 151 | 512 | 111 | 289 | 726 | 22 |
| 5 | 86 | 125 | 402 | 510 | 26 |
| 401 | 211 | 69 | 406 | 213 | 152 |
| 158 | 317 | 604 | 55 | 373 | 243 |
| 29 | 21 | 423 | 303 | 142 | 39 |

(d)

| | | |
|---|---|---|
| 0.3 | 8.2 | 6.5 |
| 4.6 | 8.4 | 2.3 |
| 2.7 | 5.1 | 3.2 |
| 1.3 | 0.8 | 2.6 |
| 7.9 | 6.5 | 7.4 |
| 6.4 | 5.3 | 4.8 |
| 3.2 | 2.2 | 5.2 |
| 1.2 | 8.8 | 3.6 |
| 0.4 | 7.0 | 1.6 |
| 9.9 | 9.8 | 1.7 |

(e)

| | | |
|---|---|---|
| 2 | 2 | 2 |
| 4 | 4 | 2 |
| 6 | 10 | 4 |
| 2 | 2 | 4 |
| 4 | 2 | 2 |
| 8 | 6 | 2 |
| 2 | 6 | 2 |
| 4 | 2 | 4 |
| 8 | 2 | 2 |
| 10 | 6 | 2 |

2. Use the methods of this chapter to form an opinion about the bivariate normal as an adequate model to describe the following data:

| $x$ | $y$ | $x$ | $y$ | $x$ | $y$ |
|---|---|---|---|---|---|
| 13.25 | 106 | 11.32 | 102 | 8.09 | 87 |
| 10.60 | 98 | 12.67 | 104 | 9.82 | 92 |
| 12.13 | 107 | 10.69 | 96 | 10.81 | 96 |
| 12.22 | 105 | 10.89 | 96 | 10.34 | 93 |
| 10.84 | 97 | 11.01 | 101 | 11.26 | 98 |
| 9.63 | 91 | 12.02 | 108 | 13.22 | 111 |
| 10.38 | 94 | 13.26 | 112 | 9.93 | 91 |
| 11.17 | 103 | 14.09 | 117 | 10.40 | 95 |
| 13.51 | 107 | 11.55 | 99 | 11.00 | 97 |
| 10.63 | 95 | 9.91 | 93 | 10.19 | 94 |

3. Use the $\chi^2$ test of significance to judge the adequacy of a Poisson model with $\lambda = 5.0$ to represent the following data:

| | | | | | |
|---|---|---|---|---|---|
| 22 | 3 | 2 | 8 | 7 | 16 |
| 5 | 5 | 6 | 5 | 2 | 9 |
| 3 | 6 | 8 | 2 | 6 | 3 |
| 7 | 8 | 4 | 9 | 5 | 5 |
| 0 | 10 | 3 | 7 | 2 | 5 |
| 4 | 1 | 5 | 11 | 4 | 6 |

4. Use a $\chi^2$ test to form an opinion about the following data representing a random sample from a negative exponential distribution with a mean of 4.3.

| | | |
|---|---|---|
| 10.6 | 5.3 | 7.1 |
| 0.9 | 5.4 | 8.2 |
| 7.1 | 2.5 | 10.0 |
| 1.6 | 6.3 | 5.8 |
| 2.3 | 1.9 | 0.6 |
| 3.2 | 0.3 | 0.4 |
| 0.2 | 4.4 | 1.6 |
| 2.1 | 4.2 | 1.7 |
| 4.3 | 5.0 | 8.4 |
| 8.4 | 6.2 | 9.3 |

5. For the data of Problems 3 and 4 plot $S_n(x)$ and $F(x)$ versus $x_{(i)}$ on the same graph.
6. For the data of Problems 3 and 4 plot $S_n(x)$ versus $F(x)$.
7. Construct probability paper for the negative exponential distribution and for the gamma distribution.
8. Show that $\sum_1^n (O_i - E_i)^2/E_i = \sum_i O_i^2 - n$ when $E_i = 1$ for all $i$.
9. Verify the formula given in Equation (9.18).
10. Suppose we have $k$ multinomial classes with probabilities $p_1, p_2, \ldots, p_k$ and that a sample of $n$ observations gives $n_i$ in the $i$th class. Determine

$$E\left[\sum_i \frac{(n_i - np_i)^2}{np_i}\right]$$

11. Determine the variance of the $\chi^2$ criterion in the previous problem.
12. Suppose that $k$ multinomial classes have equal probabilities which are $1/k$. Then the distribution of numbers in the classes $n_i$, $i = 1, 2, \ldots, k$ is given by the multinomial distribution

$$\frac{n!}{\prod_i n_i!}\left(\frac{1}{k}\right)^n$$

Obtain the factorial moments. Then obtain the first, second, and third moments of

$$\sum_i \frac{k}{n}\left(n_i - \frac{n}{k}\right)^2$$

13. Express the binomial density function in terms of $\theta_1 = p/(1 - p)$ and $\theta_2 = 1 - p$ and give the range of $\theta_1$ and $\theta_2$.

14. Express the normal density function in terms of the parameter $\theta_1 = \mu$ and $\theta_2 = e^{-(1/2\sigma^2)}$, giving the range of $\theta_1$ and $\theta_2$.

15. Express the negative exponential density function in terms of the parameter $\mu = e^{-1/\theta}$, giving the range of $\mu$.

16. A researcher counts the time required to observe 50 particles emitted from blood samples of rats exposed to a radioactive source. The count is performed automatically with an expensive piece of equipment. However the researcher suspects that background radioactivity may be interfering with his data. Thirty successive times are recorded below:

    38, 43, 46, 51, 47, 49, 53, 55, 57, 61, 62, 56, 46, 43, 38, 36, 47, 72, 46, 40, 49, 52, 54, 63, 67, 69, 70, 42, 35, 30

    Use the run test to test that these data represent a random sample from a single continuous distribution. Use the normal approximation and the exact formula to evaluate the significance level.

17. Prove that the sum of $n$ consecutive integers is $n(n + 1)/2$.

18. Prove that

$$\sum_{i=1}^{n} i^2 = n(n + 1)(2n + 1)/6$$

19. Prove that

$$\sum_{i=1}^{n} (X_i - \bar{X})^2 = n(n^2 - 1)/12$$

    for $n$ consecutive integers.

20. Devise a random number generator to generate zero's and one's with cycle "time" greater than 100. Using this generator, generate 100 numbers and check for randomness.

21. Toss a coin 100 times, assigning one's and zero's to heads and to tails and apply the same checks for randomness as in Problem 20.

22. Write down a sequence of 100 zero's and one's and check the randomness of your mental generator.

# 10

# Parametric Models and Likelihood Theory

## 10.1 Introduction

We now consider questions of the following sort: Can a sequence of $A$'s and $B$'s—say $A$, $A$, $B$, $B$, $A$, $B$, $A$, $B$, $B$, $A$, $B$, $B$, $A$—be regarded as a realization of a set of independent Bernoulli trials with $P(A) = p$, $P(B) = 1 - p$, in which $p$ is *not* specified? Can a set of numbers $x_1, x_2, \ldots, x_n$ be regarded as a realization of a sequence of $n$ independent Poisson random variables with $P(x_i) = e^{-m}m^{x_i}/x_i!$ for some value of $m$?

There are two distinct aspects to this type of question. Suppose the observed set of numbers $x_1, x_2, \ldots, x_n$ have an order of occurrence in time and/or space. We can ask if the observed sequence is like a realization of a random sequence. We shall examine this by some procedure of the sort described in Sect. 9.7. If the $x_i$ were obtained in some time order, and that order is suppressed we merely have the set $x_1, x_2, \ldots, x_n$ as an unordered set. It is this latter case which we now discuss.

It is a critical part of this type of problem that we envisage the observations as being like a random sample from a parametric class of models, such as $\mathrm{Bi}(n, p)$ with $p$ unspecified, or $N(\mu, \sigma^2)$ with $(\mu, \sigma^2)$ unspecified. This chapter is concerned with the case of an unstructured set of numbers $x_1, x_2, \ldots, x_n$ being representable as a random sample from some class of distributions $F(x)$ that is indexed or labeled by parameters. It is first necessary, however, to discuss how we can arrive at a class of models for consideration.

## 10.2 Indicators of Distribution

What are the steps in the choice of a mathematical distribution as a reporting device? In some respects they are very elementary. We will always ask how the data were collected. In some cases this will determine what distributions are candidates. In many cases, however, the mode of collection of the data is uninformative. It may consist merely of looking at $n$ individuals that are accessible. It is then necessary to examine the data themselves for indicators of distribution. This field is little developed, but there are some elementary examples. Suppose the data consist of $n_0$ 0's, $n_1$ 1's, $n_2$ 2's, $n_3$ 3's, $n_4$ 4's, and $n_5$ 5's, where $n_0 + n_1 + n_2 + n_3 + n_4 + n_5 = n$. This suggests a discrete distribution over the integers 0, 1, 2, 3, 4, 5, and the simplest example is the class of binomials Bi(5, $p$). Now we know that the average of this distribution is $5p$ and its variance is $5p(1 - p)$. We therefore compute the data average and the data variance. Suppose we found $\bar{x} = 3.6$. This would suggest a $p$ of 0.72, and a variance consonant with this model would be $5(0.72)(0.28) = 1.008$. Suppose that the data variance is 0.95. Then the model Bi(5, $p$) for some $p$ is plausible. Suppose we observe $n_0$ 0's, $n_1$ 1's, etc., and there is no reason to propose that the count must be less than a particular number. Then a candidate distribution is the Poisson, which has some mean $m$, and also a variance equal to $m$. If we noted that the sample average and sample variance are close we would consider the Poisson distribution as a candidate. There is *at this stage* no quantification of degree of closeness; we shall take this up later. As a third example suppose the data look like a random sample from a normal distribution $N(\mu, \sigma^2)$. Then we should find that the sample skewness is near zero and sample kurtosis is near 3.

The upshot of this is clear. The general idea is that particular classes of mathematical distributions have different shapes, using the word in a very general sense, and we pick a class of distributions which conform in shape to some extent to the corresponding shape attributes of the set of data. The normal distribution has much appeal as a mode of representing measurement data. This is due to the frequency of occurrence of data which look normal, to the fact that the central limit theorem tells us that a resultant of a large number of additively acting variable small forces will generate a distribution for the sum to be close to a normal distribution, and to the fact that statistical techniques are well developed for normal distributions. A simple way of getting an opinion of whether data are like a sample from a normal distribution is to prepare a histogram and see if this histogram can be approximated by the familiar bell-shaped density function of the normal distribution. We can make normal probability plots. These may be followed by more objective modes of examining whether the approximation by such a bell-shaped curve is acceptable. Alternatively, these plots may suggest some other distribution. The need for familiarity with mathematical distributions is clear.

Finally, we must discuss briefly a critical aspect of choice of distribution. We have mentioned mode of collection and pure inspection of the data. A third way is extremely important. This consists of thinking about the phenomena leading to the data and developing a model by a probability argument. We have included only a short chapter on stochastic processes, but the reader will have seen how a mathematical form for the probability of the observations may be developed by combining what are thought to be elementary and plausible random events to yield a probability of any potential observation.

In general the choice of possible distribution must be made on the basis of data examination. A main basis for choosing a family of distributions, given that account has been taken of the range of the distribution, is to search for one that has relationships between low moments like those between corresponding moments of the data. If, for instance, $\sum (x_i - \bar{x})^3$ is near zero and $\sum (x_i - \bar{x})^4/(n - 1)$ is near $3[\sum (x_i - \bar{x})^2/(n - 1)]^2$, the set of $x$-values is like a normal distribution. We note that Pearson developed a system of univariate distributions which are differentiated by relationships among the first four moments. See Elderton (1938), for example.

We now suppose that by one route or another we are led to consider the applicability of a particular class of models and wish to judge the tenability of the *class*. It seems intuitively reasonable that a test of goodness of fit of the class of models will be based in some way on the degree of agreement of the data with what would be expected from random sampling from a distribution fitted to the data. Going further, we have to be concerned in some way with the deviations of the observed values from expected values. We shall get expected values by some sort of fitting process, by which we mean the choice of numbers, say $\hat{\theta}_1, \hat{\theta}_2, \ldots, \hat{\theta}_r$ for the unspecified parameters $\theta_1, \theta_2, \ldots, \theta_r$. Sets of data which give different values for $\hat{\theta}_1, \hat{\theta}_2, \ldots, \hat{\theta}_r$ should not be expected to be similar. It seems reasonable, therefore, to consider our actual set of data as a member of possible sets which would give the same value for $\hat{\theta}_1, \hat{\theta}_2, \ldots, \hat{\theta}_r$.

## 10.3 A Basic Principle

We state as a completely binding principle that the goodness of fit of a model or a class of models should be based on the probability of the data under the model. Certainly, if the probability of a set of data for a certain probability model is zero, we will reject that model completely. Certainly also, if the probability of a set of data from a model is unity, the model cannot be discarded without additional thought. To explain this a little, we can always construct a probability model which gives probability one to a particular unique, necessarily finite set of data. But a model which achieves this may be completely *ad hoc*, i.e., suggested totally by the data with no possibility of extension to other sets of data like the existent set which we can imagine ourselves obtaining (e.g., by repeating an experiment).

We take then as an axiom that the statistical assessment of a model should be based on ideas of probability of the data. Scientific assessment of a model will be based on many other considerations: utility as representing a broad spectrum of empirical knowledge, utility with regard to suggesting lines of additional study, economy insofar as it is derivable from assumed empirical or elementary model laws, etc.

## 10.4 Factorization of Probability of Data

We consider some examples to motivate the reader to a general idea.

*Example 1*

Consider the data $x_1, x_2, \ldots, x_k$, which are integers in the sets $\{0, 1, \ldots, n_i\}$, when we have the tentative model that the $X_i$ are independent binomial variables with density functions $\mathrm{Bi}(n_i, p)$. Then, writing $X_i = x_i$, $i = 1, 2, \ldots, k$ as $\{X_i = x_i\}$

$$P[\{X_i = x_i\}|\{n_i\}, p] = \prod_{i=1}^{k} \binom{n_i}{x_i} p^{x_i}(1 - p)^{n_i - x_i}$$

$$= \left[\binom{n}{\sum x_i} p^{\Sigma x_i}(1 - p)^{n - \Sigma x_i}\right] \frac{\prod_{i=1}^{k}\binom{n_i}{x_i}}{\binom{n}{\sum x_i}} \tag{10.1}$$

where $n = \sum n_i$.

We factor the probability of the data into two parts. The first factor is the probability that we would get $\sum x_i$ from $\mathrm{Bi}(n, p)$. The second factor is the probability that we have the data configuration $x_1, x_2, \ldots, x_k$ given that $\sum X_i = \sum x_i$. Let us examine an example in detail. Let $n_1 = 4$, $n_2 = 3$, $x_1 = 3$, $x_2 = 1$. Then $n = 7$, $\sum x_i = 4$. The possible data configurations conditional on $\sum x_i = 4$ and their probabilities as given by the second term in Equation (10.1) are as follows:

| $x_1$ | $x_2$ | Probability |
|---|---|---|
| 4 | 0 | $\binom{4}{4}\binom{3}{0}\Big/\binom{7}{4} = 1/35$ |
| 3 | 1 | $\binom{4}{3}\binom{3}{1}\Big/\binom{7}{4} = 12/35$ |
| 2 | 2 | $\binom{4}{2}\binom{3}{2}\Big/\binom{7}{4} = 18/35$ |
| 1 | 3 | $\binom{4}{1}\binom{3}{3}\Big/\binom{7}{4} = 4/35$ |

We see that the probability of getting the actual data configuration conditional on $\sum x_i = 4$ is 12/35, which is moderately large. In contrast, if we had obtained $x_1 = 4$, $x_2 = 0$, we would have a configuration with probability equal to 1/35, which we might well regard as being rather small and such that we would regard the model as somewhat untenable.

To make a test of significance we have to order the possible samples. This means that we have to consider a single attribute of each possible sample. What are reasonable candidates? We expect $x_j$ to be close to $(\sum x_i)n_j/n$, so we could consider

$$C = \sum_{i=1}^{k} \left(\frac{x_i}{n_i} - \bar{x}\right)^2, \text{ where } \bar{x} = \frac{\sum x_i}{n} \tag{10.2}$$

The conditional distribution of $C$ given $\sum X_i = \sum x_i$ can be tabulated, and we can then compute $P(C \geq C_O)$, where $C_O$ is the observed value as a significance level of the goodness of fit test of significance based on this ordering of possible samples.

*Example 2*

Let us now turn to another example which is very much like the one just discussed. Suppose we have count data $x_1, x_2, \ldots, x_k$ and are led by one route or another to consider the model of independent negative binomial variables with parameters $r_i$ and $p$, with $r_i$ known. Then

$$P[\{X_i = x_i\}]$$

$$= \prod_{i=1}^{k} \frac{(x_i - 1)!}{(r_i - 1)!(x_i - r_i)!} p^{r_i}(1 - p)^{x_i - r_i}, \qquad x_i = r_i, r_i + 1, \ldots$$

$$= \left[\prod_{i=1}^{k} \frac{(x_i - 1)!}{(r_i - 1)!(x_i - r_i)!}\right] p^{\Sigma r_i}(1 - p)^{\Sigma x_i - \Sigma r_i}$$

$$= \left[ \frac{(\sum x_i - 1)!}{(\sum r_i - 1)!(\sum x_i - \sum r_i)!} p^{\Sigma r_i}(1-p)^{\Sigma x_i - \Sigma r_i} \right] \times$$

$$\left[ \frac{(\sum r_i - 1)!(\sum x_i - \sum r_i)!}{(\sum x_i - 1)!} \prod_{i=1}^{k} \frac{(x_i - 1)!}{(r_i - 1)!(x_i - r_i)!} \right] \tag{10.3}$$

The first factor is similar to the first factor for the positive binomial case and gives $P(\sum X = \sum x)$. The second factor gives the probability of the sample conditional on $\sum X = \sum x$. Let us examine a small example. Suppose $r_1 = 3$, $r_2 = 1$, $x_1 = 4$, and $x_2 = 3$. The possible data configurations conditional on $\sum x_i = 7$ and their probabilities are as follows:

| $x_1$ | $x_2$ | Probability |
|---|---|---|
| 3 | 4 | 1/20 |
| 4 | 3 | 3/20 |
| 5 | 2 | 6/20 |
| 6 | 1 | 10/20 |

Thus the model of a negative binomial might be judged to be quite tenable, although the probabilities are certainly different than for the positive binomial model. This example is to be examined very carefully because it has deep logical implications. The content of the data with regard to a positive binomial model and a negative binomial model are quite different. We emphasize this, because the student who reads the literature may find the contrary view. There are, in fact, two configurational probabilities, neither of which depends on the parameter $p$; we cannot ignore these.

*Example 3*

A third example is the case of a Poisson model for data $x_1, x_2, \ldots, x_n$. The probability is

$$P[\{X_i = x_i\}] = e^{-nm} m^{\Sigma x_i} / \left(\prod x_i!\right)$$

which factors readily into

$$\left[ e^{-nm} \frac{(nm)^{\Sigma x_i}}{(\sum x_i)!} \right] \left[ \frac{(\sum x_i)!}{n^{\Sigma x_i} \prod x_i!} \right] \tag{10.4}$$

The first factor tells us the probability of getting a sample total equal to $\sum x_i$, and the second is a configurational probability which gives the probability of $x_1, x_2, \ldots, x_n$ given $\sum x_i$. Candidate functions for ordering the possible conditional samples are $\sum(x_i - \bar{x})^2$, $P[(X_i = x_i)|\sum X_i = \sum x_i]$, and $-\sum_i x_i \ln x_i$ with $x \ln x = 0$ if $x = 0$. These are highly related. Ordering by conditional probability is equivalent to ordering by $\prod (x_i!)$ or by $-\sum_i \ln (x_i!)$. But, by Stirling's formula

$$\ln (x_i!) \doteq x_i \ln x_i - x_i$$

so

$$-\sum_i \ln (x_i!) \doteq -\sum_i x_i \ln x_i + \sum_i x_i$$

Also note that if the $x_i$ are large and we put $\hat{m} = \bar{x}$, we can write

$$x_i = \hat{m} + \varepsilon_i,\ \varepsilon_i = x_i - \bar{x},\ \sum \varepsilon_i = 0$$

with the $\varepsilon_i$ being "small," and

$$\begin{aligned}
\sum x_i \ln x_i &= \sum (\hat{m} + \varepsilon_i) \ln (\hat{m} + \varepsilon_i) \\
&= \sum (\hat{m} + \varepsilon_i)\left[\ln \hat{m} + \ln\left(1 + \frac{\varepsilon_i}{\hat{m}}\right)\right] \\
&= \sum (\hat{m} + \varepsilon_i)\left(\ln \hat{m} + \frac{\varepsilon_i}{\hat{m}} - \frac{\varepsilon_i^2}{2\hat{m}^2}\right) + \text{a small quantity} \\
&\doteq n\hat{m} \ln \hat{m} + \ln \hat{m} \sum \varepsilon_i + \sum \frac{\varepsilon_i^2}{2\hat{m}} + \sum \varepsilon_i \\
&= \text{constant} + \frac{\sum \varepsilon_i^2}{2\hat{m}} = \text{constant} + \sum \frac{(x_i - \bar{x})^2}{2\bar{x}}
\end{aligned} \qquad (10.5)$$

Conditionally, $\bar{x}$ is constant, so ordering by conditional probability is close to ordering by $\sum (x_i - \bar{x})^2$ if $\bar{x}$ is large.

As a numerical example, consider a set of count data, $x_1, x_2, \ldots, x_{10}$ such that 8 of the $x$'s equal zero and 2 of the $x$'s equal 2. Suppose a candidate model is that the data arise by random sampling from a Poisson distribution with unspecified parameter $\lambda$. Then we can enumerate the conditional probability distribution as in Table 10.1 where the number of occurrences of a sample value is indicated in square brackets if it is greater than unity. If now we use the ordering based on conditional probability, we shall say that the significance level associated with our actual data, which is represented by 0[8], 2[2], is $28/10^3$. Because this is small, we consider there is strong evidence that the Poisson model is untenable. A "larger" example of the present type was considered by Fisher (1950b). This paper should be consulted for a discussion of the relative merits of the orderings.

**Table 10.1** Conditional Probabilities for Example 3

| Sample Values | Conditional Probability |
|---|---|
| 0[9], 4 | $1/10^3$ |
| 0[8], 2[2] | $27/10^3$ |
| 0[8], 1, 3 | $36/10^3$ |
| 0[7], 1[2], 2 | $432/10^3$ |
| 0[6], 1[4] | $504/10^3$ |

*Example 4*

Consider a set of $n$ observations which are ones and zeros in some order, and suppose we entertain the model that these have arisen by independent Bernoulli trials with parameter $p$. Then we can write

$$P[\{X_i = x_i\}] = \left[\binom{n}{\sum x_i} p^{\Sigma x_i}(1-p)^{n-\Sigma x_i}\right]\left[1\Big/\binom{n}{\sum x_i}\right] \tag{10.6}$$

The first factor is the probability that we will obtain $\sum X_i = \sum x_i$ and the second is the probability that we will get the sequence $x_1, x_2, \ldots, x_n$ given that $\sum X_i = \sum x_i$. Goodness of fit of the model will be based on this conditional distribution, which states that all sequences with $\sum x_i$ successes are equiprobable. To evaluate goodness of fit, we have to choose a criterion on the points of the $n$-dimensional hyperplane $X_i$ equal zero or one, $\sum X_i = \sum x_i$. Under the conditional distribution the order numbers of the $x$'s which are one is a random sample of size $\sum x_i$ from the integers $1, 2, \ldots, n$. Suppose therefore that we define

$$\begin{aligned} T_i &= i \qquad \text{if } X_i = 1 \\ &= 0 \qquad \text{if } X_i = 0 \end{aligned} \tag{10.7}$$

Then a reasonable ordering criterion is $\sum_{i=1}^{n} T_i$. What can we say about the distribution of this, which is induced by the proposed conditional distribution? The complete distribution is discrete and can be enumerated easily for small values of $n$ and $\sum x_i$. We use the material of Chapter 6 to describe it approximately. We know that if $Z_i$ is a random member of the set $z_1, z_2, \ldots, z_n$, then

$$\begin{aligned} E(Z_i) &= \textstyle\sum z_i/n = \bar{z} \\ E(Z_i^2) &= \textstyle\sum z_i^2/n \\ V(Z_i) &= \textstyle\sum (z_i - \bar{z})^2/n = \sigma^2 \end{aligned} \tag{10.8}$$

Also, we know that the correlation of two members, randomly selected without replacement, is $-1/(n-1)$, so the covariance of two members is $-\sigma^2/(n-1)$. The expectation of the sum of $r$ members chosen at random without replacement is therefore $r\bar{z}$, and the variance is

$$
\begin{aligned}
r\sigma^2 - r(r-1)\left(\frac{\sigma^2}{n-1}\right) &= r\sigma^2\left(\frac{n-r}{n-1}\right) \\
&= r\sigma^2\left(\frac{n}{n-1}\right)\left(1-\frac{r}{n}\right) \\
&\doteq r\sigma^2\left(1-\frac{r}{n}\right) \qquad (10.9)
\end{aligned}
$$

In our case

$$
\begin{aligned}
\sum z_i &= 1 + 2 + \cdots + n = n(n+1)/2 \\
\sum z_i^2 &= 1^2 + 2^2 + \cdots + n^2 = n(n+1)(2n+1)/6
\end{aligned}
$$

So

$$\sigma^2 = (n^2-1)/12$$

Therefore

$$
\begin{aligned}
E(\sum T_j) &= r(n+1)/2 \\
V(\sum T_j) &\doteq r\left(1-\frac{r}{n}\right)(n^2-1)/12 = \phi^2 \qquad (10.10)
\end{aligned}
$$

We can therefore use a normal approximation to state that an approximate significance level with regard to the criterion $\sum T_j$ is

$$2\int_{|u|-1/2\phi}^{\infty} \frac{1}{\sqrt{2\pi}} \exp\left(-\frac{1}{2}t^2\right) dt = 2\int_{|u|-1/2\phi}^{\infty} dN(0,1) \qquad (10.11)$$

where

$$u = \left[\sum t_j - \frac{r(n+1)}{2}\right]\Big/\phi$$

in which we have $r$ equal to $\sum x_i$. If the data or related knowledge suggest that there might be a preponderance of ones early in the sequence, we could use the same criterion but use as the significance level

$$\int_{-\infty}^{u+1/2\phi} dN(0,1)$$

If an alternative candidate were a preponderance of ones toward the end of the sequence, we could use

$$\int_{u-1/2\phi}^{\infty} dN(0,1)$$

as a level of significance. There are many other interesting alternative candidate models.

*Example 5*

As a final example, consider a set of nonnegative integers $x_1, x_2, \ldots, x_n$ which are hypothesized to be a random sample from $U[0(1)\theta]$ where $\theta$ is a positive integer. So the probability of any $x_1, x_2, \ldots, x_n$ is $1/(\theta + 1)^n$, provided $0 \leq x_i \leq \theta$, $i = 1, 2, \ldots, n$, and zero, otherwise. Let $u$ be the maximum of the $x$'s. Then

$$\begin{aligned} P(u) &= \frac{(u + 1)^n - u^n}{(1 + \theta)^n}, \qquad u > 0 \\ &= \frac{1}{(1 + \theta)^n}, \qquad u = 0 \end{aligned} \tag{10.12}$$

and the probability of any set $x_1, x_2, \ldots, x_n$ conditional on the maximum being $u$ is $1/[(u + 1)^n - u^n]$, $u > 0$, and is unity if $u = 0$, over the points in the lattice defined by $x_i$ being a nonnegative integer and maximum $\{x_i\}$ equal to $u$. Any test of goodness of fit of the class of models would consist of ordering these conditional samples by a criterion $C$ and then obtaining $P(C \geq C_O)$.

The idea that we used in these examples of testing goodness of fit was that the probability of the data could be factored into parts, one part being the distribution of a statistic and the second part being the distribution of the observations given the statistic. So that the distribution of possible data sets given a statistic $S$, say, will be useful as a basis for testing goodness of fit, it is necessary that Prob(data$|S$) be mathematically independent of the unspecified parameter $\theta$ ($S$ and $\theta$ may be vectors). If Prob(data$|S$) depended on the parameter value, then the goodness of fit test would depend upon the parameter values in our class of distributions. This is undesirable, and it is natural to ask under what conditions such a factorization is possible.

For all the examples above it was possible to obtain a statistic which in an intuitive sense contains "all the information" about the parameter. Furthermore, the statistic had a distribution dependent on the parameter and is intuitively irreducible in the sense that no reduction of the statistic—such as taking its absolute value if the statistic takes both positive and negative values—retains "all the information." It is necessary now to try to give logical content to ideas of reduction of a statistic, condensation of data, and information in a statistic. This leads us directly into likelihood theory. We shall then return briefly to the question of goodness of fit.

## 10.5 Likelihood

Suppose we have a data set $D$ and a class of probability models $M(\theta)$ indexed or labeled by a parameter $\theta$ (which may be a vector). Then we are led to the view that the inferential content of the data with regard to the parameter $\theta$ of the class of models $M(\theta)$ is given by $\text{Prob}[D; M(\theta)]$.

*Definition 10.1. A function $h(D; \theta)$ is an exhaustive function for $\theta$ with the class of models $M(\theta)$ if the distribution of the data $D$ given that $h(D; \theta)$ equals $H(\theta)$ does not depend on $\theta$.*

If we are informed that $h(D; \theta) = H(\theta)$, of what additional use is it to be told that the data were actually $D$? The conditional distribution of $D$ given that $h(D; \theta) = H(\theta)$ does not depend on $\theta$ and has no relevance to making any judgment about $\theta$. This line of thought suggests that any judgment about $\theta$ should be based on the distribution of an exhaustive function.

***Theorem 10.1*** The function $L(\theta) = cP(D; \theta)$ is an exhaustive function, where $c$ is an arbitrary scalar independent of $\theta$.

*Proof:* Consider the case of a denumerable set of possible data sets $D_1, D_2, \ldots$. Let the possible functions for $L(\theta)$ be $l_1(\theta), l_2(\theta), \ldots$. Suppose data sets $D_{ij}, j = 1, 2, \ldots$, give the function $l_i(\theta)$ so that

$$P(D_{ij}; \theta) = C_{ij} l_i(\theta) \tag{10.13}$$

where $C_{ij}$ depends on $D_{ij}$, but not on $\theta$. Then

$$\sum_j P(D_{ij}; \theta) = P[L(\theta) = l_i(\theta)] = l_i(\theta) \sum_j C_{ij}$$

or

$$P[(D_{ij}; \theta) | L(\theta) = l_i(\theta)] = C_{ij} / \sum_j C_{ij} \tag{10.14}$$

which is independent of $\theta$.

Any function $L(\theta)$ as defined in the theorem is called the likelihood function; the import of the theorem is that it is an exhaustive function, and we need know only the observed likelihood function and the class of possible likelihood functions to make judgments about $\theta$. Note however, and this is very important, we do not yet say how one is to use the likelihood function for this purpose. To anticipate a little, we do note that a method of fitting the model that is widely used is to determine $\theta^*$ such that $L(\theta^*)$ is the maximum with respect to $\theta$ of $L(\theta)$. This is the method of maximum likelihood.

In describing this function, we find it logically essential to make a dichotomy of situations. There are two distinct classes of observational situations, discrete and continuous. In the former the possible observations are categories, such as eye color, defective or not defective, or counts. In the continuous case a value is observed only with a grouping error of measurement, which is specified by the measurement process. We therefore present a discussion of likelihood separately for these two cases. This has a further advantage in that the ideas of conditional distributions are very simple and clear-cut with discrete distributions.

### 10.5.1 The Discrete Case

We have thought of the probability density function $f(x;\theta)$ as a function of $x$ for each $\theta$. Obviously we can also think of it as a function of $x$ *and* $\theta$ or as a function of $\theta$ for each fixed $x$. In general, the joint density function of the observations in a data set is given by

$$P(x_1, x_2, \ldots, x_n; \theta_1, \theta_2, \ldots, \theta_k) = P(D; \theta) \tag{10.15}$$

where we use $D$ to denote the data vector $(x_1, x_2, \ldots, x_n)$. When this is regarded as a function of $\theta$ for a fixed sample, it is referred to as the likelihood function of the sample. The likelihood function is not unique because $cP(D;\theta)$ is also called the likelihood function for any $c$ not involving $\theta$. To present the idea in a very general setting, we suppose that under the model $M(\theta)$, with $\theta$ belonging to a set $\Theta$, the possible data sets are $D$ belonging to a set $\mathscr{D}$. Then the likelihood function is $cP(D;\theta)$, with $c$ arbitrary, defined for $D$ belonging to $\mathscr{D}$ and $\theta$ belonging to $\Theta$. If we wish, we can remove the indeterminancy due to $c$ by taking $1/c$ equal to the maximum over $D$ belonging to $\mathscr{D}$ and maximum (or supremum) over $\theta$ belonging to $\Theta$, of $P(D;\theta)$. It is very important to note that if one is sure of the probability model, the realized likelihood function is a random member of a class of functions.

Let us suppose that we have observed the number of visual defects on ten silicon crystals and obtained the data: 2, 3, 1, 4, 0, 5, 1, 2, 1, 0. On the basis of certain experience we surmise that the data might have arisen by independent Poisson trials. Suppose the model is judged adequate by goodness of fit tests. Then the likelihood function for this data set is given by

$$L(\lambda) = \frac{e^{-n\lambda}\lambda^{\Sigma x_i}}{\prod x_i!} = \frac{e^{-10\lambda}\lambda^{19}}{69{,}120} \tag{10.16}$$

Note that $n$ and $\sum x_i$ specify the likelihood function; if $n$ is considered to be always known, then $\sum x_i$ specifies the likelihood function.

It is appropriate to introduce here some basic definitions. Here, and often in the future, we shall use $\{h_i(X) = h_i(x)\}$ to denote $h_i(X) = h_i(x)$, for all values of the subscript $i$.

*Definition 10.2.* Statistic. *Given a random variable $X$ which may be a vector, a statistic is a set of functions $h_i(X)$, $i = 1, 2, \ldots, r$. The variable $X$ may be and very often is a label of an observational class, perhaps in vector form.*

*Definition 10.3.* Sufficient statistic. *Given a random variable $X$ with realizable values labeled by $x$ and following a distribution dependent on a parameter $\theta$ from a set of values $\Theta$ ($X$, $x$, and $\theta$ may be vectors), a statistic $h_i(X)$, $i = 1, 2, \ldots, r$, is sufficient if and only if*

$$P(X = x; \theta) = P[\{h_i(X) = h_i(x)\}; \theta] \times P[X = x | \{h_i(X) = h_i(x)\}] \qquad (10.17)$$

*in which the second factor on the right-hand side does not depend on $\theta$ in $\Theta$.*

Obviously the original data are sufficient, but this is not useful and we are interested in condensing or reducing the data.

Before turning to this, we note that $\{h_i(X), i = 1, 2, \ldots, r\}$ is sufficient if and only if

$$P(X = x; \theta) = f[\{h_i(x)\}; \theta]g(x) \qquad (10.18)$$

for *all* values of $x$, with positive probability for some value of $\theta$ belonging to $\Theta$, where the second factor does not depend on $\theta$. If Equation (10.17) holds, then Equation (10.18) holds. If Equation (10.18) holds, then

$$P[\{h_i(X) = h_i(x)\}; \theta] = \sum\nolimits^* P(X = x^*; \theta) \qquad (10.19)$$

where $\sum^*$ denotes summation over $x^*$ such that $h_i(x^*) = h_i(x)$, $i = 1, 2, \ldots, r$. But

$$\sum\nolimits^* P(X = x^*; \theta) = f[\{h_i(x)\}; \theta] \sum\nolimits^* g(x^*) \qquad (10.20)$$

Hence

$$f[\{h_i(x)\}; \theta] = \frac{P[\{h_i(X) = h_i(x)\}; \theta]}{\sum^* g(x^*)} \qquad (10.21)$$

and the factorization Equation (10.17) results.

If $h_i(X)$, $i = 1, 2, \ldots, r$ form a sufficient statistic, the conditional distribution of any other statistic given the sufficient statistic does not depend on $\theta$. This is sometimes used as a definition of a sufficient statistic. For this reason we are interested in finding a "smallest" sufficient statistic.

*Definition 10.4.* Minimal sufficient statistic. *A statistic is a minimal sufficient statistic if it is a sufficient statistic and is a function of every sufficient statistic.*

This definition was given by Lehmann and Scheffé (1950). Previously such a statistic had been called exhaustive by Fisher. Throughout, we shall use the symbol $T$ for a statistic, without regard to whether it is a scalar, a vector, or merely a label of class in a partition of the sample space. (See below.)

***Theorem 10.2*** A sufficient statistic $T$ is a minimal sufficient statistic if for no pair of values of $T$, $T_1$ and $T_2$, it is the case that

$$P(T_1; \theta) = kP(T_2; \theta) \tag{10.22}$$

for all relevant $\theta$, with $k$ depending on $T_1$ and $T_2$ and not on $\theta$.

*Proof:* If it were the case that Equation (10.22) held for some $T_1$ and $T_2$, we could define a new statistic

$$U = T \text{ when } T \neq T_1, T \neq T_2$$

and

$$U = T_1 \text{ when } T = T_1 \text{ or } T = T_2 \tag{10.23}$$

The statistic $U$ is a function of $T$ and is obviously sufficient. However, $T$ is not a function of $U$ and hence is not minimal sufficient. If Equation (10.22) holds for two "values" $T_1$ and $T_2$ of the statistic $T$, we say that these two values are poolable as in Equations (10.23).

*Corollary 10.1.* With a space of two parameter values, say $\theta_1$, $\theta_2$, the likelihood ratio statistic $P(x; \theta_1)/P(x; \theta_2)$ is minimal sufficient.

*Proof:* Let $r = r(x) = P(x; \theta_1)/P(x; \theta_2)$. Then $r$ is the realized value of a random variable $R$, say. Also

$$P(R = r; \theta_2) = \textstyle\sum_x^* P(x; \theta_2) = P_R(r; \theta_2)$$

say, where $\sum_x^*$ denotes summation over values of $x$ such that $r(x) = r$;

$$\begin{aligned} P(R = r; \theta_1) &= \sum_x{}^* P(x; \theta_1) \\ &= \sum_x{}^* P(x; \theta_2) \frac{P(x; \theta_1)}{P(x; \theta_2)} \\ &= rP_R(r; \theta_2) \end{aligned}$$

Hence

$$P(X = x|R = r;\theta_2) = \frac{P(x;\theta_2)}{P_R(r;\theta_2)}$$

$$\begin{aligned} P(X = x|R = r;\theta_1) &= \frac{P(x;\theta_1)}{P_R(r;\theta_1)} \\ &= \frac{rP(x;\theta_2)}{rP_R(r;\theta_2)} \\ &= P(x;\theta_2)/P_R(r;\theta_2) \\ &= P(X = x|R = r;\theta_2) \end{aligned}$$

Hence the conditional distribution of $X$ given $R = r$ is the same for the 2 values of $\theta$. Hence $R = r(X)$ is sufficient.

To show minimality, suppose $R$ is not a minimal sufficient statistic so that there exist values $r_1$, $r_2$ with $r_1 \neq r_2$ such that

$$\frac{P(r_1;\theta_1)}{P(r_2;\theta_1)} = \frac{P(r_1;\theta_2)}{P(r_2;\theta_2)}$$

We drop the subscript $R$ on $P$ because no confusion will result. Then

$$\frac{P(r_1;\theta_1)}{P(r_1;\theta_2)} = \frac{P(r_2;\theta_1)}{P(r_2;\theta_2)}$$

But

$$\frac{P(r_1;\theta_1)}{P(r_1;\theta_2)} = r_1, \frac{P(r_2;\theta_1)}{P(r_2;\theta_2)} = r_2$$

Hence $r_1 = r_2$, which is a contradiction.

***Theorem 10.3*** A minimal sufficient statistic is unique, except for 1 to 1 transformations.

*Proof:* Suppose $T$ and $W$ are both minimal sufficient statistics and the pairs of realized values $T_1$, $W_1$ and $T_1$, $W_2$ each occur with nonzero probability. Then

$$\begin{aligned} P(T_1, W_1;\theta) &= P(T_1;\theta)P(W_1|T_1) \\ P(T_1, W_2;\theta) &= P(T_1;\theta)P(W_2|T_1) \\ P(T_1, W_1;\theta) &= P(W_1;\theta)P(T_1|W_1) \\ P(T_1, W_2;\theta) &= P(W_2;\theta)P(T_1|W_2) \end{aligned} \tag{10.24}$$

It follows that

$$\begin{aligned} P(T_1;\theta) &= P(T_1, W_1;\theta)/P(W_1|T_1) \\ &= P(W_1;\theta)P(T_1|W_1)/P(W_1|T_1) \end{aligned} \tag{10.25}$$

and, similarly,

$$P(T_1; \theta) = P(W_2; \theta)P(T_1 | W_2)/P(W_2 | T_1) \tag{10.26}$$

so that $P(W_1; \theta) = kP(W_2; \theta)$, where $k$ does not involve $\theta$. This is a contradiction to the assumption that the statistic $W$ is minimal sufficient.

The following theorem is almost a tautology.

***Theorem 10.4*** The likelihood function is determined completely, up to the arbitrary multiplying constant, by a minimal sufficient statistic.

One could start with the definition that a sufficient statistic must enable one to recover the likelihood function, given the class of models. If one has a set of data $x_1, x_2, \ldots, x_n$ with no ties and the model is that the $x_i$ are independent realizations of a random sampling from a distribution, the order statistic $X_{(1)} < X_{(2)} < \ldots < X_{(n)}$ is sufficient. It is intuitively obvious but not easy to prove with complete rigor that if $F$ is arbitrary, the order statistic is minimal sufficient.

We now introduce a definition, which seems to be original and to convey partially what R. A. Fisher meant by sufficient in his writings. It is very convenient if there is, in one sense or another, a 1 to 1 correspondence of the values taken by the statistic and the values taken by the parameter.

*Definition 10.5*. F-sufficient statistic. *A statistic $T$ is called F-sufficient if* (a) *$T$ is minimal sufficient and* (b) *the relation between $T$ and the value of $\theta$ which maximizes $P(T; \theta)$ is 1 to 1.*

The desirability of a definition of this sort arises from the confusion that exists in the literature. In parts, "sufficient" means sufficient as we have defined it, in other parts, it means minimal sufficient, and occasionally (particularly in Fisher's writings) it appears to mean what we call $F$-sufficient.

In the case of existence of an $F$-sufficient statistic we are in the fortunate position of being able to reduce the data to a single statistic by maximizing the likelihood function. The maximum likelihood statistic is minimal sufficient and conveys the impact of the data on our knowledge or opinions with regard to $\theta$. Just how we use an $F$-sufficient statistic will depend on our aims. If, for instance, we have a scalar parameter $\theta$ and an $F$-sufficient statistic $h(X)$, we may decide to use that function of $h(X)$ whose expectation is $\theta$ as an "estimate" or nominated guess of the value of $\theta$. Or we may use that function of $h(X)$, say $g[h(X)]$, whose mean square error, i.e., $E\{g[h(X)] - \theta\}^2$ is a minimum.

In the case of a discrete sample space, a statistic is a partitioning of the sample space—regions in the sample space being given by values of the statistic. One may define the partitioning by the statistic, or the statistic by the partitioning. Let $X = (X_1, X_2)$ be a two-dimensional random variable. Then, for example,

$$A_1 = \{X = (X_1, X_2): X_1 - X_2 = 0\}$$
$$A_2 = \{X = (X_1, X_2): X_1 - X_2 \neq 0\}$$

is a partitioning of the sample space. We could define a statistic $S$ by saying $S = 1$ if $X \in A_1$, and $S = 2$ if $X \in A_2$. For many purposes it is more useful and more intuitive to think about a partition of the sample space. A partition of the sample space $A$, say, is a division of $A$ into disjoint sets of points $A_1, A_2, \ldots$ . The utility arises because we can associate a number with each of the disjoint sets in any way we like, and the number of the observed set is a statistic.

We present some useful definitions and ideas associated with the idea of partitions.

*Definition 10.6. A partition of the sample space $A_1, A_2, \ldots, A_k$ is wider or coarser than a partition $B_1, B_2, \ldots, B_m$ if and only if every B region is contained in an A region and at least one A region contains at least two B regions. A statistic corresponding to the A regions is said to be a condensation of a statistic corresponding to the B regions.*

*Definition 10.7. A partition $A_1, A_2, \ldots, A_k$ is sufficient for a parametric class of models indexed by a parameter $\theta$ if and only if $P(X = x|X \in A_j)$ is independent of $\theta$ for all x having nonzero probability for at least one value of $\theta$ and for all $j = 1, 2, \ldots, k$.*

*Definition 10.8. A partition $A_1, A_2, \ldots, A_k$ is minimal sufficient if and only if any wider or coarser partition is not sufficient.*

One of the useful aspects of this mode of thought is that one constructs statistics, minimal statistics, and so on, by pooling observations and classes of observation by a strictly aggregative process. It is also worth noting that a partition induces an algebra of sets.

It is useful for Chapter 14 to adjoin another definition.

*Definition 10.9. A partition $A_1, A_2, \ldots, A_k$ is complete if*

$$\sum_{j=1}^{k} g(A_j)P(A_j; \theta) = 0$$

*for all $\theta$ in the class of $\theta$ under consideration implies $g(A_j) = 0$ for all $j$ with $P(A_j; \theta) > 0$ for some $\theta$.*

These definitions are strictly applicable only to the discrete case. Extension to continuous probability distributions is discussed below.

### 10.5.2 The Continuous Case

The theory of measurement is very complex. We cannot observe a continuous variable exactly. The consequences of this fact are quite obscure and entail extensive discussion. We refer the reader to the book by Campbell (1957), *Foundations of Science: The Philosophy of Theory and Experiment*. This contains a detailed examination of what we call inevitable grouping error of measurement. It also contains (pp. 487–88) a highly persuasive argument to the effect that "errors of measurement" do not follow the normal or Gaussian curve of error. Campbell also discusses the fact that if a measurement is a nonlinear function of other measurements, as is often the case, it is mathematically inconsistent to suppose that a Gaussian distribution holds exactly for all the measurements, including the derived one.

Let us suppose we have observed heights of 10 individuals and find the values (in inches): 68, 73, 69, 70, 67, 68, 69, 70, 71, 74. Also suppose we wish to summarize these data with the model that they arise from a normal distribution, $N(\mu, \sigma^2)$, with a grouping error of $\pm 1/2$ inch. The probability that we will obtain the value 69 is then

$$\int_{69-1/2}^{69+1/2} \frac{1}{\sqrt{2\pi}\sigma} \exp\left[-\frac{(t-\mu)^2}{2\sigma^2}\right] dt \tag{10.27}$$

In general, if a recorded observation $x_i$ means that the actual observation is somewhere between $x_i - \Delta/2$ and $x_i + \Delta/2$, and the continuous probability model has density $f(t; \theta_1, \theta_2, \ldots, \theta_p)$, the probability of the data $x_1, x_2, \ldots, x_n$ is

$$\prod_{i=1}^{n} \int_{x_i-\Delta/2}^{x_i+\Delta/2} f(t; \theta_1, \theta_2, \ldots, \theta_p) dt \tag{10.28}$$

Now suppose $f(t; \theta_1, \theta_2, \ldots, \theta_p)$ satisfies the Lipschitz condition that for all $\varepsilon \neq 0$ in the interval $(-\Delta/2, \Delta/2)$,

$$\left|\frac{f(t+\varepsilon; \theta_1, \theta_2, \ldots, \theta_p) - f(t; \theta_1, \theta_2, \ldots, \theta_p)}{\varepsilon}\right| \tag{10.29}$$

is less than some positive number $h(t; \theta_1, \theta_2, \ldots, \theta_p)$. Then, for $\varepsilon$ in the interval $(-\Delta/2, \Delta/2)$, $f(x + \varepsilon; \theta_1, \theta_2, \ldots, \theta_p)$ is bounded above by $f(x; \theta_1, \theta_2, \ldots, \theta_p) + h|\varepsilon|$ and below by $f(x; \theta_1, \theta_2, \ldots, \theta_p) - h|\varepsilon|$. So the probability of any recorded $x$ is between

$$\int_{-\Delta/2}^{+\Delta/2} [f(x; \theta_1, \theta_2, \ldots, \theta_p) - h(x; \theta_1, \theta_2, \ldots, \theta_p)|\varepsilon|] d\varepsilon$$

and

$$\int_{-\Delta/2}^{+\Delta/2} [f(x; \theta_1, \theta_2, \ldots, \theta_p) + h(x; \theta_1, \theta_2, \ldots, \theta_p)|\varepsilon|] d\varepsilon \tag{10.30}$$

Hence, it lies between

$$f(x; \theta_1, \theta_2, \ldots, \theta_p)\Delta - h(x; \theta_1, \theta_2, \ldots, \theta_p)\frac{\Delta^2}{4}$$

and

$$f(x; \theta_1, \theta_2, \ldots, \theta_p)\Delta + h(x; \theta_1, \theta_2, \ldots, \theta_p)\frac{\Delta^2}{4} \quad (10.31)$$

If, furthermore, $\Delta$ is small, each of these is close to

$$f(x; \theta_1, \theta_2, \ldots, \theta_p)\Delta \quad (10.32)$$

so that the probability of the data $x_1, x_2, \ldots, x_n$ under the model is close to

$$\Delta^n \prod_{i=1}^{n} f(x_i; \theta_1, \theta_2, \ldots, \theta_p) \quad (10.33)$$

In addition we have agreed that we are concerned only with the dependence of the probability on the unknown parameters. We take therefore, as the likelihood of the observations $x_1, x_2, \ldots, x_n$, the function

$$\prod_{i=1}^{n} f(x_i; \theta_1, \theta_2, \ldots, \theta_p) \quad (10.34)$$

In writing the likelihood in this form, we have assumed a definite magnitude for $\Delta$, the grouping error of measurement, in relation to special properties of the continuous density which we use as a model. We may note, for instance, that if the density does not satisfy the Lipschitz condition, the approximation may break down.

Suppose we have data $x_1, x_2, \ldots, x_n$, and preliminary examination suggests that these data are like a random sample from a normal distribution $N(\mu, \sigma^2)$ with a small grouping error $\Delta$. Then the probability of the data is equal to

$$\Delta^n \prod_{i=1}^{n} \frac{1}{\sqrt{2\pi\,\sigma^2}} \exp\left[\frac{-1}{2\sigma^2}(x_i - \mu)^2\right] \quad (10.35)$$

Note that if observations had different grouping errors, we would alter the factor $\Delta^n$. This probability, with the $x_i$ as the actual data and regarded as a function of $\mu$ and $\sigma^2$, is the likelihood. It is easier to look at the logarithm:

$$\ln L = \text{constant} - \frac{1}{2\sigma^2}\sum_i (x_i - \mu)^2 - \frac{n}{2}\ln\sigma^2$$

$$= \text{constant} - \frac{\sum x_i^2}{2\sigma^2} + \mu\frac{\sum x_i}{\sigma^2} - \frac{n\mu^2}{2\sigma^2} - \frac{n}{2}\ln\sigma^2 \quad (10.36)$$

The constant is $n \ln \Delta - (n/2)\ln(2\pi)$, which does not depend on the data. This function is specified by two functions of the data $\sum x_i$ and $\sum x_i^2$. These two statistics comprise a minimal sufficient statistic. If the sample size were not fixed, we would say that the triplet $(n, \sum x_i, \sum x_i^2)$ is a minimal sufficient statistic.

We can factor approximately the total probability, as we did for the Poisson case. We know that if $\Delta$ is negligibly small, $\bar{X}$ is distributed as $N(\mu, \sigma^2/n)$ and $\sum (X_i - \bar{X})^2/\sigma^2 = (n-1)s^2/\sigma^2$ is distributed as $\chi^2$ with $n - 1$ degrees of freedom. This suggests the approximate factorization:

$$L \doteq \left\{\frac{\sqrt{n}}{\sqrt{2\pi}}\frac{1}{\sigma}\exp\left[-\frac{n}{2\sigma^2}(\bar{x}-\mu)^2\right]\Delta\right\}\times$$
$$\left\{\frac{1}{2^{(n-1)/2}\Gamma\left(\frac{n-1}{2}\right)}\exp\left[-\frac{\sum(x-\bar{x})^2}{2\sigma^2}\right]\left[\frac{\sum(x-\bar{x})^2}{\sigma^2}\right]^{(n-3)/2}\frac{\Delta}{\sigma^2}\right\}\times$$
$$\left\{\frac{\Gamma\left(\frac{n-1}{2}\right)}{\sqrt{n}\pi^{(n-1)/2}}\frac{\Delta^{n-2}}{\left[\sum(x-\bar{x})^2\right]^{(n-3)/2}}\right\} \tag{10.37}$$

The first factor is close to the probability that a random sample mean $\bar{X}$ lies in the interval $(\bar{x} \pm \Delta/2)$, the second is the probability that $\sqrt{\sum (X_i - \bar{X})^2}$ is in the interval $\sqrt{\sum (x_i - \bar{x})^2} \pm \Delta/2$, and the third is the conditional probability that $X_i$ lie in $(x_i \pm \Delta/2)$ given the conditions on $\bar{X}$ and $\sum (X_i - \bar{X})^2$. As in the Poisson case, goodness of fit should be evaluated by means of the third factor, though there is no totally compelling way. We can examine the likelihood function or its logarithm without this factoring. Some problems require the student to sketch a graph of the logarithm for some sets of data.

The extensions of the definitions of statistic, sufficient statistic, minimal sufficient statistic, and partition to the case of a continuous model are simple from one point of view. All probabilities are replaced by probability densities, and the dependence on grouping error of measurement $\Delta$ is concealed. The idea of a finite or enumerable partition extends immediately. But to have a useful, workable, approximative theory, it is convenient to suppress the dependence on $\Delta$ and to use probability densities with associated processes of the integral calculus. The underlying logic with regard to data analysis is that while actual data and distributions of properties of potential data sets are discrete, we approximate properties of these distributions by integrals in the way that a sum can be approximated by an integral. In particular, the Euler summation formula

$$f(1) + f(2) + \cdots + f(x) = \int_1^x f(t)dt + (1/2)f(x) + R$$

where $R$ is a remainder approximates a sum by an integral (e.g., see Bromwich, 1965). This replacement results in much ease of computation and causes little trouble because the logic of processes of statistics usually involves probabilities over regions and integrals.

A strict development of these notions requires advanced mathematics and, in particular, measure theory. Furthermore, the idea of a partition of the sample space has to be extended to the idea of a generalized class of sets, as discussed in Sect. 3.13.

At the level of this text the passage from the discrete to the continuous case is accomplished at an intuitive level by envisaging a grouping error of measurement $\Delta$, say, and letting $\Delta$ tend to zero. The definitions given for the discrete case are then translated verbatim, by replacing "probability" by "probability density." We warn the reader, however, that there are deep problems, as illustrated by Pitcher (1957).

It is in the continuized case that the idea of dimensionality of a statistic has some relevance. Fisher conceived of a sufficient statistic as having the same dimensionality as the parameter, and it might be thought that $F$-sufficiency could be defined in terms of the spaces of the statistic and of the parameters. However, situations may be defined in which the dimensionality of the statistic is unclear. An example arising from a case considered by Barankin and Katz, with $0 < \theta \leq 2$, is:

$$\begin{aligned} f(x) &= \theta/2, \qquad 0 \leq x < 1 \\ &= \left(1 - \frac{\theta}{2}\right)\frac{1}{x^2}, \qquad 1 \leq x < \infty \end{aligned}$$

Consider a sample of size 2. All observations in the square $[0, 1) \times [0, 1)$ are poolable, having equal probability. Observations with a particular $x_2 \geq 1$ and $x_1$ in $[0, 1)$ are poolable, etc. The dimensionality of the resulting minimal statistic is obscure. Note here, as always, that a statistic is a labeling of classes of poolable data.

It is important to note that problems can arise by using the probability density of the observations as the likelihood (apart from an arbitrary multiplier). This difficulty is exemplified by the case of $\log(y - \alpha)$ being $N(\mu, \sigma^2)$, where $\alpha$, $\mu$, $\sigma^2$ are parameters with unspecified values. One will find that the "continuous" likelihood can go to infinity as $\alpha$ tends to the minimum value of $y$ in the set of observations. Thus the maximum likelihood statistic defined as that value for the parameter which maximizes the probability density of the observations is useless. If, however, one assumes that there are definite grouping errors of measurement, an assumption which seems quite unassailable, this problem does not arise. The distribution then becomes multinomial, and the probability of the observations is less than unity for all parameter values. Our view is that the continuous likelihood is to be regarded as an approximation enabling us to use continuous mathematics, an approximation which usually works well and gives answers which are correct if grouping errors are small, no matter what they may be.

## 10.6 Theory of the Likelihood Function

We have seen that the likelihood function is essentially the probability of the actual data as a function of the parameters. To consider this function, we have to recognize several different cases. It may happen that (1) the probability of an observation is a differentiable function of the parameter $\theta$ for all possible observations and (2) the range of possible observations does not depend on the parameter $\theta$. More explicitly, by (2) we mean that there is no region of the sample space which has zero probability for some values of $\theta$ and nonzero probability for other values of $\theta$. An example for which this condition does not hold is that of the uniform distribution from zero to $\theta$. When we refer to "regularity conditions," we shall mean (1), (2), and that the distribution is multinomial. We shall call this the "regular" case. Our development will emphasize the case when $\theta$ is scalar, and we will merely state extensions to the case when $\theta$ is a vector.

It might be thought that, for the case of a scalar parameter, if there is a minimal sufficient statistic which is a scalar and the likelihood has a unique maximum for every possible observation, the maximum likelihood statistic would be sufficient. To see that this is not the case, consider a random variable $X$ which takes the values $-2$, $-1$, 1, 2 with probabilities

$$\begin{aligned} P(X = -2) = 1/4 - \theta,\ P(X = -1) = 1/4 - \theta/2 \\ P(X = 1) = 1/4 + \theta/2,\ P(X = 2) = 1/4 + \theta \end{aligned} \tag{10.38}$$

in which $-1/4 \le \theta \le 1/4$. We see that $X$ is minimal sufficient. The maximum likelihood statistic $\theta^*$ is given by

$$\theta^*(-2) = -1/4,\ \theta^*(-1) = -1/4,\ \theta^*(1) = 1/4,\ \theta^*(2) = 1/4 \tag{10.39}$$

However $\theta^*$ is not sufficient because

$$P(X = -2|\theta^* = -1/4) = (1/4 - \theta)/(1/2 - 3\theta/2)$$

which depends on $\theta$. However, Theorem 10.5 follows almost trivially.

***Theorem 10.5*** If the likelihood function has a unique maximum at $\theta = W(D)$, then $W(D)$ is a function of the minimal sufficient statistic.

This theorem suggests that we may initiate a search for a minimal sufficient statistic by examining the maximum likelihood statistic. Let us consider, for example, the continuous observation case

$$f(x; \theta)\, dx = \frac{1}{\theta}\, \delta(x, \theta)\delta(0, x)\, dx$$

where $\delta(a, b) = 1$ if $a \leq b$, and $= 0$, otherwise. The likelihood function with data $x_1, x_2, \ldots, x_n$ is

$$\frac{1}{\theta^n}\, \delta(x_{\max}, \theta)\delta(0, x_{\min})$$

This has a maximum at $\theta = x_{\max}$. As can be seen, $X_{\max}$ is $F$-sufficient.

In view of our previous discussion it is appropriate to consider the variability in repeated sampling of the realized likelihood function. There is essentially no general theory. To form an idea of the complexity, the reader should consider the case represented by Equation (10.38) and the above case with probability density equal to $\delta(x, \theta)\delta(0, x)/\theta$ for different possible values of a single observation $x$.

We shall treat only the case of discrete data, because then we have an enumerable collection of possible likelihood functions if a defined procedure with a termination rule for determining possible data sets has been used or specified. In that case the space of possible likelihood functions is discrete. So we go to the case of a multinomial distribution in which the data consist of numbers of occupancies of each of $k$ cells, say $x_1, x_2, \ldots, x_k$, and the probability of the data is

$$\frac{n!}{\prod x_i!} \prod_{i=1}^{k} [p_i(\theta)]^{x_i} \tag{10.40}$$

For the most part we take $\theta$ to be scalar. The restriction to multinomial data is in a certain sense not critical. We are concerned with the interpretation of achievable data, and an essential consequence is that even though we may have a model with a continuum of possible observations, our actual observations will be given by using a grid of measurement. The likelihood function can be written as

$$L(\theta) = L(\theta; x_1, x_2, \ldots, x_k) = \prod_{i=1}^{k} [p_i(\theta)]^{x_i} \tag{10.41}$$

If the space of possible "values" for $\theta$, the parameter, is such that for some values of $\theta$, say $\theta \in \Theta_1 \subset \Theta$, the probability $p_i(\theta)$ is zero, then a corresponding observation $x_i$ which is not equal to 0 automatically and absolutely rules out values of $\theta$ in $\Theta_1$. The probability of ruling out with a random sample of size $n$ such a value, say $\theta_1$, when $\theta = \theta_2$ is

$$1 - \left[1 - \sum_i{}^{*}\, p_i(\theta_2)\right]^n$$

where $\sum^*$ denotes summation over $i$ such that $p_i(\theta_1)$ is zero. This increases exponentially with $n$, and we mention for the advanced reader that this is one of the causes of so-called "superefficiency," in which variance of an estimate decreases faster than $1/n$ with sample size $n$.

Because the likelihood function under random sampling is essentially an exponential function in the observations, it is natural to attempt to scale it down by taking logarithms, and this we now consider. So we define

$$\begin{aligned}\mathscr{L}(\theta) &= \mathscr{L}(\theta; x_1, x_2, \ldots, x_k) = \ln L(\theta; x_1, x_2, \ldots, x_k) \\ &= \sum_{i=1}^{k} x_i \ln p_i(\theta) \qquad (10.42)\end{aligned}$$

Note that we use the script $\mathscr{L}$ to denote the logarithm of the likelihood function. We consider the variability of this function in repetitions of sampling from the model indexed by $\theta$.

If we are sampling from a model with $\theta$ equal to $\theta_T$, which is such that $p_i(\theta_T)$ is equal to zero for some $i$, the function $\mathscr{L}(\theta)$ will take the value $-\infty$ for some values of $x_1, x_2, \ldots, x_k$, which have nonzero probability under other values of $\theta$. This makes the behavior of $\mathscr{L}(\theta)$, the whole function defined over $\theta$ belonging to some $\Theta$, complex from some points of view. From other viewpoints the behavior is fairly simple in that the expected log-likelihood function under sampling from the model labeled by $\theta_T$ takes the value $-\infty$ for some values of $\theta$.

No general theory of the likelihood function for such cases exists it appears (though it is possible that we have failed to see relevant work).

We pass therefore to the theory of the log-likelihood function for the case in which $p_i(\theta) > 0$ for all $\theta$ belonging to $\Theta$ and for all $i$. This has the effect that the relevant range of summation for $i$, denoted by $k$ above, is the same for all $\theta$ belonging to $\Theta$. Taking the expectation of $\mathscr{L}(\theta_0)$ under sampling from $\theta_1$, we have

$$E_{\theta_1}[\mathscr{L}(\theta_0)] = n \sum_{i=1}^{k} p_i(\theta_1) \ln p_i(\theta_0)$$

and

$$V_{\theta_1}[\mathscr{L}(\theta_0)] = n\left\{ \sum_{i=1}^{k} p_i(\theta_1)[\ln p_i(\theta_0)]^2 - \left[ \sum_{i=1}^{k} p_i(\theta_1) \ln p_i(\theta_0) \right]^2 \right\}$$

where the subscript $\theta_1$ indicates that we are sampling from the distribution specified by $\theta = \theta_1$. Hence

$$E_{\theta_1}\left[\frac{1}{n} \mathscr{L}(\theta_0)\right] = \sum_{i=1}^{k} p_i(\theta_1) \ln p_i(\theta_0) \qquad (10.43)$$

and

$$V_{\theta_1}\left[\frac{1}{n}\mathscr{L}(\theta_0)\right] = \frac{1}{n}\left\{\sum_{i=1}^{k} p_i(\theta_1)\ln^2 p_i(\theta_0) - \left[\sum_{i=1}^{k} p_i(\theta_1)\ln p_i(\theta_0)\right]^2\right\}$$

It is intuitively plausible (and an elementary moment generating function argument shows) that as $n$ tends to $\infty$, a form $(1/n)\sum_{i=1}^{k} a_i x_i$ in multinomial variables has a limiting normal distribution with mean $\sum a_i p_i(\theta)$ and variance $K/n$ equal to

$$\frac{1}{n}\left\{\left[\sum_i a_i^2 p_i(\theta)\right] - \left[\sum_i a_i p_i(\theta)\right]^2\right\}$$

Hence $(1/n)\ \mathscr{L}(\theta_0)$ has a distribution which tends with increasing $n$ to the distribution $N(\mu, K/n)$. So $(1/n)\ \mathscr{L}(\theta_0)$ tends in distribution to a constant which depends on $\theta_0$, the value for which $\mathscr{L}(\theta)$ is calculated and $\theta_T$, the true value. We note that in the above $(1/n)\ \mathscr{L}(\theta_0)$ has finite mean and variance as long as $p_i(\theta_0)$ is not zero for any $i$.

In general we do not know $\theta_T$. The whole purpose of getting data is to form an idea of the $p_i(\theta)$ and of the $\theta_T$ which represent the situation. Consider now two candidate values for $\theta$, say $\theta_1$ and $\theta_2$, and suppose our data originate from $\theta_T$.

***Theorem 10.6*** The quantity

$$\frac{1}{n}[\mathscr{L}(\theta_1) - \mathscr{L}(\theta_2)] = \frac{1}{n}\sum x_i \ln\frac{p_i(\theta_1)}{p_i(\theta_2)}$$

is asymptotically distributed as $N(\mu, K/n)$, where

$$\mu = \sum_{i=1}^{k} p_i(\theta_T)\ln\frac{p_i(\theta_1)}{p_i(\theta_2)} \tag{10.44a}$$

and

$$K = \left\{\sum_{i=1}^{k} p_i(\theta_T)\ln^2\frac{p_i(\theta_1)}{p_i(\theta_2)} - \left[\sum_{i=1}^{k} p_i(\theta_T)\ln\frac{p_i(\theta_1)}{p_i(\theta_2)}\right]^2\right\} \tag{10.44b}$$

Going further, $(1/n)[\mathscr{L}(\theta_1) - \mathscr{L}(\theta_2)]$ tends as $n \to \infty$ to the constant $\mu$. A basic inequality in information theory is that if $\sum a_i = \sum b_i = 1$, all $a_i > 0$, $b_i > 0$,

$$\sum_{i=1}^{k} a_i \ln\frac{a_i}{b_i} \geq 0 \tag{10.45}$$

with equality only if $a_i = b_i$. By applying Taylor's expansion with Lagrange remainder we have

$$\ln(1 + z) = z - \frac{z^2}{2(1 + \phi z)^2}, \qquad 0 < \phi < 1 \tag{10.46}$$

Using this with

$$z_i = \frac{b_i}{a_i} - 1 = \frac{b_i - a_i}{a_i}$$

we have

$$\sum_i a_i \ln \frac{b_i}{a_i} = \sum_i a_i \frac{b_i - a_i}{a_i} - (1/2) \sum_i a_i \frac{(b_i - a_i)^2}{a_i^2 \left[1 + \phi_i\left(\frac{b_i}{a_i} - 1\right)\right]^2}$$

$$= -(1/2) \sum_i \frac{a_i(b_i - a_i)^2}{[(1 - \phi_i)a_i + \phi_i b_i]^2} \tag{10.47}$$

But $a_i$ and $b_i$ are in $(0, 1)$ and $(1 - \phi_i)a_i + \phi_i b_i$ is in $(a_i, b_i)$, so $(1 - \phi_i)a_i + \phi_i b_i < 1$. Hence

$$\sum_i a_i \ln \frac{b_i}{a_i} \leq -\sum_i a_i \frac{(b_i - a_i)^2}{2} \tag{10.48}$$

or

$$\sum_i a_i \ln \frac{a_i}{b_i} \geq (1/2) \sum_i a_i(b_i - a_i)^2$$

We can also see (Problem 28) that

$$\sum_i a_i \ln \frac{a_i}{b_i} < \sum_i (a_i - b_i)^2/b_i$$

It follows that $[\mathscr{L}(\theta_1) - \mathscr{L}(\theta_T)]/n$, where $\theta_T$ is the true value of $\theta$ in the assumed sampling, tends in distribution with increasing $n$ to the constant

$$\sum_i p_i(\theta_T) \ln \frac{p_i(\theta_1)}{p_i(\theta_T)} \tag{10.49}$$

which is zero if $\theta_1 = \theta_T$ and is negative if $\theta_1 \neq \theta_T$.

We may note in passing that if there were only two candidate values $\theta_1$, $\theta_2$ for $\theta$, we would consider $[\mathscr{L}(\theta_1) - \mathscr{L}(\theta_2)]/n$, which would have asymptotically a normal distribution with known mean and variance in sampling from either distribution. If $\theta_1$ is the true value of $\theta$, this mean is positive; if $\theta_2$ is the true value, it is negative. A positive value for $[\mathscr{L}(\theta_1) - \mathscr{L}(\theta_2)]/n$ suggests, therefore, that $\theta_1$ is the true value.

The theory of maximization of likelihood is very complex, and we refer the reader to Rao (1965) and references cited therein. At an elementary heuristic level we find the following sequence of statements convincing for the case of a scalar parameter:

1. $(1/n)\mathscr{L}(\theta; x_1, x_2, \ldots, x_k) = (1/n)\sum x_i \ln p_i(\theta)$ is a bounded function of $\theta$ and is in fact less than $\max_i \ln p_i(\theta)$.
2. $(1/n)\mathscr{L}(\theta)$ must have a maximum value, which may occur at several $\theta$ values (perhaps at infinitely many) and perhaps at $\theta = \infty$.
3. Suppose $\theta^*(x_1, x_2, \ldots, x_k)$ denotes the maximum value of $\theta$ which gives the maximum value of $(1/n)\mathscr{L}(\theta)$. Then $\theta^*(x_1, x_2, \ldots, x_k)$ is a statistic which exists for every possible sample from the hypothesized probability model.
4. For every $n$ this statistic $\theta^*$ has a distribution $F_n$, say.
5. This distribution must become increasingly concentrated about the true value $\theta_T$. Suppose that for $\varepsilon_2 > \varepsilon_1 > 0$

$$P[\theta^* \in (\theta_T + \varepsilon_1, \theta_T + \varepsilon_2)] \to p > 0$$

Then for a proportion $p$ of samples

$$(1/n)\mathscr{L}(\theta; x_1, \ldots, x_k) - (1/n)\mathscr{L}(\theta_T; x_1, \ldots, x_k) > 0$$

for some $\theta$ in $(\theta_T + \varepsilon_1,\ \theta_T + \varepsilon_2)$. But this quantity is distributed asymptotically for any such $\theta$ with negative mean and variance tending to zero. So the frequency with which

$$(1/n)\mathscr{L}(\theta; x_1, \ldots, x_k) - (1/n)\mathscr{L}(\theta_T; x_1, x_2, \ldots, x_k) < 0$$

tends to unity for any $\theta$ in $(\theta_T + \varepsilon_1,\ \theta_T + \varepsilon_2)$. A similar argument applies to any interval below $\theta_T$.

6. The same argument could be applied to the minimum of $\theta$ values which give the maximum of $\mathscr{L}(\theta; x_1, \ldots, x_k)$.
7. So it appears that any $\theta$ which gives the maximum value to $\mathscr{L}$ tends to $\theta_T$ as $n$ increases. This value of $\theta$ is therefore consistent.

The reader should refer to Rao (1965) and to Birch (1964) for a complete proof of the following theorem.

***Theorem 10.7*** Under the regularity conditions a value of $\theta$ which maximizes the log-likelihood tends in distribution to the unit spike at $\theta = \theta_T$, the true value under the hypothesized sampling. The unit spike at $\theta_T$ is the distribution with all probability mass at $\theta_T$.

## 10.7 The Likelihood Equation

It is natural in view of the above, to attempt to develop the maximization of likelihood by the use of differential calculus. We must assume that the $p_i(\theta)$ are differentiable for all $i$ such that $p_i(\theta)$ is nonzero for some $\theta$, so that $\theta$ lies in some open region of Euclidean space.

We consider, therefore

$$\frac{1}{n}\frac{\partial}{\partial\theta}\mathscr{L}(\theta) = \frac{1}{n}\mathscr{L}'(\theta) = \frac{1}{n}\sum_{i=1}^{k} x_i \frac{p_i'(\theta)}{p_i(\theta)} \tag{10.50}$$

This is again a linear function of multinomial frequencies, so if we are sampling from the distribution specified by $\theta_T$, the distribution of $(1/n)\mathscr{L}'(\theta)$ tends to $N(v, \psi/n)$, where

$$v = v(\theta_T, \theta) = \sum_{i=1}^{k} p_i(\theta_T)\frac{p_i'(\theta)}{p_i(\theta)}$$

$$\psi = \psi(\theta_T, \theta) = \sum_{i=1}^{k} p_i(\theta_T)\left[\frac{p_i'(\theta)}{p_i(\theta)}\right]^2 - \left[\sum_{i=1}^{k} p_i(\theta_T)\frac{p_i'(\theta)}{p_i(\theta)}\right]^2 \tag{10.51}$$

Now note that if $\theta$ equals $\theta_T$,

$$v(\theta_T, \theta_T) = \sum_{i=1}^{k} p_i'(\theta_T)$$

We have $\sum_{i=1}^{k} p_i(\theta) = 1$. With $k$ independent of $\theta$, we can differentiate to give $\sum_{i=1}^{k} p_i'(\theta) = 0$ for all $\theta$, and $v(\theta_T, \theta_T) = 0$.

We suppose that the distributions as indexed by $\theta$ are such that $v(\theta_1, \theta_2)$ is nonzero for every pair $\theta_1 \neq \theta_2$. Further assume that $p_i'(\theta)/p_i(\theta)$ is bounded for all $\theta$ and every $i$.

***Theorem 10.8*** Under the regularity conditions, $(1/n)\mathscr{L}'(\theta)$ is asymptotically normally distributed with mean $v(\theta_T, \theta)$ and variance $(1/n)\psi(\theta_T, \theta)$ where $v(\theta_T, \theta)$ is zero if and only if $\theta$ equals $\theta_T$, the true value of $\theta$, and is otherwise nonzero, and $\psi(\theta_T, \theta)$ is finite.

This suggests strongly that we should consider the equation

$$\mathscr{L}'(\theta) = \mathscr{L}'(\theta; x_1, x_2, \ldots, x_k) = 0 \tag{10.52}$$

This is called the *likelihood equation*. In general it will be a highly complicated equation, and explicit solution will be possible only with a computer. It is usually found to have a unique solution. If inspection of the equation suggested the existence of more than one solution, deep problems would exist; an intuitively reasonable procedure would be to choose that solution which maximizes $\mathscr{L}(\theta)$.

We assume that Equation (10.52) has a unique solution $\theta^* = \theta^*(x_1, \ldots, x_k)$ which is consistent in the sense that

$$P(|\theta^* - \theta| > \delta) \to 0 \text{ as } n \to \infty \tag{10.53}$$

This makes it possible to develop an asymptotic theory by expanding Equation (10.52) around the value $\theta_T$. We have that the distribution of

$$\frac{1}{\sqrt{n}} \mathscr{L}'(\theta) = \frac{1}{\sqrt{n}} \sum x_i \frac{p_i'(\theta)}{p_i(\theta)} \tag{10.54}$$

tends to normality with mean $\sqrt{n}\, v(\theta_T, \theta)$ and variance $\psi(\theta_T, \theta)$. Let $\theta = \theta_T + \delta$, where $\delta$ is small; then

$$v(\theta_T, \theta) = \sum p_i(\theta_T) \frac{p_i'(\theta_T + \delta)}{p_i(\theta_T + \delta)}$$

$$\doteq \sum p_i(\theta_T) \left\{ \frac{p_i'(\theta_T)}{p_i(\theta_T)} + \delta \left[ \frac{p_i''(\theta_T)}{p_i(\theta_T)} - \frac{p_i'^2(\theta_T)}{p_i^2(\theta_T)} \right] \right\} \tag{10.55}$$

by the use of a Taylor expansion. Hence

$$v(\theta_T, \theta) \doteq -\delta \sum \frac{p_i'^2(\theta_T)}{p_i(\theta_T)} \tag{10.56}$$

using $\sum p_i'(\theta_T) = \sum p_i''(\theta_T) = 0$. Similarly, sometimes suppressing the range of summation and $\theta_T$ for brevity,

$$\psi(\theta_T, \theta) \doteq \sum \frac{p_i'^2}{p_i} + 2\delta \sum p_i' \left( \frac{p_i''}{p_i} - \frac{p_i'^2}{p_i^2} \right) \tag{10.57}$$

Hence $(1/\sqrt{n})\mathscr{L}'(\theta)$ is approximately normally distributed with mean $-\sqrt{n} I_{11}(\theta_T)\delta$ and variance $I_{11}(\theta_T) + 2\delta I_{12}(\theta_T)$ where

$$I_{11}(\theta_T) = \sum \frac{p_i'^2(\theta_T)}{p_i(\theta_T)} = \sum p_i(\theta_T) \left[ \frac{p_i'(\theta_T)}{p_i(\theta_T)} \right]^2 \tag{10.58}$$

and

$$I_{12}(\theta_T) = \sum p_i(\theta_T) \frac{p_i'(\theta_T)}{p_i(\theta_T)} \left[ \frac{p_i''(\theta_T)}{p_i(\theta_T)} - \frac{p_i'^2(\theta_T)}{p_i^2(\theta_T)} \right]$$

Hence asymptotically, since $\delta$ will be small,

$$\frac{1}{n} \mathscr{L}'(\theta) \doteq -I_{11}(\theta_T)\delta + z \sqrt{\frac{I_{11}(\theta_T)}{n}} \tag{10.59}$$

where $z$ is a random variable from $N(0, 1)$. This serves to indicate that there will be only one solution to the likelihood equation. Equating $\mathscr{L}'(\theta)$ to zero, we obtain the solution $\delta^*$ where

$$\delta^* = \theta^*(x_1, \ldots, x_k) - \theta_T = \frac{1}{\sqrt{nI_{11}(\theta_T)}} z \tag{10.60}$$

***Theorem 10.9*** Under suitable regularity conditions, the unique solution to the likelihood equation is asymptotically normally distributed around $\theta_T$ with variance $1/[nI_{11}(\theta_T)]$.

The upshot of the above development, which is rather heuristic, is to suggest that there will be a single root $\theta^*(x_1, \ldots, x_k)$ to the likelihood equation, and that

$$\frac{\theta^*(x_1, \ldots, x_k) - \theta_T}{\dfrac{1}{\sqrt{nI_{11}(\theta_T)}}} \quad \text{or} \quad \frac{\theta^*(x_1, \ldots, x_k) - \theta_T}{\dfrac{1}{\sqrt{nI_{11}(\theta^*)}}}$$

will be distributed approximately according to $N(0, 1)$. We give the alternative because it will often be the more useful.

The extension of the argument to a set of $r$ parameters $\theta_1, \theta_2, \ldots, \theta_r$ is straightforward, though tight specification of regularity conditions is not easy. If the $p_i(\theta)$, where $\theta$ is now a vector $(\theta_1, \theta_2, \ldots, \theta_r)$, are differentiable, then the equations

$$\frac{\partial}{\partial \theta_j} \mathscr{L}(\theta; x_1, \ldots, x_k) = 0, \qquad j = 1, 2, \ldots, r \tag{10.61}$$

will define a statistic $(\theta_1^*, \theta_2^*, \ldots, \theta_r^*)$ and the vector $(\theta_1^* - \theta_1, \theta_2^* - \theta_2, \ldots, \theta_r^* - \theta_r)$ will be distributed approximately as the multivariate normal distribution with mean equal to the zero vector $(0, 0, \ldots, 0)$ and variance-covariance matrix which is the inverse of the information matrix $nI$ with elements

$$nI_{jk} = E\left[\frac{\partial \ln p_i}{\partial \theta_j} \frac{\partial \ln p_i}{\partial \theta_k}\right] = n \sum_i \frac{1}{p_i} \frac{\partial p_i}{\partial \theta_j} \frac{\partial p_i}{\partial \theta_k} \tag{10.62}$$

In order to use this result with actual data, the values $\theta_j^*$ are inserted in place of $\theta_j$.

## 10.8 The Concept of Statistical Information

The above material suggests that a mass of data assumed to arise as a random sample from a distribution with parameters $\theta_1, \theta_2, \ldots, \theta_r$ can be reduced to $r$ statistics $\theta_1^*, \theta_2^*, \ldots, \theta_r^*$ which are such as to make the likelihood a maximum with respect to $\theta_1, \theta_2, \ldots, \theta_r$. It also suggests strongly that if some regularity conditions are satisfied, these can be obtained from the equations

$$\frac{\partial}{\partial\theta_j}\mathscr{L}(\theta;\text{ data}) = 0, \qquad j = 1, 2, \ldots, r$$

and that the solution $(\theta_1^*, \theta_2^*, \ldots, \theta_r^*)$ has a distribution close to a multivariate normal distribution. It is critical to note that while $(\theta_1^*, \theta_2^*, \ldots, \theta_r^*)$ *can* be regarded as an estimate of $(\theta_1, \theta_2, \ldots, \theta_r)$, it may not be a good one. The theory of estimation requires some measure of loss due to error of estimation. To illustrate the point, it is not at all clear even in the case of a scalar parameter $\theta$ that $\theta^*$, the maximum likelihood statistic, is such that $E(\theta^* - \theta)^2$ is a minimum or that the maximum over $\theta$ of $E(\theta^* - \theta)^2$ is a minimum. The problem of point estimation must be approached as a problem of decision theory.

The theory of maximum likelihood should be regarded basically as a theory of data reduction. The data which relate to the parameter $(\theta_1, \theta_2, \ldots, \theta_r)$ are reduced by the use of maximum likelihood to a statistic $(\theta_1^*, \theta_2^*, \ldots, \theta_r^*)$. This may result in a loss of information. An example given by Kendall and Stuart (1966) is the following: Let $x_1, x_2, \ldots, x_n$ be a random sample from the distribution with $\theta > 0$

$$f(x)\,dx = \frac{dx}{\theta}, \qquad \theta \le x \le 2\theta$$
$$= 0, \text{ otherwise}$$

Here the likelihood is equal to $(1/\theta^n)\delta(x_{\text{max}}, 2\theta)\delta(\theta, x_{\text{min}})$. This is maximized by taking $\theta^* = x_{\text{max}}/2$. But $\theta \le x_{\text{min}}$, so that $x_{\text{min}}$ also tells us something about $\theta$.

There is no completely general theory of data reduction, except for the way that the data can be reduced to a minimal sufficient statistic. The distribution of any statistic conditional on the minimal sufficient statistic does not depend on the parameter(s). A theory of data reduction of a random sample for the case when the range of the variable does not depend on the parameter was developed partially by R. A. Fisher. We present this heuristically for the case of a multinomial distribution. In this case we have seen that under some regularity conditions $\theta^*$, the maximum likelihood statistic, is asymptotically normally distributed around $\theta$ with variance $1/nI$, where

$$I = \sum_{i=1}^{k}\frac{p_i'^2(\theta)}{p_i(\theta)} = \sum_{i=1}^{k} p_i\left[\frac{\partial}{\partial\theta}\ln p_i(\theta)\right]^2$$
$$= E_\theta\left[\frac{\partial}{\partial\theta}\ln p(x;\theta)\right]^2 \qquad (10.63)$$

The reduction of data can be regarded as the pooling of classes in which data are recorded. In the case of data from a Poisson distribution we can record which of the possible $x_1, x_2, \ldots, x_n$ occurs, we can record which of the possible $x_{(1)}, x_{(2)}, \ldots, x_{(n)}$ occurs, or we can record the value of $\sum x_i$. In each case a recorded result is that the data fall in one of a set of multinomial classes. When we record $x_{(1)}, x_{(2)}, \ldots, x_{(n)}$ we pool together $n!$ classes which give the same value for $x_1, x_2, \ldots, x_n$.

*Definition 10.10. The Fisherian amount of information on $\theta$ given by an observation which lies in one of the multinomial classes $C_1, C_2, \ldots, C_k$, whose probabilities are $p_i(\theta_T)$, $i = 1, 2, \ldots, k$ and such that $k$ does not depend on $\theta$ is*

$$I_{\theta_T} = \sum \frac{1}{p_i(\theta_T)} p_i'^2(\theta_T) = E_{\theta_T}\left[\frac{\partial}{\partial \theta_T} \ln p(x; \theta_T)\right]^2$$

$$= -E_{\theta_T}\left[\frac{\partial^2}{\partial \theta_T^2} \ln p(x; \theta_T)\right] \qquad (10.64)$$

*where $E_{\theta_T}$ denotes expectation over the multinomial specified by the true value of $\theta$.*

***Theorem 10.10*** The pooling of two classes, say $C_1$ and $C_2$, results in loss of information unless

$$\ln p_1(\theta) = \ln p_2(\theta) + \text{constant} \qquad (10.65)$$

*Proof:* All operations are with regard to the distribution specified by $\theta_T$, so we drop the subscript $T$. Consider observational classes

(a) $C_1, C_2, \ldots, C_k$ and (b) $C_1 \cup C_2, C_3, \ldots, C_k$

Then in case (a)

$$I_\theta = \frac{1}{p_1}\left(\frac{\partial p_1}{\partial \theta}\right)^2 + \frac{1}{p_2}\left(\frac{\partial p_2}{\partial \theta}\right)^2 + \sum_{i=3}^{k} \frac{1}{p_i}\left(\frac{\partial p_i}{\partial \theta}\right)^2 \qquad (10.66)$$

and in case (b)

$$I_\theta^* = \frac{1}{p_1 + p_2}\left(\frac{\partial p_1}{\partial \theta} + \frac{\partial p_2}{\partial \theta}\right)^2 + \sum_{i=3}^{k} \frac{1}{p_i}\left(\frac{\partial p_i}{\partial \theta}\right)^2 \qquad (10.67)$$

Hence

$$I_\theta - I_\theta^* = \frac{p_1 p_2}{p_1 + p_2}\left(\frac{1}{p_1}\frac{\partial p_1}{\partial \theta} - \frac{1}{p_2}\frac{\partial p_2}{\partial \theta}\right)^2 \qquad (10.68)$$

which is positive unless

$$\frac{1}{p_1}\frac{\partial p_1}{\partial \theta} = \frac{1}{p_2}\frac{\partial p_2}{\partial \theta} \tag{10.69}$$

or $\ln p_1(\theta) = \ln p_2(\theta) + \text{constant}$.

A second property is given by the following theorem.

***Theorem 10.11*** Information is additive for independent observations.

*Proof:* Let Observation 1 give a result in one of the classes $C_1, C_2, \ldots, C_k$ with probabilities $p_1(\theta), p_2(\theta), \ldots, p_k(\theta)$, and let Observation 2 give a result in one of the classes $D_1, D_2, \ldots, D_\ell$ with probabilities $p_1^*(\theta), p_2^*(\theta), \ldots, p_\ell^*(\theta)$. Then the pair of observations gives a result in classes $(C_i, D_j)$, $i = 1, 2, \ldots, k$, and $j = 1, 2, \ldots, \ell$, with probabilities $p_{ij}(\theta) = p_i(\theta)p_j^*(\theta)$. Then,

$$\begin{aligned}\sum_{ij} p_{ij}\left(\frac{\partial \ln p_{ij}(\theta)}{\partial \theta}\right)^2 &= \sum_{ij} p_{ij}\left(\frac{\partial \ln p_i}{\partial \theta} + \frac{\partial \ln p_j^*}{\partial \theta}\right)^2 \\ &= \sum_{ij} p_{ij}\left(\frac{\partial \ln p_i}{\partial \theta}\right)^2 + \sum_{ij} p_{ij}\left(\frac{\partial \ln p_j^*}{\partial \theta}\right)^2 \\ &\quad + 2\sum_{ij} p_{ij}\left(\frac{\partial \ln p_i}{\partial \theta}\right)\left(\frac{\partial \ln p_j^*}{\partial \theta}\right)\end{aligned} \tag{10.70}$$

Using

$$p_{ij} = p_i p_j^*, \; \sum_j p_{ij} = p_i, \; \sum_i p_{ij} = p_j^* \tag{10.71}$$

and

$$\sum_i p_i \frac{\partial}{\partial \theta}\ln p_i = \sum_j p_j^* \frac{\partial}{\partial \theta}\ln p_j^* = 0 \tag{10.72}$$

the result follows.

Consider now the pooling of observational classes by recording the value of a statistic $T$ which may be a vector. Many possible samples, say $x_1, x_2, \ldots, x_n$, will give the same value for the statistic $T(x_1, \ldots, x_n)$. The theorem above extends immediately to this reduction of data. Let the possible values of $T$ be $T_1, T_2, \ldots$, and the possible samples be $S_1, S_2, \ldots$. Then in reducing the data to the realized value of $T$, we have combined observational classes specified by $S_1, S_2, \ldots$. The information given by the recording procedure using $S$ classes is

$$I(S) = \sum_j P(S_j)\left[\frac{P'(S_j)}{P(S_j)}\right]^2 \tag{10.73}$$

where the prime sign denotes differentiation with regard to $\theta$, and summation is over all possible samples. The information given by the statistic $T(S_j)$ is

$$I(T) = \sum \frac{[P'(T_i)]^2}{P(T_i)} \tag{10.74}$$

with summation over the possible values for $T$. But $P(T_i)$ for any one $i$ is the sum of some of the $P(S_j)$. So let us number the samples $S_i$ within values of $T$, so that they are enumerated as $S_{iq}$.
Then

$$P(T_i) = \sum_q P(S_{iq}) \tag{10.75}$$

Then

$$I(S) - I(T) = \sum_i \left\{ \sum_q P(S_{iq}) \left[ \frac{P'(S_{iq})}{P(S_{iq})} \right]^2 - P(T_i) \left[ \frac{P'(T_i)}{P(T_i)} \right]^2 \right\}$$

$$= \sum_i P(T_i) \sum_q \frac{P(S_{iq})}{P(T_i)} \left[ \frac{P'(S_{iq})}{P(S_{iq})} - \frac{P'(T_i)}{P(T_i)} \right]^2$$

because

$$\sum_q \frac{P(S_{iq})}{P(T_i)} \cdot \frac{P'(S_{iq})}{P(S_{iq})} \cdot \frac{P'(T_i)}{P(T_i)} = \left[ \frac{P'(T_i)}{P(T_i)} \right]^2 \tag{10.76}$$

Hence $I(T)$ is less than $I(S)$ if the samples $S_{iq}$ which give any one $T_i$ do not have the same value for $P'(S_{iq})/P(S_{iq})$.

***Theorem 10.12*** Reduction of data $D$ to a statistic $T$ results in loss of information with regard to $\theta$, unless for all data sets that give the same value for the statistic $T$ the expression $(\partial/\partial\theta) \ln P(D; \theta)$ is constant with respect to $D$.

*Corollary 10.2* If a statistic $T$ is sufficient, then data reduction to $T$ results in no loss of information with regard to $\theta$.

*Corollary 10.3* Reduction of data to a minimal sufficient statistic is the maximum possible reduction without loss of information in the parameter. We relate this reduction of data to the maximum likelihood statistic by Theorem 10.13.

***Theorem 10.13*** The maximum likelihood statistic $\theta^*$ is asymptotically sufficient for the regular multinomial case. Note that it is also asymptotically $F$-sufficient.

*Proof:* In this case the total information is $nI_{\theta_T}$. But asymptotically $\theta^*$ has the normal distribution with mean $\theta$ and variance $1/nI_{\theta_T}$ and with information $nI_{\theta_T}$. Consider the equation

$$P(D;\theta) = P(\theta^*;\theta)\cdot P(D|\theta^*;\theta) \tag{10.77}$$

or

$$\ln P(D;\theta) = \ln P(\theta^*;\theta) + \ln P(D|\theta^*;\theta)$$

Differentiate, square, and take the expectation. The product term vanishes, and we have as $n$ tends to $\infty$,

$$E\left[\frac{\partial}{\partial\theta}\ln P(D|\theta^*;\theta)\right]^2 \to 0 \tag{10.78}$$

or by Theorem 4.3

$$\frac{\partial}{\partial\theta}\ln P(D|\theta^*;\theta) \to 0 \text{ in probability} \tag{10.79}$$

or $P(D|\theta^*;\ \theta)$ tends to be mathematically independent of $\theta$, or $\theta^*$ is sufficient. The following finite sample result is very important.

***Theorem 10.14*** If there is a scalar sufficient statistic $T$ for a scalar parameter $\theta$ for the regular multinomial case, the solution of the likelihood equation is a function of $T$.

*Proof:* Because $T$ is sufficient, $P(D;\theta) = P(T;\theta)P(D|T)$ and $P(D|T)$ does not depend on $\theta$. So the equation

$$\frac{\partial}{\partial\theta}\ln P(D;\theta) = 0$$

is equivalent to the equation

$$\frac{\partial}{\partial\theta}\ln P(T;\theta) = 0$$

and a solution of the former is a function of $T$.

It is important to note that the likelihood equation may have a unique solution which is not $F$-sufficient. This leads to very deep problems.

## 10.9 Other Asymptotically Sufficient Statistics

We restrict ourselves to the regular multinomial case with vector parameter $\theta = (\theta_1, \theta_2, \ldots, \theta_p)$. We have seen that the maximum likelihood statistic is given by the likelihood equations

$$\sum_i \frac{x_i}{p_i(\theta)}\frac{\partial p_i(\theta)}{\partial\theta_j} = 0, \qquad j = 1, 2, \ldots, p \tag{10.80}$$

which is equivalent to

$$\sum_i \left(\frac{x_i - np_i}{np_i}\right)\frac{\partial p_i}{\partial \theta_j} = 0, \qquad j = 1, 2, \ldots, p \tag{10.81}$$

Consider the $\chi^2$ formula

$$\sum \frac{(O_i - E_i)^2}{E_i} = \sum \frac{O_i^2}{E_i} - n \tag{10.82}$$

In the present terminology this is

$$\sum \frac{(x_i - np_i)^2}{np_i} = \sum \frac{x_i^2}{np_i} - n \tag{10.83}$$

Setting the derivatives of this with respect to $\theta_j, j = 1, 2, \ldots, p$, equal to zero gives

$$\sum_i \frac{x_i^2}{np_i^2}\frac{\partial p_i}{\partial \theta_j} = 0 \tag{10.84}$$

or

$$\sum_i \left(\frac{x_i^2 - n^2p_i^2}{n^2p_i^2}\right)\frac{\partial p_i}{\partial \theta_j} = 0 \tag{10.85}$$

or

$$\sum_i \left(\frac{x_i - np_i}{np_i}\right)\left(\frac{x_i + np_i}{np_i}\right)\frac{\partial p_i}{\partial \theta_j} = 0 \tag{10.86}$$

As $n$ tends to $\infty$, the second factor tends to two and this suggests that the minimization of $\chi^2$ as given by Equation (10.82) or Equation (10.83) will give a statistic close to that given by maximization of the likelihood. We do not pursue the matter, and merely state that the minimization of $\chi^2$ tends also to an asymptotically sufficient statistic. Consider also the minimization of

$$\sum \frac{(x_i - np_i)^2}{x_i} \tag{10.87}$$

This quantity is called modified $\chi^2$, and if we write $x_i = np_i + \delta_i$, it is approximately

$$\begin{aligned}\sum \frac{\delta_i^2}{(np_i + \delta_i)} &\doteq \sum \frac{\delta_i^2}{np_i}\left(1 - \frac{\delta_i}{np_i}\right) \\ &= \sum \frac{\delta_i^2}{np_i} - \sum \frac{\delta_i^3}{n^2p_i^2}\end{aligned} \tag{10.88}$$

For large $n$ this will be dominated by the first term which is the $\chi^2$ given in Equation (10.82) and Equation (10.83). So we may surmise that minimization of Equation (10.87) will give a statistic close to the minimum $\chi^2$ statistic. The values for $\theta_1, \ldots, \theta_r$ given by this process, like the values given by maximum likelihood and by minimizing $\chi^2$, are in the class of what Neyman (1949) called B.A.N. (best asymptotically normal) estimates. We prefer to regard the processes as yielding statistics, the use of which will be discussed in later chapters.

Because the equations $\partial(\chi^2)/\partial\theta_j = 0$ and $\partial\mathscr{L}(\theta)/\partial\theta_j = 0$ are close, a commonly used way of obtaining the likelihood statistic is to graph the $\chi^2$ quantity, particularly if only one parameter is involved. The slope at $\theta^*$, the point of minimum, is zero and the second derivative is close to $-(\partial^2/\partial\theta_j^2)[\mathscr{L}(\theta)]$. Hence if one uses minimum $\chi^2$, the matrix of second derivatives at the minimizing point is close to the estimated information matrix from the method of maximum likelihood.

## 10.10 Insufficient Statistics

We shall see later that the maximum likelihood statistic is $F$-sufficient for the regular multinomial case only if the distribution is of a particular type known widely as the Koopman-Darmois form. The maximum likelihood statistic is only asymptotically sufficient and for any particular sample size results in loss of information. The amount of information not contained in the statistic in the case of a scalar parameter is given by the following theorem, abstracted from Fisher (1925).

***Theorem 10.15*** The *number* of observations lost, in terms of information on a scalar parameter $\theta$ by reducing the data to the maximum likelihood statistic is asymptotically

$$\frac{\sum p_i\left(\frac{p_i''}{p_i} - \frac{p_i'^2}{p_i^2}\right)^2}{I^2} - \frac{1}{n} - \frac{\left[\sum p_i\left(\frac{p_i'}{p_i}\right)\left(\frac{p_i''}{p_i} - \frac{p_i'^2}{p_i^2}\right)\right]^2}{I^3} \tag{10.89}$$

where all the functions are of the true value $\theta_T$ and

$$I = \sum p_i'^2/p_i = I_{\theta_T} \tag{10.90}$$

One may consider how the loss of information due to condensation to the maximum likelihood statistic may be recovered. It seems clear that the whole likelihood function will achieve this. A partial recovery is to use as well as $\theta^*$, the statistic $\partial^2\mathscr{L}/\partial\theta_i\partial\theta_j$ evaluated at $\theta^*$. This leads directly to the approximate representation of the log-likelihood function as a quadratic expression around $\theta^*$. This is thought to be highly informative by many statisticians (Fisher, 1925).

In the case of asymptotically nonsufficient statistics a definite *proportion* of the data is lost. In the simpler cases we may have a statistic $T^*$ which is $N[\theta, V/n]$ with information on $\theta$ equal to $n/V$.

***Theorem 10.16*** A normally distributed consistent statistic $T^*$ with mean $\theta_T$ and variance $V/n$ contains a proportion $1/IV$ of the information on $\theta$ asymptotically.

An example given by Fisher is that the median of a sample which is $N(\mu, \sigma^2)$ contains a proportion $2/\pi$ of the information. The following theorem is interesting.

***Theorem 10.17*** Let $T$ be asymptotically sufficient for a scalar parameter $\theta$, and be normal $(\theta, 1/nI)$. Let $W$ be asymptotically normal $(\theta, \sigma^2/n)$, and suppose $T$ and $W$ have a bivariate normal distribution asymptotically. Then the asymptotic correlation of $T$ and $W$ is $1/\sqrt{I\sigma^2}$.

*Proof:* Consider $Z = (1 - \alpha)T + \alpha W$. Let $C = \text{Cov}(T, W)$. Then for any $\alpha$, $Z$ is asymptotically normal with mean $\theta$ and variance

$$\frac{1}{nI} - 2\alpha\left(\frac{1}{nI} - C\right) + \alpha^2\left(\frac{1}{nI} + \frac{\sigma^2}{n} - 2C\right) \qquad (10.91)$$

The coefficient of $\alpha^2 > 0$ if $T \neq W$, and the minimum of the variance with respect to $\alpha$ is

$$\frac{1}{nI} - \frac{\left(\frac{1}{nI} - C\right)^2}{\left(\frac{1}{nI} + \frac{\sigma^2}{n} - 2C\right)} \qquad (10.92)$$

If $C$ does not equal $1/nI$, the variance of $Z$ for a particular $\alpha$ is less than $1/nI$ which is impossible. Hence

$$C = \frac{1}{nI} \text{ and } \rho(T, W) = \frac{1}{nI}\Bigg/\sqrt{\frac{1}{nI}\frac{\sigma^2}{n}} = 1/\sqrt{I\sigma^2} \qquad (10.93)$$

*Corollary 10.4* If $\sigma^2 = 1/I$, then $\rho(T, W) = 1$. Hence, two asymptotically sufficient statistics are perfectly correlated in the regular case.

## 10.11 The Fisher-Koopman-Pitman-Darmois Form

***Theorem 10.18*** Suppose that in the regular multinomial case we have a random sample of size $n$ from the distribution with $P(X = x) = P(x; \theta)$ and (1) there is a scalar sufficient statistic $T = T(x_1, x_2, \ldots, x_n)$; (2) the likelihood function has a unique maximum with regard to $\theta$, which is the unique root $\theta = \theta^*$ of the equation $\partial \ln L/\partial\theta = 0$; (3) there is a 1 to 1 relation between $T$ and $\theta^*$. Then

$$P(x;\theta) = C(\theta)\exp[m(x)H(\theta)]g(x) \tag{10.94}$$

where $H(\theta)$ is of the form

$$\int h(\theta)d\theta$$

and $-\ln C(\theta)$ is of the form

$$\int \theta h(\theta)d\theta$$

*Proof:*

$$L = \prod P(x_i;\theta) = r(T,\theta)s(x_1, x_2, \ldots, x_n)$$

So

$$\ln L = \ln r(T,\theta) + \ln s(x_1, x_2, \ldots, x_n)$$

$$\frac{\partial \ln L}{\partial\theta} = \frac{1}{r(T,\theta)}\frac{\partial r(T,\theta)}{\partial\theta} = \sum_{i=1}^{n}\frac{1}{P(x_i;\theta)}\frac{\partial P(x_i;\theta)}{\partial\theta} \tag{10.95}$$

Also

$$\frac{\partial \ln L}{\partial\theta} = (\theta^* - \theta)h(\theta, x_1, x_2, \ldots, x_n) \tag{10.96}$$

in which $h(\theta, x_1, x_2, \ldots, x_n)$ cannot be zero. We may take $T$ to be $\theta^*$, because any 1 to 1 function of a sufficient statistic is sufficient. So we have that $\partial \ln L/\partial\theta$ is additive and symmetric in $x_1, x_2, \ldots, x_n$. It depends on $x_1, x_2, \ldots, x_n$ only through $\theta^*$, so $\theta^*$ must be symmetric in $x_1, x_2, \ldots, x_n$. Hence $h(\theta, x_1, x_2, \ldots, x_n)$ does not involve $x_1, x_2, \ldots, x_n$, and by the additivity must be of the form $nh(\theta)$. Also by the additivity $n\theta^*$ must be of the form $\sum m(x_i)$. Hence

$$\sum \frac{1}{P(x_i;\theta)}\frac{\partial P(x_i;\theta)}{\partial\theta} = \sum m(x_i)h(\theta) - n\theta h(\theta)$$

$$= \sum [m(x_i) - \theta]h(\theta) \tag{10.97}$$

So

$$\frac{1}{P(x_i;\theta)}\frac{\partial P(x_i;\theta)}{\partial\theta} = m(x_i)h(\theta) - \theta h(\theta)$$

or

$$\ln P(x_i; \theta) = m(x_i)H(\theta) + \ln C(\theta) + \text{constant} \tag{10.98}$$

where

$$H(\theta) = \int h(\theta)d\theta, \qquad \ln C(\theta) = -\int \theta h(\theta)d\theta$$

and the constant depends on $x_i$. Hence

$$P(x; \theta) = C(\theta)\exp[m(x)H(\theta)]g(x)$$

for some function $g(x)$ of $x$. This is commonly called the Koopman-Darmois form or the Pitman-Koopman form. It appears that it was first placed in the literature by Fisher (1934b). We therefore call it the Fisher-Koopman-Pitman-Darmois form, or for brevity the FKPD form. It is also called the exponential form.

*Corollary 10.5* The existence of an $F$-sufficient statistic in the regular multinomial case implies that the distribution is of the FKPD form. The FKPD form is particularly easy to work with. The likelihood equation is

$$\sum_{i=1}^{n} \left[\frac{C'(\theta)}{C(\theta)} + H'(\theta)m(x_i)\right] = 0$$

or

$$\sum_{i=1}^{n} [m(x_i) - \theta]h(\theta) = 0 \tag{10.99}$$

by the definitions of $H(\theta)$ and $C(\theta)$ above, so that the maximum likelihood statistic $\theta^*$ is equal to $(1/n)\sum m(x_i)$. The reader can verify that

1. $E[m(x)] = \dfrac{-C'(\theta)}{H'(\theta)C(\theta)} = \theta.$
2. $V[m(x)] = \dfrac{1}{H'(\theta)} = \dfrac{1}{h(\theta)}.$
3. $V(\theta^*) = \dfrac{1}{nh(\theta)}.$
4. The information per observation is $H'(\theta) = h(\theta)$.
5. The information in $n$ observations is $nh(\theta)$.

6. $\sum m(x)$ is $F$-sufficient, and extracts all the information.

The binomial distribution is in the FKPD class because

$$
\begin{aligned}
P(x;\theta) &= \binom{n}{x}\theta^x(1-\theta)^{n-x} \\
&= (1-\theta)^n\left(\frac{\theta}{1-\theta}\right)^x\binom{n}{x} \\
&= (1-\theta)^n\exp\left[x\ln\left(\frac{\theta}{1-\theta}\right)\right]\binom{n}{x}
\end{aligned}
$$

The reader may verify that the Poisson distribution is also in the class.

The argument above is for the discrete case. By the use of more advanced ideas, the same form can be shown to hold for the continuous case.

By using the parametrization $H(\theta) = \phi$, the FKPD form becomes

$$P(x;\phi) = D(\phi)\exp[m(x)\phi]g(x) \tag{10.100}$$

which is often used.

We shall not derive the form for the case of a vector parameter. The result is the form

$$P(x;\phi_1,\phi_2,\ldots,\phi_p) = D(\phi_1,\phi_2,\ldots,\phi_p)\exp\left[\textstyle\sum_j \phi_j T_j(x)\right]g(x) \tag{10.101}$$

which is interpreted either as a probability for the discrete case or a probability density for the continuous case. A simple example is the multinomial for which

$$P(X_i = x_i, i = 1, 2, \ldots, k) = \frac{n!}{\prod x_i!}\prod p_i^{x_i}$$

which we write as

$$(1-P)^n\exp\left[\sum_{i=1}^{k-1} x_i\ln\left(\frac{p_i}{1-P}\right)\right]\frac{n!}{\prod x_i!} \tag{10.102}$$

where $P = \sum_{i=1}^{k-1} p_i$, *provided* $p_1, p_2 \ldots, p_{k-1}$ are functionally independent.

The normal density with parameters $\mu$ and $\sigma^2$ can be written as

$$\frac{1}{\sqrt{2\pi\sigma}}\exp\left(\frac{-\mu^2}{2\sigma^2}\right)\exp\left[-x^2\left(\frac{1}{2\sigma^2}\right)+x\left(\frac{\mu}{\sigma^2}\right)\right] \tag{10.103}$$

to place it in the FKPD form.

## 10.12 Goodness of Fit

We are now in a position to discuss the general problem of goodness of fit. If there is an $F$-sufficient statistic, we will consider the distribution of possible data sets, conditional on the $F$-sufficient statistic taking its observed value, and construct a significance test over this distribution. If there is not an $F$-sufficient statistic, the problem is very complex and no general theory exists. The difficulty exists because there is no general theory of likelihood and of the maximum likelihood statistic for cases other than the regular ones we have discussed. The theory we have presented and intuitive considerations suggest the following as an approximate procedure for most situations and for general purposes.

1. Determine the realized value of the maximum likelihood statistic $\theta^*(= \theta_1^*, \theta_2^*, \ldots, \theta_r^*$, in general).
2. Construct the particular distribution whose label is $\theta^*$.
3. Partition the space of possible data sets into regions:
   (a) Ordering data sets on the basis of probability or probability density.
   (b) Partitioning the space subject to this ordering into $n$ regions each with probability content as close as possible to $1/n$.
4. Let this give regions $R_1, R_2, \ldots, R_n$ with "expected" numbers equal to $E_i$ and observed numbers $O_i$ and compute $C_O = \sum_{i=1}^{n} (O_i - E_i)^2/E_i$.
5. Obtain the significance level of $C_O$ the observed value as $P(\chi^2 \geq C_O)$, where $\chi^2$ is the mathematical $\chi^2$ distribution with degrees of freedom $n - r - 1$, where $r$ is the number of parameters fitted.

The basic theorem underlying this procedure was given by Fisher (1924). We refer the reader to Birch (1964) for a theorem which states that if the classes for computation of the $\chi^2$ quantity are fixed beforehand, it is distributed according to the $\chi^2$ distribution independently of the maximum likelihood statistic. Other work, e.g., by Steck (1957) and Monte Carlo computations, suggest that the objective rule above for making class intervals will also give the desired properties. If probabilities or probability density are constant, some other mode of partitioning must be used.

We should not leave the reader with the impression that ordering by conditional probability, which is used in the above approximate procedure, is the only way of constructing a goodness of fit test. Many orderings are possible, but this one seems highly appropriate for data evaluation and condensation. Additionally, it appears to be the only ordering for which there is general approximate distribution theory not dependent on the particular class of models under examination. It is particularly important to note that the approximate procedure is not sensitive to outliers, but the approximation in the procedure is poor when a few "wild" observations are present. One can see this very easily, for instance, in the case of Poisson data with all but one of the observations near zero (e.g., $\leq 3$) and one observation equal to 10. This particular configuration will be the extremal one and will have extremely small probability and hence a significance level very "close" to zero.

## 10.13 Inspection of Categorical Data: Contingency Tables

We take up a classic case in which there are four classes with $P_1 = \theta_1\theta_2$, $P_2 = \theta_1(1 - \theta_2)$, $P_3 = (1 - \theta_1)\theta_2$, and $P_4 = (1 - \theta_1)(1 - \theta_2)$. This arises typically with Table 10.2:

**Table 10.2** The 2 × 2 Contingency Table

| | | Criterion 2 | | |
|---|---|---|---|---|
| | | *Yes* | *No* | *Totals* |
| Criterion 1 | *Yes* | $a$ | $b$ | $a + b$ |
| | *No* | $c$ | $d$ | $c + d$ |
| | *Totals* | $a + c$ | $b + d$ | $a + b + c + d = N$ |

The idea is that Criterion 1 and Criterion 2 may not be associated. (Let Criterion 1 be brunette for husband, and let Criterion 2 be brunette for wife—a case for which we imagine the model would fail.) We write down the probability of the data, using capital letters to denote random variables. If the sample size $N$ is random, the probabilities are conditional on the realized size.

$$P(A = a, B = b, C = c, D = d)$$

$$= \frac{N!}{a!b!c!d!}(\theta_1\theta_2)^a[\theta_1(1 - \theta_2)]^b[(1 - \theta_1)\theta_2]^c[(1 - \theta_1)(1 - \theta_2)]^d$$

$$= \frac{N!}{a!b!c!d!}\theta_1^{a+b}(1 - \theta_1)^{c+d}\theta_2^{a+c}(1 - \theta_2)^{b+d} \tag{10.104}$$

It is clear that a sufficient statistic is the triplet $N$, $a + b$, $a + c$. Also, $a + b$ "tells us about" $\theta_1$, and $a + c$ "tells us about" $\theta_2$. Let us now factor this probability as follows:

$$P(A = a, B = b, C = c, D = d)$$

$$= \left[\frac{N!}{(a + b)!(c + d)!}\theta_1^{a+b}(1 - \theta_1)^{c+d}\right] \times$$

$$\left[\frac{N!}{(a+c)!(b+d)!}\theta_2^{a+c}(1-\theta_2)^{b+d}\right]\times$$

$$\left[\frac{(a+b)!(c+d)!(a+c)!(b+d)!}{N!a!b!c!d!}\right] \qquad (10.105)$$

The first factor is the probability under the model of getting $a + b$ with "yes" on Criterion 1, the second factor is the probability of getting $a + c$ with "yes" on Criterion 2, and the third probability is the probability of the whole data configuration given that these events happen. The testing of goodness of fit of the model is now straightforward. We use a procedure known as the Fisher-Irwin test, for which the third factor gives us the relevant probabilities. The only question is how we shall define an ordering in the space of conditional samples. A highly natural candidate is to order the samples on the basis of their conditional probabilities. Let us exemplify by a case which has been discussed in the literature. Let the data be the table

| | |
|---|---|
| (a) 3 | (b) 0 |
| (c) 0 | (d) 3 |

Then the possible samples in the conditional distribution are

| | | | | | | | |
|---|---|---|---|---|---|---|---|
| 3 | 0 | 2 | 1 | 1 | 2 | 0 | 3 |
| 0 | 3 | 1 | 2 | 2 | 1 | 3 | 0 |

The conditional probabilities of these are

$$\frac{(3!)^4}{6!}\left[\frac{1}{(3!)^2}, \frac{1}{(2!)^2}, \frac{1}{(2!)^2}, \frac{1}{(3!)^2}\right]$$

or (36/20) (1/36, 1/4, 1/4, 1/36), or (1/20, 9/20, 9/20, 1/20). So if we use this test of significance, the probability of getting a set of data at least as deviant in probability as 1/10, and we attach a significance level of 1/10 to the goodness of fit of the model which assumes lack of association. In this case it seems clear that any rational ordering of conditional samples related to possible association would give the same answer. Tables exist by which this goodness of fit test of significance for the $2 \times 2$ table with small entries may be made. We will note here only the "large sample" result. The maximum likelihood statistics of $\theta_1$ and $\theta_2$ are respectively $(a + b)/N$ and $(a + c)/N$. So the $\chi^2$ criterion $\sum_j (O_j - E_j)^2/E_j$, (summation over the 4 cells) leads to

$$C = \frac{(ad - bc)^2 N}{(a+b)(c+d)(a+c)(b+d)}$$

The large sample approximation is to refer $C$ to the $\chi^2$ distribution with $4 - 1 - 2$ degrees of freedom. The accuracy of estimating tail probabilities is improved by the correction (associated with the name Yates) of replacing $(ad - bc)$ with $|ad - bc| - N/2$, if $|ad - bc| > N/2$.

The same type of reasoning can be applied to contingency tables of larger size, e.g., $r \times s$ ($r$ and/or $s > 2$) and to the case of higher dimensionality (e.g., $2 \times 2 \times 2$ tables). We will not pursue these topics. We can however, state immediately a large sample approximation to the goodness of fit test of any aspect. This is to follow the general procedure outlined, of calculating expected numbers for each cell ($E$) corresponding to each observed number $O$, and forming the summation of $(O - E)^2/E$ over all the cells, referring the calculated value to the $\chi^2$ distribution with degrees of freedom equal to (number of cells minus 1 minus number of parameters fitted).

## 10.14 Special Purpose Tests of Goodness of Fit

The $\chi^2$ test we have given is a general purpose one. One may wish to have significance tests using the actual form of the probability model involved or sensitive to particular types of deviation. To test for normality there are skewness and kurtosis tests [see, e.g., Fisher (1925 on)] and specially constructed tests based on order statistics due to Shapiro and Wilk (1965). The data analyzer may wish to evaluate the significance of an extreme observation in the data set. As we have seen, if the data are in a sequence (e.g., of collection), one will wish to test for significance of nonrandomness of the sequence. The field is limitless!

## 10.15 Likelihood Ideas for Continuous Distributions

We have presented likelihood and information ideas for the class of multinomial distributions. The ideas may be carried over heuristically to the case of continuous distributions by supposing there is a grid of measurement and then by allowing the size of the grid to tend to zero. In this way, we have

$$\sum_{i=1}^{k} \frac{[p_i'(\theta)]^2}{p_i(\theta)} \to \int \left[\frac{\partial}{\partial\theta} \ln f(x;\theta)\right]^2 f(x;\theta)\,dx$$

The theorems on multinomial likelihood then go over to theorems of continuous likelihood. We recall, however, that there may be problems if the probability density is not well-behaved.

There are considerable difficulties in ideas of minimality of a statistic for very general probability structures [see, e.g., Pitcher (1957)]. However, it appears that many of these disappear if it is considered that there is a grouping error of measurement for continuous models.

## 10.16 Likelihood Ideas for More Complex Sampling

We have considered properties of the likelihood function for a random sample from a hypothetical distribution. It is to be noted, however, that at times we considered the case of a random sample of one observation from a multinomial distribution. To do this, we had to suppose that the observation could fall into one of a finite or infinite number of cells, specified by a variable $i(= 1, 2, \ldots)$ with associated probabilities $p_i(\theta)$.

Consider, now, more general cases of sampling from a distribution, which we exemplify by an example. Suppose we can obtain random observations from Bi$(5, p)$. Then the possible samples of size one are denoted by $x_i \in \{0, 1, 2, 3, 4, 5\}$. The possible samples of two are denoted by $x_1$, $x_2$, where each comes from this set, and so on. Now consider the following restricted rule of sampling: (1) If $x_1 = 4$ or 5, take no more observations; (2) if $x_1 < 4$, take another observation $x_2$; (3) if $x_1 + x_2 < 8$, take another observation $x_3$.

The possible sets of data are: one observation: $x_1$ is 4 or 5, two observations: $x_1, x_2 = (3, 5)$, three observations: $x_1, x_2, x_3$ with $x_1 < 4$, $x_1 + x_2 < 8$, $x_3$ free or unrestricted.

There are in all $2 + 1 + 138 = 141$ possible data sets under the sampling rule and model. We can write immediately the probability of any data set. The data set $x_1 = 2$, $x_2 = 4$, $x_3 = 5$ has probability $P_2 P_4 P_5$ where

$$P_i = \binom{5}{i} p^i (1 - p)^{5-i}$$

We can therefore enumerate each of the possible likelihood functions with the probability of each as a known function of the parameter $p$. We can form the maximum likelihood statistic and can study its distribution. There will, in general, be considerable difficulty in doing this because enumeration of the different possible data sets, of the maximum likelihood statistic of each, and of the probability of each will be very tedious with "small" situations. In our present case, the situation is shown in Table 10.3. The maximum likelihood statistic is not sufficient because we have $\theta^* = 0.8$ with probability $P_4$ and $\theta^* = 0.8$ with probability $P_3 P_5$, and the two probabilities are not in constant ratio independent of $p$.

**Table 10.3** Distribution of Likelihood Statistic

| Data | M.L. Statistic | Probability |
|---|---|---|
| 4 | 0.8 | $P_4$ |
| 5 | 1.0 | $P_5$ |
| 3, 5 | 0.8 | $P_3P_5$ |
| $3, i, j (i \leq 4, j$ free) | $(3 + i + j)/15$ | $P_3P_iP_j$ |
| $2, i, j (i, j$ free) | $(2 + i + j)/15$ | $P_2P_iP_j$ |
| $1, i, j (i, j$ free) | $(1 + i + j)/15$ | $P_1P_iP_j$ |
| $0, i, j (i, j$ free) | $(i + j)/15$ | $P_0P_iP_j$ |

We now briefly describe a general sampling rule. The possible sets of data are sequences $x_1, x_2, \ldots, x_n, \ldots$ in which each $x_i$ is a possible realization of a random variable defined by some distribution. Each $x_i$ may be a scalar or vector. Let the space of each $x_i$ be denoted by $E_1$. Then with unrestricted sampling the space for $x_1, x_2$ is $E_1 \times E_1$ or $E_2$, say, for $x_1, x_2, x_3$ the space is $E_1 \times E_1 \times E_1 = E_3$, etc. Let $R_1$ be a part of $E_1$, $R_2$ a part of $E_2$, etc.

Let

$$\begin{aligned} \delta_1(x_1, R_1) &= 0, \quad \text{if } x_1 \in R_1, \text{ and 1, otherwise} \\ \delta_2(x_1, x_2, R_2) &= 0, \quad \text{if } (x_1, x_2) \in R_2, \text{ and 1, otherwise, etc.} \end{aligned} \tag{10.106}$$

Then the rule of sampling is applied recursively: If

$$\delta_n(x_1, x_2, \ldots, x_n, R_n) = 0$$

take no more observations. Otherwise take another observation which is called $x_{n+1}$. This is called the *stopping rule*. If we wish to terminate certainly after $n_0$ observations, we would let $R_{n_0} = E_{n_0}$. Then the possible data sets are of the form $x_1$; $x_1, x_2$; $x_1, x_2, x_3$; etc., and

$$\begin{aligned} &P[D = (x_1, x_2, \ldots, x_n)] = P(x_1)[P(x_2)\delta_1(x_1, R_1)] \ldots \\ &[P(x_n)\delta_{n-1}(x_1, \ldots, x_{n-1}, R_{n-1})][1 - \delta_n(x_1, \ldots, x_n, R_n)] \end{aligned} \tag{10.107}$$

where each probability depends on the parameters of the distribution, $\theta$. So we have

$$\begin{aligned} &P[D = (x_1, x_2, \ldots, x_n)] \\ &\quad = P(x_1)P(x_2) \ldots P(x_n)g_n(x_1, x_2, \ldots, x_n) \end{aligned} \tag{10.108}$$

in which $g_n$ is a known function not dependent on $\theta$. If now $T_n$ is sufficient for $\theta$ with $n$ fixed,

$$\begin{aligned} &P(x_1)P(x_2) \ldots P(x_n) \\ &= P[T_n(x_1, x_2, \ldots, x_n)]P[x_1, x_2, \ldots, x_n | T_n(x_1, x_2, \ldots, x_n)] \end{aligned} \tag{10.109}$$

where the second factor does not involve $\theta$. Hence we can write

$$P_R[D = (x_1, x_2, \ldots, x_n)]$$
$$= P_U[T_n(x_1, x_2, \ldots, x_n)]h_n(x_1, x_2, \ldots, x_n) \quad (10.110)$$

where the subscript $R$ denotes restricted sampling and $U$ denotes unrestricted sampling and $h_n(x_1, x_2, \ldots, x_n)$ does not involve $\theta$. It follows from the definition of sufficiency that $(n, T_n)$ is sufficient. This is really no different from the case of fixed sample size in that complete specification of the likelihood function also requires specification of $n$, the sample size in that case.

There are, however, vast differences between the case of sampling with fixed sample size and sampling according to a scheme like the above. We make a short list of these:

1. The class of likelihood functions of which we observe a random member is quite different.
2. The properties of the maximum likelihood statistic are quite different, and are difficult to obtain in even the simplest cases.
3. The nature of asymptotic theory is quite different. In the case of fixed sample size $n$ we can envisage $n$ getting larger and under certain regularity conditions can establish limiting properties of the maximum likelihood statistic. In the case of a rule as specified above we can certainly imagine $n$ getting larger, but we have to specify the sequence of functions $g_n(x_1, x_2, \ldots, x_n)$ or the sequence of subspaces $R_n$ of $E_n$, and there is a wide variety of choice.
4. In general, any reduction of the sufficient statistic $(n, T_n)$ to some other statistic which is minimal sufficient is most obscure and certainly the existence of an $F$-sufficient statistic is even more so.
5. The distribution of $(n, T_n)$ is highly complex; even the distribution of $n$ is very complex.

Our knowledge with regard to recursively defined sampling schemes (or sequential sampling schemes, to use the more common phrase) is confined almost totally to decision aspects of the process (as for instance best point estimation with prescribed loss function or best accept-reject rules). We refer the reader to the basic treatise of Wald (1947) and to the very voluminous professional literature of the past twenty years.

It is appropriate, however, to mention one scheme of sequential sampling due to Wald that has received considerable favor. Define two completely specified distributions $F_0$ and $F_1$. Let $f(x|F_i)$ denote the probability or probability density of $x$ under random sampling from $F_i$. Then the rule is: Take another observation $x_{n+1}$ if and only if

$$A < \frac{f(x_1|F_0)}{f(x_1|F_1)} \cdot \frac{f(x_2|F_0)}{f(x_2|F_1)} \cdots \frac{f(x_n|F_0)}{f(x_n|F_1)} < B \quad (10.111)$$

A basic problem is to show that such a rule of sampling defines a space of outcomes with probability arbitrarily close to unity that the total number of observations will be less than some $N$. Any procedure that gives positive probability to $n$ being greater than any chosen $N$ would leave the sampler open to the possibility that he will never complete his observations. Derivative problems are to obtain some idea of the distribution of $n$, of the sample size, and of the operating characteristics of decision procedures. There is considerable theory given by Wald (1947) about sequential probability ratio sampling (SPRS) and the associated decision procedures.

## 10.17 Final Remarks

We have discussed how one may examine data, pick a probability model, write down the probability of the observations, factor this probability into parts which have different roles, and examine goodness of fit of the model. We have indicated that a reasonable way of condensing the likelihood function is to obtain its point of maximum and the second derivatives at that point. In the following chapters, we discuss other helpful processes. All these depend on the choice of a model which specifies completely the probability of the observations apart from some unknown parameters. It thus requires a tight specification of the hypothetical origin of the data.

Another possibility of condensing data is to reduce it to a small set of numbers which one believes to characterize the data well. For instance, with an unstructured set of data, a condensation to mean, variance, third moment about the origin, and fourth moment about the origin might be considered reasonable; or instead of using the third and fourth moments, one might use the skewness and the kurtosis. Such summary statistics reveal to a person experienced in looking at data, some attributes of the set of data being condensed. These summary statistics also have the virtue that they are easily understood, in the sense that they are obtained by very simple operations on the data. The same holds for structured data discussed in Chapter 16. In such cases, the analysis of variance, which is a partition of the total sum of squares of the observations around their mean related to the structure of the observations, is informative to the data handler who has some experience. The use of such summary statistics is crucial in the general handling of data. The problem is, of course, that without some sampling theory, the fact that a set of observations gives a value, say $z$, for a statistic, gives one no idea what value of $z$ one would obtain with a very large amount of data from the same source. A partial answer to this is to obtain the expected value of $z$, under some incompletely specified model, and its variance. Furthermore, if one can obtain an estimate of the variance of $z$ from the actual data and *if* one can form some idea of the joint distribution of the error of $z$ and the estimated variance of $z$, one can then form an opinion about what the "true value" of $z$ might be. This type of thinking leads to theoretical work on moments of statistics with incompletely workable models. Theoretical moments are sometimes used to approximate distribution of $z$, i.e., by supposing that for some constant $k$, $kz$ has a gamma distribution, the choice of mathematical distribution being made on the basis of low moments. It is noteworthy that almost no work has been done on approximation of the joint distribution of an estimate and the estimate of its variance. This will be a bivariate distribution, and at minimum one will need all moments of order up to two of the estimate and the estimated variance.

The reader might be led to conclude that the use of "ad hoc" procedures of the sort mentioned in the previous paragraph is indefensible. We cannot agree with such a view. The data exist, and the job of the statistician is to try to condense them and to form opinions in an objective way.

## PROBLEMS

1. Several sets of data are listed below. The order in which the data are listed is not necessarily that in which they were taken, so we cannot test for randomness. Assuming that in each case the data constitute a random sample from some distribution, choose a mathematical distribution to represent the data and sketch the likelihood function. In case two parameters are involved, sketch contours of the likelihood function.

| | | | | | | | |
|---|---|---|---|---|---|---|---|
| (a) | 2.06 | 7.31 | 1.76 | 6.22 | 5.55 | 5.41 | 2.15 |
| | 3.24 | 4.26 | 2.50 | 3.99 | 3.39 | 4.01 | 2.32 |
| | 2.80 | 1.01 | 2.26 | 3.08 | 5.26 | 4.81 | 1.67 |
| | 3.55 | 0.75 | 8.89 | 2.43 | 2.36 | 5.76 | 1.58 |
| | 3.71 | 0.15 | 6.43 | 1.77 | 1.21 | 4.09 | 3.78 |

| | | | | | | |
|---|---|---|---|---|---|---|
| (b) | .64 | .70 | .03 | .71 | .14 | .35 |
| | .08 | .52 | .07 | .65 | .01 | .41 |
| | .31 | .59 | .22 | .74 | .12 | .50 |
| | .25 | .41 | .44 | .42 | .26 | .49 |
| | .73 | .18 | .62 | .23 | .29 | .33 |

| | | | | | |
|---|---|---|---|---|---|
| (c) | −2.5 | 3.5 | 6.5 | 4.5 | 1.5 |
| | 13.5 | 2.5 | 5.5 | 3.5 | 2.5 |
| | 5.5 | 1.5 | 0.5 | 1.5 | 4.5 |
| | 4.5 | 2.5 | 4.5 | 0.5 | 4.5 |
| | 1.5 | 5.5 | 9.5 | 7.5 | 7.5 |
| | 0.5 | 7.5 | 11.5 | 8.5 | 3.5 |
| | −0.5 | 6.5 | 12.5 | 8.5 | 3.5 |
| | −1.5 | 10.5 | 10.5 | 3.5 | 5.5 |
| | −1.5 | 12.5 | 6.5 | 2.5 | 5.5 |
| | 4.5 | 8.5 | 5.5 | 1.5 | 2.5 |

2. Discuss briefly other possible methods of fitting the data sets in Problem 1 to the mathematical distributions you have chosen to represent the data.
3. Repeat Problem 2 but use the method of maximum likelihood for fitting the data.
4. Given a random sample of size $n$, $X_1, X_2, X_3, \ldots, X_n$, from a Poisson distribution, show that the following statistics are sufficient; i.e., show that if they are known, the likelihood function is known, apart from a factor not involving the parameter:
   (a) $\sum_1^n X_i$
   (b) $(X_1, X_2, \sum_3^n X_i)$
   (c) $(X_1, X_2^2, \sum_3^n X_i)$
   (d) $\bar{X}$
5. For the Poisson distribution show that the conditional distribution of $X_1, X_2, \ldots, X_n$ given $\sum_1^n X_i$ does not involve the parameter $\lambda$.
6. Show for discrete distributions in general that if $T$ is sufficient then the likelihood function is determined by $T$.
7. Consider a random sample of size $n$ from the population with density function
$$f(x) = p^x(1-p)^{1-x}, \qquad x = 0, 1, 0 \le p \le 1$$
$$= 0, \qquad \text{otherwise}$$
Show that $\sum_1^n X_i$ is an $F$-sufficient statistic. Show that the conditional distribution of $X_1, X_2, \ldots, X_n$, given a value for $\sum_1^n X_i$ does not involve the parameter $p$.
8. Given a random sample of size $n$ without measurement error from the uniform distribution $U(0, \theta)$, show that $X_{\max}$ is an $F$-sufficient statistic. Sketch the likelihood function. Show that the conditional distribution of $X_1, X_2, \ldots, X_n$ given a value for $X_{\max}$ does not involve $\theta$.
9. Given a random sample without measurement error from the negative exponential distribution, show that $\sum_1^n X_i$ is $F$-sufficient. Sketch the likelihood function. Find the conditional distribution of the sample given a value for $\sum_1^n X_i$ and verify that it does not depend on $\theta$.
10. Given a random sample without measurement error from the normal distribution, show that $\bar{X}$, $(n-1)s^2/n$ is an $F$-sufficient statistic.

11. Given a random sample of size two from a discrete population with values $0, 1, 2, \ldots, n$, sketch the set of points in the sample space specified by a given value for:
    (a) $\bar{X}$
    (b) $X_{\max}$
    (c) $s^2$
    (d) $X_{(1)}, X_{(2)}$
12. Repeat Problem 11 for a continuous population.
13. Consider sampling from the Poisson distribution and the sample point $x_1^o, x_2^o, \ldots, x_n^o$. Consider all sample points such that the likelihood function is given by a factor (not depending on $\lambda$) times the likelihood function at the given sample point. Show that this set is, in fact, all points such that $\sum_i x_i = \sum_i x_i^o$.
14. For the $U(0, \theta)$ distribution with perfect measurement, show that $X_{\max}$ is minimal sufficient by fixing a sample point and considering the set of all sample points where the likelihood differs by at most a factor not involving $\theta$.
15. Find a minimal sufficient statistic for $\theta$ in the Laplace distribution

$$f(x) = \frac{1}{2}\exp(-|x-\theta|), \qquad -\infty < \theta < \infty$$
$$-\infty < x < \infty$$

16. Find a minimal sufficient statistic for the uniform distribution

$$f(x) = 1, \qquad \theta - \frac{1}{2} < x < \theta + \frac{1}{2}$$
$$= 0, \qquad \text{otherwise}$$

17. Show that the Lipschitz condition mentioned in Sect. 10.5.2 holds for the normal distribution.
18. Show that
$$E\left(\frac{\partial}{\partial\theta}\ln L\right)^2 = -E\left(\frac{\partial^2}{\partial\theta^2}\ln L\right)$$
19. Verify that if the information on a single observation is $I$ that the information on $n$ observations is $nI$.
20. Determine the information in a random sample of $n$ observations from:
    (a) A binomial distribution.
    (b) A Poisson distribution.
    (c) A negative exponential distribution.
    (d) A normal distribution.
21. In simple random sampling without replacement from a finite population, show that the sample size $n$ is consistent in probability for the population size.
22. Given a random sample of size $n$, show that

$$\frac{1}{n}\left(\sum_{i=n_1}^{n} X_i + n_1 s^2\right), \; n_1 < n \text{ and fixed}$$

is consistent in probability for $\mu$.

23. Suppose

$$f(x) = \frac{1}{\theta} + \frac{x}{\theta^2}, \qquad -\theta \le x \le 0$$
$$= \frac{1}{\theta} - \frac{x}{\theta^2}, \qquad 0 < x \le \theta$$
$$= 0, \qquad \text{otherwise}$$

(a) Sketch the likelihood function for a random sample of size $n$.
(b) Use the method of maximum likelihood to fit the distribution.

24. Suppose

$$f(x) = \frac{1}{\alpha - \beta}, \qquad \alpha < x < \beta$$
$$= 0, \qquad \text{otherwise}$$

(a) Sketch the likelihood function for a random sample of size $n$.
(b) Use the method of moments to fit the distribution.
(c) Use the method of maximum likelihood to fit the distribution.

25. Suppose

$$f(x) = 1/\theta + x/\theta^2, \qquad -\theta \le x \le 0$$
$$= 1/\theta - x/\theta^2, \qquad 0 < x \le \theta$$
$$= 0, \qquad \text{otherwise}$$

(a) Obtain the minimal sufficient statistic in a sample of size $n$.
(b) Attempt to develop a goodness of fit test.

26. Consider independent Bernoulli trials with parameter $p$.
(a) Exhibit a sequential random sampling (SRS) procedure in terms of an $n$, $r$ diagram in which $n = 1, 2, \ldots; r = 0, 1, 2, \ldots, n$.
(b) Consider the stopping rule: Stop if $n > 5$ or $r > 2$. Enumerate the possible data sets and their probabilities.
(c) Evaluate the Fisherian information of this observational procedure.
(d) Obtain the distribution of the maximum likelihood statistic.

27. Prove that with SRS from independent Bernoulli trials the statistic $n$, $r$ where $n =$ total number of trials and $r =$ number of successes is minimal sufficient.

28. Use the general relationship $(x - 1)/n < \ln x < x - 1$ to show that

$$0 < \sum a_i \ln \frac{a_i}{b_i} < \sum a_i \frac{(a_i - b_i)}{b_i} = \sum \frac{(a_i - b_i)^2}{b_i}$$

You need the fact that

$$\sum b_i \frac{(a_i - b_i)}{b_i} = 0$$

This result complements that of Equation (10.48) to give the result

$$\sum a_i(b_i - a_i)^2/2 < \sum a_i \ln \frac{a_i}{b_i} < \sum \frac{(a_i - b_i)^2}{b_i}$$

29. Verify that seemingly peculiar results can occur if the parameter space $\Theta$ is discrete, by considering sufficiency, minimal sufficiency, and completeness for $X$ distributed according to $\text{Bi}(3; \theta)$ with $\Theta = (0, 1/2, 1)$.

# 11

# Inference by Likelihood and Bayes's Theorem

## 11.1 Introduction

We now discuss the situation in which the probability model is assumed to be known, apart from the value of the parameter (which may be a vector), and the problem is to form rationally an opinion about the value of the parameter. The occurrence of such a situation appears rare, but the formulation is reasonable, in that *any* interpretation of observations has to be based on assumption. It appears that there is no single logically compelling way of giving body to the phrase "form an opinion." All the present and following three chapters covering uses of likelihood, of Bayes's Theorem, of significance tests, of accept-reject rules, and of decision theory should be regarded as contributing to the wide purpose of using data intelligently.

## 11.2 Likelihood Inference

The proponents of likelihood have put forward what is called the likelihood principle. This has not been stated tightly, but it appears to be as follows. *To form opinions about parameter values from data, the only inferential content of the data is given by the realized likelihood function.* The original exponent of this appears to be R. A. Fisher, and the reader can refer to his last book (Fisher 1956); see also Barnard et al. (1962). We also suggest study of Hacking (1965) for an extensive discussion of a notion of support for hypotheses which is based strongly on likelihood. But we must caution the student that none of his development has achieved wide acceptance. Birnbaum (1962, 1969) has written two important papers on the topic; in the second he appears to question the utility of the principle.

In Chapter 10 we gave an elementary description of the basic properties of the likelihood function for a simple class of cases. This suggests that the realized likelihood function is informative. The recipe of the advocates of likelihood inference appears to be that one should "look at" the likelihood function and, additionally, that the behavior of any aspect of the data in repetitions of data under the probability model is not relevant to the formation of opinion about the parameter value. This is our understanding of what is referred to in the literature as "the likelihood principle." For instance, examination of the achieved likelihood function might yield that $\theta = 5$ has twice the likelihood of $\theta = 2$. It is easy to prepare a graph for the case of a scalar parameter. For the case of a parameter with two components a graph of contours of the likelihood function can be made by a computer and graphing device. If there are more than two components, graphical presentation of the behavior of the likelihood function is difficult. The likelihood theory of Chapter 10 indicates that in the regular case of random sampling from a single population the likelihood function will become increasingly peaked near the true value of the parameters involved.

In the regular case of Chapter 10, the likelihood function will take asymptotically the form

$$\left(\frac{1}{2\pi}\right)^{p/2} \frac{1}{\sqrt{\det V}} \exp\left[-\frac{1}{2}(\theta^* - \theta)'V^{-1}(\theta^* - \theta)\right]$$

where $\theta^*$ is the maximum likelihood statistic and $V$ is the asymptotic variance-covariance matrix of $\theta^*$, which is equal to $(nI)^{-1}$ where $I$ is the information matrix. In nonregular cases there is no similar approximate expression.

Part of the difficulty of the likelihood principle is that there are cases in which the likelihood function is uninformative. These are cases in which the probability of the sample data does not depend on the population values. An example of this is the case of random sampling without replacement from a finite population. Birnbaum (1969) gives an example in which the likelihood principle is strongly misleading.

It will have been noted that for the case of parametric hypotheses we consider the likelihood function to be exhaustive with regard to the parameter given the class of possible likelihood functions. For this case it therefore seems that the actual likelihood function and the class of likelihood functions possible under the probability model contain all the inferential content of the data, given that the model specification is appropriate.

The idea was advanced by Fisher (1956) that the likelihood function per se should be used to determine values of $\theta$ which are tenable in the light of the data. More explicitly, given a likelihood function for $\theta$ (which may be a vector) with data $D$, $L(\theta; D)$, we can take $\theta^*$, the maximum likelihood statistic, and then determine those values for $\theta$ such that

$L(\theta; D) \geq kL(\theta^*; D)$, where $k$ is some number less than unity. If we take logarithms, we would say that a set of values of $\theta$ tenable at a likelihood level of $\gamma(0 < \gamma < 1)$ is given by values of $\theta$ such that

$$\ln L(\theta; D) \geq \ln L(\theta^*; D) + \ln \gamma$$

This inequality may be transformed in the simple case that $\ln L$ is unimodal to an inequality $\theta^* - \delta_1 \leq \theta \leq \theta^* + \delta_2$. The likelihood theory for the regular case shows that $\theta^*$ is asymptotically normal with mean $\theta$ and a certain calculable variance. The probability that this inequality is satisfied, therefore, is asymptotically given by a normal integral. This has been explored by several workers. Hudson (1968) notes that if $\gamma$ is taken to be $e^{-2}$, the set of values of $\theta$ are those for which $\ln L(\theta; D) \geq \ln L(\theta^*; D) - 2$. If we have a sample of size $n$ from the population $N(\mu, \sigma^2)$ with $\sigma$ known, this gives as the interval of values for $\mu$ corresponding to a sample average, $\bar{x} - 2\sigma/\sqrt{n} \leq \mu \leq \bar{x} + 2\sigma/\sqrt{n}$. It will be seen in Chapter 13 that this is a consonance interval with coefficient 95.4% and a confidence interval with confidence coefficient of 95.4%. Hudson examines the application of this idea to some of the elementary situations such as simple Bernoulli trials.

## 11.3 Use of Bayes's Theorem

Suppose we have data assumed to come from a population whose distribution is specified except for a parameter $\theta$. We write down the function $P(D|\theta)$ for any possible set of data $D$. We write $P(D|\theta)$ here rather than $P(D; \theta)$ because we envisage $\theta$ as a chance variable. Suppose, also, that the population has been chosen at random from a population of populations which differ only in $\theta$. Let the probability of a value $\theta_0$ be $P(\theta_0)$. Then we can write the joint probability of $D$ and $\theta_0$ as

$$\begin{aligned} P(D, \theta_0) &= P(\theta_0)P(D|\theta_0) \\ &= P(D)P(\theta_0|D) \end{aligned} \tag{11.1}$$

so that

$$P(\theta_0|D) = P(\theta_0)P(D|\theta_0)/P(D) \tag{11.2}$$

The quantity $P(\theta_0|D)$ is the *a posteriori* probability of $\theta_0$ given $D$. It is critical that the logic of this argument be described precisely. We envisage the sampling of populations according to a rule, and then sampling within the chosen population. We ask: In the class of repetitions which give the data $D$, what proportion have $\theta$ equal to $\theta_0$? The answer is clearly a conditional probability.

There is, it appears, no argument about the use of this procedure in the case we have described, nor about the answer. *It is to be noted that the whole data consist of D and the knowledge of the sampling of a population from the population of populations.*

Another aspect of the above is to state that if there are two candidate values for $\theta$, say $\theta_1$ and $\theta_2$, then

$$\frac{P(\theta_1|D)}{P(\theta_2|D)} = \frac{P(\theta_1)}{P(\theta_2)} \cdot \frac{P(D|\theta_1)}{P(D|\theta_2)} \tag{11.3}$$

The left-hand side is the posterior odds for $\theta_1$ over $\theta_2$, and $P(\theta_1)/P(\theta_2)$ is the prior odds so we have

$$\text{Posterior odds} = \text{prior odds} \times \text{likelihood ratio} \tag{11.4}$$

This serves to give a role to the likelihood ratio as a factor by which prior odds, whatever they may be, are multiplied to give the posterior odds. If one wishes, one can take the logarithm of Equation (11.4) to give:

$$\log(\text{posterior odds}) = \log(\text{prior odds}) + \log(\text{likelihood ratio}) \tag{11.5}$$

Equations (11.4) and (11.5) are generally considered to carry great force.

## 11.4 Bayes, Likelihood, and Sufficiency

In Equation (11.2) we note that $P(D)$ does not depend on $\theta_0$ and is in fact the sum $\sum P(\theta)P(D|\theta)$ or integral over all $\theta$. So we can write Equation (11.2) as

$$P(\theta_0|D) \propto P(\theta_0)P(D|\theta_0) \tag{11.6}$$

where $\propto$ means "is proportional to," or as

$$P(\theta_0|D) \propto P(\theta_0)L(\theta_0; D) \tag{11.7}$$

because

$$L(\theta_0; D) \propto P(D|\theta_0) \tag{11.8}$$

Equation (11.7) can be expressed verbally as:

$$\text{Posterior probability} \propto \text{prior probability} \times \text{likelihood} \tag{11.9}$$

It is immediately clear that the data $D$ can be replaced by any sufficient condensation and indeed by a minimal sufficient statistic. But it is also clear that there is no need to bother with questions about existence of any such statistic. We can work with Equation (11.7), and a normalization will result in removal of any multiplicative factor in $P(D|\theta)$ which does not depend on $\theta$.

## 11.5 The Binomial Case

An example of antiquity is the following. Consider $n$ independent Bernoulli trials with parameter $p$, and suppose $r$ successes have been obtained. Then

$$P(r|p) = \binom{n}{r} p^r (1-p)^{n-r} \tag{11.10}$$

Now suppose that $p$ itself has arisen as a random sample of one from the beta distribution

$$f(p)\,dp = \frac{p^\alpha (1-p)^\beta}{B(\alpha+1, \beta+1)}\,dp, \qquad 0 < p < 1 \tag{11.11}$$

Then if we follow the obvious path we have, using Equation (11.7), that the posterior distribution of $p$ is given by

$$f_{\text{post}}(p)\,dp = \frac{p^{r+\alpha}(1-p)^{n-r+\beta}\,dp}{B(r+\alpha+1, n-r+\beta+1)} \tag{11.12}$$

The mean of the beta distribution with parameters $\gamma, \delta$ is $\gamma/(\gamma+\delta)$. Hence the mean of the posterior distribution given by Equation (11.12) is

$$(\alpha + r + 1)/(n + \alpha + \beta + 2)$$

Now suppose that $\alpha$ and $\beta$ are zero, so that Equation (11.11) becomes

$$f(p)\,dp = dp, \qquad 0 < p < 1 \tag{11.13}$$

Then the posterior mean is $(r+1)/(n+2)$. The above argument, using Equation (11.13), was advocated vigorously by Laplace, who appears to have considered that Equation (11.13) is the appropriate mathematical formulation of "knowing nothing about $p$." The use of $(r+1)/(n+2)$ is called the law of succession. It was very highly regarded for much of the nineteenth century as a justification of the basic philosophical problem of induction.

## 11.6 The Normal Distribution Case

Suppose we have an observation $x$ which is taken to be $N(\mu, \sigma^2)$, with $\sigma^2$ known and $\mu$ unknown. Also suppose that $\mu$ has arisen as a random member of $N(\nu, \phi^2)$, with $\nu$ and $\phi$ known. Then

$$f_{\text{post}}(\mu) \propto \frac{1}{\sqrt{2\pi}\sigma} \exp\left[-\frac{1}{2\sigma^2}(x-\mu)^2\right] \frac{1}{\sqrt{2\pi}\phi} \exp\left[-\frac{1}{2\phi^2}(\mu-\nu)^2\right] \tag{11.14}$$

Dropping factors not depending on $\mu$, we have

$$f_{\text{post}}(\mu) \propto \exp\left[-\frac{1}{2\sigma^2}(\mu^2 - 2\mu x) - \frac{1}{2\phi^2}(\mu - \nu)^2\right]$$

$$\propto \exp\left\{-\frac{1}{2}\left[\mu^2\left(\frac{1}{\sigma^2} + \frac{1}{\phi^2}\right) - 2\mu\left(\frac{x}{\sigma^2} + \frac{\nu}{\phi^2}\right)\right]\right\} \quad (11.15)$$

and completing the square,

$$f_{\text{post}}(\mu) \propto \exp\left\{-\frac{1}{2}\left(\frac{1}{\sigma^2} + \frac{1}{\phi^2}\right)\left[\mu - \frac{(x/\sigma^2 + \nu/\phi^2)}{1/\sigma^2 + 1/\phi^2}\right]^2\right\} \quad (11.16)$$

So we have

$$f_{\text{post}}(\mu)\, d\mu = \frac{1}{\sqrt{2\pi}\psi} \exp\left[-\frac{1}{2\psi^2}(\mu - \tau)^2\right] d\mu \quad (11.17)$$

or $\mu$ has a posterior distribution which is $N(\tau, \psi^2)$, where

$$\tau = \left(\frac{x}{\sigma^2} + \frac{\nu}{\phi^2}\right) \Big/ \left(\frac{1}{\sigma^2} + \frac{1}{\phi^2}\right), \quad \frac{1}{\psi^2} = \frac{1}{\sigma^2} + \frac{1}{\phi^2} \quad (11.18)$$

Note that $\tau$ is a weighted mean of $x$, an "estimate" of $\mu$ with variance $\sigma^2$, and $\nu$ an "estimate" of $\mu$ with variance $\phi^2$.

This case is interesting in the respect that if $\phi^2$ is taken to be very large, the density of the posterior distribution of $\mu$ is "close" to

$$\frac{1}{\sqrt{2\pi}\sigma} \exp[-(\mu - x)^2/2\sigma^2]$$

So if the prior distribution of $\mu$ is very "diffuse," the Bayesian argument enables us to pass from the direct probability assumption that $x$ is $N(\mu, \sigma^2)$ to the posterior probability statement that given $x$, $\mu$ is distributed according to $N(x, \sigma^2)$. The normality assumption for the distribution of $\mu$ is not critical, and the development is very simple. This general line of argument is considered by some statisticians to have considerable force. [See, e.g., Savage (1962).]

## 11.7 General Aspects of the Bayesian Argument

We have presented the basic idea of the Bayesian argument, which is that given a prior distribution for the parameter $\theta$—say $p(\theta)$, which may be a probability or probability density function—and given data $D$, we can calculate readily the posterior probability or probability density

$$p_{\text{post}}(\theta) = p(\theta)p(D; \theta)/c$$

where the constant $c$ equals the sum or integral of $p(\theta)p(D; \theta)$ over the space $\Theta$ for $\theta$. We now give a brief description of procedures that can be derived from this posterior distribution.

We can compute the posterior probability that the true value $\theta_T$ of $\theta$ lies in any region $R$ of the space $\Theta$ by the sum or integral of $p_{\text{post}}(\theta)$ over the region $R$.

We can construct a statistical interval or region for $\theta_T$. To take the case of a scalar $\theta$, for instance, we can determine $\theta_L$ and $\theta_U$ so that

$$P_{\text{post}}(\theta_T \le \theta_L) = \frac{\alpha}{2}$$

$$P_{\text{post}}(\theta_T \ge \theta_U) = \frac{\alpha}{2}$$

and

$$P_{\text{post}}(\theta_L \le \theta_T \le \theta_U) = 1 - \alpha$$

In the case of a discrete prior, some of the equality signs must be replaced, of course, by inequalities. The resulting interval can be called a $(1 - \alpha)$ Bayesian interval for $\theta_T$. We can, if we wish, find $\theta_U^*$ such that

$$P_{\text{post}}(\theta_T \le \theta_U^*) = 1 - \alpha$$

and call the interval on $\theta_T$ given by $\theta_T \le \theta_U^*$ a lower $(1 - \alpha)$ Bayesian interval for $\theta_T$.

If $\theta$ is a vector, say $(\theta_1, \theta_2)$, the Bayesian process gives a probability distribution over the space $\Theta$ for $\theta_1, \theta_2$, and we can construct a set of points $R$ in $\Theta$, with the probability that $\theta$ lies in $R$ of $1 - \alpha$, e.g., 95%. This will be a $1 - \alpha$ Bayesian region or set for $\theta$. It is not at all clear, however, how one should choose a region in such a two-space. In this case we can sum or integrate the posterior over either of the variables $\theta_1$ or $\theta_2$ to give a posterior for the component not margined out. This is totally routine, as in the computation of marginal distributions for a known population.

We can consider a procedure for estimating $\theta_T$, say $\hat{\theta}[D; p_{\text{prior}}(\theta)]$, which gives a unique number (vector) for $\theta_T$ by the use of one of many rules. We may take, for example: the mean of the posterior distribution, the median of the posterior distribution (if it exists), or the mode of the posterior distribution. Indeed, the posterior distribution is exactly like a mathematically specified population, and we may take any "measure of central tendency" we wish. We have already seen how the Laplace rule of succession arises as the mean of a posterior distribution obtained from a particular prior distribution. We can regard the whole process as giving an estimate of $\theta$ and can then consider what the distribution of the estimate is in repetitions of sampling from a distribution specified by a particular "value" for $\theta$, the true value $\theta_T$. We may in this way obtain what appears to be a "good" procedure for estimating $\theta_T$. This topic is discussed in Chapter 14. Note, however, that this use of the Bayesian process is merely a means to get a process for associating a value for $\theta$ with each sample.

We can make a prediction of a new observation, because given $p_{\text{post}}(\theta)$ and $P(X = x|\theta)$, we can calculate $P_{\text{post}}(X \in R)$, where $R$ is any region (a set of points) in the sample space as a double sum or integral. In the case of a continuous random variable $X$ and a continuous prior distribution, the answer given by the process is

$$\int_R \int_\Theta P(X = x|\theta) p_{\text{post}}(\theta) \, d\theta \, dx \tag{11.19}$$

where both $d\theta$ and $dx$ are differential elements in whatever number of dimensions is relevant for the spaces of $\theta$ and of $X$. Note in this that if we make two predictions, e.g., that one new observation $X_1$ will lie in $R_1$ and a second new observation $X_2$ will lie in $R_2$, then the two predictions are not independent, and the probability that $X_1$ lies in $R_1$ *and* $X_2$ lies in $R_2$ is not the product of the previous probabilities.

We can, in fact, perform any mathematical or computational developments on a posterior distribution that we can do with a *completely known* population. The ease of the indicated solution to all questions of inference is obvious, but it should not be allowed to influence unduly the user of statistical methods and the probability calculus. The cost is simply exhibited by the requirement, "choose a prior distribution."

Finally, we note that if the prior distribution is valid or, in other words, if we are *in fact* sampling from the prior distribution, all the results have a direct probability status as relative frequencies in repetitions. They would then be totally verifiable by repetitive observation, provided the repetitions include independent sampling from the prior distribution. If, however, the prior distribution does not have this status and is purely an idea introduced by the data interpreter with no basis in fact, the probabilities obtained are *not* verifiable in any sense. Predictions can be essentially useless, for instance.

## 11.8 Convergence of Posterior Distribution with Sample Size

Consider the case of a sample of size $n$ from a distribution which is specified by a scalar parameter $\theta$. Suppose also that the situation satisfies the regularity conditions of Chapter 10 and consider the maximum likelihood statistic $\theta^*$. This statistic is of course a function of the first $n$ observations. We saw that $\theta^* - \theta$ tends with increasing $n$ to have the normal distribution $N(0, \sigma^2/n)$, where $\sigma^2 = 1/I$ and $I$ is the Fisherian information.

Furthermore, $\theta^*$ tends to be asymptotically sufficient. Hence

$$\text{Prob}(x_1, x_2, \ldots, x_n; \theta) \to c(x_1, x_2, \ldots, x_n) \frac{\sqrt{n}}{\sqrt{2\pi}\sigma} \times \exp\left[-\frac{n}{2\sigma^2}(\theta^* - \theta)^2\right] \quad (11.20)$$

where Prob( ) here equals "probability" or "probability density."

Suppose that our prior distribution is continuous with density $p(\theta)\,d\theta$ over some interval (or collection of intervals) containing $\theta_T$, the true value of $\theta$. Then the posterior density of $\theta$ is given by

$$\frac{p(\theta)c(x_1, x_2, \ldots, x_n) \dfrac{\sqrt{n}}{\sqrt{2\pi}\sigma} \exp\left[-\dfrac{n}{2\sigma^2}(\theta^* - \theta)^2\right]}{\displaystyle\int_\theta p(\theta)c(x_1, x_2, \ldots, x_n) \frac{\sqrt{n}}{\sqrt{2\pi}\sigma} \exp\left[-\frac{n}{2\sigma^2}(\theta^* - \theta)^2\right] d\theta}$$

$$= \frac{p(\theta) \dfrac{\sqrt{n}}{\sqrt{2\pi}\sigma} \exp\left[-\dfrac{n}{2\sigma^2}(\theta^* - \theta)^2\right]}{\displaystyle\int_\theta p(\theta) \frac{\sqrt{n}}{\sqrt{2\pi}\sigma} \exp\left[-\frac{n}{2\sigma^2}(\theta^* - \theta)^2\right] d\theta} \quad (11.21)$$

It is clear intuitively that the likelihood function is a multiple of a normal density, which becomes concentrated increasingly with $n$ near the true value $\theta_T$ of the distribution being sampled because $\theta^*$ leads to $\theta_T$. In non-regular cases the rate at which this happens may be faster, for instance when the variance of $\theta^*$ decreases as $1/n^2$ or faster.

Suppose now that in the neighborhood of $\theta_T$ the prior density is nearly uniform and is given by $\alpha + \beta\theta$. Then, using a heuristic argument which can be made rigorous, we have: posterior density given $x_1, x_2, \ldots, x_n$

$$= \frac{(\alpha + \beta\theta) \dfrac{\sqrt{n}}{\sqrt{2\pi}\sigma} \exp\left[-\dfrac{n}{2\sigma^2}(\theta^* - \theta)^2\right]}{\displaystyle\int_\theta \text{numerator}\, d\theta} \quad (11.22)$$

The denominator is equal to $\alpha + \beta\theta^*$. Hence the posterior density is the same for all samples $x_1, x_2, \ldots, x_n$ which give the same maximum likelihood statistic $\theta^*$ and is close to

$$\left(\frac{\alpha + \beta\theta}{\alpha + \beta\theta^*}\right) \frac{\sqrt{n}}{\sqrt{2\pi}\sigma} \exp\left[-\frac{n}{2\sigma^2}(\theta - \theta^*)^2\right] \quad (11.23)$$

Note that if $\beta$ is zero, the posterior density is

$$\frac{\sqrt{n}}{\sqrt{2\pi}\sigma} \exp\left[-\frac{n}{2\sigma^2}(\theta - \theta^*)^2\right] \tag{11.24}$$

Consider now the distribution with density given by Equation (11.23). Its mean is equal to

$$\frac{1}{\alpha + \beta\theta^*}\left[\alpha\theta^* + \beta\left(\theta^{*2} + \frac{\sigma^2}{n}\right)\right] = \theta^* + \left(\frac{\beta}{\alpha + \beta\theta^*}\right)\frac{\sigma^2}{n} \tag{11.25}$$

and with increasing $n$ it tends to $\theta^*$, which itself tends in distribution to the unit spike at $\theta_T$, the true value for $\theta$. Consider also the variance of a particular posterior distribution. It is

$$\frac{1}{\alpha + \beta\theta^*}\left[\alpha\left(\theta^{*2} + \frac{\sigma^2}{n}\right) + 3\beta\theta^*\frac{\sigma^2}{n} + \beta\theta^{*3}\right] - \left[\theta^* + \left(\frac{\beta}{\alpha + \beta\theta^*}\right)\frac{\sigma^2}{n}\right]^2 \tag{11.26}$$

The importance of this somewhat messy expression is seen by noting that the term not involving $1/n$ is zero. Hence the variance of the posterior distribution decreases approximately as $1/n$. The posterior distribution, therefore, tends with increasing $n$ to be a unit spike at the true value of $\theta$ which we called $\theta_T$.

This argument, which can be given in a slightly different form for a discrete prior, has been regarded by many as having considerable force. It assumes that the prior distribution is "nearly" flat in a "broad" region known to include the true value $\theta_T$. However, the force has been considerably exaggerated, it would seem, because the likelihood theory above shows that the maximum likelihood statistic has a distribution which tends to the unit spike at $\theta = \theta_T$ without the introduction of any prior distribution. The relevant question is not at all that one will reach the true value of $\theta$ by indefinitely large sampling of the true distribution. This will happen with any reasonable approach to inference. Note also that if the logical basis for any sort of Bayesian argument is the above, this logical basis *does* use the idea of repeated sampling from the true population.

## 11.9 Operating Characteristics of the Bayesian Process

The Bayesian process is a procedure for calculating a posterior distribution of an unknown parameter $\theta$. This posterior distribution can be regarded as a statistic, with a finite or infinite number of components. The data are mapped, then, by the Bayesian process into a statistic. As in all cases of computing statistics it is highly natural to consider the behavior of the procedure in repetitions, which we call its operating characteristics.

Suppose there are only two candidate models $M_1$, $M_2$ for data $D$. Then if the particular model that one encounters arises by a random process with $\text{Prob}(M_1) = q_1$, $\text{Prob}(M_2) = q_2$, the prior odds-ratio, or odds-ratio in the absence of the observation for $M_1$ against $M_2$ is $q_1/q_2$. After data $D$ the posterior odds-ratio is $q_1^*/q_2^*$, where

$$\frac{q_1^*}{q_2^*} = \frac{q_1}{q_2} \frac{P(D|M_1)}{P(D|M_2)}$$

The factor by which the prior odds-ratio is multiplied to give the posterior odds-ratio is the likelihood ratio $P(D|M_1)/P(D|M_2)$, which we may denote by $LR(D; M_1, M_2)$. If we consider this as a rule for writing down posterior odds given the data $D$, it is interesting to note the expected values of the factor. We have

$$E_{M_1}[LR(D; M_1, M_2)] = \sum_D \frac{[P(D|M_1)]^2}{P(D|M_2)} \tag{11.27}$$

This will be finite only if $P(D|M_2) > 0$ for all $D$ such that $P(D|M_1) > 0$, and we consider only this case. If this did not hold, it would be possible that an observation would make the posterior odds infinite or the posterior probability of $M_1$ equal to unity. If $\sum u_i = \sum v_i = 1$, $v_i \neq 0$ then one can verify quickly that

$$\sum \frac{u_i^2}{v_i} = 1 + \sum v_i \left( \frac{u_i}{v_i} - 1 \right)^2 \tag{11.28}$$

Hence the expected value under repetitions of sampling from $M_1$ of the factor by which the prior odds-ratio of $M_1$ against $M_2$ is multiplied is greater than unity. On the other hand, the expected value under repetitions of sampling from $M_2$ is

$$\sum_D P(D|M_2) \frac{P(D|M_1)}{P(D|M_2)} = \sum_D P(D|M_1) = 1 \tag{11.29}$$

It is not comforting and is perhaps disturbing to note this result. One might hope that the factor multiplying the odds-ratio for $M_1$ against $M_2$ would be less than one on the average over random observations from $M_2$. It is not at all clear at an intuitive level that the Bayesian process has good operating characteristics except in repetitions of the *whole* process.

An example illustrating the problems follows: Suppose there are four possible observations, represented by 1, 2, 3, 4 with probabilities under

$M_1$ and under $M_2$ as follows:

| $D$ | 1 | 2 | 3 | 4 |
|---|---|---|---|---|
| $P(D\|M_1)$ | .1 | .2 | .3 | .4 |
| $P(D\|M_2)$ | .2 | .3 | .4 | .1 |

Suppose we assumed that the $M_i$ arose by sampling with equal probabilities of 1/2. Then the posterior probabilities of $M_1$ and $M_2$ are as follows:

| $D$ | 1 | 2 | 3 | 4 |
|---|---|---|---|---|
| Post Prob$(M_1)$ | $\frac{1}{3}$ | $\frac{2}{5}$ | $\frac{3}{7}$ | $\frac{4}{5}$ |
| Post Prob$(M_2)$ | $\frac{2}{3}$ | $\frac{3}{5}$ | $\frac{4}{7}$ | $\frac{1}{5}$ |

Then we see that the probability under sampling from $M_1$ that the posterior probability of $M_1$ is reduced is 0.6. On the other hand, the probability under $M_2$ that the posterior probability of $M_2$ is increased is 0.9 which may well be regarded as satisfactory. The whole result suggests, however, that the operating characteristics of the Bayesian process are not always appealing. It is natural to regard here observations for which the posterior probability of $M_1$ are less than 1/2 as providing evidence against $M_1$. The probability that an observation from $M_1$ gives a lower posterior probability for $M_1$ is 0.6. This seems rather discouraging and suggests that the Bayesian process has a high probability of being misleading. This suggests, furthermore, that the mode of approach which consists merely of writing down a prior distribution in the absence of a real random sampling of models and then performing the Bayes computations may have high probability of being misleading, as well as having no good logical basis.

A possible approach to this situation is to convert the problem of forming a probability opinion on whether the model is $M_1$ or $M_2$ to the problem of using the observation to state a conclusion that the model is $M_1$ or that the model is $M_2$. This leads, then, to the division of the observation space into two exclusive and exhaustive regions $E_1$ and $E_2$. If the observation falls in $E_i$, we give the conclusion that the model is $M_i$. This formulation converts the problem into one of formulating an accept-reject rule. The general study of any such rule falls into the domain of decision theory, to which an introduction is given in Chapter 14. We give here some initial ideas for this special case. The properties of a rule can be specified simply by the probabilities:

$$P(E_1|M_1),\ P(E_2|M_1) = 1 - P(E_1|M_1)$$
$$P(E_2|M_2),\ P(E_1|M_2) = 1 - P(E_2|M_2) \tag{11.30}$$

The erroneous conclusions here are, of course, to conclude $M_2$ when in fact the observation comes from $M_1$ and to conclude $M_1$ when the observation comes from $M_2$. There seems little recourse for the investigator but to consider the possible partitions of $E$ into $E_1$ and $E_2$ and to choose that partition which appeals to him. One can do some mathematical reasoning if one can attach values to each correct conclusion, say, $v$ dollars, and losses due to each incorrect conclusion, say of $c$ dollars. Then the expected value of a conclusion based on an observation is

$$vP(E_1|M_1) - c[1 - P(E_1|M_1)] = (v + c)P(E_1|M_1) - c \qquad (11.31)$$

when the observation comes from $M_1$, and

$$(v + c)P(E_2|M_2) - c \qquad (11.32)$$

when the observation comes from $M_2$. It is very difficult to go beyond this. One can determine $E_1$ and hence $E_2$ to maximize the minimum of these two expected values, which would amount to choosing $E_1$ and $E_2(=E - E_1)$ to maximize the minimum of $P(E_1|M_1)$ and $P(E_2|M_2) = 1 - P(E_1|M_2)$.

It may be, however, that the values of correct conclusions and losses due to incorrect conclusions have a much more general structure—e.g., $v_1$ as the value of concluding correctly that the model is $M_1$, $v_2$ as the value of concluding correctly that the model is $M_2$, $c_1$ as the loss due to concluding erroneously that the model is $M_1$, and $c_2$ as the loss due to concluding erroneously that the model is $M_2$. In that case the expected values of a rule using $E_1$ and $E_2$ are

$$v_1P(E_1|M_1) - c_1P(E_2|M_1) = (v_1 + c_1)P(E_1|M_1) - c_1 \qquad (11.33)$$

when the observation comes from $M_1$ and $(v_2 + c_2)P(E_2|M_2) - c_2$ when the observation comes from $M_2$. One will then have to consider these "values" for all partitions of $E$ and choose the partition which has the most appeal. If one had the additional information that the model was $M_1$ with probability $q_1$ and $M_2$ with probability $q_2(=1 - q_1)$, one could form an expected value

$$q_1[(v_1 + c_1)P(E_1|M_1) - c_1] + q_2[(v_2 + c_2)P(E_2|M_2) - c_2] \qquad (11.34)$$

and then choose $E_1$ so that this expected value is a maximum. A possible way of simplifying the situation is to suppose there are only losses due to incorrect conclusions and these are the same for both possibilities. Then one could choose $E_1$ and $E_2$ to maximize

$$q_1P(E_1|M_1) + q_2P(E_2|M_2) = 1 - q_1P(E_2|M_1) - q_2P(E_1|M_2) \qquad (11.35)$$

so one would minimize

$$q_1P(E_2|M_1) + q_2P(E_1|M_2) \qquad (11.36)$$

If further $q_1 = q_2 = 1/2$, one would choose $E_1$ to minimize the sum of the possible error rates $P(E_2|M_1) + P(E_1|M_2)$. The suggested basis of choosing between different procedures by attaching values to the correct and incorrect conclusions and then taking the procedure with the best expected value is not, however, highly appealing. It is remarkable, perhaps, that so simple a problem has no simple generally accepted solution.

A more general approach to this simple situation could involve a partition of the observation space into three disjoint exhaustive regions, say $E_1$ which leads to the conclusion that the model is $M_1$, $E_2$ which leads to the conclusion that the model is $M_2$, and $E_u$ which leads to the conclusion that one is uncertain. Clearly such an approach can be extended still further to regions giving "high confidence" in $M_1$, "moderate confidence" in $M_1$, "uncertain," "moderate confidence" in $M_2$, and "high confidence" in $M_2$. It appears that little work has been done on this potentially useful mode of approach.

## 11.10 Recent Bayesian Theory

We see from the above discussion that the Bayesian argument reduces to a calculation or computation problem given the prior distribution. There has been fairly extensive literature on the consequences of adjoining particular prior distributions to data probability situations, of which simple cases are the binomial and normal cases discussed above. Because of the dependence for any finite set of data of the posterior on the prior, there have been attempts to develop compelling arguments for the use of particular prior distributions. We refer the reader to Good (1950), Jeffreys (1961), Lindley (1965), Schlaifer (1959), and Savage (1954) for expositions of Bayesian arguments. Part of the development is based on what are called "improper" prior distributions, such as that given by the density $d\mu$, with $-\infty < \mu < \infty$, or in a two-parameter case the density $d\mu\, d\sigma/\sigma$ with $-\infty < \mu < \infty$, $0 < \sigma < \infty$ for normal distributions $N(\mu, \sigma^2)$. In these cases the density does not have a finite integral over the space of the random variable. See also Cornfield (1969) and L. J. Savage (all refs.).

## 11.11 Some Comments on the Bayesian Process

There is great controversy over the use of the Bayesian argument. The following remarks may be helpful:

1. There are situations in which we may wish to examine the consequences of a Bayesian argument, using a particular prior distribution.

A possible prior distribution may be obtained from a mechanistic model, as in some genetic problems, for instance.

2. It is reasonable, perhaps, to take the view that we are prepared to summarize the historical record by a prior distribution but then the question of how we obtained it has to be raised, because we can get this prior only by combining another prior distribution with the historical record. We have to note, however, that if the only possible inference is Bayesian, we have merely hidden the logical problem because we would have to "analyze" the historical record by injecting a prior distribution.
3. The use of the uniform prior as in Equation (11.13) was regarded as appropriate to characterize ignorance by Laplace but was rejected strongly by great thinkers such as Venn (1866).
4. It was pointed out that a distribution can be parametrized in many ways. In the binomial case one could reparametrize by using $\sin^2 \theta = p$ with $0 < \theta < \pi/2$, so a uniform prior would be

$$f(\theta)\, d\theta = \frac{2}{\pi}\, d\theta, \qquad 0 < \theta < \pi/2$$

This will give a different answer for the posterior distribution of $p$ than we would obtain if we used a uniform prior on $p$. Of course if we are sampling from $f(p)\, dp = dp$, we are not sampling from

$$f(\theta)\, d\theta = \frac{2}{\pi}\, d\theta$$

so there is no question. But which distribution should we use to characterize ignorance?
5. Some feel there is a case for a "natural" prior associated with a particular problem. This view has been advocated by Jeffreys (1961) and is pursued mathematically by Raiffa and Schlaifer (1961).
6. The use of the Bayesian argument for a scalar parameter is fairly easy to understand, but what seems to be a "natural" prior for a vector-parameter situation may give strange results [see, e.g., Geisser (1965)].
7. We have to give weight to the point that even in the case of sampling from a population of populations the final Bayesian statement applies to repetitions of the whole process. But in many cases we are interested in the particular population that meets our attention. The reader may verify that the mean square error of the posterior mean is greater than that of $x$ in the normal case above if $\mu$ is quite different from $\nu$. So if we have some sort of decision problem in which the loss function is mean square error for the *individual case*, we may not wish to use $\tau$.

8. R. A. Fisher (1956 and earlier) proposed a method called fiducial inference, which could be applied in certain cases, and felt that this gave *the* solution to the whole dilemma. We are among those who do not accept this view, and our comments about the process appear in Chapter 12. A clear discussion and criticism of the elementary aspects of the fiducial argument is given by Neyman (1941).
9. The ease of the Bayesian argument for the case of a vector parameter and, in particular, the simplicity of the reduction of a posterior distribution for the elements of a vector parameter to a distribution for any subset of the elements of the parameter by computing marginal posterior distributions has led some workers to regard the Bayesian process favorably. It will be seen in Chapter 12 that there are deep obscurities in "inference" for the case of vector parameters with the more traditional methods described therein.
10. We may refer the advanced reader to the papers by Pratt (1965) and Geisser (1965) in which relations of traditional inference procedures and Bayesian procedures are examined.
11. Some statisticians are of the opinion that Bayesian arguments should be the main basis for inference. Others regard them as having some force as well as the traditional methods discussed in Chapters 12 and 13. Others regard the Bayesian argument as having no force except when sampling from a known population of populations, and they use the ideas of these chapters to form opinions and to make decisions on the basis of data. We shall indicate some of the difficulties of traditional methods which have led to the resurgence of Bayesian ideas.

## PROBLEMS

1. Considering the binomial density function

$$f(x) = \binom{n}{x} p^x(1 - p)^{n-x}$$

as the conditional density of $x$ given $p$, find the conditional density of $p$ given $x$ if the marginal density of $p$ is taken to be:

(a) $g(p) = \dfrac{p^{\alpha-1}(1 - p)^{\beta-1}}{B(\alpha, \beta)}, \qquad 0 \le p \le 1$

(b) $g(p) = 1, \qquad 0 \le p \le 1$

(c) $g(p) = 4p, \qquad 0 \le p \le 1/2$

$\qquad = 4(1 - p), \qquad 1/2 < p \le 1$

2. Suppose that $X_1, X_2, \ldots, X_n$ are a random sample from a Poisson distribution with parameter $\lambda$. Find the conditional density of $\lambda$ given $x_1, x_2, \ldots, x_n$ if the density of $\lambda$ is taken to be:

   (a) $g(\lambda) = \lambda^{\alpha-1}e^{-\lambda}/\Gamma(\alpha), \qquad \lambda \geq 0$

   (b) $g(\lambda) = (1/\theta)e^{-\lambda/\theta}, \qquad \lambda \geq 0$

   (c) $g(\lambda) = \binom{k}{\lambda} p^{\lambda}(1-p)^{k-\lambda}, \qquad \lambda = 0, 1, 2, \ldots, k$

3. Consider $n$ independent observations from a normal population with mean $\mu$ and variance unity. Find the conditional density of $\mu$ given the observations if the marginal density of $\mu$ is taken to be:

   (a) $g(\mu) = \dfrac{1}{\sqrt{2\pi b}} e^{-[(\mu-a)^2/2b^2]}, \qquad -\infty < \mu < \infty$

   (b) $g(\mu) = \dfrac{1}{\theta} e^{-\mu/\theta}, \qquad \mu \geq 0$

4. Consider a random sample of $n$ observations from a normal distribution with zero mean. Find the conditional density of $\sigma$ given the observations if $(1/\sigma^2)$ is a gamma variable with parameters $\alpha$ and $\beta$.

5. Consider a random sample of $n$ observations from a normal distribution with zero mean. Consider the unknown parameter to be $1/\sigma$. Take the marginal density of $1/\sigma$ to be uniform from zero to infinity. Of course this is not a probability density function. However, carry through the algebra in a formal manner as though it were, to show that the conditional density of $1/\sigma$ given the $x$'s is

$$f(1/\sigma|x_1 \ldots x_n) = \frac{(\sum x^2)^{(n+1)/2}\exp(-\sum x^2/2\sigma^2)}{\Gamma[(n+1)/2]2^{(n-1)/2}\sigma^n}$$

6. Verify Equation (11.23).

7. Verify Equations (11.25) and (11.26).

8. Assuming normality with $\mu$ and $\sigma^2$ unknown and given the data 11.2, 10.1, 6.5, 7.4, 8.3, 9.2, 6.8, 7.5, 8.9, 9.1, sketch the region in the $\mu$, $\sigma^2$ space such that $L(\mu, \sigma^2; D) \geq .9L(\mu^*, \sigma^{2*}; D)$.

9. Determine the means, medians, and modes of the posterior distributions obtained in Problems 1, 2, and 3.

10. Determine $1 - \alpha$ Bayesian intervals for the posterior distributions obtained in Problems 1, 2, and 3.

# 12

# Statistical Tests

## 12.1 Introduction

We saw in Chapter 10 how the probability leads to the likelihood, and in Chapter 11 we saw how the likelihood would be combined with a prior distribution on the parameters to give a posterior distribution of parameter values. This computation is logically quite simple, and can always be done on a computer, though a mathematical derivation of the posterior distribution may be difficult because of the need to determine the normalizing factor.

We now take up procedures that have had wide use in science and data evaluation in the absence of a prior distribution. To understand these procedures it is essential to have an approximate picture of how knowledge accumulates and is checked. The following diagram is informative.

idea (1) ⟶ theory (2) ⟶ models for data (3) ⟶ data (4) ⟶

data analysis (5) ↗ only one of models acceptable (end) (6)

data analysis (5) ⟶ several of models partially acceptable (7)

data analysis (5) ↘ none of models acceptable (8) ⟶ return to (1)

We assume that we have reached stage 7 in that several models or perhaps a continuum of models are partially consonant with the data. By this we mean that the probability of our actual data is nonzero for the partially acceptable models. Note that the idea and theory may be very vague, for instance that raising temperature increases growth.

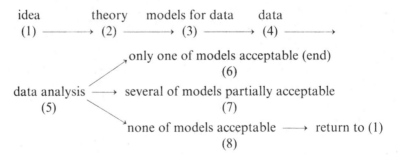

The problem is how to form some ideas of the tenability of the different models. Certainly one idea is to examine the likelihood function. The model in the set which gives the maximum of the likelihood is the most tenable. Other models which give at least 9/10 of the maximum likelihood are somewhat tenable. Models which give only 8/10 of the maximum likelihood are less tenable, etc.

In Chapter 9 we introduced tests of significance as a quantification of strength of evidence with regard to goodness of fit. In this chapter we shall attempt to give a more thorough development of testing for significance. This process goes back to antiquity in a certain sense and an early example was given by Arbuthnot (1710). However, its real development started with Karl Pearson (1900), who developed to some extent a $\chi^2$ goodness of fit test like the one of Chapter 9. The next development was due to W. S. Gossett (who published under the name of "Student") in 1908; in the period of the next 20 years R. A. Fisher developed many tests of significance. Subsequently Neyman and Pearson developed a general theory of tests of hypotheses, which seemed to be identical to tests of significance. The field of statistics has been marred by strong controversies on these matters, to the extent that similarities and differences have been obscured. Additionally, Fisher never gave a rigorous definition of what he meant by a test of significance, but produced a large number of significance tests which could be reformulated as tests of hypotheses. We will make a distinction between significance tests and tests of hypotheses and will then discuss the similarities and differences.

The starting point for tests of significance is a set of data. On the basis of prior knowledge and data analysis, one surmises that a certain class of models can be used to represent the data. Part of this data analysis consists of the tests of goodness of fit described in Chapters 9 and 10. It is essential to repeat what was said there: there is no single goodness of fit test which is compelling except under special data and prior knowledge situations. A set of data may contradict a model in many different ways. The data analyst will examine the data and will notice features which seem to him to be aberrant from the model he is inclined to use. He may apply a test of significance to any such feature that comes to his mind.

## 12.2 Tests of Significance of a Parameter Value

Suppose that we are led to envisage a model $M(\theta)$ depending on an unspecified parameter $\theta$ (which may be a vector). Very frequently we can pick out a distinctive value or set of values for $\theta$ which have scientific (not statistical) significance. For example, suppose we have $n$ animals, that we observe their weights at some point of time and apply some treatment to them (e.g., a dietary one). We shall then have weights after treatment, and we can certainly form the differences in weight, say $d_1, d_2, \ldots, d_n$. We take these to be the data of the study, though we note in passing that we have already made a condensation because the actual data consist of $n$ doublets $x_i$, $y_i$, where $x_i$ is the initial weight and $y_i$ is the weight after treatment. We now adopt the following stance. We are quite sure that the treatment has produced a change, though the change may be very small and the data may be indistinguishable from what might arise by chance. Therefore we desire some sort of test. As Fisher (1956) stated, "As early as Darwin's experiments on growth rate, the need was felt for some sort of a test of whether an apparent effect 'might reasonably be due to chance'."

We ask if the data can provide a measure of strength of evidence that there has been a change. We have to characterize what meaning and operational definition can reasonably be given to the vague phrase "strength of evidence."

A way of producing a quantification of strength of evidence is to construct what is called a *test of significance*, which we now describe in general terms. We choose a function of the observations, which in repeated sampling from a hypothesized model or class of models has a completely specified distribution, $F_0$, say. The function is to be chosen so that its distribution will shift in a systematic way from the distribution $F_0$ for an alternative class of models.

To illustrate, recall the goodness of fit tests of Chapter 9. There we stated that a particular function of the observations—one based on the hypothesized distribution—would follow (approximately) a particular known distribution, the $\chi^2$, and gave evidence that the particular function will tend to be larger for alternative distributions. Significance testing, then, consists of the following:

1. Specification of a hypothesized class of models and an alternative class of models.
2. Choice of a function of the observations $T$.
3. Evaluation of the significance level, i.e., $SL = P(T \geq t)$, where $t$ is the observed value of $T$ and where the probability is calculated for the hypothesized class of models.

In most applied writings the significance level is designated by $P$, a custom which has engendered a vast amount of confusion.

It is quite common to refer to the hypothesized class of models as the *null hypothesis* and to the alternative class of models as the *alternative hypothesis*. We shall omit the adjective "null" because it may be misleading. In either case a singleton class is called a *simple hypothesis*, whereas a class consisting of several models is called a *composite hypothesis*. In case our models are indexed by a parameter $\theta$, we can describe the situation by

$$\text{Hypothesis: } \{M(\theta)\},\ \theta \in \Omega_0$$
$$\text{Alternative: } \{M(\theta)\},\ \theta \in \Omega_1$$

If $\Omega_0$ consists of a single point $\theta_0$, the hypothesis is simple; otherwise it is composite. The same is true for $\Omega_1$.

It is essential to realize that we do not in general attach any strength to the idea that the hypothesized class of models is the true one. As stated by Fisher (1956), "In general tests of significance are based on hypothetical probabilities calculated from the null hypotheses. They do not generally lead to any probability statements about the real world, but to a rational and well-defined measure of reluctance to the acceptance of the hypotheses they test." We think about a test of significance because we wish to form an opinion of whether the data conform to the hypothesized distribution or model. We may not, in general, wish to form an opinion of whether the hypothesized distribution or model is the "true" one. Instead, we address the question of whether data conform to a particular model, and this is intrinsically an operational question. Are the data like those one would obtain from a class of repetitions related to the model? It follows that a class of repetitions must be relevant to the data situation.

We remark that there may be considerable latitude in the choice of population of repetitions, and we refer to Barnard (1947) for a discussion. The choice must be made, partially at least, on the basis of the scientific question asked. A purely mathematical principle of choice has been propounded by Fisher (1956). This may be described by a definition.

*Definition 12.1. A statistic $T(D)$ calculated from data $D$, assumed to have arisen from a class of probability models indexed by a parameter $\theta$ (which may be a vector), is said to be ancillary if the distribution of $T(D)$ in repetitions of $D$ does not depend on $\theta$.*

Fisher's principle, which is based on the idea of information discussed in Chapter 10, is that if one wishes to make a test of significance of $\theta$ and the minimal sufficient statistic is $(a, t)$ where $a$ is ancillary, then the appropriate population of repetitions is the set of possible samples for which the ancillary statistic takes its observed value. A vague heuristic argument for this is that because we have observed $a = a_0$, we know we are in a restricted population and should not base any opinion on the possibility that we are in populations for which $a \neq a_0$. Most of the cases we shall discuss do not involve this concept, except that the sample size can be regarded as ancillary. Fisher's principle is not universally accepted.

Basu (1959, 1964) has examined the Fisher idea in depth. The authors of this book accept Basu's conclusions that this type of conditioning should be done if the experiment can be reasonably regarded as a two-stage process in which the value of the ancillary occurs as a result of the first stage and the remainder of the data at the second stage. Basu calls such an experiment with the ancillary fixed at its observed value a "performable experiment." An illustration of a nonperformable conditional "experiment" is an observation from the distribution:

| $x$ | 1 | 2 | 3 | 4 |
|---|---|---|---|---|
| $P(X = x)$ | $\frac{1}{4}(1 - 2\theta)$ | $\frac{1}{4}(1 - \theta)$ | $\frac{1}{4}(1 + \theta)$ | $\frac{1}{4}(1 + 2\theta)$ |

with $-1/2 < \theta < 1/2$. It is clear that the statistic $A = 1$ if $X = 1$ or 4 and $A = 0$ if $X = 2$ or 3 is ancillary. But repetition of an actual experiment with an observed value for $A$ is not possible. Barnard and Sprott (1970) claim to have a logical procedure which resolves the dilemma. This is based on the fact that the likelihood functions for observations 1 and 4 are "similar," as are those for observations 2 and 3. However, there seems to be some doubt as to its utility. In general, several statistics may be ancillary, and no general criteria of shape of likelihood functions serves to pick out one ancillary over the others. The opinion of the authors is that in the last resort the scientist, not the statistician, must decide on populations of repetitions relevant to his interests.

It is relevant to note that significance testing can be applied to data in which the sample size is not fixed. In the case of sequential sampling [Wald (1947)], the data will have arisen by the use of a stopping rule such as: Obtain the $(n + 1)$th observation $x_{n+1}$ if $A < \sum_1^n x_i < B$, and otherwise stop. By such a rule the possible sets of data are restricted, and a significance test would consist of an ordering of the possible sets of data and the calculation of a significance level based upon this ordering.

## 12.3 Comparison of Tests of Significance

We considered the distribution of the significance level in Chapter 9 for the hypothesized class of models $M_0$ and the alternative class $M_1$. We showed that $P(SL \leq \alpha | M_0) = \alpha$, where $\alpha$ is an achievable significance level, and suggested that the corresponding probabilities under $M_1$ should be used as a basis for the comparison of different significance tests. Equivalently, the distribution of the significance level under the alternative provides a basis for the comparison of different significance orderings of the possible samples. However, there is a fundamental difficulty in making a meaningful comparison of two tests which do not have the same set of *achievable* significance levels. Although we shall give some results in Section 12.4 for the comparison of tests in this case, in general it appears that they should have the same set of achievable significance levels in order to be comparable. We state this formally as a definition.

*Definition 12.2. Two or more tests are comparable if they have the same set of achievable significance levels.*

Now if tests $T_1$ and $T_2$ are comparable, the corresponding probabilities satisfy

$$P_1(SL \leq \alpha|M_0) = P_2(SL \leq \alpha|M_0) = \alpha \tag{12.1}$$

for all achievable $\alpha$. If there is a greater probability of achieving small levels under $M_1$ for $T_1$ than for $T_2$, it seems reasonable to say that $T_1$ is more sensitive to departures from the null hypothesis than $T_2$. We state this as a definition.

*Definition 12.3. If tests $T_1$ and $T_2$ are comparable and*

$$P_1(SL \leq \alpha|M_1) \geq P_2(SL \leq \alpha|M_1) \tag{12.2}$$

*for all achievable $\alpha$, with strict inequality for some achievable $\alpha$, $T_1$ is more sensitive than $T_2$ with respect to $M_1$.*

Once we have restricted our attention to a set of achievable significance levels and have adopted a criterion for the comparison of tests, the next question is whether or not there is a best test by this criterion. We adopt as obvious the following definitions.

*Definition 12.4. $T^*$ is a most sensitive test if there is no comparable test which is more sensitive.*

*Definition 12.5. If $T^*$ is a most sensitive test for all members of the alternative class of models, it is a uniformly most sensitive test.*

In the subsequent sections we shall be concerned with finding most sensitive tests and uniformly most sensitive tests if they exist. Quite frequently, in fact usually, no uniformly most sensitive test exists. The general procedure then followed is to restrict our attention to some class of tests and to attempt to find a uniformly most sensitive test within this restricted class. For example, we have previously mentioned that it seems desirable for the distribution function for the significance level to be greater under the alternative than under the hypothesis, although we have not made this a requirement of significance tests in general. A significance test satisfying this property will be called an unbiased test.

*Definition 12.6. An unbiased test of significance is such that* $P(SL \leq \alpha | M_1) \geq \alpha$ *for all achievable* $\alpha$.

We may choose to restrict our attention to unbiased tests and attempt to find a uniformly most sensitive unbiased test.

## 12.4 Likelihood Ratio Tests of Significance

We are describing tests of significance as a way of giving operational meaning to a measure of strength of evidence. In Chapter 11 we saw that the likelihood ratio played a fundamental role in modifying the odds in favor of a given hypothesis. It has seemed to many researchers that significance tests should be based upon the likelihood ratio. We shall show that a very strong case can be made for the use of the likelihood ratio when it is a continuous statistic. However, when it is discrete, the case for using the likelihood ratio is weaker.

Consider the case where we have a simple hypothesis and a simple alternative and the data are denoted by $D$:

$$\text{Hypothesis: } P(D;\theta) = P(D;\theta_0)$$
$$\text{Alternative: } P(D;\theta) = P(D;\theta_1)$$

A very natural way of proceeding is to order the significance of samples by the ratio of likelihoods:

$$LR(D;\theta_1,\theta_0) = \frac{P(D;\theta_1)}{P(D;\theta_0)}, \text{ for the discrete case} \tag{12.3}$$

$$= \frac{f(D;\theta_1)}{f(D;\theta_0)}, \text{ for the continuous case} \tag{12.4}$$

The likelihood ratio test of significance ordering is

$$D_1 \gg D_2 \text{ if } LR(D_1;\theta_1,\theta_0) \geq LR(D_2;\theta_1,\theta_0)$$

If $LR(D_1;\theta_1,\theta_0) = LR(D_2;\theta_1,\theta_0)$, we have $D_1 \gg D_2$ and $D_2 \gg D_1$, so $D_1$ and $D_2$ are equally significant.

For brevity we use $LR$ in place of $LR(D;\theta_1,\theta_0)$. The significance level is found by calculating the probability under $\theta_0$ of a ratio as large as or larger than the observed ratio. Thus $LR$ is the test statistic.

***Theorem 12.1* (*Neyman-Pearson*)** The likelihood ratio test is a most sensitive test of significance for testing the simple hypothesis $\theta_0$ against the simple alternative $\theta_1$ in the class of tests comparable to it.

*Proof:* Let $T(D)$ denote any other statistic. We wish to prove that

$$P_{\theta_1}[SL(LR) \leq \alpha] \geq P_{\theta_1}[SL(T) \leq \alpha] \tag{12.5}$$

where $\alpha$ is an achievable significance level for both tests. Partition the sample space for $D$ into four disjoint regions, as shown in Table 12.1.

**Table 12.1** Neyman-Pearson Theorem

| | |
|---|---|
| $R_3$<br>$SL(LR) \leq \alpha$<br>$SL(T) > \alpha$ | $R_4$<br>$SL(LR) > \alpha$<br>$SL(T) > \alpha$ |
| $R_1$<br>$SL(LR) \leq \alpha$<br>$SL(T) \leq \alpha$ | $R_2$<br>$SL(LR) > \alpha$<br>$SL(T) \leq \alpha$ |

This partially represents the mappings from the sample space onto the $[0, 1]$ line segment. So we have, with an obvious notation

$$\begin{aligned} R_2 \cup R_4 &\xrightarrow{LR} (\alpha, 1] \\ R_3 \cup R_4 &\xrightarrow{T} (\alpha, 1] \\ R_1 \cup R_3 &\xrightarrow{LR} [0, \alpha] \\ R_1 \cup R_2 &\xrightarrow{T} [0, \alpha] \end{aligned} \tag{12.6}$$

We denote the probability model under hypothesis $\theta_i$ by $M_i$.

Because $\alpha$ is an achievable significance level for both $LR$ and $T$.

$$\begin{aligned} P(R_1 \cup R_3|M_0) &= P(R_1|M_0) + P(R_3|M_0) = \alpha \\ P(R_1 \cup R_2|M_0) &= P(R_1|M_0) + P(R_2|M_0) = \alpha \end{aligned} \tag{12.7}$$

So

$$P(R_2|M_0) = P(R_3|M_0) \tag{12.8}$$

Now consider possible values for the likelihood ratio in $R_2$ and $R_3$. In $R_2$ we will have small values and in $R_3$ large values.

Let

$$l_1 = \inf_D [LR(D),\ D \in R_3]$$

$$l_2 = \sup_D [LR(D),\ D \in R_2] \tag{12.9}$$

where inf and sup designate greatest lower bound and least upper bound respectively (see p. 394). Then

$$l_2 \leq l_1 \tag{12.10}$$

and

$$P(R_3|M_1) \geq l_1 P(R_3|M_0) \tag{12.11}$$

$$P(R_2|M_1) \leq l_2 P(R_2|M_0) = l_2 P(R_3|M_0) \tag{12.12}$$

Then because $l_2 \leq l_1$,

$$P(R_3|M_1) \geq P(R_2|M_1) \tag{12.13}$$

Consider now what happens for our two tests under $M_1$:

$$P[SL(LR) \leq \alpha|M_1] = P(R_1|M_1) + P(R_3|M_1)$$

and

$$P[SL(T) \leq \alpha|M_1] = P(R_1|M_1) + P(R_2|M_1) \tag{12.14}$$

but because

$$P(R_3|M_1) \geq P(R_2|M_1)$$

we have that

$$P[SL(LR) \leq \alpha|M_1] \geq P[SL(T) \leq \alpha|M_1] \tag{12.15}$$

which concludes the proof.

Although this theorem is stated in terms of some test which is comparable to the likelihood ratio test (i.e., one which has the same set of achievable significance levels), the inequality involving $P_\theta(SL \leq \alpha)$ holds for any significance level common to the two tests.

We attribute this theorem to Neyman and Pearson because their theorem which refers to accept-reject rules discussed below is closely related and may be taken over with some verbal changes to the context of significance testing.

*Corollary 12.1* $P_{\theta_1}(SL \leq \alpha)$ for the likelihood ratio test is at least as great as $P_{\theta_1}(SL \leq \alpha)$ for any other test for any $\alpha$ achievable by both tests.

To emphasize the limited nature of the preceding theorem and corollary consider the following example:

*Example 1*

$$f(x;\theta) = 1/\theta, \qquad 0 < x \leq \theta$$
$$= 0, \text{ otherwise}$$

$$\text{Hypothesis: } \theta = 10 = \theta_0$$
$$\text{Alternative: } \theta = 5 = \theta_1$$

Then for a single observation the likelihood ratio takes the value 2 for all $x$, $0 < x \leq 5$, and 0 for all $x$, $5 < x \leq 10$. Because there are only two values for $LR$, there are only two significance levels and

$$P_{\theta_0}[SL(LR) \le 1/2] = P_{\theta_0}(0 < x \le 5) = 1/2$$
$$P_{\theta_0}[SL(LR) \le 1] = P_{\theta_0}(0 < x \le 10) = 1$$
$$P_{\theta_1}[SL(LR) \le 1/2] = P_{\theta_1}(0 < x \le 5) = 1$$
$$P_{\theta_1}[SL(LR) \le 1] = P_{\theta_1}(0 < x \le 10) = 1$$

The test based on the likelihood ratio is a uniformly most sensitive test. However it has only two achievable levels, 1/2 and 1. By contrast consider the test based upon small values of $x$ where the significance level is obtained by calculating $P_{\theta_0}(X \le x)$. It is easily verified that at $\alpha = 1/2$ and 1 the probabilities $P_{\theta_0}(SL \le \alpha)$ and $P_{\theta_1}(SL \le \alpha)$ are the same as for the likelihood ratio tests. However, the second test might be preferred because it has all achievable significance levels between 0 and 1. The reader should convince himself that the anomaly noted here is not merely a result of using a continuous model but that a similar situation exists for discrete versions of this example.

In the previous example we have a case where a test not based on the likelihood ratio has achievable levels smaller than the smallest achievable level of the likelihood ratio test. This in itself does not mean that we should discard the likelihood ratio test. It simply means that the tests are not comparable. If, however, it happened that

$$P_{\theta_1}[SL(T) \le \alpha^*] \ge P_{\theta_1}[SL(LR) \le \alpha] \tag{12.16}$$

where $\alpha^*$ is an achievable level of $T$ less than the smallest achievable level $\alpha_0$ of the likelihood ratio test, we would be inclined to regard the likelihood ratio test as inferior to $T$. We show in the proof of the following theorem that this cannot happen.

***Theorem 12.2*** Consider the likelihood ratio test with achievable levels $\{a_i\}$ and any other test $T$ with achievable levels $\{b_j\}$. For no $b_j < a_i$ it is true that

$$P_{\theta_1}[SL(T) \le b_j] > P_{\theta_1}[SL(LR) \le a_i] \tag{12.17}$$

*Proof:* We shall prove the theorem by contradiction. Partition the sample space as shown in Table 12.2 for a pair of values $b_j$, $a_i$ such that $b_j < a_i$.

**Table 12.2** Partition of Sample

| | |
|---|---|
| $C$<br>$SL(LR) \le a_i$<br>$SL(T) > b_j$ | $D$<br>$SL(LR) > a_i$<br>$SL(T) > b_j$ |
| $A$<br>$SL(LR) \le a_i$<br>$SL(T) \le b_j$ | $B$<br>$SL(LR) > a_i$<br>$SL(T) \le b_j$ |

Now

$$a_i = P_{\theta_0}(A) + P_{\theta_0}(C)$$
$$b_j = P_{\theta_0}(A) + P_{\theta_0}(B)$$

Because

$$b_j < a_i$$
$$P_{\theta_0}(C) > P_{\theta_0}(B) \tag{12.18}$$

In region $B$, the likelihood ratio at each point is less than or equal to its value at any point in region $C$.
So

$$\frac{P_{\theta_0}(C)}{P_{\theta_1}(C)} \leq l_2 \leq l_1 \leq \frac{P_{\theta_0}(B)}{P_{\theta_1}(B)}$$

for some $l_1$ and $l_2$, and

$$\frac{1}{l_1} P_{\theta_0}(B) \geq P_{\theta_1}(B) \tag{12.19}$$

$$P_{\theta_1}(C) \geq \frac{1}{l_2} P_{\theta_0}(C) \tag{12.20}$$

Now suppose the conclusion of the theorem is false. That is, suppose

$$P_{\theta_1}(A \cup B) > P_{\theta_1}(A \cup C)$$

Then

$$P_{\theta_1}(B) > P_{\theta_1}(C) \tag{12.21}$$

Then from Equations (12.19) and (12.20) it follows that

$$\frac{1}{l_1} P_{\theta_0}(B) > \frac{1}{l_2} P_{\theta_0}(C) \tag{12.22}$$

and because

$$l_2 \leq l_1$$
$$P_{\theta_0}(B) > P_{\theta_0}(C) \tag{12.23}$$

which contradicts Equation (12.18). Therefore the theorem is proved.

Although we have discussed the inadequacies of the likelihood ratio test when the likelihood ratio is discrete, in most cases of practical interest we still feel that we would prefer it. It follows from the development of this section that when the likelihood ratio test yields all significance levels from zero to one, it is the most sensitive test of significance for a simple hypothesis with a simple alternative.

## 12.5 An Argument for Using the Likelihood Ratio Ordering

A mode of argumentation that leads to the *LR* ordering follows. Let us suppose that we have a discrete set of possible values for the parameter $\theta$, say $\theta_1, \theta_2, \ldots, \theta_s$, which we call the population of $\theta$ values. Suppose furthermore that the situation we are examining has arisen in a Bayesian context; i.e., suppose that the value of $\theta$ which we encounter has arisen by sampling from the population with probabilities $q_i, i = 1, 2, \ldots, s, q_i \geq 0, \sum q_i = 1$. The posterior probability of $\theta_m$ or the conditional probability (or frequency) of $\theta_m$ given that the data are $D$ is:

$$P_{\text{post}}(\theta_m) = \frac{q_m P(D; \theta_m)}{\sum_i q_i P(D; \theta_i)}$$

Now suppose that there are just two values for $\theta$, say $\theta_1$ and $\theta_2$. Then

$$P_{\text{post}}(\theta_1) = \frac{q_1 P(D; \theta_1)}{q_1 P(D; \theta_1) + q_2 P(D; \theta_2)}$$

$$P_{\text{post}}(\theta_2) = \frac{q_2 P(D; \theta_2)}{q_1 P(D; \theta_1) + q_2 P(D; \theta_2)}$$

We see, therefore, that $q_1$ is changed by the occurrence of the data $D$ to

$$q_1 \left[ \frac{1}{q_1 + q_2 \dfrac{P(D; \theta_2)}{P(D; \theta_1)}} \right] = q_1 \left[ \frac{1}{1 + q_2 \left[ \dfrac{P(D; \theta_2)}{P(D; \theta_1)} - 1 \right]} \right]$$

with a similar change for $\theta_2$. If $P(D; \theta_2)/P(D; \theta_1)$ equals unity, we would make no change from $q_1$. Hence it seems reasonable to take the view that if $P(D; \theta_2)/P(D; \theta_1)$ equals unity, we have no evidence which indicates that we should modify the opinion we had before seeing the data, whatever that opinion may have been. If, however, $P(D; \theta_2)/P(D; \theta_1) > 1$, the data $D$ indicate that we should shift the weight of our opinion away from $\theta_1$ toward $\theta_2$. Note that this argument does not depend critically on our knowing $q_1$ and $q_2$, the prior probabilities of $\theta_1$ and $\theta_2$. The likelihood ratio indicates the direction in which the data point, but not how far we should move. To ascertain this, we would have to know $q_1$ and $q_2$, and this knowledge we rarely possess. It is in this sense that a test of significance is related to the direction that the data point or indicate.

A purpose of a simple test of significance can now be seen. Even if we were sampling from the model with $\theta = \theta_1$, our data may indicate that we should entertain $\theta_2$. We may therefore ask what the probability is that our data will indicate a shift of opinion toward $\theta_2$. This is equal to the probability that $P(D; \theta_2)/P(D; \theta_1)$ is greater than unity. If this probability is low, we should regard the possible data sets $D$ as being rather insensitive to the possibility that $\theta$ is $\theta_2$ rather than $\theta_1$.

The likelihood ratio ordering significance level $\alpha$, say, is the probability beforehand that a set of data will indicate a shift of opinion from $\theta_1$ to $\theta_2$ as large as or larger than that obtained with the actual data if $\theta_1$ is the true value. Note that we do not say by how much our opinion shifts. To say this we would need to know $q_1$ and $q_2$.

The above argument illustrates another aspect of the matter. Suppose $P(D; \theta_2)/P(D; \theta_1) = 1$. Then we should not modify whatever opinion we have about whether $\theta$ is $\theta_1$ or $\theta$ is $\theta_2$ on the basis of the data. It seems, therefore, that in the case under discussion we should look at the observed likelihood ratio *and* the significance level. It may happen that the observed likelihood ratio is less than unity but the significance level is very small. This would happen, for instance, if we had two competing models $N(0, 1)$ and $N(10, 1)$, and we obtained an observation equal to four.

## 12.6 Goodness of Fit

In Chapters 9 and 10 we discussed the problem of testing whether a set of data may be considered to be a random sample from a specified distribution or class of distributions. It is possible to apply the idea of sensitivity of tests of significance to this problem. We consider only a simple case. Suppose we have multinomial data with $k$ classes and observations $n_i$, $i = 1, 2, \ldots, k$ with $n = \sum n_i$. Suppose we wish to test goodness of fit of the completely specified multinomial distribution $M_0$ with class probabilities $P_1, P_2, \ldots, P_k$, against the alternative of a completely specified multinomial model $M_1$, with probabilities $Q_1, Q_2, \ldots, Q_k$. Then the likelihood ratio ordering of possible data sets is

$$(n_1, n_2, \ldots, n_k) \gg (n'_1, n'_2, \ldots, n'_k)$$

if

$$\frac{P(n_1, n_2, \ldots, n_k; M_1)}{P(n_1, n_2, \ldots, n_k; M_0)} \geq \frac{P(n'_1, n'_2, \ldots, n'_k; M_1)}{P(n'_1, n'_2, \ldots, n'_k; M_0)}$$

This gives the ordering

$$(n_1, n_2, \ldots, n_k) \gg (n'_1, n'_2, \ldots, n'_k)$$

if

$$\left(\frac{Q_1}{P_1}\right)^{n_1}\left(\frac{Q_2}{P_2}\right)^{n_2}\cdots\left(\frac{Q_k}{P_k}\right)^{n_k} \geq \left(\frac{Q_1}{P_1}\right)^{n'_1}\left(\frac{Q_2}{P_2}\right)^{n'_2}\cdots\left(\frac{Q_k}{P_k}\right)^{n'_k}$$

or, taking logarithms, if

$$\sum_{i=1}^{k} n_i \ln \frac{Q_i}{P_i} \geq \sum_{i=1}^{k} n_i' \ln \frac{Q_i}{P_i}$$

This ordering leads to a number of achievable levels equal at most to

$$\binom{n+k-1}{n}$$

because this is the total number of possible data sets under the model.

We can use the likelihood theory of Chapter 10 to obtain an approximate idea of the behavior of this test of significance. The test of significance is equivalent to the following procedure:

1. For the data $(n_1, n_2, \ldots, n_k)$ consider the statistic or criterion

$$C = \sum_{i=1}^{k} \left(\ln \frac{Q_i}{P_i}\right) n_i$$

and let $C_O$ be the observed value.

2. Determine the distribution of the statistic in random sampling, and suppose its c.d.f. under $M_0$ is $P(C \leq c) = F(c)$.

3. Take as the significance level for the given data, the number $1 - F(C_O) = P(C > C_O)$.

Now recall the results established in Section 10.6. In sampling from the multinomial $P_1, P_2, \ldots, P_k$ with observation $(n_1, n_2, \ldots, n_k)$, a linear function $\sum a_i n_i$ is approximately normally distributed with mean $n \sum a_i P_i$ and variance $n[\sum a_i^2 P_i - (\sum a_i P_i)^2]$. Hence we have the following results, using the fact that in the present case $a_i$ equals $\ln(Q_i/P_i)$:

1. If we are sampling from $P_1, P_2, \ldots, P_k$, the distribution of $C$ is approximately normal with mean equal to $n \sum P_i \ln(Q_i/P_i)$ and variance

$$n\left[\sum P_i \left(\ln \frac{Q_i}{P_i}\right)^2 - \left(\sum P_i \ln \frac{Q_i}{P_i}\right)^2\right]$$

2. If we are sampling from $Q_1, Q_2, \ldots, Q_k$, the distribution of $C$ is approximately normal with mean $n \sum Q_i \ln(Q_i/P_i)$ and variance

$$n\left[\sum Q_i \left(\ln \frac{Q_i}{P_i}\right)^2 - \left(\sum Q_i \ln \frac{Q_i}{P_i}\right)^2\right]$$

3. The approximate significance level for the data $(n_1, n_2, \ldots, n_k)$ is therefore equal to the probability that a standard normal deviate exceeds

$$d_0 = \frac{\sum n_i \ln \frac{Q_i}{P_i} - n \sum P_i \ln \frac{Q_i}{P_i}}{\sqrt{n}\sqrt{\sum P_i \left(\ln \frac{Q_i}{P_i}\right)^2 - \left(\sum P_i \ln \frac{Q_i}{P_i}\right)^2}}$$

This is equal to $1 - \Phi(d_0)$ where

$$\Phi(u) = \frac{1}{\sqrt{2\pi}} \int_{-\infty}^{u} \exp\left(-\frac{1}{2}t^2\right) dt$$

which is given in Table A.4 of the Appendix.

An alternative way of describing the situation is that a set of data $(n_1, n_2, \ldots, n_k)$ gives significance at the level $\alpha$ or a lower level if and only if

$$C_O = \sum n_i \ln \frac{Q_i}{P_i}$$

is equal to or greater than

$$C(\alpha) = n \sum P_i \ln \frac{Q_i}{P_i} + z(\alpha) \sqrt{n\left[\sum P_i \left(\ln \frac{Q_i}{P_i}\right)^2 - \left(\sum P_i \ln \frac{Q_i}{P_i}\right)^2\right]}$$

where $P[Z \geq z(\alpha)] = 1 - \Phi[z(\alpha)] = \alpha$, where $Z$ is a standard normal variate. By the derivation it is approximately true that

$$P[C \geq C(\alpha)|M_0] = \alpha$$

4. An approximation to the c.d.f. of the significance level if we are sampling from the multinomial $Q_1, Q_2, \ldots, Q_k$ is $F_1(\alpha)$, say, where $F_1(\alpha) = P[C \geq C(\alpha)|M_1)] = 1 - \Phi[v(\alpha)]$, where

$$v(\alpha) = \frac{n \sum (P_i - Q_i) \ln \frac{Q_i}{P_i} + z(\alpha) \sqrt{n\left[\sum P_i \left(\ln \frac{Q_i}{P_i}\right)^2 - \left(\sum P_i \ln \frac{Q_i}{P_i}\right)^2\right]}}{\sqrt{n\left[\sum Q_i \left(\ln \frac{Q_i}{P_i}\right)^2 - \left(\sum Q_i \ln \frac{Q_i}{P_i}\right)^2\right]}}$$

The accuracy of these approximations is not known. Some aspects of likelihood ratio and $\chi^2$ goodness of fit tests are discussed by Wise (1963). We do know, however, from Chernoff (1952) that if we have the case of independent Bernoulli trials with probability of success equal to $P$ and if $P_n$ is the probability that the observed proportion of successes in $n$ trials is greater than $t$, then

$$\lim_{n\to\infty}\left(\frac{1}{n}\ln P_n\right) = -\left[t\ln\left(\frac{t}{P}\right) + \left(1 - t\right)\ln\left(\frac{1-t}{1-P}\right)\right]$$

See also Hoeffding (1965). This result serves to indicate that a limit distribution may give the wrong answer to a limit probability because the above limiting normal distribution does not give this answer.

The difficulty of the above is that we have a test of goodness of fit against a specified alternative. If we wish to have a test against all alternatives it is intuitively reasonable to choose as the alternative that distribution which is indicated by the data. This, then, suggests that we test the hypothesis $P_1, P_2, \ldots, P_k$ against the alternative

$$Q_1 = n_1/n, Q_2 = n_2/n, \ldots, Q_k = n_k/n$$

The likelihood ratio criterion then becomes $\sum n_i \ln(n_i/nP_i)$. The distribution of this criterion is more complex than the distribution of $C$ above. We have seen, however, that if $n$ is "large," this criterion will be close to the $\chi^2$ criterion

$$\sum \frac{(n_i - nP_i)^2}{nP_i}$$

Recent unpublished work by E. N. West shows that for small sample sizes there is little difference in the sensitivity of the $\chi^2$ and the likelihood-ratio test.

## 12.7 Distributions with Monotone Likelihood Ratio

It was seen in Section 12.1 that we are often concerned with the possibility that a parameter value $\theta$ is equal to $\theta_0$ with alternatives $\theta$ not equal to $\theta_0$. A simple case is that in which the alternatives are $\theta$ greater than $\theta_0$ or $\theta$ less than $\theta_0$, but not both of these.

Consider now the case in which the alternatives are $\theta$ greater than $\theta_0$. (We can cover the other case by considering the parameter to be $-\theta$ rather than $\theta$.) We have seen that we can compare two sets of data $D_1$ and $D_2$ for the case of $\theta$ equal to $\theta_0$ with alternative $\theta$ equals $\theta_1$ by the rule

$$D_1 \gg D_2 \text{ if } \frac{P(D_1;\theta_1)}{P(D_1;\theta_0)} \geq \frac{P(D_2;\theta_1)}{P(D_2;\theta_0)}$$

Suppose that if

$$\frac{P(D_1;\theta_1)}{P(D_1;\theta_0)} \geq \frac{P(D_2;\theta_1)}{P(D_2;\theta_0)}$$

then

$$\frac{P(D_1; \theta_2)}{P(D_1; \theta_0)} \geq \frac{P(D_2; \theta_2)}{P(D_2; \theta_0)}$$

for $\theta_2 \neq \theta_1$.

If this happens, we have the same ordering of samples whether we consider the alternative to be $\theta_2$ or $\theta_1$. We can extend this to the case of a spectrum of values of $\theta$ greater than $\theta_0$. If

$$\frac{P(D_1; \theta_1)}{P(D_1; \theta_0)} \geq \frac{P(D_2; \theta_1)}{P(D_2; \theta_0)}$$

implies

$$\frac{P(D_1; \theta)}{P(D_1; \theta_0)} \geq \frac{P(D_2; \theta)}{P(D_2; \theta_0)}$$

for all $\theta_1$ and $\theta$ greater than $\theta_0$, the ordering of samples $D$ will be the same for whatever value of $\theta > \theta_0$ that we consider for the alternative. If this happens, we can regard the ordering of significance as a statistic $T(D)$, and the likelihood ratio will be a monotone function of $T(D)$. Alternatively, we can use the definition close to that given by Lehmann (1959).

*Definition 12.7. The one-parameter family of probabilities $P(D; \theta)$ is said to have monotone likelihood ratio if there exists a scalar $T(D)$ such that for any $\theta' > \theta$ the probabilities are distinct and $P(D; \theta')/P(D; \theta)$ is an increasing function of $T(D)$.*

To illustrate the idea, consider an observation $x$ from the Poisson distribution with parameter $\lambda$ and suppose we wish to order possible observations regarding significance for the null hypothesis $\lambda = \lambda_0$ and alternative $\lambda = \lambda_1 > \lambda_0$. Then

$$\frac{P(x; \lambda_1)}{P(x; \lambda_0)} = e^{-(\lambda_1 - \lambda_0)} \left(\frac{\lambda_1}{\lambda_0}\right)^x$$

For $\lambda_1 > \lambda_0$ this is a monotone increasing function of $x$. Hence, if $x_1 \gg x_2$ for the alternative $\lambda_1 > \lambda_0$, then $x_1 \gg x_2$ for the alternative $\lambda_2 > \lambda_0$ and, in fact, $x_1 \gg x_2$ if $x_1 \geq x_2$. Similarly, if the alternatives are $\lambda < \lambda_0$ and if $x_1 \gg x_2$ for the alternative $\lambda_1 < \lambda_0$, then $x_1 \gg x_2$ for the alternative $\lambda_2 < \lambda_0$ and, in fact, $x_1 \gg x_2$ if $x_1 \leq x_2$.

Given a class of probabilities which depend upon a single parameter and which have monotone likelihood ratio, consider testing the simple hypothesis

$$\text{Hypothesis: } \theta = \theta_0$$

against the composite alternative

$$\text{Alternative: } \theta < \theta_0$$

Since a small value for $T(D)$ would seem to indicate a small value for $\theta$, it seems that the significance level should be calculated by evaluating the probability under $\theta_0$ of a value for $T$ as small or smaller than observed. This is proved to be a uniformly most sensitive test of significance in the class comparable to it.

***Theorem 12.3*** If data $D$ have probability with monotone likelihood ratio in $T(D)$, then a uniformly most sensitive test of significance for $\theta = \theta_0$ against $\theta < \theta_0$ is given by

$$SL(T) = P_{\theta_0}[T(D) \leq \text{observed } T] \tag{12.24}$$

*Proof:* Consider some simple alternative $\theta_1 < \theta_0$. A most sensitive test is given by

$$\begin{aligned} SL &= P_{\theta_0}\left[\frac{P(D;\theta_1)}{P(D;\theta_0)} \geq \text{observed } LR\right] \\ &= P_{\theta_0}[T(D) \leq \text{observed } T] \end{aligned} \tag{12.25}$$

since the likelihood ratio in this case is a decreasing function of $T$. Because the test does not depend on the value $\theta_1$ chosen, it is a uniformly most sensitive test. This result is closely related to a theorem in hypothesis testing theory [Lehmann (1959)].

It follows very simply that a uniformly most sensitive test of significance for $\theta = \theta_0$ against $\theta > \theta_0$ is given by

$$SL(D) = P_{\theta_0}[T(D) \geq \text{observed } T] \tag{12.26}$$

Consider a single observation $x$ with probabilities calculated from a Poisson model with parameter $\lambda$. Then $X$ has monotone likelihood ratio and therefore a uniformly most sensitive test of $\lambda = \lambda_0$ against $\lambda > \lambda_0$ is given by

$$SL(X) = P_{\lambda_0}(X \geq \text{observed}) \tag{12.27}$$

It is worth noting that we have required the likelihood ratio to be strictly increasing or strictly decreasing.

Many of the simple distributions we encounter are members of an exponential family and do have monotone likelihood ratio properties.

*Corollary 12.2* If $X$ has probabilities calculated under the one-parameter exponential family (or FKPD form):

$$P(x;\theta) = C(\theta)\exp[H(\theta)m(x)]g(x) \tag{12.28}$$

where $H$ is strictly monotone, then a uniformly most sensitive test for $\theta = \theta_0$ against $\theta > \theta_0$ is given by

$$SL = P_{\theta_0}[m(X) \geq \text{observed}] \tag{12.29}$$

if $H$ is increasing in $\theta$. If $H$ is decreasing, a uniformly most sensitive test is given by

$$SL = P_{\theta_0}[m(X) \leq \text{observed}] \tag{12.30}$$

## 12.8 Unbiased Tests of Significance

In Section 12.3 we defined an unbiased test of significance as one for which

$$P(SL \leq \alpha | M_1) \geq \alpha \tag{12.31}$$

for all achievable $\alpha$. This seemed to be a natural and desirable property, although the question of whether such tests exist was not dealt with at that time. In this section we wish to discuss briefly the existence of unbiased tests. When they do exist, we may want to seek the most sensitive test in the class of unbiased tests.

We shall concern ourselves entirely with the one-parameter exponential family. For the hypothesis $\theta = \theta_0$ against the one-sided alternative $\theta < \theta_0$ (or $\theta > \theta_0$), the uniformly most sensitive tests given in the preceding section are also seen to be unbiased. Therefore, we raise the question of the existence of an unbiased test for the alternative $\theta \neq \theta_0$. Since the statistic $m(X)$ mirrors the behavior of $\theta$, it seems that a reasonable way to calculate the significance level for an observed value of $X$ is to calculate the probability under $\theta_0$ of large as well as small values of $m(X)$. For the moment let us concern ourselves with obtaining conditions which must hold if there is an unbiased test (Lehmann, 1959).

***Theorem 12.4*** If an unbiased test exists for $\theta = \theta_0$ against $\theta \neq \theta_0$ for the one-parameter exponential family

$$P(D; \theta) = C(\theta)\exp[\theta m(D)]g(D) \tag{12.32}$$

then

$$E_{\theta_0}[m(D) | SL \leq \alpha] = E_{\theta_0}[m(D)] \tag{12.33}$$

for all achievable $\alpha$. Here we have replaced $H(\theta)$ by $\theta$ by a reparametrization.

*Proof:*

$$P_\theta(SL \leq \alpha) = \sum_{D \gg D_O} C(\theta)\exp[\theta m(D)]g(D) \tag{12.34}$$

where $SL(D_O) = \alpha$. Differentiating with respect to $\theta$, we obtain

$$P'_\theta(SL \leq \alpha) = \sum_{D \gg D_O} \{C'(\theta)\exp[\theta m(D)]g(D) + m(D)C(\theta)\exp[\theta m(D)]g(D)\}$$

$$= P_\theta(SL \leq \alpha)\left\{\frac{C'(\theta)}{C(\theta)} + E_\theta[m(D)|SL \leq \alpha]\right\} \quad (12.35)$$

A necessary condition for the test to be unbiased is for the derivative to vanish at $\theta = \theta_0$. Thus we have

$$0 = \frac{C'(\theta_0)}{C(\theta_0)} + E_{\theta_0}[m(D)|SL \leq \alpha] \quad (12.36)$$

or

$$E_{\theta_0}[m(D)|SL \leq \alpha] = \frac{-C'(\theta_0)}{C(\theta_0)} \quad (12.37)$$

But if we sum over all samples and differentiate with respect to $\theta$, we obtain that

$$E_{\theta_0}[m(D)] = -\frac{C'(\theta_0)}{C(\theta_0)} \quad (12.38)$$

Therefore,

$$E_{\theta_0}[m(D)|SL \leq \alpha] = E_{\theta_0}[m(D)] \quad (12.39)$$

We have thus obtained a necessary condition for a test to be unbiased. The necessary condition for unbiasedness resembles independence in that it requires the conditional expectation of $m(D)$ to equal the unconditional expectation of $m(D)$.

*Corollary 12.3* Suppose that the significance level of an unbiased test is obtained in terms of a statistic $T(D)$ by

$$SL = P_{\theta_0}[T(D) \leq \text{observed}] \quad (12.40)$$

Then $\text{Cov}[m(D), T(D)] = 0$, when $\theta = \theta_0$.

*Proof:* Because the test is unbiased,

$$\begin{aligned} E_{\theta_0}[m(D)|SL \leq \alpha] &= E_{\theta_0}[m(D)] \\ &= E_{\theta_0}[m(D)|T(D) \leq t] \text{ for all observable } t \end{aligned} \quad (12.41)$$

Therefore,

$$E_{\theta_0}[m(D)] = E_{\theta_0}[m(D)|T(D) = t] \quad (12.42)$$

Now, under $\theta = \theta_0$,

$$\begin{aligned} \text{Cov}[m(D), T(D)] &= E[T(D)m(D)] - E[m(D)]E[T(D)] \\ &= E\{E[T(D)m(D)|T(D)]\} - E[m(D)]E[T(D)] \\ &= E[T(D)]E[m(D)] - E[m(D)]E[T(D)] \\ &= 0 \end{aligned} \quad (12.43)$$

Then a necessary condition for a test to be unbiased is that the test statistic be uncorrelated with $m(D)$. The same result holds if the significance level is obtained by calculating the probability under $\theta_0$ of large values of $T(D)$.

*Example 2*

It is important to realize that unbiased tests of significance do not exist for many very simple situations, even ones of exponential form. In the discrete case, existence is the exception rather than the rule. To illustrate the problem, consider the case of Bi(5, 0.3); i.e., the hypothesized model is that the observation which takes one of the values 0, 1, 2, 3, 4, 5 is binomially distributed with $n = 5$ and $p = 0.3$. Consider the alternative to be $p \neq 0.3$. The possible data sets are the numbers 0, 1, 2, 3, 4, 5, and any test of significance is an ordering of these data sets with a computation of success probabilities under the model. Obviously there are $6! = 720$ orderings if ties are not allowed. A fairly reasonable one for our null hypothesis is $5 \gg 4 \gg 0 \gg 1 \gg 3 \gg 2$. If we use this ordering, the achievable significance levels are $P(5)$, $P(5 \text{ or } 4)$, $P(5 \text{ or } 4 \text{ or } 0)$, $P(5 \text{ or } 4 \text{ or } 0 \text{ or } 1)$, $P(5 \text{ or } 4 \text{ or } 0 \text{ or } 1 \text{ or } 3)$, and finally 1; these probabilities being calculated with $p = 0.3$. It is easy to see that $P(5|p)$ is not greater than or equal to $P(5|0.3)$ for all $p$ and similarly for the other probabilities. It is also easy to see that no ordering of the possible observations gives an unbiased test of significance. For this binomial situation no unbiased test for $p = p_0$ against $p \neq p_0$ exists except when $p_0 = 1/2$, 0, or 1. For $p_0 = 1/2$, the reader can verify that an unbiased test is given by $SL = P_{1/2}(X \leq x) + P_{1/2}(X \geq n - x)$ when $x \leq n/2$ and by $SL = P_{1/2}(X \geq x) + P_{1/2}(X \leq n - x)$ when $x \geq n/2$.

*Example 3*

As an example of an unbiased test, consider the hypothesis $X_i \sim N(\mu, \sigma_0^2)$ against the alternative $X_i \sim N(\mu, \sigma^2)$ with $\sigma^2 \neq \sigma_0^2$ and $\mu$ known. Lehmann (1959) shows that an unbiased test is given by

$$SL = P_{\sigma_0}(T \leq \text{observed})$$

where

$$T = \left[\frac{\sum (X_i - \mu)^2}{\sigma_0^2}\right]^{n/2} \exp\left[\frac{-\sum (X_i - \mu)^2}{2\sigma_0^2}\right] \tag{12.44}$$

It is worthwhile verifying that $T$ is uncorrelated with $\sum (X_i - \mu)^2/\sigma_0^2$.

We saw in Chapter 10 that under certain regularity conditions the maximum likelihood statistic $\theta^*$ has asymptotically a normal (or multivariate normal) distribution with mean $\theta$ and variance-covariance matrix obtainable from the information matrix. Consider the case of a scalar parameter $\theta$ and suppose that the information does not depend on $\theta$. It follows for such cases that $\theta^*$ is asymptotically $F$-sufficient and has a normal distribution of exponential form. Asymptotically, most sensitive one-sided tests of significance of $\theta = \theta_0$ are given, therefore, by calculating $P(\theta^* \geq \theta^*_{\text{observed}}|\theta_0)$ or $P(\theta^* \leq \theta^*_{\text{observed}}|\theta_0)$, each of which will be simply a tail area of a known normal distribution. Also, an asymptotically unbiased test is given by $P(|\theta^* - \theta_0| \geq |\theta^*_{\text{observed}} - \theta_0|)$.

## 12.9 The Generalized Likelihood Ratio Ordering for Significance Testing

We have seen that a plausible ordering of significance of samples $D_1, D_2, \ldots$ for the case of testing the significance of a parameter value $\theta = \theta_0$ with regard to an alternative value $\theta = \theta_1$ for the case of a distribution depending only on $\theta$ is given by $D_1 \gg D_2$, meaning $D_1$ is at least as significant as $D_2$ if and only if

$$\frac{P(D_1; \theta_1)}{P(D_1; \theta_0)} \geq \frac{P(D_2; \theta_1)}{P(D_2; \theta_0)} \tag{12.45}$$

This gives a partial ordering of all possible samples $D_1 \gg D_2 \gg D_3 \gg \cdots$ and the significance level of an observed sample $D_O$ is

$$\sum_{D \gg D_O} P(D; \theta_0) \tag{12.46}$$

We have seen that in special cases the ordering of samples for testing $\theta = \theta_0$ with alternative $\theta = \theta_1$ is the same as the ordering for testing $\theta = \theta_0$ with another alternative (or spectrum of alternatives) $\theta = \theta_2$. This happens, in fact, in the case of a monotone likelihood ratio with one-sided alternatives such as $\theta > \theta_0$. What is one to do in the absence of a monotone likelihood ratio? Suppose we wish to test $\theta = \theta_0$ with the class of alternatives $\theta \neq \theta_0$ or, more specifically, $\theta$ in some set $\Omega - \theta_0$. A natural candidate for ordering of samples is the generalized likelihood ratio, due to Neyman and Pearson (1933), defined as follows. Consider a sample $D$ with probability $P(D; \theta)$. Then the generalized likelihood ratio ordering is as follows: $D_1 \gg D_2$ if and only if

$$\frac{\max_{\theta \in \Omega - \theta_0} P(D_1; \theta)}{P(D_1; \theta_0)} \geq \frac{\max_{\theta \in \Omega - \theta_0} P(D_2; \theta)}{P(D_2; \theta_0)} \tag{12.47}$$

As usual, we say that $D_1$ and $D_2$ are equally significant if equality holds in Equation (12.47).

The theory of significance testing by this route is very complex. We give a "small" example. Consider the testing of $p = 1/3$ for the case of five independent Bernoulli trials; i.e., Bi$(n, p)$ with $n = 5$. The possible observations are indexed by $r$, the number of successes, and

$$P(r|p) = \binom{n}{r} p^r q^{n-r}, \qquad q = 1 - p$$

The maximum of this with regard to $p$ occurs at $p = r/n$, and has the value

$$\binom{n}{r}\left(\frac{r}{n}\right)^r\left(\frac{n-r}{n}\right)^{n-r}$$

The computations for the generalized likelihood ratio ordering are given in Table 12.3, in which $p^*$ denotes the maximizing $p$. So the ratios for $r = 0, 1, 2, 3, 4, 5$ are equal to $3^5/2^5$ times 1, 512/3125, 432/3125, 864/3125, 4096/3125, 32 respectively. Hence the generalized likelihood ratio ordering is $5 \gg 4 \gg 0 \gg 3 \gg 1 \gg 2$. The associated significance levels are

$$SL(5) = \frac{1}{243}, SL(4) = \frac{11}{243}, SL(0) = \frac{43}{243}$$

$$SL(3) = \frac{83}{243}, SL(1) = \frac{163}{243}, SL(2) = 1$$

**Table 12.3** Computation of Ordering of Samples

| $r$ | Prob $p = p_0 = 1/3$ | $p^*$ | max(Prob) | Ratio |
|---|---|---|---|---|
| 0 | $\left(\frac{2}{3}\right)^5$ | 0 | 1 | $3^5/2^5$ |
| 1 | $5\left(\frac{1}{3}\right)\left(\frac{2}{3}\right)^4$ | $\frac{1}{5}$ | $5\left(\frac{1}{5}\right)\left(\frac{4}{5}\right)^4$ | $\frac{4^4}{5^5}\cdot\frac{3^5}{2^4} = \frac{3^5}{2^5}\cdot\frac{2\cdot 4^4}{5^5}$ |
| 2 | $10\left(\frac{1}{3}\right)^2\left(\frac{2}{3}\right)^3$ | $\frac{2}{5}$ | $10\left(\frac{2}{5}\right)^2\left(\frac{3}{5}\right)^3$ | $\frac{2^2 3^3}{5^5}\cdot\frac{3^5}{2^3} = \frac{3^5}{2^5}\cdot\frac{2^4 3^3}{5^5}$ |
| 3 | $10\left(\frac{1}{3}\right)^3\left(\frac{2}{3}\right)^2$ | $\frac{3}{5}$ | $10\left(\frac{3}{5}\right)^3\left(\frac{2}{5}\right)^2$ | $\frac{2^2 3^3}{5^5}\cdot\frac{3^5}{2^2} = \frac{3^5}{2^5}\cdot\frac{2^5 3^3}{5^5}$ |
| 4 | $5\left(\frac{1}{3}\right)^4\left(\frac{2}{3}\right)$ | $\frac{4}{5}$ | $5\left(\frac{4}{5}\right)^4\left(\frac{1}{5}\right)$ | $\frac{4^4}{5^5}\cdot\frac{3^5}{2} = \frac{3^5}{2^5}\cdot\frac{2^4 4^4}{5^5}$ |
| 5 | $\left(\frac{1}{3}\right)^5$ | 1 | 1 | $3^5 = \frac{3^5}{2^5}\cdot 2^5$ |

This example is extremely important. We have used a particular procedure to obtain the ordering of significance $5 \gg 4 \gg 0 \gg 3 \gg 1 \gg 2$. It is natural to ask if this ordering has the property of unbiasedness, namely, that for the achievable levels $\alpha_1, \alpha_2, \ldots, 1$,

$$P(SL \leq \alpha | p \neq p_0) \geq P(SL \leq \alpha | p = p_0) \tag{12.48}$$

It will be recalled that this is a valuable requirement for testing $p = p_0$ with a single alternative $p = p_1$, say. It is easy to examine whether Equation (12.48) holds, and one sees immediately that it does not. For instance,

$$P\left(SL \leq \frac{1}{243} | p\right) = P(r = 5) = p^5$$

and if $p$ is less than 1/3

$$P\left(SL \leq \frac{1}{243} | p\right) < P\left(SL \leq \frac{1}{243} | p = \frac{1}{3}\right) = \frac{1}{243}$$

So the test of significance of $p = 1/3$ is not unbiased. However, we should not take this seriously because no unbiased test of significance exists. In fact, excluding ties in ordering significance of data points, there are $6!(= 720)$ possible orderings of possible data points. For instance, the ordering $3 \gg 0 \gg 5 \gg 1 \gg 2 \gg 4$ is possible, though it is difficult to imagine circumstances under which one would use it. It is obvious that no ordering is unbiased. Consider for example the case in which $r = 3$ is most significant. Then the significance level of the observation $r = 3$ is 40/243, and the probability that the significance level is less than or equal to 40/243 is $10p^2q^3$. If $p = 0$, a value certainly different from $p = 1/3$, this probability is 0, so there is an alternative hypothesis to that of $p = 1/3$, for which $\text{Prob}(SL \leq 40/243)$ is less than 40/243 and is in fact zero.

The generalized likelihood ratio ordering is difficult to compute. There is an ordering which is easy to obtain, namely, order by $|p^* - p_0|$ or by $|(r/n) - p_0|$. It is interesting to note that this principle *also* gives the ordering $5 \gg 4 \gg 0 \gg 3 \gg 1 \gg 2$.

The extension of the generalized likelihood ratio procedure to a test of significance of a composite hypothesis with a composite alternative is direct. Let the hypothesis under test specify that the parameter $\theta$ is in a region $\omega$ of parameter space, and let the alternative hypothesis be that $\theta$ is in the region $\Omega - \omega$. Then we say that $D_1$ is more significant than $D_2$ if

$$\frac{\max_{\theta \in \Omega - \omega} P(D_1; \theta)}{\max_{\theta \in \omega} P(D_1; \theta)} > \frac{\max_{\theta \in \Omega - \omega} P(D_2; \theta)}{\max_{\theta \in \omega} P(D_2; \theta)} \tag{12.49}$$

In principle this may be applied to any situation, but the mathematical computations may be very onerous. A major difficulty arises when we wish to attach a significance level to an observed sample $D_O$. Following the general ideas given above we would say that

$$SL(D_o) = \sum_{D \gg D_O} P(D; \theta) \tag{12.50}$$

But this no longer works because the R.H.S. of Equation (12.50) is a function of $\theta$. We have to nominate a value of $\theta$ in $\omega$ at which this is computed.

At first sight it would appear reasonable to take $SL(D_O)$ to be the minimum over $\theta$ in $\omega$ of Equation (12.50). We shall then have the difficulty that there can be two samples $D_1$ and $D_2$, such that by the ordering $D_1 \gg D_2$ the significance level of $D_1$ is greater than the significance level of $D_2$. For this reason the idea of pivotal quantities discussed in Sect. 12.10 may be useful. The general conclusion is that there is no "best" test of significance in the composite case (cf. Sect. 12.16). It is also clear that the generalized likelihood ratio cannot always be used for testing significance. This is a critical point.

## 12.10 Pivotal Quantities

We have seen that a significance test of a model or class of models $\omega$ consists of a mapping of possible data sets $D$ into the real line and a computation of probabilities under the model being tested. A data set $D_1$, say, is mapped into a point $x_1$. Many sets of data may be mapped to a single $x$ value. If our observed set of data $D_O$ is mapped into $x_O$, the significance level is the probability that the random variable $X$ defined by the mapping is less than or equal to $x_O$ under $\omega$. The mapping must be such that the distribution of $X$ is the same for all models in the class $\omega$.

Such a mapping exists only in quite special circumstances. Consider the case in which $\omega$ consists of the models Bi(3, $p$), with $p$ equal to 1/4 or 2/3. Any pair of values for $p$ can be taken. Consider now a possible ordering (in this case, a complete ordering) of the possible observations, such as 3, 0, 1, 2. Then we must satisfy the requirements

$$P(3; p = 1/4 \text{ or } 2/3) = \alpha_1$$
$$P(3 \text{ or } 0; p = 1/4 \text{ or } 2/3) = \alpha_2$$
$$P(3 \text{ or } 0 \text{ or } 1; p = 1/4 \text{ or } 2/3) = \alpha_3$$

But this is not achievable. The only partial ordering that satisfies the requirement is $3 = 0 = 1 = 2$ with an achievable significance level equal to unity. We take the viewpoint that there is no test of significance for this problem.

For some classes of distribution and for the testing of some composite models relating to the class of distribution, tests of significance are possible. These are derivable from the existence of pivotal quantities.

*Definition 12.8. Suppose we have a class of models for data D such that the probability of D depends on a scalar parameter $\theta$, with $\theta$ belonging to a space $\Theta$. Then a function of the data D and the parameter $\theta$, say $h(D, \theta)$, is said to be a pivotal quantity for $\theta$ if the distribution of $h(D, \theta)$ is the same completely known distribution for all $\theta$ belonging to $\Theta$.*

*Definition 12.9. Suppose we have a class of models for data D such that the probability of D depends on a vector parameter $\theta = (\theta_1, \theta_2, \ldots, \theta_r)$, with $\theta$ belonging to some space $\Theta$. Let $\phi$ be a subset of $\theta$, say $(\theta_{\alpha_1}, \theta_{\alpha_2}, \ldots, \theta_{\alpha_s})$. Then a function of the data D and $\phi$, say $h(D, \phi)$, is said to be a pivotal quantity for $\phi$ if the distribution of $h(D, \phi)$ is the same completely known distribution for all $\theta$ belonging to $\Theta$.*

*Definition 12.10. Suppose we have a class of models for data D such that the probability of D depends on a vector parameter $\theta = (\theta_1, \theta_2, \ldots, \theta_r)$ with $\theta$ belonging to some space $\Theta$. Let $\phi$ be a subset of s of the $\theta_i$. Then $h_1(D, \phi)$, $h_2(D, \phi)$, $\ldots$, $h_w(D, \phi)$ is a set of w pivotals for $\phi$ if the joint distribution of $h_1, h_2, \ldots, h_w$ is a completely known distribution for all $\theta$ belonging to $\Theta$.*

The existence of pivotal quantities is most obscure. We give below a few such quantities that are associated with the normal distribution, and later we give some which do not depend on complete specification of the probability of any data set $D$.

Consider the case of data $D = (x_1, x_2, \ldots, x_n)$, being regarded as a random sample from $N(\mu, \sigma^2)$. Then we have:

1. $\sum (x_i - \mu)^2/\sigma^2$ is distributed as $\chi^2$ with $n$ degrees of freedom. Hence this is a pivotal for $\mu$ and $\sigma$.
2. $\sum (x_i - \bar{x})^2/\sigma^2$ is distributed as $\chi^2$ with $n - 1$ degrees of freedom for any $\mu$. Hence $\sum (x_i - \bar{x})^2/\sigma^2$ is a pivotal for $\sigma$.
3. $t = \sqrt{n}(\bar{x} - \mu)/s$, where $s^2 = \sum (x - \bar{x})^2/(n - 1)$, is distributed according to the $t$ distribution with $(n - 1)$ degrees of freedom for any $\sigma$. Hence this is a pivotal for $\mu$ alone.
4. The quantity $\sqrt{n}(\bar{x} - \mu)/\sigma$ is distributed as $N(0, 1)$; hence this is a pivotal quantity for $\mu$ and $\sigma$.
5. The number of $x$ values such that $x - \mu$ is positive has the binomial distribution $\mathrm{Bi}(n, 1/2)$. Hence $\sum_i \mathrm{sgn}(x_i - \mu)$ is a pivotal quantity for $\mu$.

Consider now $D = (x_1, x_2, \ldots, x_m, y_1, y_2, \ldots, y_n)$, with $x_1, x_2, \ldots, x_m$ as a random sample of size $m$ from $N(\mu, \sigma^2)$ and $y_1, y_2, \ldots, y_n$ as a random sample of size $n$ from $N(\mu + \Delta, \sigma^2)$. There are many pivotals and we mention only one, namely, $c(\bar{y} - \bar{x} - \Delta)/s$, where

$$c = \sqrt{\frac{1}{n} + \frac{1}{m}}$$

$$s^2 = [\sum (x - \bar{x})^2 + \sum (y - \bar{y})^2]/(n + m - 2)$$

This function of the observations and $\Delta$ has the $t$ distribution with $n + m - 2$ d.f. for any value of $\mu$ and $\sigma$ and hence is a pivotal for $\Delta$.

To exemplify the case of two pivotals, consider $\sqrt{n}(\bar{x} - \mu)/\sigma$ and $\sum (x - \bar{x})^2/\sigma^2$ for a random sample $x_1, x_2, \ldots, x_n$ from $N(\mu, \sigma^2)$.

It seems that all exact tests of significance are based on the idea of pivotals. There is a whole array of tests of significance associated with samples from normal distributions which are derived in this way. The logic of the process is not clear and well established, however. It seems that there should be ideas of best pivotals for particular purposes. For tests of significance that $\mu = \mu_0$ with a random sample from $N(\mu, \sigma^2)$, the $t$ pivotal given above seems to exhaust the data. Just how one will use the pivotal and its distribution is also unclear in general. In the case of testing $\mu = \mu_0$, with alternative $\mu > \mu_0$, the common usage is to calculate the probability that Student's $t$ is greater than its observed value $\sqrt{n}(\bar{x} - \mu_0)/s$. It seems also that a test of significance of $\sigma^2$, with $\mu$ unknown, should be based on the pivotal $\sum (x - \bar{x})^2/\sigma^2$. These ideas carry over to the case of several normal populations with unknown means and the same unknown $\sigma^2$.

It should be noted, however, that there are great complexities in the case of a test of significance involving two parameters. If, for instance, we suppose that we have a random sample, $x_1, x_2, \ldots, x_n$ from $N(\mu, \sigma^2)$ and we wish to make a test of significance of $\mu = \mu_0$ and $\sigma^2 = \sigma_0^2$, there is a variety of tests available. We can use the single pivotal $\sum (x - \mu_0)^2/\sigma_0^2$ or the pair of pivotals $\sqrt{n}(\bar{x} - \mu_0)/\sigma_0$ and $\sum (x - \bar{x})^2/\sigma_0^2$; we suspect there are other possibilities. Furthermore, if we use two pivotals, we have to decide what weight to give to each. The problem can also be viewed as that of making a choice of direction in the parameter space along which one wishes to have greatest sensitivity. It is obvious that no single test of significance has maximum sensitivity in all directions. There is no way out of the problem except for the scientist to specify directions that interest him in the parameter space. Also, one might wish to consider several directions with a given data situation.

The neo-Bayesian solution (Chapter 11) to data evaluation removes this type of indeterminacy by making an all-powerful assumption of a prior distribution.

The asymptotic theory of likelihood indicates that, under certain regularity conditions, quantities that are asymptotically pivotal exist. Let the maximum likelihood statistic for $\theta = (\theta_1, \theta_2, \ldots, \theta_p)$ be $\theta^* = (\theta_1^*, \theta_2^*, \ldots, \theta_p^*)$. Let the information matrix (10.62) be

$$I = \begin{pmatrix} I_{11} & I_{12} & \cdots & I_{1p} \\ I_{21} & I_{22} & \cdots & I_{2p} \\ \cdots & \cdots & \cdots & \cdots \\ I_{p1} & I_{p2} & \cdots & I_{pp} \end{pmatrix} \tag{12.51}$$

Now we know that asymptotically $\theta^*$ is $F$-sufficient and has a multivariate normal distribution and that the quadratic form

$$Q = (\theta^* - \theta)'(nI)(\theta^* - \theta) \tag{12.52}$$

is distributed as $\chi^2$ with $p$ degrees of freedom. Hence, $Q$ is a pivotal quantity asymptotically, and if we let

$$C_O = (\theta_O^* - \theta_0)'[nI(\theta_0)](\theta_O^* - \theta_0) \tag{12.53}$$

where $\theta_0$ is the hypothesized value, then a good significance test is given by $P(\chi_p^2 \geq C_O)$.

For a composite hypothesis $\theta_1 = \theta_{10}, \theta_2 = \theta_{20}, \ldots, \theta_r = \theta_{r0}$ with $\theta_{r+1}, \ldots, \theta_p$ unspecified, we can form the pivotal

$$\begin{pmatrix} \theta_1^* - \theta_{10} \\ \theta_2^* - \theta_{20} \\ \cdots\cdots \\ \theta_r^* - \theta_{r0} \end{pmatrix}' V^{-1} \begin{pmatrix} \theta_1^* - \theta_{10} \\ \theta_2^* - \theta_{20} \\ \cdots\cdots \\ \theta_r^* - \theta_{r0} \end{pmatrix} \tag{12.54}$$

where $V$ is the leading $r \times r$ diagonal block of the inverse of $nI$, and $\theta_{r+1}, \ldots, \theta_p$ are replaced in this by their maximum likelihood statistics under the restriction $\theta_1 = \theta_{10}, \theta_2 = \theta_{20}, \ldots, \theta_r = \theta_{r0}$. This pivotal will have the $\chi^2$ distribution with $r$ degrees of freedom. This result is closely related to the following theorem which we state without proof.

***Theorem 12.5*** Let $\lambda$ be the generalized likelihood ratio criterion of Equation (12.49) for testing $\theta_1 = \theta_{10}, \theta_2 = \theta_{20}, \ldots, \theta_r = \theta_{r0}$. Then $-2 \ln \lambda$ is distributed asymptotically as $\chi^2$ with $r$ degrees of freedom.

The problem in a nonasymptotic context is that there may be several different pivotal quantities. In the area of Chapter 16, for instance, there is a pivotal quantity related to the $F$ distribution, and also one related to range/(estimate of $\sigma$). These give different tests of significance with differing sensitivities.

## 12.11 Distribution-free Tests of Significance

The tests of significance described so far in this chapter have all been concerned with the case of a fully defined class of parametric distributions, e.g., normal, uniform. It is clear that any set of data may be approximated well by any of a large number of such parametric distributions which will look quite different, at least in mathematical form and possibly in shape. This raises the question of how strongly development of a significance level associated with a hypothesized value and a class of alternative values for a parameter depends on the exact mathematical distributional form that is assumed in the development.

We consider in this section some tests of significance which use a less-defined distributional assumption. The oldest and simplest is the *sign test* directed to the testing of significance of a mean, using the model that the data are like a random sample from a distribution which is specified only to be symmetrical about a value $\theta$. The parameter $\theta$ is called a location parameter. Suppose we wish to examine the tenability of the value $\theta_0$ for $\theta$ and that the data are $x_1, x_2, \ldots, x_n$. Then, under the model the modified observations $x_i - \theta_0$ are symmetrically distributed around zero. A naive approach is to consider merely the number of the $x_i - \theta_0$ which are positive. Let this number be $r$. Then, under the model this number $r$ is distributed as $\mathrm{Bi}(n, 1/2)$; and a test of significance of the binomial frequency equal to 1/2—either against alternatives greater than 1/2, less than 1/2, or unequal to 1/2—is obtained just as in the example of a binomial test. The achievable levels of significance are $K/2^n$, where $K$ is of the form

$$\binom{n}{0} + \binom{n}{1} + \binom{n}{2} + \cdots + \binom{n}{r}$$

or twice this for a two-sided test. This amounts to using the pivotal $\sum_i \mathrm{sgn}(x_i - \theta_0)$.

A second method [Wilcoxon (1947)], is the *signed rank test*. Rank the $|x_i - \theta_0|$ from one for the smallest to $n$ for the largest. Denote the rank of $|x_i - \theta_0|$ by $r_i$ and consider the pivotal

$$W = \sum_i r_i \,\mathrm{sgn}(x_i - \theta_0) = \sum_i r_i \delta_i \qquad (12.55)$$

where

$$\begin{aligned} \delta_i &= +1 \text{ if } x_i - \theta_0 > 0 \\ &= -1 \text{ if } x_i - \theta_0 < 0 \end{aligned}$$

We consider the distribution of $W$ conditional on the observed $\{|X_i - \theta_0|\}$.

**Table 12.4** Complete Distribution of $W$

| $i$ | 1 | 2 | 3 | 4 | |
|---|---|---|---|---|---|
| $x_i$ | $-2$ | 1 | 3 | 4 | |
| $x_i - 1/4$ | $-9/4$ | 3/4 | 11/4 | 15/4 | |
| $\|x_i - 1/4\|$ | 9/4 | 3/4 | 11/4 | 15/4 | |
| $r_i$ | 2 | 1 | 3 | 4 | |
| | $\delta_1$ | $\delta_2$ | $\delta_3$ | $\delta_4$ | $W$ |
| 1 | +1 | +1 | +1 | +1 | 10 |
| 2 | +1 | +1 | +1 | −1 | 2 |
| 3 | +1 | +1 | −1 | +1 | 4 |
| 4 | +1 | +1 | −1 | −1 | −4 |
| 5 | +1 | −1 | +1 | +1 | 8 |
| 6 | +1 | −1 | +1 | −1 | 0 |
| 7 | +1 | −1 | −1 | +1 | 2 |
| 8 | +1 | −1 | −1 | −1 | −6 |
| 9 | −1 | +1 | +1 | +1 | 6 |
| 10 | −1 | +1 | +1 | −1 | −2 |
| 11 | −1 | +1 | −1 | +1 | 0 |
| 12 | −1 | +1 | −1 | −1 | −8 |
| 13 | −1 | −1 | +1 | +1 | 4 |
| 14 | −1 | −1 | +1 | −1 | −4 |
| 15 | −1 | −1 | −1 | +1 | −2 |
| 16 | −1 | −1 | −1 | −1 | −10 |

There are $2^n$ points in this distribution determined by

$$\{(\delta_1, \delta_2, \ldots, \delta_n) : \delta_i = \pm 1\}$$

If the model is appropriate, the $\delta_i$ are independently distributed, each with probability 1/2 of being +1 and probability 1/2 of being −1. Therefore each of the $2^n$ points has probability of $1/2^n$. We illustrate the computations with four observations, $x_1 = -2$, $x_2 = 1$, $x_3 = 3$, and $x_4 = 4$, in Table 12.4 for $\theta_0 = 1/4$. We have demonstrated how to obtain the distribution of $\sum r_i\delta_i$ under the hypothesis of $\theta_0$. Let us consider evaluating for the same set of data the null hypothesis that $\theta = \theta_0 + \Delta$, $\Delta > 0$. Now $x_i - (\theta_0 + \Delta) < x_i - \theta_0$. So with $r\{|x_i - \theta_0 - \Delta|\}$ denoting the rank of $|x_i - \theta_0 - \Delta|$ in the set $\{|x_i - \theta_0 - \Delta|\}$

if $x_i - \theta_0 - \Delta > 0$ and $x_i - \theta_0 > 0$, $r\{|x_i - \theta_0 - \Delta|\} \leq r\{|x_i - \theta_0|\}$

if $x_i - \theta_0 - \Delta \leq 0$ and $x_i - \theta_0 \leq 0$, $r\{|x_i - \theta_0 - \Delta|\} \geq r\{|x_i - \theta_0|\}$

if $x_i - \theta_0 - \Delta \leq 0$ and $x_i - \theta_0 > 0$

$$r\{|x_i - \theta_0 - \Delta|\}\,\mathrm{sgn}(x_i - \theta_0 - \Delta) < 0$$

and
$$r\{|x_i - \theta_0|\}\,\mathrm{sgn}(x_i - \theta_0) > 0 \qquad (12.56)$$

In all cases

$$r\{|x_i - \theta_0 - \Delta|\}\operatorname{sgn}(x_i - \theta_0 - \Delta) \leq r\{|x_i - \theta_0|\}\operatorname{sgn}(x_i - \theta_0) \quad (12.57)$$

Hence

$$W(\theta_0 + \Delta) \leq W(\theta_0) \quad (12.58)$$

Hence smaller values of $W$ are obtained for larger hypothesized values.

Thus for a given set of data obtained from an unknown $\theta$, the further $\theta_0$ is to the left of $\theta$, the larger $W$ will be and therefore large values of the criterion are indicative of $\theta > \theta_0$. It is reasonable then to take as the strength of evidence against $\theta = \theta_0$, with respect to alternatives $\theta > \theta_0$, the probability under $\theta_0$ of a criterion value equal to or greater than the observed value. The significance level in our example for $\theta = 1/4$ with alternatives $\theta > 1/4$ is $1/8$.

A third partially distribution-free test, which we call the *Fisher randomization test*, was given by Fisher (1935a). The idea again is that we have a set of data from a distribution symmetric about $\theta$. Consider the pivotal

$$\begin{aligned} R &= \sum_i |x_i - \theta_0|\operatorname{sgn}(x_i - \theta_0) \\ &= \sum_i |x_i - \theta_0|\delta_i \end{aligned} \quad (12.59)$$

If the model is appropriate, the distribution for this random variable conditional on the observed $|x_i - \theta_0|$ is obtained in the same way as the distribution of $W$. There are $2^n$ possible configurations of the $\delta_i$'s, each with probability $1/2^n$. Against the alternative $\theta < \theta_0$, the significance level is obtained by calculating the probability of a value for $R$ as small or smaller than the observed value for $R$. Against $\theta > \theta_0$, the significance level is the probability of a value for $R$ as large or larger than the observed value of $R$. We state without proof that the significance level of $\theta_0$ with regard to alternatives $\theta < \theta_0$ is less than the significance level of $\theta_1$ with regard to alternatives $\theta < \theta_1$ if $\theta_0 > \theta_1$.

A closely related test of significance is the *Wilcoxon-Mann-Whitney two-sample test*. Suppose we have data $x_1, x_2, \ldots, x_m$ assumed to be a random sample from $F(x)$ and $y_1, y_2, \ldots, y_n$ assumed to be a random sample from $G(y)$, and we wish to form an idea of consonance of the data with the hypothesis that $F(x) = G(x)$ for all $x$. A simple procedure is to rank the full set of $m + n$ observations and to consider the sum of the ranks of the $x$ observations $\sum_{i=1}^{m} r(x_i)$. If the distributions $F$ and $G$ are the same, this sum of the ranks $T$ has a known distribution. Alternatively, one can consider all $mn$ pairs $(x_i, y_j)$, $i = 1, 2, \ldots, m$; $j = 1, 2, \ldots, n$ and let $u(x_i, y_j) = 1$ if $x_i < y_j$ and $0$ if $x_i > y_j$. Let

$U = \sum_{i,j} u(x_i, y_j)$. Then it can be shown that $E(U) = mn/2$,

$$V(U) = [mn(m + n + 1)]/12$$

if $F(x) = G(x)$. Furthermore, $U$ is asymptotically normally distributed with these parameters.

It can be shown that $U + T = mn + [m(m + 1)]/2$, so the distribution of $T$ can also be obtained. For finite $m$, $n$, the exact distribution of $U$ can be obtained and a table is given by Fix and Hodges (1955).

## 12.12 Relation of Fisher Randomization Test to Student's *t* Test

In the Fisher randomization test for $\theta = 0$ we obtain the percentile point of $\sum x$ in the discrete distribution of $R = \sum_i |x_i|\delta_i$, where each $\delta_i$ is either $+1$ or $-1$. It may be seen that

$$E(R) = 0$$

$$E(R^2) = \sum_{i=1}^{n} x_i^2 = \sigma^2, \quad \text{say}$$

$$E(R^3) = 0$$

$$E(R^4) = \sum_{i=1}^{n} x_i^4 + 3 \sum_{i,i'=1}^{n\ \ddagger} x_i^2 x_{i'}^2$$

$$= 3\left(\sum_{i=1}^{n} x_i^2\right)^2 - 2\sum_{i=1}^{n} x_i^4 \tag{12.60}$$

So if

$$R^* = R/\sqrt{\sum x_i^2}$$

$$E(R^*) = 0$$

$$E(R^{*2}) = \text{Var}(R^*) = 1$$

$$E(R^{*3}) = 0$$

$$E(R^{*4}) = 3 - 2\left(\sum_{i=1}^{n} x_i^4\right)\Big/\left(\sum_{i=1}^{n} x_i^2\right)^2 \tag{12.61}$$

The maximum value of $\sum x_i^4/(\sum x_i^2)^2$ is unity and its minimum value is $1/n$. If we envisage sets $\{x_i\}$ of increasing sizes $n$ with $\sum x_i^4/(\sum x_i^2)^2$ decreasing as $n$ increases, the fourth moment would tend to three. This type of argument can be made rigorous to show that $R^*$ may be expected to be approximately normally distributed with zero mean and variance unity. Hence an approximate Fisher randomization test significance level against the alternative that $\theta > 0$ is the probability that a standard normal deviate will exceed $\sum x_i/\sqrt{\sum x_i^2}$.

We now compare this with the $t$ test. If we use the $t$ test, we say that the significance level associated with $\sum x_i$ is

$$P\left(t \geq \frac{\bar{x}}{\left[\frac{\sum (x_i - \bar{x})^2}{n(n-1)}\right]^{1/2}}\right) \tag{12.62}$$

As $n$ gets large, the $t$ distribution approaches the standard normal distribution, so with large $n$ the level of significance is approximately

$$P\left(Z \geq \frac{\sum x_i}{\sqrt{\sum x_i^2 - \frac{(\sum x_i)^2}{n}}}\right) \tag{12.63}$$

where $Z$ is a standard normal variable. We see therefore that we shall get somewhat similar significance levels by the $t$ test and the $R$ test, provided $(\sum x_i)^2/[n \sum x_i^2]$ is small. If this is large, the $t$ test will give for positive $\sum x_i$ a higher significance, i.e., a smaller level of significance. This smaller level is based strongly, however, on the accuracy of representation of the parent distribution by the normal distribution *in the tails* and is therefore questionable because our data will not be able to indicate that normality does hold in the extreme tails. The $R$ test depends *only* on symmetry of the parent distribution. It is said to be more robust.

The ordering of possible samples by the Fisher randomization test is ordering approximately by $(\sum x_i)/\sqrt{\sum x_i^2}$, and it can be seen that this approximate ordering is the same ordering as that given by the $t$ statistic. The significance level in the former is a frequency in a restricted population of samples giving the same set of $|x_i|$, while in the latter it is in the population of all possible samples.

The three tests presented above for testing a location parameter are called permutation tests because they are based on the set of all permutations of the data. Since the sign test ($S$) uses only the signs of $x_i - \theta_0$, the Wilcoxon test ($W$) uses only the signs and the ranks of $|x_i - \theta_0|$, and the Fisher test ($R$) uses the $x_i - \theta_0$ without condensation, the Fisher test is superior as a significance test.

In terms of earlier ideas, the situation is that if the sets of achievable levels are denoted by $\{S\}$, $\{W\}$, and $\{R\}$ for the three tests then $\{S\} \subset \{W\} \subset \{R\}$. It is therefore possible to achieve a lower level with $R$ than with $W$ and with $W$ than with $S$. We have seen that $W$ and $R$ have the monotonicity property, so that $SL(\theta_0 + \Delta) \leq SL(\theta_0)$. It can happen with the sign test that $SL(\theta_0 + \Delta) = SL(\theta_0)$, but with the Wilcoxon test $SL(\theta_0 + \Delta) < SL(\theta_0)$. Also it can happen that with the Wilcoxon test $SL(\theta_0 + \Delta) = SL(\theta_0)$, but with the Fisher test $SL(\theta_0 + \Delta) < SL(\theta_0)$. If any one of the tests gives significance at an extremal level of $1/2^n$ or one, the others will do likewise. On an intuitive basis it seems there is a clear order of sensitivity

$$R > W > S \tag{12.64}$$

The tests are of somewhat different computational difficulty. The $S$ test is trivially easy. Significance levels of the values of the $W$ criterion can be constructed easily. The $R$ test, in contrast, requires computation and ordering of all the possible $2^n$ forms:

$$\sum_{i=1}^{n} \delta_i |x_i|, \qquad \delta_i = +1 \text{ or } -1 \tag{12.65}$$

Actually only 1/2 of these need be computed because each absolute value occurs twice. If $n$ is say 20, this is a formidable task. However, if $(\sum x_i^4)/(\sum x_i^2)^2$ is "small," one can use the $N(0, 1)$ approximation for $\sum x_i/\sqrt{\sum x_i^2}$; and if $\sum x_i^4/(\sum x_i^2)^2$ is large, it will not be a difficult task to enumerate the more discrepant configurations. We suggest that the Fisher procedure will be shown to be based on a "best" pivotal. Some discussion of the various tests is given by Kempthorne and Doerfler (1969).

## 12.13 General Remarks on Tests of Significance

We have seen that the formulation of a test of significance is fairly straightforward if the problem is to test a completely specified model $M_0$. In the case of a vector parameter $\theta = (\theta_1, \theta_2, \ldots, \theta_p)$, however, there are considerable difficulties. To illustrate approaches we take $\theta$ to be two dimensional and suppose we wish to test $\theta_1 = \theta_{10}$ without making an assumption or restriction about the value of $\theta_2$.

An approach that comes readily to mind is the following. We may assume that if $\theta_2$ is known to have the value $\theta_{20}$, we can formulate a test of significance, which we will regard as a weight of evidence against the value $\theta_{10}$ of $\theta_1$. To accommodate the fact that we do not know the value of $\theta_2$, we can follow the Bayesian approach of supposing that $\theta_2$ is a random variable with a particular prior distribution. We can use the data to obtain a posterior distribution for $\theta_2$ and can then form a weighted average of the significance for fixed $\theta_2$ over this posterior distribution of $\theta_2$. The procedure is relatively easy to imagine, though the mechanical aspects may not be. The difficulties with this mode of approach are that there are no reasonable rules for choosing a prior distribution for $\theta_2$ and the significance level will *not* have the frequency properties which occur in the simple case.

A second approach, due to R. A. Fisher (1935b) and called fiducial inference, is to use the data to obtain a sort of posterior distribution for $\theta_2$ without the injection of a prior distribution. The logic of this mode of development is still very unclear, and fiducial inference is not at all widely accepted. To illustrate how the data can be used (but not necessarily should be used) to give a sort of posterior distribution for $\theta_2$, suppose there is a statistic $T_2$ and a parametrization from $\theta_2$ to another parameter $\phi$ such that $\text{Prob}(T_2 - \phi \leq k) = F(k)$, where $k$ ranges over the real line,. and $F(k)$ is the c.d.f. of $T_2 - \phi$. This is a direct probability statement. The Fisher process is to invert the relationship to give $\text{Prob}_{\text{fid}}(\phi \geq T_2 - k) = F(k)$, in which $T_2$ is taken to be fixed at its observed value, the random variable is now $\phi$ rather than $T_2$, and the subscript fid on Prob denotes fiducial. This fiducial distribution for $\phi$ gives a quasi-posterior distribution for $\theta_2$, which is then used to make an average of a procedure defined for fixed $\theta_2$, such as calculation of a significance level.

In this text it is possible to comment only briefly on the processes. A significance test for $\theta_1 = \theta_{10}$ obtained by either a Bayesian or fiducial argument does *not* have the basic property $\text{Prob}(SL \leq \alpha | \theta_1 = \theta_{10}) = \alpha$. Neither procedure should be rejected because it does not possess a property that was not sought. However, it is not at all clear that either procedure possesses other properties which are compelling to some extent. If a reasonable basis for making the fiducial argument unique could be obtained, the fiducial approach would be superior to the Bayesian approach based on choice of a prior which has no real relation to the situation at hand. But a unique fiducial type of approach has not been specified and developed. The controversy here appears to be a stand-off.

## 12.14 Tests of Significance of a Vector Parameter

Consider the general situation of data $D$ which we envisage as random from some distribution $F(D; \theta_1, \theta_2, \ldots, \theta_r)$. We have to envisage a multiplicity of models with regard to the parameter which we would wish to test; e.g., we might wish to examine $\theta_1 = \theta_{10}, \theta_2 = \theta_{20} \ldots, \theta_r = \theta_{r0}$ or $\theta_1 = \theta_{10}, \theta_2 = \theta_{20}, \ldots, \theta_s = \theta_{s0}, s < r$. The simplest of the latter is a test on a single $\theta_j$.

In the case of a test of a single $\theta_j$ we can envisage the possibility of testing $\theta_j = \theta_{j0}$ with alternative $\theta_j > \theta_{j0}$ or $\theta_j < \theta_{j0}$ or $\theta_j \neq \theta_{j0}$. So we have the possibility of an upper-tail test, a lower-tail test or a two-tailed test. Consider now a test of $\theta_j = \theta_{j0}, \theta_k = \theta_{k0}$ involving two components of the whole vector parameter. The idea of one-sided or two-sided tests now becomes impossible. In two or more dimensions, there are an "infinity of sides." We can envisage a test with sensitivity in any specified direction in the two-dimensional space. Because a test of significance is an ordering of samples, each possible sample that is envisaged under the model is mapped into a point of the line segment (0, 1). To each ordering there corresponds a mapping and vice versa. The distribution of significance level under the hypothesis $\theta_j = \theta_{j0}, \theta_k = \theta_{k0}$ is by necessity uniform. The distribution under a different hypothesis will be nonuniform with a shift toward the lower end dependent on some function of $\theta_j - \theta_{j0}$ and $\theta_k - \theta_{k0}$. This function is in a certain sense the noncentrality aspect of the test of significance. Intuitively there cannot be a test of significance that is more sensitive in all directions of the parameter space; therefore, the data analyzer has to choose what directions are of interest to him. Alternatively, he may select a particular test with the full realization that this test is sensitive in some respects and insensitive in others. Consider the case of three means $\bar{x}$, $\bar{y}$, $\bar{z}$, based on $n_1$, $n_2$, and $n_3$ observations respectively, envisaged as arising from three normal populations with the same variance $\sigma^2$ but different means $\mu_x$, $\mu_y$, $\mu_z$. The usual analysis of variance test (discussed in Chapter 16) for differences of the means has sensitivity dependent on

$$(1/2\sigma^2)[n_1(\mu_x - \theta)^2 + n_2(\mu_y - \theta)^2 + n_3(\mu_z - \theta)^2]$$

where $\theta$ is $(n_1\mu_x + n_2\mu_y + n_3\mu_z)/(n_1 + n_2 + n_3)$. This test has sensitivity toward $\mu_x \neq \mu_y$, but a more sensitive test in this direction is given by the $t$ test on $\bar{x} - \bar{y}$. There is no way out of the dilemma posed here from the viewpoint of assessment of data. If one wishes to develop a best accept-reject rule (see below) with power dependent on $f(\mu_x - \mu_y, \mu_x - \mu_z)$ for some specified function $f$, one can in theory, at least, do so. But to consider that such an accept-reject rule gives a most sensitive test of equality of $\mu_x$, $\mu_y$, and $\mu_z$ in all directions is not reasonable.

## 12.15 Tests of Hypotheses: Accept-Reject Rules

We have developed a theory of significance testing as a means of quantifying the weight of evidence against a model or class of models. We have not attached any strength of opinion to the possibility that either the hypothesized model or the alternative model was the true one. We believe that in most applications the hypothesized model is unacceptable as a true model. Rather it is set up as a "straw man" to determine whether the data suggest that it is false. For example, a common experiment consists of giving a treatment to one group of subjects and another treatment to another group. The hypothesis that the treatments have equal effects is then tested. In most cases the experimenter knows this hypothesis is not true. Yet he wishes to determine if the data provide such evidence. We have followed Fisher in discussing tests of significance as evaluation techniques rather than accept-reject rules.

Hypothesis testing, on the other hand, is concerned with accept-reject rules. Upon the basis of the data we either accept or reject the hypothesis (rejecting or accepting the alternative). Although tests of significance are due largely to Fisher, the theory of tests of hypotheses was originally developed by Neyman and Pearson.

In the literature on this topic, the hypothesis under test is usually denoted by $H_0$ and the alternative by $H_A$. If we do accept or reject the hypothesis $H_0$, there are two possible decision errors which we might make. We are certainly in error if we reject $H_0$ when it is true or if we accept $H_0$ when it is false. By convention these errors are referred to as Type I and Type II errors respectively. The situation can be conveniently represented in a 2 × 2 table as in Table 12.5.

**Table 12.5** Types of Error

| | Accept $H_0$ | Reject $H_0$ |
|---|---|---|
| $H_0$ True | No error | Type I error |
| $H_0$ False | Type II error | No error |

As far as the possibility that a hypothesized model is true is concerned, our point of view is that no mathematical model is true in the sense that it *exactly* describes the data-producing mechanism. However, when we accept $H_0$, we recognize this difficulty and presumably act in the way we would behave if it were true. With this stipulation we never "know" whether or not our decision is in error, but we can determine the probabilities of Type I and Type II errors. In general there is not *a* probability of a Type I error or *a* probability of a Type II error. Rather, the probability of rejecting $H_0$ will depend upon $\theta$; and if $H_0$ is composite, the probability of a Type I error will be a function of $\theta$. Similarly, the probability of a Type II error is a function of $\theta$.

From the above viewpoint the problem is one of choosing an accept-reject rule. The choice is usually made by considering for each rule the probability of rejecting $H_0$ as a function of $\theta$. This function is referred to as the power function, $P(\theta)$, of the test (rule).

For

$$H_0: \theta \leq 10$$
$$H_A: \theta > 10$$

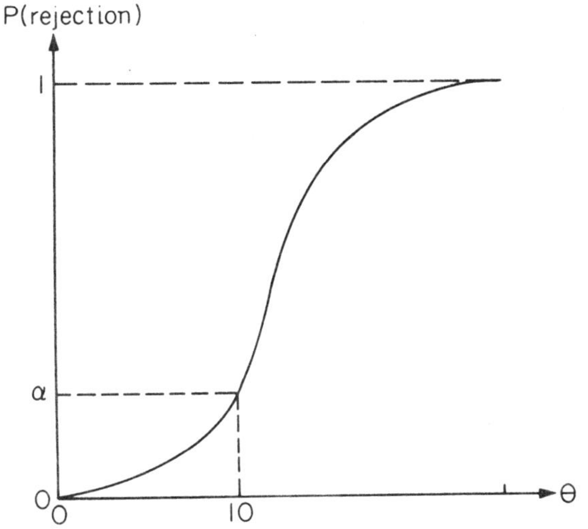


**Fig. 12.1** Typical Power Curve

the power function for a test might be as indicated in Figure 12.1. Note that when $H_0$ is true, the probability of rejecting $H_0$ is less than or equal to $\alpha$. The quantity $\alpha$, i.e., the least upper bound for $P(\theta)$ when $H_0$ is true, is referred to as the size of the test and often as the significance level of the test. The latter is unfortunate because it has created confusion between tests of significance and tests of hypotheses. In tests of significance the significance level $\alpha$ is calculated from the data and is a random variable, while in testing hypotheses the size of the test is a property of the decision rule and is known prior to collecting the observations. In all future discussion of hypothesis testing we shall refer to the size, not the significance level. If $\alpha$ is actually the maximum of $P(\theta)$ under $H_0$, then the test is said to be of exact size $\alpha$.

For various reasons, it is sometimes more convenient to graph the probability of accepting $H_0$ versus $\theta$ rather than the probability of rejecting $H_0$. Such a graph is known as an operating characteristic (O.C.) curve. For both O.C. curves and power functions, which may be spoken of as power curves, some function of $\theta$ rather than $\theta$ may be displayed on the horizontal axis.

It would appear that given an array of tests and their respective power curves, one would simply choose the one he prefers. This is essentially the case. However, a common modus operandi seems to be that the choice is made only from the rules which have the same size $\alpha$, arbitrarily chosen in advance. If a test of size $\alpha$ has a power curve above the power curve of every other test of size $\alpha$ for all $\theta$ in the region of $H_A$, it is referred to as a uniformly most powerful test of size $\alpha$, and presumably we would choose such a test. It appears that a resolution of the question of choice of $\alpha$ has to be based on decision theory, with introduction of prior opinion and loss functions.

## 12.16 Randomized Test Functions

Much of the mathematical elegance of the theory developed on Neyman and Pearson lines results from the introduction of randomized decision rules. Such procedures are generally described in terms of a critical function which maps the sample observations into the probability of rejecting the hypothesis. A random device is then used to decide whether to accept or reject the hypothesis. This formulation is reasonable if we must make one of the two decisions and are concerned with the operating characteristics of our decision rule.

To illustrate the idea of a critical function, suppose we are using the binomial model Bi$(n, p)$ with

$$H_0: p = 1/2$$
$$H_A: p \neq 1/2$$

and we have chosen the critical function

$$\phi(D) = e^{-\Sigma_1^n x_i}$$

We are using this merely as an example, not necessarily a desirable test. Then if $\sum x_i = 0$, i.e., if all $x$'s are zeros, $\phi(D) = 1$ and we reject $H_0$. If we observed only one success, i.e., $\sum x_i = 1$, $\phi(D) = e^{-1}$ and we reject $H_0$ with probability $e^{-1}$. To achieve this, we call on a chance mechanism which will give a specified result with probability $e^{-1}$. For instance, we can draw a four-digit random number between zero and one and reject $H_0$ if we obtain a number less than $e^{-1}$.

Certain critical functions deserve special notice. If $\phi(D)$ takes only the values zero and one, the test is nonrandomized. In this case the sample space is divided into two regions, the rejection region and the acceptance region. The rejection region is also commonly called the critical region. If $\phi(D) = \alpha$ for all $D$, the test is a purely randomized test of size $\alpha$. Regardless of the data the hypothesis is rejected with probability $\alpha$.

The power $P(\theta)$ can be expressed very nicely by means of the critical function. We have

$$\begin{aligned} P(\theta) &= P_\theta(\text{rejecting } H_0) \\ &= \sum_D P(\text{rejecting } H_0|D)P(D;\theta) \end{aligned}$$

$$= \sum_D \phi(D)P(D;\theta)$$

$$= E_\theta[\phi(D)] \tag{12.66}$$

Thus the power of the test is given by the expected value of the critical function.

The introduction of a randomized rule allows us to restrict our attention to critical functions which are functions of sufficient statistics.

***Theorem 12.6*** Given a critical function $\phi(D)$ and a sufficient statistic $T$, there exists a critical function $\psi(T)$ with

$$E_\theta[\phi(D)] = E_\theta[\psi(T)] \tag{12.67}$$

*Proof:* Because the conditional distribution of $D$ given $T$ does not depend upon $\theta$, we can let

$$\psi(T) = E[\phi(D)|T] \tag{12.68}$$

and

$$E_\theta[\psi(T)] = E_\theta\{E[\phi(D)|T]\} = E_\theta[\phi(D)] \tag{12.69}$$

This theorem plays a fundamental role in the theory of hypothesis testing.

## 12.17 Most Powerful Tests

In the development of significance tests we were forced to consider common achievable significance levels in order to compare tests. Here we do not have that difficulty because there is always a test of size $\alpha$ for any $\alpha$, $0 < \alpha < 1$ (namely the randomized test of size $\alpha$). In hypothesis testing the likelihood ratio plays an important role. We shall now state and prove the Neyman-Pearson theorem in terms of hypothesis testing.

***Theorem 12.7*** A test which has the form

$$\begin{aligned} \phi(D) &= 1 \qquad \text{if } \frac{P(D;\theta_1)}{P(D;\theta_0)} > c \\ &= a \qquad \text{if } \frac{P(D;\theta_1)}{P(D;\theta_0)} = c \\ &= 0 \qquad \text{if } \frac{P(D;\theta_1)}{P(D;\theta_0)} < c \end{aligned} \tag{12.70}$$

where $a$ and $c$ are chosen so that the test is of exact size $\alpha$, is most powerful for testing the simple hypothesis $\theta = \theta_0$ against the simple alternative $\theta = \theta_1$.

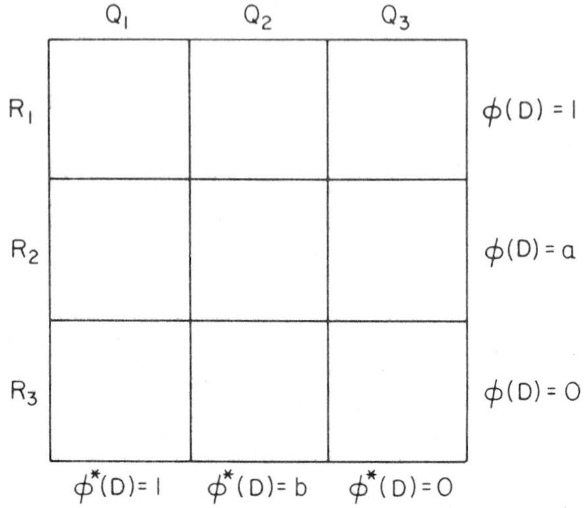


**Fig. 12.2** For Proof of Neyman-Pearson Theorem

*Proof:* The Neyman-Pearson test divides the sample space into three regions $R_1$, $R_2$, $R_3$, specified by $\phi(D) = 1, a, 0$ respectively. Consider a competing test $\phi^*$ of size $\alpha$ (i.e., power under $H_0 \leq \alpha$) which divides the sample space into three regions $Q_1$, $Q_2$, $Q_3$ specified by $\phi^*(D) = 1, b, 0$ respectively. This situation is depicted in Figure 12.2.

Consider the difference in power of the two tests equal to the following, where $P_0(\ )$ denotes probability under $H_0$ and $P_1(\ )$ probability under $H_A$:

$$
\begin{aligned}
&E_{\theta_1}[\phi(D)] - E_{\theta_1}[\phi^*(D)] \\
&\quad = P_1(R_1) + aP_1(R_2) - P_1(Q_1) - bP_1(Q_2) \\
&\quad = P_1(R_1Q_1) + P_1(R_1Q_2) + P_1(R_1Q_3) + aP_1(R_2Q_1) + aP_1(R_2Q_2) \\
&\qquad + aP_1(R_2Q_3) - P_1(R_1Q_1) - P_1(R_2Q_1) - P_1(R_3Q_1) - bP_1(R_1Q_2) \\
&\qquad - bP_1(R_2Q_2) - bP_1(R_3Q_2) \\
&\quad = (1-1)P_1(R_1Q_1) + (1-b)P_1(R_1Q_2) + P_1(R_1Q_3) \\
&\qquad + (a-1)P_1(R_2Q_1) + (a-b)P_1(R_2Q_2) + aP_1(R_2Q_3) \\
&\qquad - P_1(R_3Q_1) - bP_1(R_3Q_2) + (1-1)P_1(R_3Q_3) \qquad (12.71)
\end{aligned}
$$

Now, in $R_1$,

$$P(D;\theta_1) > cP(D;\theta_0)$$

in $R_2$,

$$P(D;\theta_1) = cP(D;\theta_0)$$

and in $R_3$,

$$P(D; \theta_1) < cP(D; \theta_0) \tag{12.72}$$

So the difference in powers exceeds

$$\begin{aligned}
&c[(1-1)P_0(R_1Q_1) + (1-b)P_0(R_1Q_2) + P_0(R_1Q_3)\\
&+ (a-1)P_0(R_2Q_1) + (a-b)P_0(R_2Q_2) + aP_0(R_2Q_3)\\
&- P_0(R_3Q_1) - bP_0(R_3Q_2) + (1-1)P_0(R_3Q_3)]\\
&\qquad = c[P_0(R_1) + aP_0(R_2) - P_0(Q_1) - bP_0(Q_2)]\\
&\qquad = c[E_{\theta_0}\phi(D) - E_{\theta_0}\phi^*(D)] \qquad (12.73)\\
&\qquad \geq 0
\end{aligned}$$

because $\phi$ is of exact size $\alpha$ and $\phi^*$ is only of size $\alpha$.

It should be noted that we have not given a proof here that a most powerful test must have the form specified; however, that is the case. We now illustrate the use of Theorem 12.7 by direct application to problems involving simple hypotheses and also its use when the hypotheses are composite. Suppose that we have $n$ observations considered to be normal variables and that the hypotheses are

$$\begin{aligned}
H_0&: \mu = \mu_0, \sigma^2 = 1\\
H_A&: \mu = \mu_1(>\mu_0), \sigma^2 = 1
\end{aligned}$$

According to the theorem the most powerful test must have the form: Reject $H_0$ if

$$\frac{(1/\sqrt{2\pi})^n \exp[-\sum (x_i - \mu_1)^2/2]}{(1/\sqrt{2\pi})^n \exp[-\sum (x_i - \mu_0)^2/2]} > c \tag{12.74}$$

Operating on the inequality and retaining the same symbol $c$ to denote a different constant, we see that the criterion for rejection is given by any term in the following sequence of terms:

$$\begin{aligned}
\exp[\textstyle\sum (x_i - \mu_0)^2/2]\exp[-\sum (x_i - \mu_1)^2/2] &> c\\
\exp[\textstyle\sum x_i(\mu_1 - \mu_0)] &> c\\
\sqrt{n}\,\bar{x} &> c\\
\sqrt{n}\,(\bar{x} - \mu_0) &> c \qquad (12.75)
\end{aligned}$$

Note that we use the fact that $\mu_1 - \mu_0 > 0$.

The rejection criteria given by this sequence of inequalities are equivalent only if it is noted that $c$ does not represent the same quantity in all cases. Using the last term in the sequence as a test criterion, let us choose $c$ so that the test has size $\alpha$. Since $\sqrt{n}\,(\bar{x} - \mu_0)$

is a normal variable with mean zero and variance unity when $\mu = \mu_0$, we see that $c$ can be taken to be $z_\alpha$ such that $P(Z \geq z_\alpha) = \alpha$, where $Z$ is a standard normal variable.

Let us now consider the simple null hypothesis

$$H_0: \mu = \mu_0, \sigma = 1$$

and the composite alternative hypothesis

$$H_A: \mu > \mu_0, \sigma = 1$$

Note that this is a composite hypothesis because there are many parameter values specified by the statement $\mu > \mu_0$. The procedure usually followed is to choose a single point in the region specified by the alternative and to use the Neyman-Pearson theorem to find a most powerful test against this simple alternative. If the test does not, in fact, depend upon which point is chosen, it is most powerful against all parameter values of the alternative; i.e., it is uniformly most powerful. In the present example any simple alternative $\mu = \mu_1(>\mu_0)$, $\sigma = 1$ leads to the same test, which is therefore uniformly most powerful.

In case the null hypothesis is composite, we mention two possible approaches:

1. Choose a simple null hypothesis contained in the composite null hypothesis and proceed to develop a test. Then check to see that the test is of size $\alpha$, i.e., that it has power less than or equal to $\alpha$ under the composite null hypothesis.
2. Choose a weighted average of the parameter values covered by the null hypothesis as a simple null hypothesis and proceed as in (1).

We refer the reader to Lehmann (1959) for a complete development of the theory of testing hypotheses.

## 12.18 Sequential Hypothesis Testing

We have covered the concept of a population of samples which could be generated by repeatedly drawing samples from a population. In general, we considered the sample to consist of a specified number $n$ of observations; i.e., the samples in the conceptual population of samples are all of size $n$.

In sequential sampling the sample size is not specified in advance, and the samples in the conceptual population of samples are of varying sizes. The observations are drawn sequentially until it is decided to stop sampling in accordance with some stopping rule. For instance, we might stop drawing observations when (1) the data look favorable, (2) we accept or reject some hypothesis, (3) we reach an upper limit for sample size, or (4) the mean is twice the standard deviation. Note that we do not claim these to be "good" stopping rules. We are merely indicating that there are many possible stopping rules.

When the sample is drawn sequentially, we still have the problems of inference: making sense of the data, condensation of the data, tests of significance, etc. We mention here only the decision problem of testing a hypothesis. The rule for accepting or rejecting is generally based upon the likelihood ratio; and after each observation the decision is made to accept the hypothesis, reject the hypothesis, or to draw another observation.

The advantage of such a procedure over the fixed sample size approach is that a specified O.C. curve can be achieved with an average sample size smaller than the sample size that is specified in advance. There is, of course, the disadvantage that drawing observations sequentially is troublesome, particularly when one does not know how many one is going to take.

The sequential hypothesis testing procedure developed by Wald (1947) calls for specifying two numbers $A$ and $B$ such that $0 < B < A < 1$. The observations are taken sequentially; and after each observation is taken, the likelihood ratio $\lambda$ is computed and (a) if $\lambda \leq B$, $H_0$ is accepted; (b) if $B < \lambda < A$, an additional observation is taken; and (c) if $\lambda \geq A$, $H_A$ is accepted.

## 12.19 Comparison of Hypothesis Testing and Significance Testing

Tests of significance and tests of hypotheses are very closely related, and the theory developed for significance tests closely parallels that developed for tests of hypotheses. In most textbooks, only the theory of hypothesis testing or accept-reject rules is presented. We feel that formal accept-reject rules are too rigid for most data analysis situations and in fact are not strictly followed by practicing statisticians. Perhaps the most notable exception to this comment is that of acceptance sampling of manufacturing lots. In the simplest case a sample is taken from each lot, and the lot is either accepted or rejected on the basis of the sample.

A simple correspondence can be established so that a significance test gives a test of hypothesis. However, a test of hypothesis does not always yield a significance test. To illustrate, suppose we have a significance test of a hypothesis against an alternative and that $\alpha_0$ is an achievable significance level. Let us establish the rule

$$\begin{aligned} &\text{Reject } H_0 \text{ if } SL \leq \alpha_0 \\ &\text{Accept } H_0 \text{ if } SL > \alpha_0 \end{aligned} \tag{12.76}$$

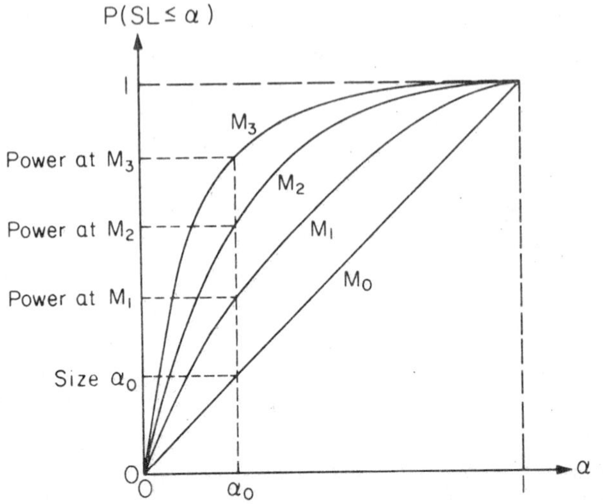


**Fig. 12.3** Correspondence between Significance Tests and Tests of Hypotheses

This rule gives an exact size $\alpha_0$ test of hypothesis because

$$P(\text{reject } H_0|H_0) = P(SL \leq \alpha_0|H_0) \tag{12.77}$$

and we know from the preceding sections that

$$P(SL \leq \alpha_0|H_0) = \alpha_0 \tag{12.78}$$

(If we decided to reject $H_0$ if $SL < \alpha_0$, we would have only a size $\alpha_0$ test, not exact size $\alpha_0$.) The power of this test is

$$P(\text{reject } H_0|H_1) = P(SL \leq \alpha_0|H_1) \tag{12.79}$$

and is obtained from the distribution function of the significance level under the alternative hypothesis. This correspondence is shown in Figure 12.3 which illustrates the continuous case where all values $0 \leq \alpha \leq 1$ are achievable. A similar sketch holds for the discrete case but involves a family of step functions. This figure serves to emphasize one major distinction between the development of significance tests and hypothesis tests. In the first case we are concerned with the behavior of a family of functions over the entire [0, 1] interval. In the second case we are concerned with the behavior of this family at only a single point $\alpha_0$.

Let us suppose now that we start with an accept-reject rule of size $\alpha_0$ and ask whether the rule specifies a test of significance. If the rule is non-randomized we can partially order the samples as follows: Samples in critical region $\gg$ samples in acceptance region. This ordering then specifies a significance test with two achievable levels $\alpha_0$ and 1. Furthermore, if the test of hypothesis is a uniformly most powerful test, the corresponding test of significance is a uniformly most sensitive test of significance. However, we would prefer a test of significance to have many achievable levels. If we have a sequence of nonrandomized accept-reject rules of size $\alpha_1, \alpha_2, \ldots$, we may be able to obtain a test of significance with the same set of achievable significance levels. If it is true for all $\alpha_i$, $\alpha_j$, with $\alpha_i < \alpha_j$ that

$$\begin{pmatrix}\text{critical region for} \\ \text{size } \alpha_i \text{ test}\end{pmatrix} \subset \begin{pmatrix}\text{critical region for} \\ \text{size } \alpha_j \text{ test}\end{pmatrix} \tag{12.80}$$

then we obtain a test of significance by letting

$$SL(D) = \min(\alpha_i | D \text{ is in critical region of size } \alpha_i) \tag{12.81}$$

It is not at all clear that a randomized critical function has any utility for the construction of a significance test. Use of the rule $\phi(D) = \alpha$ may be a good way of reaching decisions in certain instances, but it appears to have no role in the weighing of evidence.

A notable difference in the two theories we are discussing arises in the construction of unbiased tests. A test of hypothesis of size $\alpha$ is unbiased if

$$P(\theta) \leq \alpha \text{ for } H_0$$

and

$$P(\theta) \geq \alpha \text{ for } H_A \tag{12.82}$$

An unbiased test of hypothesis always exists, namely $\phi(D) = \alpha$. However, we pointed out in a preceding section that unbiased significance tests may not exist; in fact, they rarely exist for discrete probability laws and only under rather special circumstances for continuous probability laws.

## 12.20 Goodness of Fit Tests of Parameter Values

Consider a model that observations $x_1, x_2, \ldots, x_n$ are like a random sample from some distribution $f(x; \theta)$. Suppose that the model has been judged acceptable and we wish to form an idea of whether a particular value $\theta_0$ is tenable. A test of significance which has considerable intuitive appeal for many cases is based on the idea of goodness of fit. Consider the statistic $GF(\theta_0)$ = probability of possible samples whose probability is less than or equal to the probability of the observed sample.

In symbols

$$GF(\theta_0; x_1, x_2, \ldots, x_n)$$

$$= \int_{u_1, u_2, \ldots, u_n} f(u_1; \theta_0) f(u_2; \theta_0) \ldots f(u_n; \theta_0) \delta(u_1, u_2, \ldots, u_n; x_1, x_2, \ldots, x_n) \prod_{i=1}^{n} du_i$$

where

$$\delta(u_1, u_2, \ldots, u_n; x_1, x_2, \ldots, x_n) = 1$$

if

$$\prod_{i=1}^{n} f(u_i; \theta_0) \leq \prod_{i=1}^{n} f(x_i; \theta_0)$$

and is zero otherwise.

The statistic $GF(\theta_0; x_1, x_2, \ldots, x_n)$ will be under the model a random number between 0 and 1, and the value obtained may be used as a measure of the tenability of the value $\theta_0$. The scale of tenability is fixed by the fact that if $x_1, x_2, \ldots, x_n$ is the most probable observation set with the parameter value $\theta_0$, $GF(\theta_0)$ will be unity. If at the other extreme the probability of $x_1, x_2, \ldots, x_n$ with the parameter value $\theta_0$ is zero, $GF(\theta_0)$ will be zero.

This mode of evaluating tenability is very close to likelihood ratio tests of significance except that an alternative to $\theta_0$ is not explicitly stated. It has virtue in that it can be applied to a situation of any dimensionality for the parameter, in which concepts of most powerful or most sensitive tests are essentially tangential. This is so because one can envisage a multitude of alternatives, and a test which is highly sensitive to deviations from $\theta_0$ in one direction of the parameter space will usually be quite insensitive in other directions from $\theta_0$.

Some of the difficulties of the Neyman-Pearson theory of tests of hypotheses and confidence intervals are discussed by Dempster (1964), who suggests that the word "confidence" in the phrase "confidence interval" should be replaced by indiffidence. The paper by Dempster gives a long list of references that the interested reader may pursue to advantage.

## PROBLEMS

1. A metallurgist has collected some data on the thermal expansion of two different alloys $A$ and $B$. From previous experience he is willing to accept the normal distribution as adequately describing the data. A casual inspection of the data leads him to suspect that $\mu_A > \mu_B$, but he wishes to quantify this suspicion. Assuming equality of variances, determine the significance level for the model $\mu_A = \mu_B$ against the alternative that $\mu_A > \mu_B$.

| *A* | | | | *B* | | |
|---|---|---|---|---|---|---|
| 18.1 | 19.7 | 18.4 | 13.2 | 16.4 | 15.6 | 18.3 |
| 13.6 | 15.4 | 19.5 | 15.8 | 17.6 | 16.2 | 17.2 |
| 19.2 | 22.3 | 17.0 | 22.6 | 16.9 | 23.8 | 16.1 |
| 21.0 | 20.6 | 20.0 | 24.8 | 19.2 | 18.8 | 22.7 |
| 25.6 | 19.9 | 18.4 | 21.1 | 21.3 | 17.2 | 12.0 |
| 21.2 | 21.7 | 22.3 | 14.2 | 18.4 | 19.4 | 15.3 |

2. A chemist who has been working on the development of a new type of battery believes that he has developed a process which will yield batteries with mean efficiency of 50% or better. However, his opinion is based upon the rather limited set of data:

| Percent Efficiency | | | |
|---|---|---|---|
| 44.3 | 47.2 | 51.2 | 46.3 |
| 58.3 | 49.0 | 54.7 | 46.2 |
| 49.1 | 53.5 | 52.0 | 49.3 |

Assume that the data may be represented by a normal population and determine the significance level for the model that $\mu = 50$ against the alternative that $\mu > 50$.

3. Suppose we use the upper-tailed standard normal deviate test to test the model $\mu = 0$ against the alternative that $\mu > 0$. Graph the c.d.f. of the significance level for $\mu = 0, 1, 2, 3$. Take $n = 10$, $\sigma = 1$.

4. Suppose we use the two-tailed Student's $t$ test to test the model $\mu = 0$ against the model $\mu \neq 0$. Graph the c.d.f. of the significance level for $\mu/\sigma = 0, 1, 2, 3$. Take $n = 4$.

5. A gasoline distributor has been bothered by complaints from retailers that the grade of gasoline being supplied is extremely variable. Thirty tank trucks are sampled and the octane rating determined for each sample. The ratings are

| | | |
|---|---|---|
| 31.2 | 33.3 | 25.4 |
| 29.8 | 37.1 | 42.3 |
| 41.6 | 32.4 | 42.2 |
| 35.4 | 33.6 | 25.8 |
| 27.7 | 36.7 | 26.2 |
| 36.3 | 34.2 | 37.6 |
| 39.0 | 40.3 | 33.4 |
| 42.3 | 43.2 | 37.1 |
| 39.6 | 27.3 | 35.9 |
| 38.2 | 28.2 | 35.2 |

In the past, the octane ratings have been well described by a normal distribution with variance 4. Use the upper-tailed $\chi^2$ test to test the model $\sigma^2 = 4$ against the alternative model that $\sigma^2 > 4$.

6. For the upper-tailed $\chi^2$ test of the model that $\sigma^2 = \sigma_0^2$ against the alternative that $\sigma^2 > \sigma_0^2$, graph the c.d.f. of the significance level for $\sigma^2/\sigma_0^2 = 1, 2, 3, 4$. Use $n = 10$.

7. A project engineer has been considering the adoption of a new process for etching silicon wafers to size. He is interested in the variability, which for the new process was .25. A sample of 25 wafers is selected and the maximum dimension measured. The data are:

| | | | | |
|---|---|---|---|---|
| 2.3 | 4.6 | 6.0 | 2.7 | 3.9 |
| 4.3 | 5.7 | 4.0 | 3.6 | 4.2 |
| 7.2 | 3.9 | 6.2 | 3.8 | 4.0 |
| 1.9 | 3.6 | 5.8 | 2.6 | 4.1 |
| 5.4 | 5.3 | 4.4 | 2.5 | 3.7 |

Assuming normality, evaluate the significance level for the model $\sigma^2 = .25$ against the alternative model that $\sigma^2 \neq .25$.

8. Graph the c.d.f. of the significance level for the test used in Problem 7 for values of $\sigma^2 = .15, .20, .25, .30, .35$.

9. If $n$ independent binomial trials yield $x_0$ successes, let us determine the significance level for $p = p_0$ against the alternative $p > p_0$ by calculating $P(x \geq x_0)$. Show that

$$\sum_{x=x_0}^{n} \binom{n}{x} p^x (1-p)^{n-x} = \int_0^p \frac{n!}{(x_0-1)!(n-x_0)!} z^{x_0-1}(1-z)^{n-x_0}\, dz$$

**Hint:** Use integration by parts.

10. Show that the significance level for the binomial test of significance in Problem 9 can be determined from the $F$ distribution.

11. If an observation from a Poisson distribution yields $x = x_0$, let us determine the significance level for the model $\lambda = \lambda_0$ against the alternative $\lambda > \lambda_0$ by calculating $P(x \geq x_0)$. Show that

$$\sum_{x=x_0}^{\infty} \frac{\lambda^x e^{-\lambda}}{x!} = \int_0^\lambda \frac{z^{x_0-1} e^{-z}}{\Gamma(x_0)}\, dz$$

12. Show that the significance level for the Poisson test of significance in Problem 11 can be determined from the $\chi^2$ distribution.

13. A sequential sampling model calls for stopping at the following pairs of $(n, r)$: (1, 1), (3, 2), (4, 0), (4, 1). Order these results in a reasonable way for testing $p = 0.5$ against $p \neq 0.5$. Graph the c.d.f. of the significance level for $p = 0.1$, 0.5, and 0.9.

14. A sequential sampling model calls for stopping at the following pairs of $(n, r)$: (3, 0), (4, 1), (5, 2), (6, 3), (6, 4), (6, 5), (6, 6). For testing significance of the model

$p = .5$ against the alternative $p \neq .5$, order the seven possible results in a reasonable way and graph the c.d.f. of the resulting significance level for $p = .1, .3, .5, .7, .9$.

15. Obtain the likelihood ratio criterion to test the model $\mu = 0$ against the alternative $\mu \neq 0$ for the normal distribution.

16. Obtain the likelihood ratio criterion to test the model $\sigma^2 = \sigma_0^2$ against the alternative $\sigma^2 \neq \sigma_0^2$ for the normal distribution.

17. In the case of two normal distributions with means $\mu_1$, $\mu_2$ and variances $\sigma_1^2, \sigma_2^2$, devise a significance test for the model that $(\sigma_1^2 + \sigma_2^2)/\sigma_1^2 = h_0$ against the alternative that $(\sigma_1^2 + \sigma_2^2)/\sigma_1^2 > h_0$.

18. In a development laboratory of a large company, a test has been developed which is much cheaper to perform than the standard American Society for Testing Materials test. The developers of the test believe that it is measuring the same thing as the ASTM test and have used both in an experiment. Specimens of material were submitted in pairs and the two tests performed with the following results:

| Pair | New | Standard |
|---|---|---|
| 1 | .46 | .32 |
| 2 | .51 | .49 |
| 3 | .58 | .62 |
| 4 | .43 | .38 |
| 5 | .47 | .54 |
| 6 | .65 | .56 |
| 7 | .32 | .28 |
| 8 | .59 | .63 |
| 9 | .87 | .84 |

(a) Use the sign test to determine the significance level for the model that the tests are equivalent.
(b) Use the signed-rank test to determine the significance of the model that the tests are equivalent.

19. Given $x$ successes in $n$ Bernoulli trials, find a uniformly most powerful test of size $\alpha$ for testing

$$H_0: p \leq p_0$$

against

$$H_A: p > p_0$$

20. Given a random sample of size $n$ from the Poisson distribution, find a uniformly most powerful test of size $\alpha$ for testing

$$H_0: \lambda \leq \lambda_0$$

against

$$H_A: \lambda > \lambda_0$$

21. Given a random sample of size $n$ from the normal distribution $N(\mu, \sigma^2)$, find a uniformly most powerful test of size $\alpha$ for testing

$$H_0: \sigma^2 \leq \sigma_0^2$$

against

$$H_A: \sigma^2 > \sigma_0^2$$

22. Given that we use the uniformly most powerful test of size .05 for testing

$$H_0: \mu \leq \mu_0$$

against

$$H_A: \mu > \mu_0$$

for the normal distribution with known $\sigma$, construct the power curve, graphing Prob(rejection) against $(\mu - \mu_0)/\sigma$.

23. The specifications for a cough syrup call for 32 mg of ammonium chloride. If the content exceeds 32 mg by as much as 2 mg, it is desired to detect it with probability .95. If the content does not exceed 32 mg, it is desired to accept the batch with probability .99.
    (a) How many observations should be taken if the standard deviation is no greater than 0.5 mg if we assume normality?
    (b) If the sample of the size indicated by (a) is taken and $\bar{x} = 30.8$, $\sum (x_i - \bar{x})^2 = 40$, should the batch of cough syrup be accepted or rejected?

24. A chemical engineering graduate student has proposed the development of a new process for his research topic. The members of his committee give their approval on the condition that the standard deviation of the process actually is 20% smaller than the existing process standard deviation of 15. He decides to collect some experimental evidence. If the standard deviation is greater than or equal to 15 he wishes to accept the new process with probability no greater than .05. If the standard deviation is as small as 12, he wishes to accept the new process with probability .90.
    (a) How large a sample should he take if he assumes normality?
    (b) Suppose he takes the sample size as dictated by (a) and that $\sum (x_i - \bar{x})^2 = 200$. Should he reject or accept the new process?

25. The manufacturer of a household cleanser consults a statistician. The specifications call for 12% trisodium phosphate. If the concentration in a batch is no greater than 12%, it is desired to accept the batch with probability .95. If the concentration is as high as 15%, it is desired to reject the batch with probability .90. The manufacturer asks how large the sample should be if the standard deviation is believed to be no greater than 1%. Determine the required sample size if we assume normality.

# 13

# Statistical Intervals

## 13.1 Introduction

The specification of a margin of error is a well-established practice in science, engineering, and industry. It is quite common for a researcher to report a constant determined to within plus or minus 10%. Engineering specifications are commonly expressed in this manner, and data sheets supplied by manufacturers often give an interval of values within which the purchaser assumes most devices will fall.

There have been several formulations of this imprecise but very important concept of an interval of values as an expression of uncertainty. In this chapter we present an elementary introduction to this rich and detailed body of literature.

If we are able to incorporate a Bayesian argument as described in Chapter 11, the conclusion is a probability distribution for the unknown parameter, the posterior distribution. This is generally regarded as a complete solution to the inference problem when it is applicable. It leads to the possibility of calculating, for any interval $A$, the probability that $\theta$ lies in $A$. The array of such probability statements can be used to make wagers. Furthermore, the probability statements are unbiased in the sense that they will withstand challenges by an opponent of the person making them. To illustrate this, suppose individual $I$ states that the probability of heads is 1/2. Then an opponent $O$ can challenge $I$ in one of two ways:

1. $O$ says, "I will place \$1 on heads and if the outcome is heads, you are to pay me \$2."
2. $O$ says "I will place \$1 on tails and if the outcome is tails, you are to pay me \$2."

It may be the case that opponent $O$ can make a choice between challenges (1) and (2) for each set of data in an agreed on set of repetitions which has the consequence that he will win *on the average*. If this can be done, it would seem that $I$'s original probability statement is not good. The idea of successfully withstanding such challenges, a requirement of a statement of probability, on the basis of data has been discussed in the literature, particularly by Buehler (1959). It appears to be the case that it is impossible for an opponent to win from a challenger on the *average* if the conditions for validity of the Bayesian argument are completely satisfied.

Because of the absence of a single totally compelling prior distribution for most of the data interpretation problems that arise, the ideas of significance tests and of tests of hypotheses were developed. The former give a measure of weight of evidence against a hypothesized value for a parameter. The latter give an accept-reject rule of such a hypothesized value. We now take up the specification of tenability of sets of points in the parameter space by non-Bayesian arguments.

## 13.2 The Inversion of Significance Tests: Consonance Intervals

Frequently one of our jobs is to form an opinion of the degree of agreement of parameters of a particular distribution with the data. The basic difficulty is to decide what meaning can be attached to the statement: "The data are consonant with a definite model to a particular degree" or to the statement "A certain group of models is consonant with the data." There are elementary situations for which a meaning might be accepted by everyone. If, for instance, we have a model for coin tossing that heads and tails are equally probable and upon examining the coin find that it has two heads, the model is obviously not consonant with our observation. But suppose our model is the same, and we have made 1000 tosses and have observed 408 heads. Suppose also that examination of the sequence of observed results indicates acceptability of the model of independent Bernoulli trials. The data are surely consonant with the particular model that $p$ of the Bernoulli trials is close to 0.41. Also we imagine that the reader would not object to a value for $p$ of 0.40, or even 0.39 or 0.42. The fact is that no value of $p$ except 0 or 1 can be ruled out by the processes of deductive logic, just as a run of 15 heads in coin tossing can happen with a "reasonable" penny and tossing mechanism. We must, therefore, have some means of forming an opinion as to what values of the parameter are plausible in the light of the data.

The ideas of this chapter enable us to make an approach to the matter. Let us take what is perhaps the easiest mathematically because it has been "continuized" and leads to easy continuous mathematics, namely the case of a sample $x_1, x_2, \ldots, x_n$ supposedly from a normal distribution with unknown mean $\mu$ and with variance unity. Suppose we base our test of significance upon $\sqrt{n}(\bar{x} - \mu_0)$ which, if $\mu = \mu_0$, is distributed normally with mean zero and variance unity. As indicated in Chapter 12, the significance level of $\mu_0$ with regard to alternatives $\mu > \mu_0$ is calculated by determining the probability that a standard normal deviate exceeds $\sqrt{n}(\bar{x} - \mu_0)$. To decide what values of $\mu$ are consonant with the data, it is reasonable to determine the hypothesized values for $\mu$ for which the significance level exceeds specified values.

This process will be called inversion of the significance test. Since the level for $\mu$ will be less than .025 if $\sqrt{n}(\bar{x} - \mu) \geq 1.96$, we can say that all values of $\mu$ less than $\bar{x} - 1.96/\sqrt{n}$ would have significance level less than .025. Since large values for the significance level indicate consonance with the data, we therefore say that the interval of $\mu$ values $[\bar{x} - 1.96/\sqrt{n}, \infty)$ is consonant with the data by a one-sided normal deviate test, at the level of significance .025.

We need not be restricted to consideration of the consonant values of $\mu$ at a given level of significance but can and should consider the entire family of consonance intervals obtained by varying the significance level. A convenient way of doing this is to plot the significance level versus $\mu$. To continue with the present example, suppose that $n = 9$ and $\bar{x} = 10$. The significance levels for various values of $\mu$ are shown in Table 13.1.

**Table 13.1**
Significance Levels for Range of $\mu$ Values $\sigma = 1$, $\bar{x} = 10$, $n = 9$

| $\mu$ | $\sqrt{n}(\bar{x} - \mu)$ | $SL = P[Z \geq \sqrt{n}(\bar{x} - \mu)]$ |
|---|---|---|
| 9.0 | 3.0 | .001 |
| 9.1 | 2.7 | .003 |
| 9.2 | 2.4 | .008 |
| 9.3 | 2.1 | .018 |
| 9.4 | 1.8 | .036 |
| 9.5 | 1.5 | .067 |
| 9.6 | 1.2 | .115 |
| 9.7 | 0.9 | .184 |
| 9.8 | 0.6 | .274 |
| 9.9 | 0.3 | .382 |
| 10.0 | 0.0 | .500 |
| 10.1 | −0.3 | .618 |
| 10.2 | −0.6 | .726 |
| 10.3 | −0.9 | .816 |
| 10.4 | −1.2 | .885 |
| 10.5 | −1.5 | .933 |
| 10.6 | −1.8 | .964 |
| 10.7 | −2.1 | .982 |
| 10.8 | −2.4 | .992 |
| 10.9 | −2.7 | .997 |
| 11.0 | −3.0 | .999 |

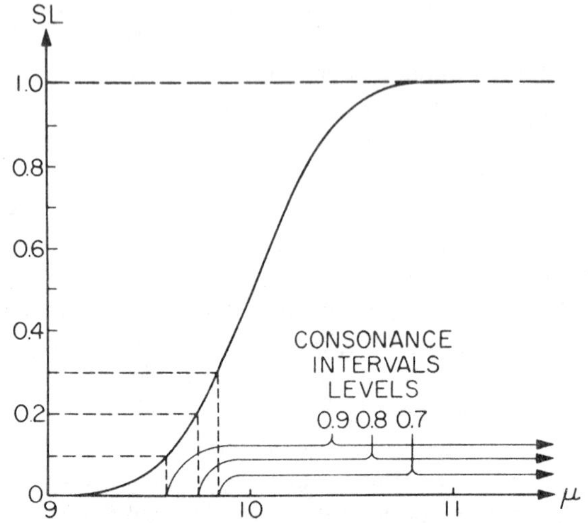


**Fig. 13.1** Family of Consonance Intervals

Figure 13.1 shows the significance level graphed versus the values for $\mu$ and displays the entire family of consonance intervals. That is, it shows the entire family of intervals which are consonant with the data by the one-sided normal deviate test at the possible levels.

In the above situation we could also consider the values of $\mu$ consonant with the data by the two-sided normal deviate test at the $\alpha$ level of significance. Since the significance level in such a case is determined by $P[|Z| > |\sqrt{n}(\bar{x} - \mu_0)|]$, all values of $\mu$ such that $-z_{\alpha/2} \leq \sqrt{n}(\bar{x} - \mu) \leq z_{\alpha/2}$ are consonant with the data at the $\alpha$ level of significance. Thus a consonance interval for $\mu$ at the $\alpha$ level of significance is given by

$$[\bar{x} - z_{\alpha/2}\sqrt{1/n}, \bar{x} + z_{\alpha/2}\sqrt{1/n}]$$

It is again useful to display the entire family of consonance intervals. This is done in Figure 13.2 for the case $n = 9$, $\bar{x} = 10$.

We consider it valid to perform as many significance tests on as many hypothesized values as we like, and the entire family of consonance intervals with a test procedure is obtained by considering all possible significance levels. A family of consonance intervals is based on a particular test of significance. How can we compare two families of consonance intervals? This is a complex subject which we cannot fully discuss here. An elementary answer, however, is to compare the sensitivities of the two tests which by inversion lead to the intervals.

We have illustrated the inversion of a significance test in the case of a test of significance for the mean of a normal distribution. In general we will have a test of significance for the hypothesis $\theta_1 = \theta_{10}$, $\theta_2 = \theta_{20}$,

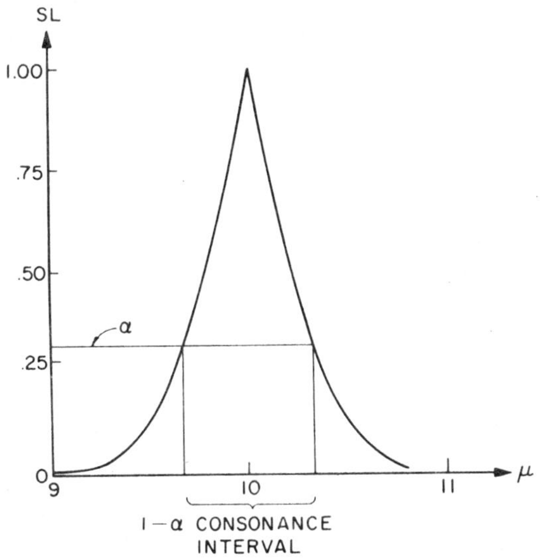


**Fig. 13.2** Family of Consonance Intervals: Two-sided Standard Normal Deviate Test

$\theta_3 = \theta_{30}, \ldots, \theta_k = \theta_{k0}$ and the consonance region for the $\alpha$ level of significance is the region in the space of $\theta$'s for which the significance level is greater than or equal to $\alpha$. We shall now give some further examples.

Consider the test of significance for the binomial model of the hypothesis $p = p_0$ against the alternative $p > p_0$ given by $SL = P(X \geq x_O)$, where $x_O$ is the observed number of successes. We want to determine all values of $p$ that give a higher value than $\alpha$. That is, we want to find all values of $p$ such that

$$A = \sum_{x=x_O}^{n} \binom{n}{x} p^x (1-p)^{n-x}$$

is greater than or equal to $\alpha$. Now $A$ is a continuous function of $p$. Further, we find the derivative with respect to $p$ to be

$$\frac{dA}{dp} = \sum_{x=x_O}^{n} \binom{n}{x} (x - np) p^{x-1} (1-p)^{n-x-1} \tag{13.1}$$

This is positive, and therefore $A$ is an increasing function of $p$. So the values of $p$ consonant with $x_O$ at the $\alpha$ level of significance are all values such that $p_1 < p \leq 1$ where

$$\sum_{x=x_O}^{n} \binom{n}{x} p_1^x (1-p_1)^{n-x} = \alpha \tag{13.2}$$

We can solve for $p_1$ explicitly because

$$\sum_{x=x_O}^{n} \binom{n}{x} p_1^x (1 - p_1)^{n-x} = \int_0^{p_1} \frac{n!}{(x_O - 1)!(n - x_O)!} y^{x_O - 1}(1 - y)^{n - x_O}\, dy \tag{13.3}$$

This can be proved by repeated integration by parts. Then by transforming the beta variable to an $F$ variable, we find

$$p_1 = \frac{x_O}{x_O + (n - x_O + 1)F_{\alpha;\, 2(n - x_O + 1),\, 2x_O}} \tag{13.4}$$

where $F_{\alpha;m,n}$ is the value exceeded with probability $\alpha$ for the $F$ distribution with $m$ and $n$ degrees of freedom.

A common way of testing significance of a binomial parameter being equal to $p_0$ against the alternative that $p \neq p_0$ is to take the significance level $SL(p_0)$ as $P[|X - np_0| \geq |x_O - np_0|]$.

To obtain a set of consonance intervals on $p$ given the observation of $x_O$ successes, we can determine the values of $p$ such that $SL(p) \geq \alpha$ for $0 \leq \alpha \leq 1$. Because of the normal approximation to the binomial, an approximation is to calculate $SL(p_0)$ by

$$P[|Z| \geq |x_O - np_0|/\sqrt{np_0(1 - p_0)}]$$

where $Z$ is a standard normal variate. This leads easily to an inversion. The values of $p$ such that $SL(p) \geq \alpha$ are the values in the interval

$$[p_L(x_O, \alpha), p_U(x_O, \alpha)]$$

where $p_L$ and $p_U$ are the roots of

$$(x_O - np)^2/np(1 - p) = z_\alpha^2$$

where $z_\alpha$ is the value of a standard normal variable exceeded with probability $\alpha$. A simpler approximation is to take $p_L$ and $p_U$ as given by

$$p^* - z_{\alpha/2}\sqrt{\frac{p^*(1 - p^*)}{n}}$$

and

$$p^* + z_{\alpha/2}\sqrt{\frac{p^*(1 - p^*)}{n}}$$

respectively, where $p^* = x_O/n$. Numerous tables and graphs are given for two-sided intervals on $p$ [e.g., Dixon and Massey (1969)]. See also Clopper and Pearson (1934).

Suppose that we have a sample $x_1, x_2, \ldots, x_m$ from a normal population with mean $\mu_x$ and variance $\sigma^2$, and a sample $y_1, y_2, \ldots, y_n$ from a normal population with mean $\mu_y$ and variance $\sigma^2$. Recall now some distributional facts:

1. $\sqrt{m}(\bar{x} - \mu_x)/\sigma$ is $N(0, 1)$.
2. $\sqrt{n}(\bar{y} - \mu_y)/\sigma$ is $N(0, 1)$.
3. $\sum (x - \bar{x})^2/\sigma^2$ and $\sum (y - \bar{y})^2/\sigma^2$ are distributed as $\chi^2$ with $m - 1$ and $n - 1$ degrees of freedom respectively and are independent, so that by the additive property of independent $\chi^2$ variables
$$\left[\sum (x - \bar{x})^2 + \sum (y - \bar{y})^2\right]/\sigma^2$$
is distributed as $\chi^2$ with $m + n - 2$ degrees of freedom.
4. The random variables (1), (2), and the last one in (3) are independent.
5. $[m(\bar{x} - \mu_x)^2 + n(\bar{y} - \mu_y)^2]/\sigma^2$ is distributed as $\chi^2$ with 2 degrees of freedom.
6. Because of the independence,
$$\left[\frac{m(\bar{x} - \mu_x)^2 + n(\bar{y} - \mu_y)^2}{2}\right] \bigg/ \left[\frac{\sum(x - \bar{x})^2 + \sum(y - \bar{y})^2}{m + n - 2}\right]$$
is distributed as $F$ with 2 and $m + n - 2$ degrees of freedom.

Hence we can use this last quantity for a test of significance about $\mu_x$ and $\mu_y$. We calculate the significance level for $\mu_x$ and $\mu_y$ as follows:

$$SL = P\left\{F_{2,\,m+n-2} \geq \frac{(m + n - 2)[m(\bar{x} - \mu_x)^2 + n(\bar{y} - \mu_y)^2]}{2[\sum(x - \bar{x})^2 + \sum(y - \bar{y})^2]}\right\} \quad (13.5)$$

Then the values of $\mu_x$ and $\mu_y$ for which the significance level exceeds a specified value, $\alpha$ say, are given by the values for which

$$\frac{(m + n - 2)[m(\bar{x} - \mu_x)^2 + n(\bar{y} - \mu_y)^2]}{2[\sum(x - \bar{x})^2 + \sum(y - \bar{y})^2]} \leq F_{\alpha;\,2,\,m+n-2} \quad (13.6)$$

This is equivalent to saying that the consonance region for $\mu_x$ and $\mu_y$ at the level of significance consists of the points on or inside the ellipse

$$\frac{(\mu_x - \bar{x})^2}{1/m} + \frac{(\mu_y - \bar{y})^2}{1/n} = \frac{2F_{\alpha;\,2,m+n-2}[\sum(x - \bar{x})^2 + \sum(y - \bar{y})^2]}{m + n - 2} \quad (13.7)$$

Suppose we wish to form an opinion about what the value of $\mu_x - \mu_y$ might be. Let us note that $\bar{x} - \mu_x$ is $N(0, \sigma^2/m)$ and $\bar{y} - \mu_y$ is $N(0, \sigma^2/n)$

and these are independent. Hence

$$\frac{(\bar{x} - \bar{y}) - (\mu_x - \mu_y)}{\left[\left(\frac{1}{m} + \frac{1}{n}\right) \frac{\sum(x - \bar{x})^2 + \sum(y - \bar{y})^2}{m + n - 2}\right]^{1/2}}$$

is distributed as Student's $t$. We can therefore evaluate the significance level for any hypothesized value of $\mu_x - \mu_y$ by the $t$ test. Also we can invert this test to give a family of consonance intervals just as in the case of one mean $\mu$. We obtain the symmetrical intervals, for instance

$$[\bar{x} - \bar{y} - t_{\alpha/2} SE(\bar{x} - \bar{y}), \bar{x} - \bar{y} + t_{\alpha/2} SE(\bar{x} - \bar{y})]$$

where

$$SE(\bar{x} - \bar{y}) = \left[\left(\frac{1}{m} + \frac{1}{n}\right) \frac{\sum(x - \bar{x})^2 + \sum(y - \bar{y})^2}{m + n - 2}\right]^{1/2} \quad (13.8)$$

the notation $SE$ denoting "standard error." By varying values of $\alpha$, this gives us a family of consonance intervals for the unknown quantity $\mu_x - \mu_y$, and this family enables us to form an opinion about what values of $\mu_x - \mu_y$ are consonant with the data. Of course "consonant with the data" means, by virtue of the derivation, "consonant with the data by the particular form of the test which is inverted." In general, the consonance interval of level $\alpha$ for a parameter $\theta$ is the set of values of $\theta$ for which given the data and the test of significance $SL(\theta) \geq \alpha$.

## 13.3 The Inversion of Nonrandomized Tests of Hypotheses: Confidence Intervals

In the preceding section we developed a method of giving operational meaning to the question, What class of models is consonant with the data? The method consisted of determining the set of parameter values for which a given significance test gives levels in excess of $\alpha$ for a given set of data. We recommended that this be done for all achievable significance levels of the particular test being used. This provides a useful way of quantifying the agreement between data and model. Furthermore, the usefulness does not hinge upon any assumption or belief on our part that the class of models is the true class (in the sense of an actual data-generating mechanism). Suppose, however, that we are willing to make this assumption (except for the usual problem of assuming a continuous model to describe discrete measurement data). Having obtained a consonance interval for a given level of significance, we might well ask how likely it is that the interval we have obtained contains the true parameter value. The way to proceed, it seems, is to suppose that we repeatedly sample from the same distribution and determine an interval for each sample. This process will generate a population of intervals. The interval corresponding to a random sample is then a random interval with at least one of its end points a random variable.

Consider the example developed in Section 13.2 for a random sample from a population approximately normally distributed with unknown mean $\mu$ and variance one. The consonance interval of $\mu$ values at the .025 level of significance was $[\bar{x} - 1.96/\sqrt{n}, \infty)$. We wish to determine

$$P_\mu(\bar{X} - 1.96/\sqrt{n} \leq \mu) = P_\mu\left(\frac{\bar{X} - \mu}{1/\sqrt{n}} \leq 1.96\right) = P(Z \leq 1.96)$$

where $Z$ is a normal variable with mean zero and variance unity. So in repeated sampling the probability that we will obtain a consonance interval by this test at the .025 level which contains $\mu$ is .975. The reader may quickly surmise that this result can be generalized, and we shall state a general theorem in this section.

For the present, however, let us ask what our attitude should be toward the interval obtained from a given sample which, before the sample was obtained, had a probability of .975 of containing $\mu$. Obviously we are more confident that the given interval contains $\mu$ than if we were unaware of the probability statement. However, "confidence" is not a quantity which is ordinarily expressed on a quantitative scale. Thus its use as it occurs in statistical literature requires some explanation. It is quite common in the present example to refer to the interval $[\bar{x} - 1.96/\sqrt{n}, \infty)$ as a .975 confidence interval on $\mu$. Since as we have already said, confidence is not ordinarily measured quantitatively, this is essentially a definition of .975 confidence. We now give the following definition.

*Definition 13.1. A $(1 - \alpha)$ confidence interval on the parameter $\theta$ is an interval obtained by a procedure which has probability $(1 - \alpha)$ of giving an interval containing $\theta$.*

Will the consonance interval for significance level $\alpha$ always be a $(1 - \alpha)$ confidence interval? The answer, as the reader may surmise, is in the affirmative. We state this result as a theorem.

***Theorem 13.1*** A consonance region for a given significance test at the $\alpha$ level of significance is a $(1 - \alpha)$ confidence region.

*Proof:* The proof is almost immediate. The parameter $\theta$ will be in the consonance region if the significance level for $\theta \geq \alpha$. But

$$\begin{aligned} P_\theta[SL(\theta) \geq \alpha] &= 1 - P_\theta[SL(\theta) \leq \alpha] \\ &= (1 - \alpha) \end{aligned} \tag{13.9}$$

when $\alpha$ is an achievable level because of the fundamental property of significance tests.

In the previous chapter we stated the close connection between tests of hypotheses and significance tests and that a test of size $\alpha$ is given by rejecting if $SL \leq \alpha$, otherwise accepting. This simple correspondence allows us to state a means of obtaining $(1 - \alpha)$ confidence regions from size $\alpha$ tests. We state the following corollary without proof.

*Corollary 13.1* For a given sample a $(1 - \alpha)$ confidence region is given by the parameter values which would be accepted by a size $\alpha$ accept-reject rule.

The statement in this corollary represents the more usual presentation of confidence intervals. For this reason, we have referred to confidence intervals, in the title of this section, as the inversion of tests of hypotheses.

When the viewpoint that we are dealing with the appropriate class of models is accepted, the idea of talking about the prior probability that a random interval will contain the parameter value seems perfectly reasonable. Very quickly, however, we face a fundamental difficulty. We can construct an infinity of different $(1 - \alpha)$ intervals by inverting different tests and even using the same test. Other properties must be introduced. When we are considering bounded intervals, the length has some appeal; since we are considering random intervals the length may be a random variable. Then, subject to the requirement that the prior probability of containing the parameter exceeds a given number, we may attempt to minimize the expected length of the interval. When we consider one-sided intervals, that is, intervals bounded only from one side, for example $[\bar{x} - 1.96/\sqrt{n}, \infty)$, length of the interval is no longer a defined quantity, so we must look for some other property of interest.

A question of some interest is, With what probability does a random interval contain false values for the parameter? It seems reasonable that we would desire to contain false values with lower probability than the true value. In general such an interval will be said to be unbiased. However, we must distinguish between one-sided and two-sided intervals.

*Definition 13.2.* $\underline{\theta}$ *is a* $(1 - \alpha)$ *unbiased lower bound for* $\theta$ *if*

$$P_\theta[\underline{\theta} \leq \theta] = (1 - \alpha) \tag{13.10}$$

*and*

$$P_\theta[\underline{\theta} \leq \theta'] \leq (1 - \alpha) \text{ for } \theta' < \theta \tag{13.11}$$

In this case the interval $[\underline{\theta}, \infty)$ will obviously contain $\theta'' > \theta$ if it contains $\theta$. Therefore we concern ourselves with the probability of containing false values less than $\theta$. A corresponding definition holds for upper bounds. In the case of a bounded interval, we can state a definition which is closer to the intuitive concept.

*Definition 13.3.* $[\underline{\theta}, \bar{\theta}]$ *is a* $(1 - \alpha)$ *unbiased confidence interval for* $\theta$ *if*

$$P_\theta[\underline{\theta} \leq \theta \leq \bar{\theta}] = 1 - \alpha \tag{13.12}$$

*and*

$$P_\theta[\underline{\theta} \leq \theta' \leq \bar{\theta}] \leq 1 - \alpha \text{ for } \theta' \neq \theta \tag{13.13}$$

Not only does it seem reasonable that the probability of containing false values should be less than the probability of containing the true value but the idea of making the probability of containing false values as small as possible also seems compelling. Such intervals are called most accurate.

*Definition 13.4.* $\underline{\theta}$ *is a* $(1 - \alpha)$ *most accurate lower bound for* $\theta$ *if*

$$P_\theta[\underline{\theta} \leq \theta] = 1 - \alpha \tag{13.14}$$

*and*

$$P_\theta[\underline{\theta} \leq \theta'] \leq P_\theta[\underline{\theta}^* \leq \theta'] \tag{13.15}$$

*where* $\theta' < \theta$ *and* $\underline{\theta}^*$ *is any other* $(1 - \alpha)$ *lower bound for* $\theta$.

*Definition 13.5.* $[\underline{\theta}, \bar{\theta}]$ *is a* $(1 - \alpha)$ *most accurate confidence interval for* $\theta$ *if*

$$P_\theta[\underline{\theta} \leq \theta \leq \bar{\theta}] = 1 - \alpha \tag{13.16}$$

*and*

$$P_\theta[\underline{\theta} \leq \theta' \leq \bar{\theta}] \leq P_\theta[\underline{\theta}^* \leq \theta' \leq \bar{\theta}^*] \tag{13.17}$$

*where* $\theta' \neq \theta$ *and* $[\underline{\theta}^*, \bar{\theta}^*]$ *is any other* $(1 - \alpha)$ *confidence interval for* $\theta$.

How can we obtain unbiased confidence intervals? It should not be surprising that we obtain them from the inversion of unbiased tests. Also, we obtain most accurate confidence intervals from the inversion of most powerful tests. The theory is complex and we refer the reader to Lehmann (1959).

Consider the problem of finding an upper confidence bound for the binomial parameter $p, p \neq 0$, i.e., a confidence interval of the form $[0, \bar{p}]$. A $(1 - \alpha)$ confidence interval of this form is obtained by taking $\bar{p} = 1$ with probability $(1 - \alpha)$ and $\bar{p} = 0$ with probability $\alpha$. Then for all $p$ other than $p = 0$

$$P_p[p \leq \bar{p}] = 1 - \alpha \tag{13.18}$$

and for $p = 0$

$$P_p[p \leq \bar{p}] = 1 \tag{13.19}$$

In terms of the concepts of consonance intervals this procedure, which ignores the data, provides no information about parameter values consonant with the data. From the standpoint of confidence interval theory one would suspect that the procedure is quite poor in that it yields intervals containing false parameter values with high probability.

Lehmann (1959) presents a uniformly most accurate procedure for obtaining a $(1 - \alpha)$ upper confidence bound. Given the binomial random variable $X$, an extraneous variable $U$ uniform on $[0, 1]$ is observed independently, and the sum $T = X + U$ is formed. Then the upper bound $\bar{p}$ is the solution, if it exists, of the equation

$$P_p[T \leq t] = \alpha \tag{13.20}$$

where $t$ is an observed value for $T$. In the cases where a solution does not exist, $\bar{p}$ is taken to be either 1 or 0. For example, if $X = 0$, $U = u < \alpha$

$$\begin{aligned} P_p(X + U \leq u) &= P_p(X = 0)P(U \leq u) \\ &= (1 - p)^n u \end{aligned} \tag{13.21}$$

For no $p$ is this equal to $\alpha$. In this case we take $\bar{p} = 0$. If $u \geq \alpha$, we solve for $\bar{p}$ to obtain

$$\bar{p} = 1 - \left(\frac{\alpha}{u}\right)^{1/n} \tag{13.22}$$

Despite the fact that this procedure yields a uniformly most accurate upper confidence bound on $p$, we surmise that there would often be a reluctance to introduce an extraneous randomization.

## 13.4 A Possible Method for Obtaining Confidence Intervals

The idea of confidence intervals was originated by Neyman (1935) at a time when there was deep confusion about the nature of fiducial intervals, discussed in Section 13.6. The following procedure was presented. Let $\theta^*$ be a statistic with distribution depending only on $\theta$, and suppose we can find functions $u(\theta)$ and $l(\theta)$ such that

$$\begin{aligned} P[\theta^* \geq u(\theta); \theta] &= \gamma_1 \\ P[\theta^* \leq l(\theta); \theta] &= \gamma_2 \end{aligned} \tag{13.23}$$

Here $u(\theta)$ and $l(\theta)$ depend on $\theta$ because the distribution of $\theta^*$ depends on $\theta$. Now consider the graph of $u(\theta)$ and $l(\theta)$ plotted against $\theta$ given in Figure 13.3. Then

$$P[l(\theta) \leq \theta^* \leq u(\theta); \theta] = 1 - \gamma_1 - \gamma_2 \tag{13.24}$$

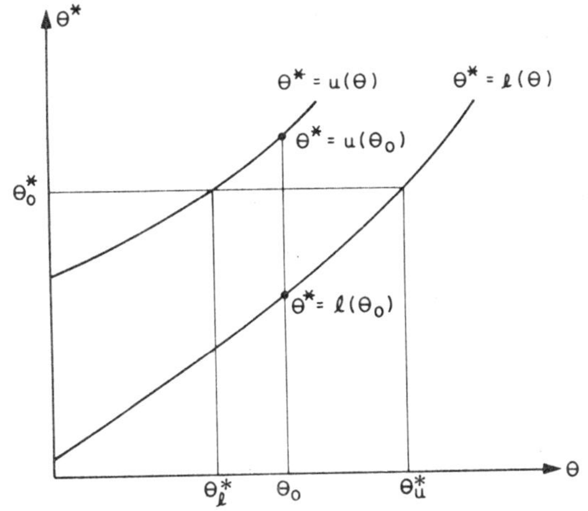


**Fig. 13.3** Graph Underlying Procedure for Obtaining a Confidence Interval

provided $\gamma_1 + \gamma_2 \leq 1$. Let $\theta_0^*$ be the observed value for $\theta^*$. Draw the horizontal line $\theta^* = \theta_0^*$ and let the abscissa corresponding to the intersection of this line with the curves $\theta^* = u(\theta)$ and $\theta^* = l(\theta)$ be $\theta_l^*$ and $\theta_u^*$ respectively. Note the inversion of subscripts here. In repetitions of the sampling process $\theta_0^*$ varies and hence $\theta_l^*$ and $\theta_u^*$ vary, so $\theta_l^*$ and $\theta_u^*$ are random variables. Now consider the statement

$$\theta_l^* \leq \theta \leq \theta_u^* \tag{13.25}$$

In a single repetition of the sampling this statement is either true or false. Consider, however, the family of such statements generated by repetitions of the sampling. It is natural to ask what proportion of the statements in the family is correct. The family of statements (13.25) is logically equivalent to the statement that the point $(\theta^*, \theta)$ lies in the region bounded by the curves $\theta^* = u(\theta)$ and $\theta^* = l(\theta)$. But the relative frequency of correctness of the latter statement is $1 - \gamma_1 - \gamma_2$ in repetitions of sampling with any fixed $\theta$. Alternatively, if we are sampling from $\theta_0$ the resulting interval contains $\theta_0$ if $\theta^*$ lies between $l(\theta_0)$ and $u(\theta_0)$ and does not contain $\theta_0$ if $\theta^*$ falls outside this interval. So the relative frequency with which statement (13.25) is true is $1 - \gamma_1 - \gamma_2$.

Figure 13.3 has been drawn for the case in which $l(\theta)$ and $u(\theta)$ are monotonic in $\theta$. If this property does not hold, the procedure will give more than one interval and may give infinite intervals or the whole real line.

It is very important to note that the relative frequency of correctness is based on the distribution of $\theta^*$. With particular sets of data $D$ say, we may be able to find two statistics $\theta^* = \theta^*(D)$ and $\phi^* = \phi^*(D)$ and then apply the procedure to give $\theta_l^*$ and $\theta_u^*$ such that

$$P(\theta_l^* \leq \theta \leq \theta_u^*) = 1 - \gamma_1 - \gamma_2$$

but also find a set, say $C$, of the real line such that $P(\theta_l^* \leq \theta \leq \theta_u^* | \phi^* \text{ is in } C) \neq 1 - \gamma_1 - \gamma_2$ identically in $\theta$. If such a statistic $\phi$ and a set $C$ can be determined, it is not reasonable to ignore the fact and to attach confidence $1 - \gamma_1 - \gamma_2$ to the statement (13.25). [See Buehler (1959) and Buehler and Feddersen (1963).]

The properties of the procedure if applicable in relation to Definitions 13.1 to 13.5 have to be examined in each individual case. The procedure has great appeal if $\theta^*$ is sufficient for $\theta$, because then $\theta^*$ contains "all the information in the data on $\theta$." This suggests further that one should use the maximum likelihood statistic for $\theta^*$. However, the matter of conditioning as in the use of a supplementary statistic $\phi$ is very obscure.

We may note that the above procedure if applicable is generally used with $\gamma_1 = \gamma_2$. Also we note that if $\gamma_1$ is taken to be zero, the statistic $\theta_u^*$ can be taken as an upper bound for $\theta$, while if $\gamma_2$ is taken as zero, the statistic $\theta_l^*$ can be taken as a lower bound on $\theta$.

It should be noted that if the probability model is discrete, it will be impossible to draw Figure 13.3 in exactly the form presented. The problem is that any statistic $\theta^*$ will have a discrete set of achievable values, and one will not be able to find $\theta^*$, $u(\theta)$, and $l(\theta)$ such that the required properties hold. The purpose of the introduction of the extraneous variable $U$ in the previous section is precisely to allow probabilities to be achieved exactly. The difficulties of constructing confidence intervals with a discrete model are presented by Neyman (1935).

We may note finally that the theory of confidence intervals for a vector parameter is very obscure. The commonly used procedure is that of using Theorem 13.1 and inverting a test of significance.

## 13.5 Some Comments on Intervals

We have made a distinction between intervals obtained by a test of significance and intervals obtained by a test of hypothesis. The example of the previous section is compelling, we think. If one wishes to have a rule of constructing a *single* interval for each set of data that is envisaged as possible under the probability model which has the desired properties, the solution is unquestionable. If, however, one wishes to use the computed interval to form an opinion as to what values of $\theta$ are tenable on the basis of the data, the introduction of the observed value of an extraneous random variable is not appealing. We are of the opinion that an investigator would object strongly to the process. The process is convincing only as an accept-reject rule. But even in this respect the imposition of the requirement of being correct in $(1 - \alpha)$ of the repetitions and incorrect in $\alpha$ of the repetitions may be questioned. It is our opinion, then, that randomized confidence intervals are not relevant to forming ideas of tenability of parameter values in the light of the data.

We now take up another aspect of statistical intervals. If these intervals are to give us an ordering of tenability of parameter values, it seems essential to require the following:

1. Values in the center of the interval should be more tenable than values at the edges (we use terms loosely here).
2. Values in a 95% interval should be in a 99% interval, say; and more generally if the $(1 - \alpha_1)$ interval with data $D$ is $I(D, 1 - \alpha_1)$, then $I(D, 1 - \alpha_1)$ is a proper subset of $I(D, 1 - \alpha_2)$ if $1 - \alpha_1 < 1 - \alpha_2$.

Requirement (2) is satisfied by many of the rules for constructing intervals. However, a general basis for constructing a rule must be questioned if in particular cases it fails to have a property we desire. Therefore we present the following example for which the confidence theory was developed by Welch (1939). The problem is that we have two observations $x_1, x_2$ assumed to be independent realizations from the uniform distribution over the interval described by $\theta \pm 1/2$. Welch (1937) obtains the following $(1 - \alpha)$ confidence interval for $\theta$:

$$
\begin{aligned}
I_1(1-\alpha)& \\
&= [x_G - 1/2, x_L + 1/2] \qquad \text{if } x_G - x_L > \sqrt{\alpha/2} \\
&= [x_L - 1/2 + \sqrt{\alpha/2}, x_G + 1/2 - \sqrt{\alpha/2}] \qquad \text{if } x_G - x_L \le \sqrt{\alpha/2}
\end{aligned}
\tag{13.26}
$$

where $x_L = \min(x_1, x_2)$ and $x_G = \max(x_1, x_2)$.

Suppose that we observe $x_1 = 1.9$, $x_2 = 2.4$ and compute intervals for some values of $(1 - \alpha)$. We get

| Confidence $(1 - \alpha)$ | Interval |
|---|---|
| .95 | 1.9 to 2.4 |
| .90 | 1.9 to 2.4 |
| .80 | 1.9 to 2.4 |
| .70 | 1.9 to 2.4 |

The system of intervals does not satisfy property (2) above and appears to be ineffective as an ordering of tenability of values for $\theta$ for *the given set of data*.

This example is interesting in that Fisher's idea of basing a test of significance on the conditional distribution of the data given the observed value of an ancillary statistic can be applied. The reader may verify that a minimal sufficient statistic is $X_1 - X_2, X_1 + X_2$ and that $X_1 - X_2$ is ancillary because its distribution does not depend on $\theta$. Welch shows that the Fisherian process leads to a test of significance conditional on $|x_1 - x_2|$ and to the system of intervals

$$I_2(1 - \alpha) = [z_1 - (1 - \alpha)(1/2 - z_2), z_1 + (1 - \alpha)(1/2 - z_2)] \quad (13.27)$$

where $z_1 = (x_1 + x_2)/2$ and $z_2 = |x_1 - x_2|/2$.

This set of intervals has property (2), and in our opinion is better for quantifying the tenability of $\theta$ values in the light of the data. Welch proves that the system $I_1(1 - \alpha)$ has a lower probability of containing an untrue value of $\theta$ than the system $I_2(1 - \alpha)$. He concludes that the Fisherian process is poor; we differ. The reader must form his own opinion.

## 13.6 Fiducial Intervals

Any discussion of statistical intervals would be incomplete without some mention of fiducial intervals because the subject occupies an important place in the history of statistics. The fiducial argument lies at the heart of one of the major controversies of statistics, and the distinction between confidence intervals and fiducial intervals is an item of great confusion.

Although the fiducial argument was presented vigorously by Fisher, many of the ideas remain obscure. In a formal manner, however, we can present the basic idea at least. Consider normal distribution theory with the variance known. Then for a sample of size $n$,

$$P\left(\frac{\bar{X} - \mu}{\sigma/\sqrt{n}} \leq z_\alpha\right) = 1 - \alpha \quad (13.28)$$

The fiducial argument consists of taking this as a probability statement about $\mu$ as a random variable with $\bar{X}$ fixed at its observed value. If we adopt this viewpoint, we can then obtain a distribution function and density function for $\mu$. Subscripting $P$ with $F$ to indicate that the probabilities are based upon the fiducial argument, we have

$$P_F\left[\frac{\sqrt{n}(\bar{X} - \mu)}{\sigma} \geq k\right] = 1 - F_Z(k) \quad (13.29)$$

where $F_Z$ is the cumulative distribution function for a standard normal variable. Then

$$P_F\left(\mu \leq \bar{X} - k\frac{\sigma}{\sqrt{n}}\right) = 1 - F_Z(k) \tag{13.30}$$

$$P_F(\mu \leq c) = 1 - F_Z\left(\frac{\bar{X} - c}{\sigma/\sqrt{n}}\right) \tag{13.31}$$

The last statement gives the fiducial distribution function for $\mu$ which we denote by $G_\mu$. Thus

$$G_\mu(c) = \int_{(\bar{x}-c)/(\sigma/\sqrt{n})}^{\infty} \frac{1}{\sqrt{2\pi}} e^{-y^2/2}\, dy \tag{13.32}$$

Using Leibnitz's rule of differentiation, we differentiate $G_\mu(c)$ with respect to $c$ to find the fiducial density function of $\mu$. Thus

$$g_\mu(c) = \frac{d}{dc} G_\mu(c) = \frac{\sqrt{n}}{\sigma\sqrt{2\pi}} \exp\left[\frac{-n(c - \bar{x})^2}{2\sigma^2}\right] \tag{13.33}$$

We have arrived at a probability density function on $\mu$ which is normal with mean $\bar{x}$ and variance $\sigma^2/n$. Then an interval symmetric about $\bar{x}$ containing $(1 - \alpha)$ of the distribution has end-points given by

$$\bar{x} \pm z_{\alpha/2}\sigma/\sqrt{n} \tag{13.34}$$

which is exactly the same interval we would have obtained from the consonance interval perspective or from the confidence interval viewpoint. So even if one had reservations about the fiducial argument, one would be inclined to accept the result in this simple case because it agrees with results obtained from other arguments. For a time it was believed that fiducial intervals and confidence intervals were the same and the names were used interchangeably. The reader may note the similarity of the procedure in Section 13.4 to that used in Equations (13.30) and (13.31) to see how confusion can arise. In some simple cases the distinction is partly one of semantics. Fisher insisted on the term "fiducial probability," while Neyman objected (correctly, it is felt) to the use of the word probability. Even today the expression fiducial interval is sometimes used to describe a confidence interval.

However, different results were obtained by these two methods for the celebrated Behrens-Fisher problem. We will not trace the argument but wish to inform the reader of the role of the problem in this important controversy. The Behrens-Fisher problem is concerned with the means of two normal distributions with unknown and unequal variances. The fiducial argument yields different results than does the Neyman-Pearson theory.

What attitude should one have toward a distribution on a parameter such as $\mu$? The attitude of statisticians trained in the classic frequentist viewpoint of probability has been a reluctance to accept such a concept. In more recent years, however, the Bayesian viewpoint has been more popular, and the idea of a distribution on $\mu$ has been more acceptable. However, most of those who accept the Bayesian viewpoint find the fiducial argument an unacceptable way of arriving at a probability distribution on $\mu$. Instead, they argue that one should start with a prior distribution on $\mu$, $g(\mu)$ and invoke Bayes's theorem to obtain the posterior distribution on $\mu$, i.e., the conditional distribution of $\mu$ given the sample, $g(\mu|D)$. Fisher completely rejected this approach and staunchly defended the fiducial argument as a means of obtaining a distribution on parameter values without the use of a prior distribution or the use of Bayes' theorem. The controversies in this general area are not at all resolved and will undoubtedly continue for decades.

It is appropriate, we think, to give the reader our views:

1. If one has a history of phenomena which one judges to be like the phenomenon under study, one may be able to make a guess of a prior distribution based on this history, and a computation of a posterior distribution using this guess is interesting.
2. If a fiducial distribution is merely a restatement of a test of significance, we see no need for it.
3. If one could use a fiducial distribution as a prior distribution for interpretation of subsequent data, we might judge it to be useful; but Lindley (1958) showed that inconsistencies will occur except in special cases.
4. There seem to be basic mathematical difficulties of a fiducial development in the case of a vector parameter.
5. Also, there seem to be basic difficulties with a Bayesian argument in this case since a natural prior gives results which seem to conflict with common sense [Geisser and Cornfield (1963)].
6. As a consequence we are of the opinion that there is no unique way of summarizing the inferential content of data, and one informative mode of attack is to apply significance tests which seem to be appropriate and to consider inversions of these; but it is a mistake to suppose there is *one* best test.

Finally we refer the reader to Fisher (1956) and to Hacking (1965) for discussion of the fiducial argument, to Fraser (1968) for a treatment which seems very close but attempts to avoid the fiducial label, and to Neyman (1941) for a criticism of fiducial intervals.

## 13.7 Statistical Tolerance Intervals

A very common situation encountered in manufacturing is that of trying to manufacture items in conformity with specification limits established by a design engineer. Professional pride may lead to the claim that all items meet the specification limits; however, a realistic process engineer knows that some items will not, and he may make a more conservative statement that most, say 90%, of the items are within "specs." A dubious customer may ask for the basis of such a statement.

The theory of statistical tolerance intervals is relevant to problems of this type. In the manufacturing problem the specification limits are fixed, and we would be interested in the fraction of items in the population which meet "specs." The approach of statistical tolerance intervals on the other hand is to calculate a random interval $[L(D), U(D)]$ where $L(D)$ and $U(D)$ are random variables depending upon the sample. The fraction of the population contained in such a random interval is a random variable which we shall denote by $C$ (for content). Given a scheme for calculating a random interval, we can consider the probability distribution of $C$ as in Figure 13.4.

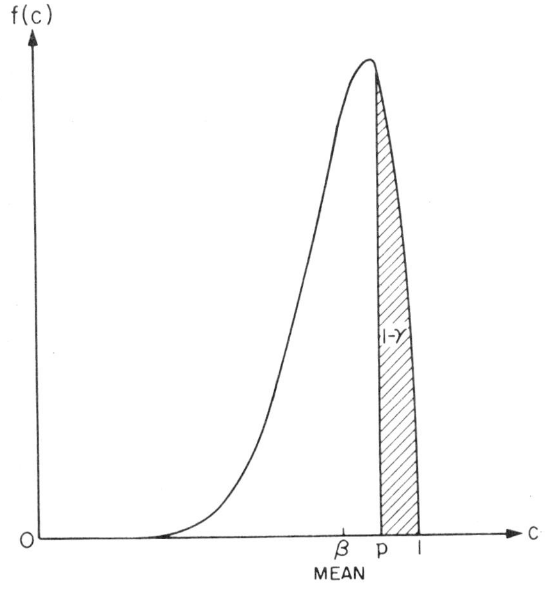


**Fig. 13.4** Distribution of Content

The theory of statistical tolerance intervals has developed along two lines:

1. Development of procedures with a specified mean content, i.e., $E(C) = \beta$.
2. Development of procedures with a specified $1 - \gamma$ percentile $p$, i.e., $P(C \geq p) \geq 1 - \gamma$.

These intervals are called $\beta$-expectation tolerance intervals and $(1 - \gamma)$ confidence $p$-content tolerance intervals respectively. Both types of intervals play prominent roles in literature on statistical tolerance intervals and we will discuss each briefly.

For normal distribution theory the factor $K$ has been calculated so that the probability that the content contained in the interval

$$[\bar{X} - Ks, \bar{X} + Ks]$$

exceeds $p$ is $(1 - \gamma)$. See Bowker and Lieberman (1959) for tables. Tables are also given for one-sided intervals; i.e., intervals of the form $(-\infty, \bar{X} + Ks]$ and $[\bar{X} - Ks, \infty)$. In the case of normal distribution theory, $(1 - \alpha)$ expectation tolerance intervals are given by

$$\bar{X} \pm t_{\alpha/2} s \sqrt{\frac{n + 1}{n}}$$

This is a consequence of the following theorem which we state without proof.

***Theorem 13.2*** A $(1 - \alpha)$ confidence interval for $X_{n+1}$ based on $X_1, X_2, \ldots, X_n$ is a $(1 - \alpha)$ expectation tolerance interval.

As an application of this theorem, let us consider $n + 1$ observations from a normal distribution with $\bar{X}$ and $s^2$ computed from the first $n$ observations. Then $X_{n+1} - \bar{X}$ is a normal variable with mean zero and variance $\sigma^2(n + 1)/n$. Also it is independent of $s^2$. So

$$\frac{X_{n+1} - \bar{X}}{s\sqrt{\dfrac{n + 1}{n}}} \tag{13.35}$$

is a Student's $t$ variable with $n - 1$ degrees of freedom. So

$$P\left(-t_{\alpha/2} \leq \frac{X_{n+1} - \bar{X}}{s\sqrt{\dfrac{n + 1}{n}}} \leq t_{\alpha/2}\right) = 1 - \alpha$$

$$= P\left(\bar{X} - t_{\alpha/2} s\sqrt{\frac{n + 1}{n}} \leq X_{n+1} \leq \bar{X} + t_{\alpha/2} s\sqrt{\frac{n + 1}{n}}\right) \tag{13.36}$$

where $t_{\alpha/2}$ is the point of the $t$ distribution exceeded with probability $\alpha/2$.

Distribution-free tolerance intervals are easily obtained by using the order statistics. Suppose that we have a sample of size $n$ and assume a continuous density function $f(x)$ with a distribution function $F(x)$. Consider the content of the random interval $[X, Y]$ where $X$ and $Y$ denote the smallest and largest values, respectively. From Chapter 6 we know that the joint density function of $X$ and $Y$ is given by

$$g(x, y) = n(n-1)f(x)f(y)[F(y) - F(x)]^{n-2} \tag{13.37}$$

Let $U = F(X)$, and $C = F(Y) - F(X)$. Then the joint density function of $U$ and $C$ is given by

$$g(u, c) = n(n-1)c^{n-2}, \qquad 0 \le c \le 1, \quad 0 \le u \le 1 - c \tag{13.38}$$

So the probability density function of the content $C$ is given by

$$\begin{aligned} h(c) &= \int_0^{1-c} n(n-1)c^{n-2}\,du \\ &= n(n-1)c^{n-2}(1-c) \\ &= \frac{1}{B(n-1, 2)} c^{n-2}(1-c) \end{aligned} \tag{13.39}$$

So in this case the distribution of the content sketched in Figure 13.4 is known completely and is a beta distribution with parameters $n - 1$ and 2. We can use this distribution to obtain formulas for tolerance intervals of both types that we have discussed. For example

$$E(C) = (n-1)/(n+1)$$

for this distribution, so we can determine the necessary sample size to give a beta-expectation tolerance interval for any specified $\beta$. If we want $P(C \ge p) \ge 1 - \gamma$, we require

$$\int_0^p n(n-1)(c^{n-2} - c^{n-1})\,dc \le \gamma$$

$$n(n-1)\left(\frac{p^{n-1}}{n-1} - \frac{p^n}{n}\right) \le \gamma$$

$$p^{n-1}[n - (n-1)p] \le \gamma \tag{13.40}$$

Then for a specified $\gamma$ and $p$ we can determine the sample size required.

In general we need not use the smallest and largest observations in the sample but can use $[X_{(i)}, X_{(j)}]$. In this case $C = F[X_{(j)}] - F[X_{(i)}]$ and

$$h(c) = \frac{1}{B(j-i, n-j+i+1)} c^{j-i-1}(1-c)^{n-j+i}$$

## 13.8 The Theory of Random Sampling from a Finite Population

We have seen how we may attach consonance limits with regard to the mean of a population when we accept the normality assumption as reasonable. Suppose, however, that we have an existent finite population of $N$ elements and wish to develop an opinion about the mean. We have seen in Chapter 8 that the expectation of the sample mean $\bar{x}$ of $n$ observations is the population mean $\mu$ and that the variance of the sample mean is $[(N - n)/N] \cdot S^2/n$ where $S^2$ is the sum of squares of deviations divided by $N - 1$. If we know $S^2$, we could use Chebyshev's inequality to state that

$$P\left(|\bar{x} - \mu| \geq k\sqrt{\frac{N-n}{N}\frac{S^2}{n}}\right) \leq \frac{1}{k^2}$$

and this would give us an invertible test of significance, valid for any underlying population of $x$ values. Of course, the bound given by the inequality may be very poor. In addition we do not know $S^2$. To overcome this defect, an objective guess of $S^2$ is usually taken to be the mean square deviation around the mean of the sample, say $s^2$. This is unbiased, though this is irrelevant to the ensuing argument because tests of significance for finite populations are based upon the hope that the pivotal

$$\frac{\bar{x} - \mu}{\sqrt{\dfrac{N-n}{N}\dfrac{s^2}{n}}}$$

has a distribution like the $t$ distribution. This is the critical point, not whether $s^2$ is biased or unbiased.

Here the deficiencies of sampling theory become readily apparent. The central limit theorem is frequently quoted, but it is not strictly relevant because the sample members are not independent. However, even if $\bar{x} - \mu$ has a normal-like distribution and $(n - 1)s^2/S^2$ has a $\chi^2$-like distribution, we still have no assurance that the above pivotal will have a $t$-like distribution. We need this property so that we can make an inversion. If we in fact have an "interval of estimation," we have a significance test!

We thus see that the procedures of random sampling of finite populations are objective, since one does not depend upon the ability of any individual to pick a representative sample but insists upon random sampling. They are subjective to the extent that a judgment is made that a $t$-like pivotal will have a distribution like Student's $t$ distribution. We commend to the reader exercises which indicate the nature of the problem. The problem is a quantification of the extent to which distributional theory for discrete distributions can be approximated by distributional theory for continuous distributions.

## 13.9 Concluding Remarks on Data Analysis

Let us summarize the status we have achieved with regard to the interpretation of data, at least in the type of situations we have discussed. We have examined the data for consonance with a class of models. The consonance is measured by computational procedures which have a basis in terms of their behavior in repetitions which one would encounter from the class of models. We have developed procedures by which we measure the degree of consonance of the data with particular subclasses of the class of models by calculational processes of the same type. Data can be consonant with a model in one respect and not in another, so we cannot insist on only one test.

The upshot of this line of examination is that we quantify the degree of consonance of the data with a class of models by a particular test of significance.

We emphasize that we do not talk about the probability of a particular model or subclass of models. In our framework, probability refers to frequency in an infinite set of repetitions; we do not think of the true model for our situation (which we do not know) as being a random member of some infinite population of models (unless as sometimes, but infrequently, a scientific model says this is the case).

The conclusion is *only* that the data are consonant with certain models. It would be foolish to assert that we have in any way exhausted the possible models which are consonant with the data. The choice of a class of models with which we say the data are consonant is based on our own experience and that of other people as to what models have been found useful in situations which seem to us to be like the situation we are facing. It is totally within the bounds of realism for some individual to discover that our data are consonant with a model that looks mathematically very different from the one we have reached.

It is essential to emphasize that the measurement of consonance of data and a model (or class of models) can be done in many ways. Developing a measure of distance of the data from the model is involved. In a small class of situations we have seen that one measure of consonance (or distance) may be best in a particular respect, but no measure is best in all respects. We recognize that we are vulnerable to the possibility of someone saying, "You are not interpreting your data well because if you had used such and such a test you would have concluded that the data are not consonant with the model."

The contingency described in the previous paragraph is no different from what might happen to a biologist with a taxonomic guide, looking at an insect and reporting: "I noted that the body structure, the pattern on the wings, and the antennae are like those illustrated in the guide for species X, and I conclude that the insect belongs to species X." Another biologist may say, on hearing this statement: "But did you look at the placement of the eyes? I believe that if you had done so, you would have been forced to the conclusion that the insect does not belong to species X." To hope that statistical methodology can avoid totally this type of criticism of one's conclusions is fallacious.

We have given emphasis to measures of consonance based on significance tests. Another measure which has been pursued extensively to the point of obtaining partial answers for a broad selection of situations is based on ideas of information theory (Kullback, 1959). It is interesting that similar results are obtained, but it appears to be accepted that there are arbitrary elements in this general approach.

## PROBLEMS

1. A curing process requires an evenly maintained temperature over a long period of time. The ovens are supplied by two competing companies. A random sample of readings was taken from competing ovens during a period when the temperature was supposed to be maintained at 500 ° F. The readings were:

| Oven $A$ | | Oven $B$ | |
|---|---|---|---|
| 502 | 498 | 499 | 509 |
| 500 | 500 | 493 | 497 |
| 501 | 492 | 501 | 502 |
| 497 | 501 | 502 | 503 |
| 495 | 507 | 500 | 504 |
| 502 | 498 | 496 | 492 |

Assuming normality, what values of $\sigma_A^2/\sigma_B^2$ are consonant with the model; $\sigma_A^2 = \sigma_B^2$ against the alternative models; $\sigma_A^2 > \sigma_B^2$ at the 5% level of significance?

2. Consider inversion of the two-tailed $t$ test of significance for the mean of a normal distribution. Show that values of $\mu$ within the interval

$$(\bar{x} - t_{\alpha/2}s_{\bar{x}}, \bar{x} + t_{\alpha/2}s_{\bar{x}})$$

yield a higher significance level than $\alpha$, while values outside the interval yield a significance level lower than $\alpha$.

3. Consider inversion of the upper-tailed test of significance for the mean of a normal distribution. Show that the values of $\mu$ yielding a level of significance greater than $\alpha$ are the values which exceed $\bar{x} - t_\alpha s_{\bar{x}}$.

4. Consider the test of significance for the model $\sigma^2 = \sigma_0^2$ against the model $\sigma^2 \neq \sigma_0^2$ for a normal distribution. Obtain the consonance interval at the $\alpha$ level of significance for each of the following tests of significance:

   (a) $SL = P[|\chi^2 - (n-1)| \geq |\sum (x_i - \bar{x})^2/\sigma_0^2 - (n-1)|]$

   (b) $SL = \min\{2P[\chi^2 \geq \sum (x_i - \bar{x})^2/\sigma_0^2], 2P[\chi^2 < \sum (x_i - \bar{x})^2/\sigma_0^2]\}$

5. To evaluate the level of significance for the binomial model that $p = p_0$ against the alternative $p > p_0$, use the formula

$$SL = P(x \geq x_0)$$
$$= \sum_{x=x_0}^{n} \binom{n}{x} p^x(1-p)^{n-x}$$

   For the inversion of this significance test, consider all values of $p$ that give a higher level of significance than a specified value $\alpha$. Show that this gives all values of $p$ greater than $p_1$ where

$$\sum_{x=x_0}^{n} \binom{n}{x} p_1^x(1-p_1)^{n-x} = \alpha$$

6. In Problem 5 show that $p_1$ can be determined from the cumulative beta distribution by

$$\int_0^{p_1} \frac{n!}{(x_0-1)!(n-x_0)!} z^{x_0-1}(1-z)^{n-x_0}\,dz$$

7. In Problems 5 and 6 show that $p_1$ is given by

$$p_1 = \frac{x_0}{x_0 + (n - x_0 + 1)F_{\alpha;\,2(n-x_0+1),\,2x_0}}$$

8. If we evaluate the significance for the binomial model that $p = p_0$ against the alternative that $p < p_0$, show that the values of $p$ for which the level of significance is greater than $\alpha$ are those values of $p$ less than $p_2$ where

$$\sum_{x=0}^{x_0} \binom{n}{x} p_2^x(1-p_2)^{n-x} = \alpha$$

9. In Problem 8 show that $p_2$ is given by

$$\frac{(x_0+1)F_{\alpha;\,2(x_0+1),\,2(n-x_0)}}{(n-x_0) + (x_0+1)F_{\alpha;\,2(x_0+1),\,2(n-x_0)}}$$

10. To evaluate the significance level for the Poisson model that $\lambda = \lambda_0$ against the model that $\lambda > \lambda_0$, use the formula $SL = P(x \geq x_0)$. For the inversion of this significance test consider all values of $\lambda$ that give a higher level of significance than a specified value $\alpha$. Show that these are the values of $\lambda$ greater than $\lambda_1$ where

$$\sum_{x=x_0}^{\infty} \frac{e^{-\lambda_1}\lambda_1^x}{x!} = \alpha$$

11. In Problem 10 show that $\lambda_1$ can be determined from the cumulative gamma distribution by

$$\int_0^{\lambda_1} \frac{z^{x_0 - 1} e^{-z}}{\Gamma(x_0)} \, dz = \alpha$$

12. Show that in Problems 10 and 11, $\lambda_1$ is given by $\lambda_1 = \chi^2_{1-\alpha,\, 2x_0}/2$.

13. If we evaluate the significance level for the Poisson model that $\lambda = \lambda_0$ against the alternative that $\lambda < \lambda_0$, show that the values of $\lambda$ for which the level of significance is greater than $\alpha$ are those values of $\lambda$ less than $\lambda_2$ where

$$\sum_{x=0}^{x_0} \frac{\lambda_2^x e^{-\lambda_2}}{x!} = \alpha$$

14. In Problem 13 show that $\lambda_2$ is given by

$$\int_{\lambda_2}^{\infty} \frac{z^{x_0} e^{-z}}{\Gamma(x_0 + 1)} \, dz = \alpha$$

15. In Problems 13 and 14 show that $\lambda_2 = \chi^2_{\alpha,\, 2(x_0 + 1)}/2$.

# 14
# Decision Making

## 14.1 Introduction

In the preceding chapters on statistical inference we attempted to distinguish between decision making and the formulation of opinion. We realize that from some points of view all intellectual activity consists of sequences of decisions, including the process of forming opinions so that the distinction we are making requires further discussion. Nevertheless, we believe that the distinction is worth making and is in fact, fundamental to our philosophy about statistical inference.

Let us attempt to express more clearly what we mean by decision and opinions. It is doubtful that any formal definition will serve our purposes completely. However, the definition of a decision given by Webster which is close to what we have in mind is "A settling or deciding, as of a controversy, by giving a judgment on the matter." We also want to quote Webster's definition of an opinion as "A belief stronger than an impression but not based on positive knowledge." When we use the word decision, we are thinking of a situation where a chain of actions or consequences flows from that decision regardless of our belief. For example, the verdict of a jury is a settling of a controversy which determines to some extent the future of the individuals involved based on the individual beliefs and opinions of the jury members.

Often our decisions reflect our opinions about matters, but not always. For example, some travelers as a matter of course carry large life insurance policies on air travel. The decision to buy a large policy prior to a plane trip does not necessarily indicate a strong belief that the plane will crash. As another example, the decision to wager against the home team in a football game does not necessarily indicate a lack of faith in the team. Another way of expressing these ideas is to say that acting as though we believe $A$ is true does not necessarily mean that we believe $A$ is true. There is a difference between behavior and belief.

We feel that much scientific activity is concerned with formulation of opinion. In particular, much of statistics is concerned not with decision making but with the formulation of beliefs and opinions; in other words, not with inductive behavior, but with inductive inference. For example, we believe it more common that a statistical analysis results in forming an opinion than in accepting or rejecting a hypothesis.

We do believe that statistics plays an important role in decision making. However, we do not believe that decisions always flow, as an immediate consequence, from the formal application of a statistical procedure. The matter is discussed by Tukey (1960).

Despite this discussion we recognize the fact that many situations do require definite actions or decisions, and the need for a formal theory of decision making is a very real one. In this chapter we give an elementary exposition of statistical decision theory. A very general formulation and development has been given by Wald (1950). We illustrate the application of this general formulation to some specific problems and then consider the special problem of estimation of parameters.

## 14.2 Formulation of a Model for Decision-making Problems

As a basis for discussing the matter in abstract terms let us consider a few decision problems:

1. Acceptance sampling—lots of manufactured items are submitted to outgoing inspection. Each lot has an unknown fraction defective. Upon the basis of a sample a decision is made either to accept or to reject the lot.
2. Petroleum exploration—the manager of a petroleum production department must decide upon the basis of seismograph data whether or not to drill a well in a certain location.
3. Estimation of parameters—a statistician has already decided upon a parametric class of models. He now wishes to decide, upon the basis of his data, which parameter value to use.
4. Sample size determination—a market analyst must decide how large a sample to take in a survey of consumer preference for brand $X$.
5. Production expansion—an expanding company has decided to build a new plant and must decide upon the plant location.

The reader can add to this list if he so desires. In all cases we are concerned with a decision rather than the formulation of opinion.

What characteristics do these problems have in common? It seems that in all examples there is a decision space $\mathscr{D}$ and a state space $\Theta$ and that a decision $d$ must be selected from $\mathscr{D}$ without knowledge of the state of the system $\theta$. An explicit description of $\mathscr{D}$ and $\Theta$ for the five examples considered might be as follows:

1. Acceptance sampling
   $\mathscr{D} = \{\text{accept, reject}\}$
   $\Theta = \{p: 0 \leq p \leq 1\}$
   where $p$ is the unknown fraction defective
2. Petroleum exploration
   $\mathscr{D} = \{\text{drill, don't drill}\}$
   $\Theta = \{\text{productive region, nonproductive region}\}$
3. Estimation of parameters
   $\mathscr{D} = \{\hat{\theta}: -\infty < \hat{\theta} < \infty\}$
   $\Theta = \{\theta: -\infty < \theta < \infty\}$
4. Sample size determination
   $\mathscr{D} = \{n: n = 1, 2, 3, \ldots\}$
   $\Theta = \{p: 0 \leq p \leq 1\}$
   where $p$ is the unknown fraction favoring brand $X$
5. Production expansion
   $\mathscr{D}$ = set of locations under consideration
   $\Theta$ = possible states of the economy

It is to be noted that the precision with which we can describe or define $\mathscr{D}$ and $\Theta$ for these five examples varies. This is a difficulty with the theory we shall subsequently present. However, we shall assume that the spaces $\mathscr{D}$ and $\Theta$ are completely specified, although the state of the system $\theta(\theta \in \Theta)$ is unknown.

Given the decision space $\mathscr{D}$ and the state space $\Theta$, we still do not have a model for decision making. We must specify in some way the consequences of making any particular decision for each possible state of the system. It is conventional to formulate this in terms of a loss function $W$, a real-valued function of $\theta$ and $d$. $W(\theta, d)$ gives the loss when the system is in state $\theta$ and decision $d$ is chosen. A negative value, of course, represents a gain rather than a loss. In most real situations there will be great difficulty in specifying $W(\theta, d)$, because the consequences of terminal decisions extend into the indefinite future. It is generally assumed that this difficulty has been solved by the user of the procedures.

## 14.3 Decision-making Rules

We do not hope to say how people should make decisions. Rather we wish to present some decision-making rules which have had considerable appeal. To simplify the presentation, let us suppose that both the decision space and the state space are finite. It is also convenient to think of the decision-making problem as a game against nature, where $\theta$ represents the state of nature. Then we can represent the situation in an array as shown in Table 14.1.

**Table 14.1** Game against Nature

| | | Player II Decision Maker | | | |
|---|---|---|---|---|---|
| | | $d_1$ | $d_2$ | ... | $d_n$ |
| Player I Nature | $\theta_1$ | $W(\theta_1, d_1)$ | $W(\theta_1, d_2)$ | ... | $W(\theta_1, d_n)$ |
| | $\theta_2$ | $W(\theta_2, d_1)$ | $W(\theta_2, d_2)$ | ... | $W(\theta_2, d_n)$ |
| | . . . | ... | ... | ... | ... |
| | $\theta_m$ | $W(\theta_m, d_1)$ | $W(\theta_m, d_2)$ | ... | $W(\theta_m, d_n)$ |

If the state of nature is known to us, we simply make the decision which minimizes our loss. What decision do we make when we do not know the state of nature? A principle which has considerable appeal is the so-called minimax principle. For each decision we consider the worst possible outcome, i.e., the maximum loss, and then choose the decision for which the maximum loss is smallest. That is, we minimize our maximum possible loss.

Consider the artificial example represented by the loss function in Table 14.2. Then $d_2$ is the decision given by the minimax rule because it achieves the smallest maximum loss.

**Table 14.2** Loss Function

| | $d_1$ | $d_2$ | $d_3$ | $d_4$ |
|---|---|---|---|---|
| $\theta_1$ | −10 | 2 | 10 | 5 |
| $\theta_2$ | 0 | 4 | −10,000 | 5 |
| $\theta_3$ | 10 | 1 | 5 | 5 |
| Max Loss | 10 | 4 | 10 | 5 |

Another way of tackling the problem is to form a weighted average of the loss for each decision and then to choose the decision for which this average loss is minimized. This is called a Bayes rule. The weights need not be probabilities, although they are generally described as such. The Bayes decision is the decision which minimizes $\sum_i P(\theta_i) W(\theta_i, d_j)$. That is, a Bayes decision is the decision which minimizes the expected loss where the expectation is taken over the distribution of $\theta$.

Let us use the same numerical example we used to illustrate the minimax principle. Suppose that the probabilities $P(\theta_i)$ were 1/3, 1/6, 1/2. Then the expected losses are given by:

| Decision | $d_1$ | $d_2$ | $d_3$ | $d_4$ |
|---|---|---|---|---|
| Expected Loss | 10/6 | 11/6 | $-9965/6$ | 5 |

Then $d_3$ is the Bayes decision for this distribution on $\theta$ because it results in the smallest expected loss.

The two preceding examples serve to illustrate the fact that the minimax rule is generally much more conservative than a Bayes rule because it concentrates upon the worst possible result. This conservative attitude might in fact be proper if we were playing against an antagonistic opponent, but there is a philosophical question about whether or not nature is trying to beat us.

We have not thus far incorporated one feature possessed by many decision-making problems, namely, the existence of data. Let us now suppose that the decision maker can obtain data giving some information about the state of nature. Obviously this could be incorporated into the model in many ways, but a simple approach is to suppose that the data represents a random sample $D$ from a distribution depending upon the state of the system $\theta$. That is, we suppose that we have the class of models $P(D; \theta)$, $\theta \in \Theta$.

Suppose now that a decision rule $\delta$ maps the data $D$ into one of the decisions in $\mathscr{D}$. That is $d = \delta(D)$. If the decision maker chooses to make his decisions in such a manner, his problem is not to choose a decision from $\mathscr{D}$ but rather to choose a decision rule $\delta$ from a set of possible rules, or rule space $\Delta$. Since $\delta(D)$ is itself a random quantity, the loss $W(\theta, \delta(D))$ is a random variable, and it is usual to consider the expected loss as a basis for choosing the decision rule $\delta$. This expected value is called the risk and is denoted by $R(\theta, \delta)$ or $R_\delta(\theta)$.

Now the formal situation is precisely the same as where we started. Instead of a loss function we have a risk function; and instead of making a decision, the decision maker must choose a decision function $\delta$. For instance, he can choose the decision function in accordance with a Bayes rule, i.e., choose the decision function which minimizes the expected risk.

In Chapter 12, we introduced the idea of a randomized decision rule when we discussed randomized accept-reject rules for tests of hypotheses. We now want to generalize this idea. Suppose that we have a randomized decision rule $\delta_R$ which maps the data, not into a decision, but into a probability distribution on the decision space, so that $d_1$ is chosen with probability $Q(d_1)$ and $d_2$ with probability $Q(d_2)$, etc. Then the risk is given by taking expectation not only with respect to the distribution of the sample $P(D; \theta)$ but also with respect to the distribution on the decision space $Q(d)$.

It may be that the decision maker will wish to use a formulation such that maxima and minima do not exist over the spaces of interest.

*Definition 14.1. A decision rule $\delta^*$ is said to be minimax if*

$$\sup_{\theta \in \Theta} R(\theta, \delta^*) = \inf_{\delta \in \Delta} \sup_{\theta \in \Theta} R(\theta, \delta) \tag{14.1}$$

*Definition 14.2. A decision rule $\delta^*$ is said to be Bayes with respect to $P(\theta)$ if*

$$E[R(\theta, \delta^*)] = \inf_{\delta \in \Delta} E[R(\theta, \delta)] \tag{14.2}$$

The notions of "sup" and "inf" are possibly advanced for the reader. We recall therefore that if we have a function $g(\alpha)$ of $\alpha$ where $\alpha$ belongs to some set $\mathscr{A}$ then

$$\sup_{\alpha \in \mathscr{A}} g(\alpha) = c$$

if and only if $g(\alpha) \leq c$ for all $\alpha \in \mathscr{A}$ and for any $\varepsilon > 0$ there exists an $\alpha$ such that $g(\alpha) > c - \varepsilon$. Similarly

$$\inf_{\alpha \in \mathscr{A}} g(\alpha) = d$$

if and only if $g(\alpha) \geq d$ for all $\alpha \in \mathscr{A}$ and for any $\varepsilon > 0$ there exists an $\alpha$ such that $g(\alpha) < d + \varepsilon$. If the set $\mathscr{A}$ is finite, the "sup" is the maximum and the "inf" is the minimum.

We now wish to illustrate the ideas of this section by giving a simplified version of an important decision problem. The manager of a petroleum production department must decide upon the basis of seismograph data whether or not to drill a well in a certain location. The seismograph data $X$ indicate either that oil is present ($X = 1$) or not present ($X = 0$). The manager believes the data to have come from one of three possible structures: $\theta_1$, a nonproducing structure; $\theta_2$, a high-producing structure; $\theta_3$, a low-producing structure. Further, from historical data he knows the probability distributions $P(X; \theta)$ as given in Table 14.3. The desirability of drilling or not drilling is measured by the

**Table 14.3** Historical Seismograph Probabilities

| | $P(X; \theta)$ | | |
|---|---|---|---|
| $X$ | $\theta_1$ | $\theta_2$ | $\theta_3$ |
| 1 | .2 | .5 | .9 |
| 0 | .8 | .5 | .1 |

**Table 14.4** Drilling Loss Function

| | | Drill | Don't Drill |
|---|---|---|---|
| Structure | $\theta_1$ | 10 | 0 |
| | $\theta_2$ | $-5$ | 0 |
| | $\theta_3$ | $-15$ | 0 |

losses in Table 14.4. Of course, in the absence of data the minimax rule leads to the decision not to drill since the maximum loss in that case is zero and the maximum loss in the other case is ten. This indicates the sterility of the minimax rule in some cases. The Bayes rule in the case of no data amounts to choosing the action which corresponds to the minimum of two elements: $10p_1 - 5p_2 - 15p_3$, 0, where $p_i$ is the $P(\theta = \theta_i)$.

Now consider the situation in which we bring data into the choice of decision. The four possible nonrandomized decision functions are:

| $X$ | $\delta_1$ | $\delta_2$ | $\delta_3$ | $\delta_4$ |
|---|---|---|---|---|
| 0 | don't drill | don't drill | drill | drill |
| 1 | don't drill | drill | don't drill | drill |

The expected losses or risks are easily obtained. For example,

$$\begin{aligned} R_{\delta_3}(\theta_2) &= (-5)\ \text{Prob}(\text{drill}|\theta_2) \\ &= (-5)\ \text{Prob}(X = 0|\theta_2) \\ &= -2.5 \end{aligned}$$

In such a fashion we obtain the risk functions in Table 14.5. As in the case with no data the minimax rule leads to the decision not to drill.

To determine the Bayes rule, we must first obtain the expected risk. This is also shown in Table 14.5. The Bayes rule, then, depends upon the values of $p_1$, $p_2$, and $p_3$. For $p_1 = p_2 = p_3$, the Bayes rule is $\delta_2$, for example. The reader is encouraged to determine the Bayes rule for various other values of $p_1$, $p_2$, and $p_3$. A clear discussion of risks for reaching a decision with a table like Table 14.1 is given by Luce and Raiffa (1957).

**Table 14.5** Drilling Risk Function

| | $\theta_1$ | $\theta_2$ | $\theta_3$ | Expected Risk |
|---|---|---|---|---|
| $\delta_1$ | 0 | 0 | 0 | 0 |
| $\delta_2$ | 2 | $-2.5$ | $-13.5$ | $2p_1 - 2.5p_2 - 13.5p_3$ |
| $\delta_3$ | 8 | $-2.5$ | $-1.5$ | $8p_1 - 2.5p_2 - 1.5p_3$ |
| $\delta_4$ | 10 | $-5$ | $-15$ | $10p_1 - 5p_2 - 15p_3$ |

## 14.4 Estimation of Parameters

In Chapters 8–13 we discussed the general idea of estimation, by which we mean the development of an opinion, in an objective way, about values for unknown parameters which are consonant with the data. Thus we really mean what is more commonly called interval estimation, though we envisage, not just one interval, but the totality of intervals arising by varying a probability from zero to unity.

We now discuss a rather different concept of estimation, commonly called point estimation. By this is meant the choice of one numerical value for each unknown parameter. The only way to phrase this problem realistically appears to be in terms of decision theory. Suppose the unknown parameter is $\theta$ and that we have a rule for associating with each sample a number $\hat{\theta}$. Then the process we should follow in picking just one number $\hat{\theta}$ has to be dependent in some way, perhaps rather vaguely, on losses arising from incorrect choice.

Given a probability distribution on $\theta$ and a loss function, how can we go about finding a Bayes estimator? The question is not trivial and there is no general answer. However, for many loss functions the answer is obtained by considering the consequences of rewriting the expression for the expected risk. Now

$$R_{\hat{\theta}}(\theta) = \sum_{D} W(\theta, \hat{\theta}) P(D|\theta) \tag{14.3}$$

Then, if the prior density is $h(\theta)$,

$$\begin{aligned} E[R_{\hat{\theta}}(\theta)] &= \int_{\theta} \left[ \sum_{D} W(\theta, \hat{\theta}) P(D|\theta) \right] h(\theta) d\theta \\ &= \int_{\theta} \sum_{D} W(\theta, \hat{\theta}) P(D, \theta) d\theta \\ &= \sum_{D} \left[ \int_{\theta} W(\theta, \hat{\theta}) P(\theta|D) d\theta \right] P(D) \end{aligned} \tag{14.4}$$

We assume that inversion of order of integration and summation is permissible. Thus the expected risk can be minimized by minimizing the quantity inside the brackets for each sample $D$. In fact, we need only minimize this quantity for the $D$ we actually observe, so that we can obtain the Bayes estimator by minimizing the expected loss with respect to the posterior distribution on $\theta$. This can be done quite simply for special loss functions. For example: (1) If $W(\theta, \hat{\theta}) = (\theta - \hat{\theta})^2$, the

Bayes estimator is the mean of $P(\theta|D)$, the posterior density of $\theta$; and (2) if $W(\theta, \hat{\theta}) = |\hat{\theta} - \theta|$, the Bayes estimator is the median of $P(\theta|D)$.

*Example 1*

Consider the problem of estimation of the binomial parameter $p$ with $x$ successes in $n$ trials. Suppose the prior distribution of $p$ is given by

$$h(p) = 3p^2, \qquad 0 \leq p \leq 1$$

Then

$$P(x, p) = 3\binom{n}{x}p^{x+2}(1 - p)^{n-x}$$

So the marginal probability of $x$ is

$$g(x) = 3\int_0^1 \binom{n}{x}p^{x+2}(1 - p)^{n-x}\,dp$$

$$= 3\binom{n}{x}B(x + 3, n + 1 - x)$$

and

$$P(p|x) = \frac{p^{x+2}(1 - p)^{n-x}}{B(x + 3, n + 1 - x)}$$

That is, the posterior distribution of $p$ is a beta distribution with parameters $x + 3$ and $n + 1 - x$. If the loss function is squared error; that is, if $W(p, \hat{p}) = (\hat{p} - p)^2$, the Bayes estimator is given by the mean of the posterior distribution which is readily found to be $(x + 3)/(n + 4)$.

It is curious that the single most useful method of finding a minimax estimator is provided by the Bayes rule.

***Theorem 14.1*** If the risk function for an estimator $\hat{\theta}$, $R_{\hat{\theta}}(\theta)$, is constant for all values of $\theta$ and there is a probability distribution on $\theta$ such that $\hat{\theta}$ is the Bayes estimator for the probability distribution, $\hat{\theta}$ is a minimax estimator.

*Proof:* Suppose $h(\theta)$ is the density on $\theta$ for which $\hat{\theta}$ is the Bayes estimator. Let $\theta^*$ be any other estimator. Then, because $\hat{\theta}$ has constant risk, the maximum risk equals the average risk. So

$$\begin{aligned} \sup_\theta R_{\hat{\theta}}(\theta) &= \int R_{\hat{\theta}}(\theta)h(\theta)\,d\theta \\ &\leq \int R_{\theta^*}(\theta)h(\theta)\,d\theta \\ &\leq \sup_\theta R_{\theta^*}(\theta) \end{aligned} \qquad (14.5)$$

That is, $\hat{\theta}$ has a smaller maximum risk than any other estimator.

The minimax principle is sometimes used because we are unwilling to discuss a probability distribution on $\theta$. It is curious that we are led to do so by this theorem purely as a mechanical means of obtaining a minimax estimator.

*Example 2* (*Hodges and Lehmann, 1950*)
We wish to find the minimax estimator of the binomial parameter $p$ where the loss function is $W(p, \hat{p}) = (\hat{p} - p)^2$. We must find an estimator with constant risk and then try to find a prior distribution on $p$ for which it is the Bayes estimator.

In searching for an estimator with constant risk, it seems natural to try simple functions of $X$ first of all. So we try the estimator $\hat{p} = aX + b$ and evaluate the risk

$$\begin{aligned} R(p, \hat{p}) &= E(aX + b - p)^2 \\ &= [(an - 1)^2 - a^2n]p^2 + [a^2n + 2b(an - 1)]p + b^2 \end{aligned}$$

We want to find $a$ and $b$ for which this is constant, i.e., for which $(an - 1)^2 - a^2n = 0$, $a^2n + 2b(an - 1) = 0$. Solving for $a$ and $b$ we obtain

$$a = \frac{1}{\sqrt{n}(1 + \sqrt{n})}, \qquad b = \frac{1}{2(1 + \sqrt{n})}$$

Then the risk for these values of $a$ and $b$ is $b^2$.

Next we need a prior distribution for which the Bayes estimator is the one we have already obtained. A very natural prior distribution to try for the binomial parameter is the beta distribution,

$$\frac{1}{B(\alpha, \beta)} p^{\alpha - 1}(1 - p)^{\beta - 1}$$

Then the posterior density of $p$ given $x$ is

$$\frac{1}{B(x + \alpha, n - x + \beta)} p^{x + \alpha - 1}(1 - p)^{n - x + \beta - 1}$$

and the Bayes estimator is the mean of the posterior distribution, namely $(x + \alpha)/(n + \alpha + \beta)$. Now we can see that by letting

$$\frac{1}{n + \alpha + \beta} = \frac{1}{\sqrt{n}(1 + \sqrt{n})}$$

and

$$\frac{\alpha}{n + \alpha + \beta} = \frac{1}{2(1 + \sqrt{n})}$$

this coincides with the estimator we already have. Thus the Bayes estimator for a prior distribution with parameters $\alpha = \sqrt{n}/2$ and $\beta = \sqrt{n}/2$ has constant risk and is therefore the minimax estimator.

We note, parenthetically, that this is a procedure that a Bayesian (cf. Chapter 11) would probably not use because it has the objectionable feature that the prior distribution depends on $n$, the amount of data.

We stated at the beginning of this section that estimation of parameters appears to be a decision problem. It would follow that a realistic way to assess the merits of a particular estimator is to use general decision theory. In any particular application presumably this would be done. However, a great deal of the theory of estimation has developed along lines tangential to decision theory. Instead, various properties for estimators have been considered.

A property that is often useful in searching for estimators that are good with regard to some criterion of merit is that of consistency. There is obscurity even in this matter. There are two definitions of consistency in the literature; the first is widely used and the second is not. [See Rao (1965) for a detailed and precise definition.]

*Definition 14.3. Consistency: An estimator $T_n$ of $\theta$ defined for all sample sizes n is said to be consistent if $T_n$ tends to $\theta$ with probability 1 as n tends to $\infty$.*

*Definition 14.4. Fisher-Consistency: For the case of a sample $x_1, x_2, \ldots, x_k$, $\sum_1^k x_i = n$ from the multinomial distribution with probabilities $p_i(\theta)$, a statistic $T(x_1, x_2, \ldots, x_k)$ is Fisher-consistent for $g(\theta)$ if*

$$T[p_1(\theta), p_2(\theta), \ldots, p_k(\theta)] = g(\theta) \tag{14.6}$$

The first definition is very widely used in the mathematical theory of statistics in the development of asymptotic results. The hope underlying this is that a statistic which is consistent will prove to have good properties with small samples. The second definition is restricted to the multinomial case but may, perhaps, have more force, although little work has been done on it.

Undoubtedly one of the most commonly used properties is that of unbiasedness. An estimator $\hat{\theta}$ of $\theta$ is unbiased if in repetitions of sampling the average value of $\hat{\theta}$ is $\theta$; i.e., if $E(\hat{\theta}) = \theta$. In our opinion unbiasedness is not an essential property. Nevertheless it has wide acceptance and the theory of unbiased estimation is interesting in its own right. A procedure for estimation is to choose the estimator with smallest variance in the class of unbiased estimators, if one exists.

We may note incidentally that if an unbiased estimator has minimum variance, it is unique. (In mathematically advanced texts a phrase, "except for sets of measure zero" is included for exactness.) Suppose

$\hat{\theta}_1$ is unbiased and has the minimum possible variance $V$. Suppose there is another unbiased estimator $\hat{\theta}_2$ also with variance $V$. Consider

$$\alpha\hat{\theta}_1 + (1 - \alpha)\hat{\theta}_2 \tag{14.7}$$

It is also unbiased for any choice of $\alpha$. Its variance is

$$V[\alpha^2 + (1 - \alpha)^2 + 2\alpha(1 - \alpha)\rho] \tag{14.8}$$

where $\rho$ is the correlation of $\hat{\theta}_1$ and $\hat{\theta}_2$. But this equals

$$V[1 - 2\alpha(1 - \alpha)(1 - \rho)]$$

Thus, if $\rho$ is different from unity, we can find a value for $\alpha$ so that $\alpha\hat{\theta}_1 + (1 - \alpha)\hat{\theta}_2$ has variance less than $V$. But this is impossible because $V$ is the least possible variance. Hence $\rho$ equals 1, or $\hat{\theta}_1$ and $\hat{\theta}_2$ are perfectly correlated, and $\hat{\theta}_1 - \hat{\theta}_2$ has zero mean and variance. Thus if two unbiased estimators $\hat{\theta}_1$ and $\hat{\theta}_2$ achieve minimum variance, they are identical.

The idea of minimum variance unbiasedness is interesting mathematically. One reason is given by the following theorem.

***Theorem 14.2* (*Cramér-Rao*)** If the sample space does not depend on the parameter $\theta$ (a scalar), then with a random sample of size $n$ and any estimate $\hat{\theta}$ of $\theta$, $V(\hat{\theta}) \geq [1 + b'(\theta)]^2/nI(\theta)$, where $E(\hat{\theta}) = \theta + b(\theta)$, and $I(\theta)$ is the Fisherian information on $\theta$.

*Proof:* We take the case of a discrete sample space. Then $\hat{\theta} = \hat{\theta}(D)$, where $D$ is a possible sample. So we have

$$E(\hat{\theta}) = \sum \hat{\theta}P(D; \theta) = \theta + b(\theta) \tag{14.9}$$

Differentiating with respect to $\theta$ because the range of summation does not depend on $\theta$, we obtain

$$\sum \hat{\theta}P'(D; \theta) = 1 + b'(\theta) = \sum \hat{\theta}\left[\frac{P'(D; \theta)}{P(D; \theta)}\right]P(D; \theta)$$

$$= E\left[\hat{\theta}\frac{\partial}{\partial\theta}\ln L(\theta)\right] \tag{14.10}$$

But under regularity conditions,

$$E\left[\frac{\partial}{\partial\theta}\ln L(\theta)\right] = 0 \tag{14.11}$$

and

$$V\left[\frac{\partial}{\partial\theta}\ln L(\theta)\right] = nI(\theta) \tag{14.12}$$

So

$$\operatorname{Cov}\left[\hat{\theta}, \frac{\partial}{\partial\theta}\ln L(\theta)\right] = 1 + b'(\theta) \tag{14.13}$$

But

$$\operatorname{Cov}^2(X, Y) \le V(X)V(Y)$$

So

$$V(\hat{\theta}) \ge [1 + b'(\theta)]^2/nI(\theta) \tag{14.14}$$

with equality if and only if

$$\frac{\partial \ln L(\theta)}{\partial\theta} = c + d\hat{\theta} \tag{14.15}$$

We have, therefore, the following corollaries.

*Corollary 14.1* In the regular case, the variance of an unbiased estimate is not less than $1/nI(\theta)$.

*Corollary 14.2* In the regular case, if $(\partial/\partial\theta)\ln L(\theta) = (\hat{\theta} - \theta)g(\theta)$ then $\hat{\theta}$ has variance $1/nI(\theta)$.

*Corollary 14.3* In the regular case for which the maximum likelihood statistic $\hat{\theta}$ is asymptotically $N(\theta, 1/nI(\theta))$, $\hat{\theta}$ is asymptotically minimum variance unbiased.

The generalization to a vector parameter requires the additional idea of a concentration ellipsoid and we refer the reader to Cramér (1946, Chapter 32). The upshot is that under mild regularity conditions, the most crucial of which appears to be that the sample space does not depend on the parameter value, the maximum likelihood statistic is asymptotically best unbiased (with a reasonable definition of best, of course).

We now give a method due to Rao (1945) and Blackwell (1947), independently, which is useful in the search for a minimum variance unbiased estimator. We present it for the case of a scalar parameter.

***Theorem 14.3*** If $\hat{\theta}$ is an unbiased estimator of $\theta$ and $T$ is a sufficient statistic for $\theta$, then $E(\hat{\theta}|T) = g(T)$ is an unbiased estimator for $\theta$ and

$$\operatorname{Var}[g(T)] \le \operatorname{Var}(\hat{\theta}) \tag{14.16}$$

*Proof:* Since $T$ is sufficient, the conditional expectation of $\hat{\theta}$ given $T$ does not involve $\theta$ and is a function of $T$. It can therefore be used as an estimator. The expectation of a conditional expectation gives an unconditional expectation. That is, $E[E(X|Y)] = E(X)$. Therefore $g(T)$ is unbiased since

$$E[g(T)] = E[E(\hat{\theta}|T)] = E(\hat{\theta}) = \theta \tag{14.17}$$

We know that the variance of $\hat{\theta}$ given $T = t$ is nonnegative. That is

$$E(\hat{\theta}^2|t) \geq [E(\hat{\theta}|t)]^2 \tag{14.18}$$

Since this inequality holds for every value of the random variable $T$,

$$\begin{aligned} E[E(\hat{\theta}^2|T)] &\geq E[E(\hat{\theta}|T)]^2 \\ E(\hat{\theta}^2) &\geq E[g(T)]^2 \\ E(\hat{\theta}^2) - \theta^2 &\geq E[g(T)]^2 - \theta^2 \end{aligned} \tag{14.19}$$

or

$$\text{Var}(\hat{\theta}) \geq \text{Var}[g(T)] \tag{14.20}$$

So given any sufficient statistic for $\theta$ and any set of unbiased estimators of $\theta$, $\hat{\theta}_1, \hat{\theta}_2, \ldots$, we know that $E(\hat{\theta}_i|T)$, $i = 1, 2, \ldots$ are unbiased and

$$V[E(\hat{\theta}_i|T)] \leq V(\hat{\theta}_i), i = 1, 2, \ldots \tag{14.21}$$

Suppose that $E(\hat{\theta}_i|T)$ is a unique function of $T$, $g(T)$ for all unbiased estimators $\hat{\theta}_i$. Then $g(T)$ is the minimum variance unbiased estimator of $\theta$. There could be no other unbiased estimate, say $\theta^*$, with smaller variance than $g(T)$ because by the Rao-Blackwell theorem,

$$V[E(\theta^*|T)] = V[g(T)] \leq V(\theta^*) \tag{14.22}$$

It is too much to hope that $E(\hat{\theta}_i|T)$ will give a unique answer $g(T)$ for all sufficient statistics $T$, and it is reasonable therefore to try to determine what property the distribution of $T$ must have for this to occur. The answer appears almost cyclical in that we require for unbiased estimators $\hat{\theta}_1$ and $\hat{\theta}_2$ the equality of $g_1(T) = E(\hat{\theta}_1|T)$ and $g_2(T) = E(\hat{\theta}_2|T)$. Stated another way, we require that if $E[g_1(T)] = E[g_2(T)]$, then $g_1(T) = g_2(T)$. This is the property referred to as completeness in Chapter 10 and which we repeat in a slightly different form.

*Definition 14.3. The distribution of $X$ is complete if there is no nontrivial unbiased estimator of 0; i.e., if $E[g(X)] = 0$ implies $g(X) = 0$.*

Lehmann and Scheffé (1950) proved the following theorem which results from the previous discussion.

***Theorem 14.4*** If $T$ is sufficient for $\theta$ and the distribution of $T$ is complete, the minimum variance unbiased estimate of $\theta$ is given by $E(\hat{\theta}|T)$, where $\hat{\theta}$ is any unbiased estimate of $\theta$.

*Example 3*
Suppose we are working on a reliability problem, and we are willing to assume that the number of failures in a given time interval is a Poisson variable. We wish to estimate the probability of no failures in a given time period; i.e., we want to estimate $e^{-\lambda}$.

Now $T = \sum_1^n X_i$ is sufficient for $\lambda$; furthermore it is complete because $E[g(T)] = \sum_{t=0}^{\infty} g(t)e^{-n\lambda t}(n\lambda)^t/t!$, which is a power series in $\lambda$. If the series is zero for all $\lambda$, each coefficient must be equal to zero, so that $g(t) = 0$. To find the minimum variance unbiased estimator of $e^{-\lambda}$, we need only find the conditional expectation, given $t$, of any unbiased estimator.

An unbiased estimator of $\theta = e^{-\lambda}$ is given by

$$\begin{aligned} \hat{\theta} &= 1 \text{ if } X_1 = 0 \\ &= 0 \text{ if } X_1 > 0 \end{aligned}$$

because $E(\hat{\theta}) = P(X_1 = 0) = e^{-\lambda}$. Now

$$\begin{aligned} E(\hat{\theta}|T = t) &= P(\hat{\theta} = 1|T = t) \\ &= P(X_1 = 0|T = t) \\ &= \frac{P(X_1 = 0, T = t)}{P(T = t)} \\ &= \frac{P\left(X_1 = 0, \sum_2^n X_i = t\right)}{P\left(\sum_1^n X_i = t\right)} \\ &= \frac{e^{-\lambda}[(n-1)\lambda]^t e^{-(n-1)\lambda}/t!}{e^{-n\lambda}(n\lambda)^t/t!} \\ &= \left(\frac{n-1}{n}\right)^t \end{aligned}$$

So the minimum variance unbiased estimator of $e^{-\lambda}$ is $[(n-1)/n]^T$.

*Example 4*
Finally, it is important to note that a minimum variance unbiased estimator may give a result which seems useless for a particular sample even though it has good or best properties on the average. Consider a

single observation from a Poisson in which zero is unavailable to us. That is, suppose

$$P(x) = \frac{e^{-\lambda}\lambda^x/x!}{1 - e^{-\lambda}}, \qquad x = 1, 2, 3, \ldots$$

Further suppose we wish to find an unbiased estimator, $a(x)$, of $e^{-\lambda}$. In order for $a(x)$ to be unbiased, we must have

$$\frac{\sum_1^\infty a(x)e^{-\lambda}\lambda^x/x!}{1 - e^{-\lambda}} = e^{-\lambda}$$

or

$$\sum_1^\infty a(x)\lambda^x/x! = 1 - e^{-\lambda}$$

$$= \sum_1^\infty (-1)^{x+1}\lambda^x/x!$$

Therefore

$$a(x) = +1, \text{ if } x \text{ is odd}$$
$$= -1, \text{ if } x \text{ is even}$$

Because this is the only unbiased estimator, it is the minimum variance unbiased estimator. This example illustrates a weakness of minimum variance unbiased estimation.

An aspect of minimum variance unbiased estimation is given by the following theorem, the origin of which is obscure, though it is certainly given by Lehmann and Scheffé (1950).

***Theorem 14.5*** Let $\hat{\theta}$ be an unbiased estimator of $\theta$. Suppose that $\hat{\theta}$ is uncorrelated with every statistic $Z$, whose expected value is zero. Then $\hat{\theta}$ is the minimum variance unbiased estimator of $\theta$.

*Proof:* The proof is by contradiction. Let $V(\hat{\theta}) = \sigma^2$ and suppose there is an unbiased estimator $\theta^*$ with variance $k\sigma^2$, $k < 1$. Then

$$\text{Cov}(\hat{\theta}, \theta^*) = l\sigma^2, \ l < \sqrt{k} < 1 \tag{14.23}$$

But $\theta^* - \hat{\theta}$ has zero expectation and

$$\text{Cov}(\theta^* - \hat{\theta}, \hat{\theta}) = (l - 1)\sigma^2 \neq 0 \tag{14.24}$$

So $\hat{\theta}$ is correlated with an unbiased estimator of zero, contradictory to the hypothesis of the theorem.

It is worthwhile to mention that the best estimate of a parameter $\theta$ depends on the loss function, and sometimes very strongly. Also there is no necessary relation of the best estimate of a function of a parameter, say $g(\theta)$, to the best estimate of $\theta$.

## 14.5 Ad Hoc Methods of Point Estimation

We have discussed above the elementary aspects of point estimation, and we have seen that this problem requires a specification of a risk function. We now present some methods of estimation which are based solely on intuitive grounds. These methods may have good properties but not necessarily so.

### 14.5.1 The Method of Moments

The classical case of application of this method is that of estimating the parameters $\theta_1, \theta_2, \ldots, \theta_p$ from a random sample drawn from a population with distribution $F(x; \theta_1, \theta_2, \ldots, \theta_p)$. The procedure is to obtain low sample moments and corresponding low population moments in terms of the parameters and then to determine the values for $\theta_1, \theta_2, \ldots, \theta_p$ which result in population moments being equal to sample moments. The number of moments used is the number of parameters. We give three examples.

*Example 5*
Given a random sample $x_1, x_2, \ldots, x_n$ from $N(\mu, \sigma^2)$, we have:

$$\text{1st moment: in sample} = \bar{x}\text{: in population} = \mu$$

$$\text{2nd moment: in sample} = \frac{\sum x_i^2}{n}\text{: in population} = \mu^2 + \sigma^2$$

Hence, the method of moments gives

$$\hat{\mu} = \bar{x}$$

$$\hat{\sigma}^2 = \frac{\sum x_i^2}{n} - \bar{x}^2 = \frac{\sum (x_i - \bar{x})^2}{n}$$

*Example 6*
Given a random sample $x_1, x_2, \ldots, x_n$ from Poisson $(m)$, we have:

$$\text{1st moment: in sample} = \bar{x}\text{: in population} = m$$

Hence the method of moments gives $\hat{m} = \bar{x}$.

*Example 7*
Given a random sample of doublets, $(x_i, y_i)$, $i = 1, 2, \ldots, n$, from the bivariate normal with parameters $\mu_x, \mu_y, \sigma_x^2, \sigma_y^2, \sigma_{xy}$

$$\text{1st } x \text{ moment: in sample} = \bar{x}\text{: in population} = \mu_x$$

$$\text{1st } y \text{ moment: in sample} = \bar{y}\text{: in population} = \mu_y$$

$$\text{2nd } x \text{ moment: in sample} = \frac{\sum x_i^2}{n} : \text{in population} = \mu_x^2 + \sigma_x^2$$

$$\text{2nd } y \text{ moment: in sample} = \frac{\sum y_i^2}{n} : \text{in population} = \mu_y^2 + \sigma_y^2$$

$$\text{2nd } xy \text{ moment: in sample} = \frac{\sum x_i y_i}{n} : \text{in population} = \mu_x \mu_y + \sigma_{xy}$$

Hence the method of moments gives the estimates

$$\hat{\mu}_x = \bar{x}, \hat{\mu}_y = \bar{y}$$

$$\hat{\sigma}_x^2 = \frac{\sum (x_i - \bar{x})^2}{n}, \hat{\sigma}_y^2 = \frac{\sum (y_i - \bar{y})^2}{n}, \hat{\sigma}_{xy} = \frac{\sum (x_i - \bar{x})(y_i - \bar{y})}{n}$$

This method cannot be extended simply to any situation, because it is not clear what moments of a complex set of data should be used.

### 14.5.2 The Method of Maximum Likelihood

This is a method of point estimation which is completely related to the likelihood theory presented in Chapter 10. The idea is to use as the estimate of the parameter $\theta$ the maximum likelihood statistic for $\theta$. In all three examples given above it results in the same estimate as the method of moments. The method can be applied to any situation, no matter how complicated the structure. There may be calculational and computational difficulties. Frequently the maximum likelihood estimate can be obtained only numerically for a given set of data by a computer routine for maximization. In such cases the properties of the maximum likelihood estimate are difficult to obtain. In general the procedure followed is to use the asymptotic likelihood theory of Chapter 10 to yield approximate distributional properties of the estimators. We note that under special regularity conditions the maximum likelihood estimate has minimum variance for any sample size, and that under certain regularity conditions the maximum likelihood estimate has minimum asymptotic variance. In the case that the maximum likelihood statistic is asymptotically normally distributed, it may also be asymptotically complete and then give asymptotically uniform minimum variance unbiased estimators. It seems that the maximum likelihood estimator has the edge over all competitors both from the viewpoint of sufficiency and efficiency, i.e., variance.

### 14.5.3 The Method of Minimum $\chi^2$

Consider multinomial data $x_1, x_2, \ldots, x_k$ assumed to be a random sample of size $n$ and

$$\chi^2 = \sum_{i=1}^{k} \frac{[x_i - np_i(\theta)]^2}{np_i(\theta)}$$

The estimation method known as the method of minimum $\chi^2$ is to determine the value of $\theta$ (which may be a vector) which gives the minimum value of $\chi^2$.

The method of modified minimum $\chi^2$ is to choose the value $\hat{\theta}$ which gives the minimum with respect to $\theta$ of

$$\chi_m^2 = \sum_{i=1}^{k} \frac{[x_i - np_i(\theta)]^2}{x_i} \tag{14.25}$$

A broad class of estimators has been developed by Neyman (1949) called best asymptotically normal (B.A.N.) estimators.

## 14.6 The Role of Sufficiency

It is natural for the reader to speculate on the relation of the material of Chapters 10 and 11 which makes extensive use of sufficiency, either in terms of a sufficient statistic or, equivalently, in terms of a sufficient partition of the space of possible observations, to the ideas of the present chapter. The relationship can be summarized by a general argument which includes decision theory application as a special case.

Suppose that we envisage discrete observations which we label by $(ij)$ with the labeling such that the index $i$ is a sufficient statistic, or alternatively speaking, that the partition of the possible observations by the first subscript is a sufficient partition. The subscript $j$ indexes possible observations within a sufficiency class. Now suppose that $\Omega$ is the class of discrete distributions defined over the closed interval $[0, 1]$, and let $\Omega_{ij}$ be a distribution from $\Omega$ which we associate with the observation $(ij)$. We suppose that we have a physical process, e.g., a coin-tossing apparatus, which permits us to obtain a realization of a random variable which follows any one of the distributions in $\Omega$. We now suppose that given an observation $(ij)$ and a realization $u_{ij}$ of a random variable having the distribution $\Omega_{ij}$, we choose an entity $c$ from a discrete class $\mathscr{C}$ of entities. That is, the observations $(ij)$ and $u_{ij}$ determine a $c$. The entity which we choose can be a point estimate, a decision from a space of decisions, an interval for a parameter or evidential content of the data or whatever. So we have a rule which maps a combination $[(ij), u_{ij}]$ into the space $\mathscr{C}$. If we repeat indefinitely the sampling of data and the sampling of the resultant distribution $\Omega_{ij}$, we shall have a resultant probability or relative frequency that each entity $c$ of $\mathscr{C}$ is obtained. Because the distribution of the data will be indexed by a parameter $\theta$, the overall operating characteristic of the procedure for picking an element $c$ of $\mathscr{C}$ is given by $P(c; \theta)$.

Now consider as an alternate procedure reducing the data to a sufficient statistic, which in the above context means suppressing the index $j$. Consider picking a discrete distribution $\Omega_i^*$, and denote a realization of the random variable having distribution $\Omega_i^*$ by $u_i^*$. Suppose also that we have a rule which maps a combination $[i, u_i^*]$ into the space $\mathscr{C}$. Repetition of this process will lead to an operating characteristic of the procedure of picking an element $c$ of $\mathscr{C}$, which we denote by $P^*(c; \theta)$. We now state the following theorem.

***Theorem 14.6*** For any choice of the distributions $\{\Omega_{ij}\}$ and a rule $R:[(ij), u_{ij}] \to c$, there is a choice of distributions $\Omega_i^*$ and a rule $R^*:[i, u_i^*] \to c$ such that the operating characteristics of $R^*$ is identical to that of $R$.

*Proof:* In repetitions of the variable $u_{ij}$ with fixed $(ij)$ the combinations $[(ij), u_{ij}]$ gives any particular $c$, say $c_k$, with a probability which we denote by $P_k^{ij}$, with $\sum_k P_k^{ij} = 1$. Hence from the fact that the statistic $i$ is sufficient, the overall probability of $c_k$ is

$$\begin{aligned}\sum_{ij} P(ij; \theta)P_k^{ij} &= \sum_{ij} P(i; \theta)P(ij|i)P_k^{ij} \\ &= \sum_{i} P(i; \theta) \sum_{\substack{j \\ \text{given } i}} P(ij|i)P_k^{ij} \\ &= \sum_{i} P(i; \theta)w_k^i, \text{ say} \end{aligned} \tag{14.26}$$

We now note that the $w_k^i$ are probabilities with $\sum_k w_k^i = 1$, because $w_k^i \geq 0$ clearly, and

$$\begin{aligned}\sum_{k} w_k^i &= \sum_{k} \sum_{\substack{j \\ \text{given } i}} P(ij|i)P_k^{ij} \\ &= \sum_{\substack{j \\ \text{given } i}} P(ij|i) \sum_{k} P_k^{ij} \\ &= \sum_{\substack{j \\ \text{given } i}} P(ij|i) = 1\end{aligned}$$

Now let $\{z_k; k = 1, 2, \ldots\}$ be a discrete set which has 1 to 1 correspondence with $\mathscr{C}$. Define the discrete distribution $\Omega_i^*$ to be that which gives $u_i^* = z_k$ with probability $w_k^i$. Finally, say that our rule $R^*$ gives $c_k$ if the observation is $i$ and if $u_i^* = z_k$. Then we have constructed a rule $R^*$ based on the sufficient statistic which has the same operating characteristic as that based on the total observation.

We now comment briefly on the above result. The rule $R$ is based on the actual observation and the realized value of a random variable having known distribution for each observation. These random variables are called auxiliary random variables. They correspond to the use of a coin toss to achieve a choice of an entity $c$ in $\mathscr{C}$. Similarly, $R^*$ is based on the sufficient statistic and the realized value of a random variable having a known distribution for each value of the sufficient statistic.

We now supplement the above by a theorem which tells us about the use of insufficient statistics. As before we treat the discrete sample case. From the above we know that for any procedure using the full data there is a procedure based on the sufficient partition with the same operating characteristic. If then there is a way of attaching a value to operating characteristics such that we can say that the particular procedure has maximum value, we can find this by examining the function

$$\text{Value}\left\{\sum_i P(i;\theta)\sum_k w_k^i c_k\right\}$$

with the restrictions, $w_k^i \geq 0$ and $\sum_k w_k^i = 1$ for each $i$. Suppose that probabilities $w_k^{*i}$ give the best operating characteristic.

Now consider a procedure based on an insufficient statistic. We suppose that this insufficient statistic takes values $1, 2, \ldots, W$. Suppose that the value $\alpha$ for the insufficient statistic for the data $(ij)$ is determined by the array of numbers $\delta_\alpha^{ij}$, which equals 1 if $(ij)$ gives $\alpha$ and zero otherwise. Suppose furthermore that we associate with the statistic $\alpha$ the auxiliary random variable $R_\alpha$, and that in repetitions of fixed $\alpha$ and varying $R_\alpha$, the rule of picking a $c$ gives the probability array $\sum_k q_k^\alpha c_k$. Then the operating characteristic of this procedure is

$$\begin{aligned}\sum_{ij} P(ij;\theta)\sum_\alpha \delta_\alpha^{ij}\sum_k q_k^\alpha c_k &= \sum_{ij} P(i;\theta)P(ij|i)\sum_\alpha \delta_\alpha^{ij}\sum_k q_k^\alpha c_k\\ &= \sum_i P(i;\theta)\sum_k\left[\sum_{j\alpha} P(ij|i)\delta_\alpha^{ij}q_k^\alpha\right]c_k\\ &= \sum_i P(i;\theta)\sum_k \tilde{w}_k^i c_k, \text{ say} \qquad (14.27)\end{aligned}$$

This tells us that the use of any statistic and a rule using auxiliary random variables for determining a $c$ from $\mathscr{C}$ has an operating characteristic equal to that of one of the class of procedures based on the use of a sufficient statistic and auxiliary random variables. By the definition of $\{w_k^{*i}\}$ and the assumption that there is a best sufficient procedure, the procedure based on the insufficient statistic cannot be superior to the best one based on the sufficient statistic. This indicates that if one is concerned with determining a best operating characteristic, one need concern oneself only with those based on a sufficient statistic. So we have the following theorem.

***Theorem 14.7*** Any search for an optimal procedure with any definition of optimality based on the operating characteristic of the procedure can be confined to a search among procedures based on a sufficient partition and auxiliary random variables for each class of a sufficient partition.

*Corollary 14.4* The statement of Theorem 14.7 can be modified by replacing the phrase "sufficient partition" by "minimal sufficient partition."

The relevance of this result is that for any procedure based on the observation and auxiliary random variables there is a procedure based on the sufficient statistic and auxiliary random variables which has identical performance in the totality of possible repetitions. Hence if we are searching for a procedure using auxiliary random variables to map the data into a space $\mathscr{C}$ which has an optimal operating characteristic in any sense of the word "optimal," we can restrict our search to procedures based on the sufficient statistic and auxiliary random variables. Furthermore, no procedure can be superior to a procedure which uses a sufficient condensation of the data plus auxiliary random variables. It may be noted that in our proof we have used no regularity conditions, though we have used discreteness of all observations intimately. Continuization of the whole argument is a matter of mathematical technique, not one of intrinsically different concepts.

It is important to comment on another aspect of the process described above. We may agree readily that if we have to make a terminal decision on the basis of data and if the data above do not indicate a unique one of the entities in $\mathscr{C}$, we may use the result of a coin toss to make a unique choice. But we should recognize that the use of an auxiliary random variable should not add to the evidential content of the data, whatever meaning we attach to this idea. To make the point in explicit form, can we "know" more about the proportion of defective items in a lot over and above what we know from our actual observation by being told also that the toss of a fair coin gave head as the result? It is failure to recognize that the answer to this question must be in the negative which has led to much controversy in statistics. For example, confidence intervals based on auxiliary random variables are not regarded by some as having valid evidential content.

The role of auxiliary random devices in the playing of games is obvious—one has merely to follow with a little attention any game, such as tennis, football, or bridge. There has been very extensive mathematical treatment of games going back to Borel early in this century, and von Neumann in the twenties, leading up to the definitive volume of von Neumann and Morgenstern (1947), with statistically oriented descriptions by Blackwell and Girshick (1954), and others. A very informative review is given by Luce and Raiffa (1957). It is of some historical interest that Fisher (1934a), apparently in ignorance of other work, wrote a short simple paper which discusses an old enigma of card playing which he resolves by the use of an auxiliary random variable. This four-page paper is rather easy to read and presents the essential basic idea.

## 14.7 Empirical Bayes Ideas

The following constitutes an application of the Bayesian idea which has validity in the sense of leading to predictions that may be expected to have verifiability properties. We use the class of normal distributions because it is so easy to manipulate. Suppose we have observations $x_1, x_2, \ldots, x_n$ which we know to be normally and independently distributed with unknown means $\mu_1, \mu_2, \ldots, \mu_n$ and with common known variance $\sigma^2$. Suppose also that we know or are willing to assume that the $\mu_i$ are independent realizations of a normal random variable with unknown mean $\varphi$ and unknown variance $\psi^2$. Hence we can write

$$x_i = \mu_i + f_i$$

$$\mu_i = \varphi + e_i$$

or

$$x_i = \varphi + e_i + f_i$$

in which $e_i \sim IN(0, \psi^2)$, and $f_i \sim IN(0, \sigma^2)$. Hence $x_i \sim IN(\varphi, \sigma^2 + \psi^2)$. Therefore we can use our ideas of the present chapter to estimate $\varphi$ and $\sigma^2 + \psi^2$. Conventional estimates are

$$\hat{\varphi}_n = \bar{x}_n = \left(\sum_{i=1}^{n} x_i\right)\Big/ n$$

and

$$\widehat{(\sigma^2 + \psi^2)}_n = \sum_{i=1}^{n} (x_i - \bar{x}_n)^2/(n-1)$$

where we use the subscript "$n$" to indicate that the estimates are based on the first $n$ $x$'s. It follows that a conventional estimate for $\psi^2$ is

$$\hat{\psi}^2 = \left[\sum_{i=1}^{n} (x_i - \bar{x}_n)^2/(n-1)\right] - \sigma^2 \tag{14.28}$$

Suppose now we have a new observation $x_{n+1}$ which is assumed to be a realization of a normal random variable with mean $\mu_{n+1}$ and variance $\sigma^2$, and we again can assume that $\mu_{n+1}$ is an independent realization of a normal random variable with mean $\varphi$ and variance $\psi^2$. Then if we know $\varphi$ and $\psi^2$, we would use as an estimator of $\mu_{n+1}$ with minimum square error, the quantity

$$\hat{\mu}_{n+1} = \left(\frac{x_{n+1}}{\sigma^2} + \frac{\varphi}{\psi^2}\right)\Big/\left(\frac{1}{\sigma^2} + \frac{1}{\psi^2}\right) \tag{14.29}$$

It is natural then to use as an estimator of $\mu$, in the absence of precise knowledge of $\varphi$ and $\psi^2$, the function

$$\mu^*_{n+1} = \left(\frac{x_{n+1}}{\sigma^2} + \frac{\hat{\varphi}_n}{\hat{\psi}_n^2}\right)\Big/\left(\frac{1}{\sigma^2} + \frac{1}{\hat{\psi}_n^2}\right) \tag{14.30}$$

There are slight technical difficulties here because, for example, $\hat{\psi}_n^2$ above might be a negative number and would not be appropriate. It seems clear intuitively that as $n \to \infty$, $\hat{\varphi}_n$ tends in probability to the unknown fixed number $\varphi$ and $\hat{\psi}_n^2$ tends in probability to the unknown number $\psi^2$. Hence as $n \to \infty$, $\mu^*_{n+1}$ tends to the best (in the sense of mean square error) predictor of $\mu_{n+1}$, which would be given by the correct prior distribution for $\mu_{n+1}$.

Going further, we may take the view that if $x_i$ is symmetrically distributed around $\mu_i$ and if the distribution of $\mu_i$ changes slowly in relation to the spread of the conditional distribution of each $x_i$ given $\mu_i$, then the actual cumulative distribution function of the observed $x_i$, $i = 1, 2, \ldots, n$, will be a reasonable estimate of the prior distribution of the $\mu_i$. It is natural then to do the following.

Let the known probability (or probability density function) of $X_i$ given $\mu_i$ be $f(x_i; \mu_i)$. Estimate the prior distribution of the $\mu_i$ by the discrete distribution

$$P(\mu = x_i) = 1/n, \qquad i = 1, 2, \ldots, n$$

Then the joint distribution of $(X_{n+1}, \mu_{n+1})$ is estimated to have the probability (or probability density function)

$$(1/n) f(x_{n+1}; x_i), \qquad \text{for } i = 1, 2, \ldots, n$$

and 0, otherwise. The marginal distribution of $X_{n+1}$ is estimated to have the probability (or probability density)

$$(1/n)\sum_{i=1}^{n} f(x_{n+1};\, x_i) \tag{14.31}$$

Hence the posterior probability distribution of $\mu_{n+1}$, given $x_1, x_2, \ldots, x_n, x_{n+1}$, is discrete with probability mass

$$f(x_{n+1};\, x_i)\bigg/\sum_{i=1}^{n} f(x_{n+1};\, x_i) \tag{14.32}$$

at each of the points $x_1, x_2, \ldots, x_n$. Furthermore, the mean of this posterior distribution is

$$\frac{\sum_{i=1}^{n} x_i\, f(x_{n+1};\, x_i)}{\sum_{i=1}^{n} f(x_{n+1};\, x_i)} \tag{14.33}$$

This simple result has been given by Krutchkoff (1970).

The idea of a recursive procedure like the above was initiated in the context of terminal decision theory by Robbins (1955). The initial example was the following. Suppose we have the discrete conditional distribution given the parameter $\lambda$ with probability $(1 - \lambda)\lambda^{x-1}$, $x = 1, 2, \ldots$. Suppose the prior for $\lambda$ is $f(\lambda)d\lambda$. Then the joint probability of $x$ and $\lambda$ is

$$(1 - \lambda)\lambda^{x-1} f(\lambda)d\lambda \tag{14.34}$$

The marginal probability of $x$ is

$$\int_\lambda (1 - \lambda)\lambda^{x-1} f(\lambda)d\lambda \tag{14.35}$$

and the conditional distribution of $\lambda$ given $x$ or posterior distribution of $\lambda$ given $x$ has density

$$\frac{(1 - \lambda)\lambda^{x-1} f(\lambda)}{\int_\lambda (1 - \lambda)\lambda^{x-1} f(\lambda)d\lambda} \tag{14.36}$$

Hence the posterior mean for $\lambda$ is

$$\frac{\int_\lambda (1 - \lambda)\lambda^{x} f(\lambda)d\lambda}{\int_\lambda (1 - \lambda)\lambda^{x-1} f(\lambda)d\lambda} \tag{14.37}$$

But the numerator is the marginal probability that the random variable $X$ takes the value $x + 1$ and the denominator is the marginal probability that it takes the value $x$. Hence the posterior mean of the distribution of $\lambda$ given the observation $x$ is equal to $P(X = x + 1)/P(X = x)$. Furthermore, the sequence of observed values $x_1, x_2, \ldots, x_n$ is a random sample from the marginal distribution of the random variable $X$. Clearly, the actual sequence $x_1, x_2, \ldots, x_n$ provides an estimate of this marginal distribution. Hence a procedure for estimating the particular $\lambda$, say $\lambda_{n+1}$, associated with the observation $x_{n+1}$ is to smooth the empirical distribution of $x_1, x_2, \ldots, x_n$ to yield a distribution $P_n(X = u)$ for all integral $u$ greater than zero, and then to use $P_n(X = x_{n+1} + 1)/P_n(X = x_{n+1})$ to estimate $\lambda_{n+1}$. This example is curious in that estimation of $f(\lambda)$ is not needed when one is interested solely in the posterior means. It seems interesting to note that if we write $G_n(t) = \sum_u P_n(X = u)t^u$, then the estimated prior distribution $f_n(\lambda)$ is the solution of the integral equation

$$\int \frac{(1-\lambda)t}{(1-\lambda t)} f_n(\lambda) d\lambda = G_n(t)$$

The general idea has been the subject of several workers under the leadership of Robbins. A good review of developments to 1969 is given by Copas (1969), but we should warn the student that because development of properties of the procedure are aimed at asymptotic ($n \to \infty$) properties, the mathematics is distinctly nonelementary. Our own preference is to regard the problem as simply one of data interpretation of a readily understandable form involving the conditional distribution given the parameter values and the distribution of the parameter values.

Note that if instead of an observation $x_i$, we had observations $x_{ij}, j = 1, 2, \ldots, r$ which are distributed around $\mu$ with *unknown* variance $\sigma^2$, we would have the data problem of components of variance discussed in Chapter 16, and the formulation would have to be extended. Note also that the method of empirical Bayes estimation is totally dependent on independence of sampling of parameter values. This assumption seems to be precisely what is in question in many cases.

It appears that the developers of the empirical Bayes idea and the developers of the neo-Bayesian idea mentioned in Section 11.10 regard the two processes as totally distinct (c.f. Lindley in the discussion of the Copas paper). However, it seems justified to take the view that the empirical Bayes procedures are a rational formulation of a legitimate Bayesian approach. To lend weight to this, we simply ask the question, How does the neo-Bayesian obtain his prior, if not by examining the historical record, which will include both observational data and theoretical knowledge (c.f. Cox in the discussion of the Copas paper)?

## PROBLEMS

1. A product is being sold at \$15 per item with a "double your money back" guarantee for defective items. It is manufactured in lots of ten, and it is desired to devise a screening procedure where the test is a destructive test. The following decision rules are being considered:

$\delta_1$: Accept the lot without testing.
$\delta_2$: Reject the lot without testing.
$\delta_3$: Test one item and reject the lot if the item is defective, otherwise accept.

(a) If the cost of rejecting the lot is \$20, find the loss function $W(k, \delta_i)$ where $k$, the number of defective items in the lot, may range from 0 to 10.
(b) Find the risk function $R_{\delta_i}(k)$.
(c) Find the minimax decision rule.
(d) If the lots of ten items can be regarded as samples from a binomial distribution with $p = .1$, find the Bayesian decision rule.

2. Give a proof of the Cramér-Rao lower bound on variance for continuous variables.

3. Find the Cramér-Rao lower bound for an unbiased estimator of $p^2$ based on $X$ where

$$f(x) = \binom{n}{x} p^x q^{n-x}, \qquad x = 0, 1, 2, \ldots, n$$

Show that $[X(X-1)]/[n(n-1)]$ is an unbiased estimator of $p^2$ and its variance is that given by the Cramér-Rao lower bound.

4. For the binomial distribution of the previous problem, find the Cramér-Rao lower bound for an unbiased estimator of $p^k$, $k \leq n$ and find an unbiased estimator which has this variance.

5. Find the Cramér-Rao lower bound for an unbiased estimator of $\lambda^k$ based on a random sample of size $n$ from the Poisson distribution with parameter $\lambda$. Find an unbiased estimator with this variance.

6. Find the Cramér-Rao lower bound for the variance of an unbiased estimator of $\mu$ for $N(\mu, \sigma^2)$ with $\sigma^2$ known. Find an unbiased estimator with this variance.

7. Find the Cramér-Rao lower bound for the variance of an unbiased estimator of $\sigma^2$ for $N(\mu, \sigma^2)$ with $\mu$ known. Find an unbiased estimator with this variance.

8. Given a random sample of size $n$ from the Poisson distribution with parameter $\lambda$, show that

$$\begin{aligned} g(X_1, X_2, \ldots, X_n) &= 1 \text{ if } X_1 = 0 \\ &= 0 \text{ if } X_1 \neq 0 \end{aligned}$$

is an unbiased estimator of $e^{-\lambda}$ and find its variance.

9. Given a random sample of size $k$ from the binomial distribution with density function

$$f(x) = \binom{n}{x} p^x q^{n-x}, \qquad x = 0, 1, 2, \ldots, n$$

show that an unbiased estimator of $P(X = 0) = (1 - p)^n$ is given by

$$\begin{aligned} g(X_1, X_2, \ldots X_k) &= 1 \text{ if } X_1 = 0 \\ &= 0, \text{ otherwise} \end{aligned}$$

Find the variance of this estimator.

10. Show that the conditional expectation of the estimator in Problem 8, given that $\sum_1^n X_i = t$, is $[(n-1)/n]^t$. Then show that $[(n-1)/n]^T$ is an unbiased estimator of $e^{-\lambda}$ and find its variance. Show that it has smaller variance than the other estimator, thereby illustrating the Rao-Blackwell theorem.

11. Show that the conditional expectation of the estimator in Problem 9, given that $\sum_1^k X_i = t$, is

$$\binom{n}{0}\binom{(k-1)n}{t}\Big/\binom{kn}{t}$$

12. Given a single observation from the Poisson distribution with parameter $\lambda$, show that the only unbiased estimator of $e^{-\lambda}$ is given by

$$\begin{aligned} a(X) &= 1 \text{ if } X = 0 \\ &= 0 \text{ if } X \neq 0 \end{aligned}$$

13. Show that the Poisson distribution is complete by expressing $E[g(X)]$ as a power series in $\lambda$ and show that this can be zero for all $\lambda$ if and only if $g(x) = 0$ for all $x$.

14. Show that the binomial distribution is complete.

15. Find the minimum variance unbiased estimator of $F(x_0)$ given a random sample of size $n$ from the Poisson distribution.

16. Find the minimum variance unbiased estimator of $F(x_0)$ given a random sample of size $k$ from the binomial distribution.

17. Mice have been exposed to radioactive material, and blood samples are being analyzed by counting the particles emitted. It is assumed that the number of particles emitted per hour is a Poisson random variable. We wish to estimate the probability of observing no more than 20 particles in an hour. The numbers of particles counted in six one-hour periods were: 15, 10, 25, 22, 30, 6.
(a) Obtain the minimum variance unbiased estimate of the desired probability.
(b) Set a 95% confidence interval on the mean number of particles per hour $\lambda$.

18. It is believed that the number of defects in rolls of camera film is a Poisson variable. Ten rolls were selected at random from a production lot and the number of defects counted. The data are: 5, 10, 6, 12, 4, 15, 3, 8, 10, 12.
(a) Obtain a minimum variance unbiased estimate of the probability that a roll selected at random will have no more than five defects.
(b) Set a 95% upper confidence limit on the mean number of defects.

19. Diodes are manufactured in large production lots. From each lot a random sample of 50 diodes is taken, and the number of defective diodes is noted. If there are three or more defective the lot is submitted to 100% inspection. The following numbers are the numbers of defective diodes in seven samples: 1, 0, 2, 1, 4, 3, 2.
(a) Obtain the minimum variance unbiased estimate of the probability that a lot will be completely inspected.
(b) Set a 99% confidence interval on the mean number of defective diodes per sample.

20. Consider a random sample of size $n$ from a normal distribution $N(\mu, \sigma^2)$. Since $X_1$ and $\bar{X}$ are both linear combinations of $X_1, X_2, \ldots, X_n$, $(X_1, \bar{X})$ has the bivariate normal distribution. Find the joint density function $f(x_1, \bar{x})$ and show that the conditional density function $f(x_1|\bar{x})$ is normal with mean $\bar{x}$ and variance $\sigma^2\,[(n-1)/n]$.

21. Using the result that $\bar{X}$ is a sufficient statistic for $\mu$ in the case of a normal distribution with known $\sigma^2$ and that the distribution of $\bar{X}$ is complete, show that a minimum variance unbiased estimator of $F(x_0)$ is given by

$$F_Z\left(\frac{x_0 - \bar{x}}{\sigma\sqrt{\dfrac{n-1}{n}}}\right)$$

where $F_Z$ denotes the cumulative distribution of a standard normal variable.

22. An extrusion process is used to produce a plastic fitting for which the inside diameter is critical. The manufacturing specifications call for .20 $\pm$ .001. The standard deviation of the process is rather constant from production run to production run and has been set at .0003. However, the mean of the process varies from run to run. A random sample of five was taken and the dimensions measured as .021, .0208, .022, .0201, .0202. Assuming that the dimension is approximately normally distributed:
    (a) Obtain the minimum variance unbiased estimate of the fraction of parts within specification limits.
    (b) Set a 95% confidence interval on the mean $\mu$.
    (c) Test the hypothesis that $\mu = .020$ against the alternative that $\mu \neq .020$ with a test of size .01.

23. Show that $\hat{\theta}$ is a Bayes estimator if and only if it minimizes the expected loss with respect to the posterior distribution of $\theta$ given $x$.

24. Show that $E(X - a)^2$ is minimized by taking $a = E(X)$.

25. Show $E|X - a|$ is minimized by taking $a =$ median.

# 15

# Relationships of Two Variables and Curve Fitting

## 15.1 Introduction

The bulk of statistical techniques so far presented has been aimed at the class of situations in which we have a set of $n$ observations, each of which consists of a single number. We now turn to the class of situations in which each observation consists of a pair of numbers $(x, y)$ and the totality of observations consists of the $n$ pairs $(x_i, y_i)$ with $i$ running from 1 to $n$. We have discussed in Section 4.10 the case in which both $x$ and $y$ values are taken to be random variables, for instance with $x_i$ as the height of the $i$th individual and $y_i$ as the weight of the $i$th individual. We now turn to a more detailed discussion of the whole matter.

It is essential to spend some time describing and discussing possible situations because the occurrence simply of $n$ pairs of numbers $(x_i, y_i)$ is quite insufficient to suggest definite methodology. We have to think about the status of the variables which are quantified by $x$ and $y$.

Let us ask first whether we "know" $x$ for each individual or have a measurement of $x$ which is subject to error of measurement. If the number $x$ is a classificatory label, the values for $x$ are presumably not subject to error. If, however, the number $x$ for an individual is the result of a measurement process which has sizable error, it will be reasonable to regard $x$ as the sum of a true value, say $X$, and an error of measurement $e$, i.e.,

$$x = X + e \tag{15.1}$$

Similarly, we may have the case that

$$y = Y + f \tag{15.2}$$

Next we ask whether $X$ is a random variable or a controlled value. For example, $X$ might represent a temperature reading, subject to control by an experimenter, or it might represent the uncontrolled temperature of a random specimen. The situation is the same with $Y$.

A case of particular interest, although of the greatest difficulty, is where both $X$ and $Y$ are random. Thus we may have the situation that the values $(X, Y)$ have a bivariate distribution. If we suppose that $X$ and $Y$ have a bivariate normal distribution with means $\mu_X$, $\mu_Y$, with variances $\sigma_X^2$, $\sigma_Y^2$ and covariance $\rho\sigma_X\sigma_Y$, and if we also suppose that the measurement errors $e$ and $f$ are independent normally distributed variables with means zero in both cases, and variances $\sigma_e^2$, $\sigma_f^2$ respectively, then $(x, y)$ will have a bivariate normal distribution with means $\mu_X$, $\mu_Y$, variances $\sigma_X^2 + \sigma_e^2$, $\sigma_Y^2 + \sigma_f^2$ respectively, and covariance $\rho\sigma_X\sigma_Y$. The statistical problems would then revolve around forming opinions about $\sigma_X^2$, $\sigma_e^2$, $\sigma_Y^2$, $\sigma_f^2$, and $\rho$. This topic is not elementary.

We make a 4 × 4 table of possibilities with regard to the status of the two components $x$, $y$ of the observation. These are displayed in Table 15.1.

**Table 15.1** Possible Status of Each of Two Variables in a Doublet Observation

| | | $x$ | | | |
|---|---|---|---|---|---|
| | | Random | | Controlled | |
| | | Error of Measurement | | Error of Measurement | |
| | | Yes *RE* | No *RN* | Yes *CE* | No *CN* |
| $y$ | *RE* | $xRE\|yRE$ | $xRN\|yRE$ | $xCE\|yRE$ | $xCN\|yRE$ |
| | *RN* | $xRE\|yRN$ | $xRN\|yRN$ | $xCE\|yRN$ | $xCN\|yRN$ |
| | *CE* | $xRE\|yCE$ | $xRN\|yCE$ | $xCE\|yCE$ | $xCN\|yCE$ |
| | *CN* | $xRE\|yCN$ | $xRN\|yCN$ | $xCE\|yCN$ | $xCN\|yCN$ |

Let us consider each of the 16 possibilities which we designate with a shorthand exemplified by $xRE|yRN$, indicating that $x$ is random—measured with error—and $y$ is random—measured without error.

$xRE|yRE$: This could arise when we have a production process leading to individuals, e.g., transistors, which have two attributes. Both attributes $x$ and $y$ are measured with error. Here probably the interest will be in discovering the magnitude of the measurement errors and in determining aspects of the covariability of the underlying true attributes.

*xRE|yRN:* This could arise when we have a process leading to individuals with two attributes $x$, $y$ such that $x$ is measured with error, but $y$ with unappreciable error. This might arise in the examination of students—the attribute $y$ being age, and the attribute $x$ being intelligence as measured by some test. All such tests have, as is well known, an error of moderate magnitude in the sense that ability to perform tasks that are closely alike varies from occasion to occasion of testing.

*xRE|yCE:* Here the idea is that $y$ is a controlled variable which cannot be controlled precisely. For example, consider administering a dosage of silver iodide crystals to clouds. The delivering of a specified dose $y$ is very difficult. The variable $x$ might be precipitation from the cloud in the 24 hours after dosage. This, it will be admitted, would be difficult to measure without error. Furthermore, the clouds treated are variable entities in themselves so that $x$ is random with error.

*xRE|yCN:* Suppose we have specimens of an alloy which we subject to an annealing process in a very carefully controlled environment for $y$ hours. Suppose that each specimen has a true corrosive resistance $x$ which we can measure only with error.

*xRN|yRN:* This could be the specification of a situation in which one measures height and weight of human male adults of a particular age with devices customary to an anthropometric laboratory.

*xRN|yCE:* The example which occurs to us is that we are producing physical specimens which are given various treatments, labeled by $y$, which are difficult to reproduce, and a property of the specimens is measured without error. The nature of the situation is, however, that the specimens are variable.

*xRN|yCN:* Here the situation is like the previous except that the treatments, labeled by $y$, are easy to reproduce.

*xCE|yCE:* This case may arise when we are considering two variables which are believed to be mathematically related, but both are measured with error.

*xCE|yCN:* This is the situation in which a variable $y$ is controlled without error, and $x$ is a "response" measured with error. We can suppose, for instance, that $y$ is a dosage of heat treatment applied to specimens which are very much alike and that $x$ is an attribute measured with error.

*xCN|yCN:* This may be regarded as the mathematical case of two related variables $x$ and $y$.

The remaining six cases differ from those described only in interchange of $x$ and $y$.

It might be hoped that statistical techniques would exist for all the ten structurally different cases. In fact, this is not the case. The whole situation is very complex, but all the structurally different cases do occur, and one elementary contribution a statistician can make is to recognize the different situations. We shall remark about some of the cases and give some definite techniques for only a few. Our primary interest in all the cases is to study the joint variability and to form ideas about how variation in one variable can be explained in terms of variability of another. There is no logical basis for envisaging an explanation of a controlled variable in terms of a random variable, so with a case like $xRN|yCN$ the only issue is that of explaining $x$ in terms of $y$ and not vice versa.

The simplest case is that designated as $xCN|yCN$. Indeed as the subject is commonly envisaged, there is no question of statistics being involved. All one can do is to plot $x$ against $y$, or some function of $x$ against some function of $y$, and hope that a simple relationship is perceived. It is implicit in our specifications that there is no variability in the $x$, $y$ relationship because if there were variability, one or the other of $x$ and $y$ would be random.

## 15.2 The Bivariate Normal Model

A model frequently used for several of the cases described in the previous section is the bivariate normal distribution which we introduced in Chapter 4. We recall that if a bivariate normal density function is of the form $C\exp(-Q/2)$, then $Q$ has the $\chi^2$ distribution with 2 degrees of freedom. This is an example of $xRN|yRN$.

For a partial representation of the bivariate distribution we may use a succession of ellipses

$$\frac{(x-\mu_x)^2}{\sigma_x^2} - 2\rho\frac{(x-\mu_x)(y-\mu_y)}{\sigma_x\sigma_y} + \frac{(y-\mu_y)^2}{\sigma_y^2} = \chi_\gamma^2(1-\rho^2) \quad (15.3)$$

where $\chi_\gamma^2$ is such that the proportion of the distribution contained within the ellipse is $1-\gamma$.

Let us now review quickly some inferential aspects of data consisting of $n$ doublet observations $(x_i, y_i)$ which are assumed to be a random sample from a bivariate normal distribution. The first aspect is the likelihood. In conformity with the description of likelihood for continuous distributions given in Chapter 8, we assume that there is a grouping error of $x = \Delta x$, the same for all $x$, and a grouping error of $y = \Delta y$. Then the probability of the sample is approximately

$$\prod_{i=1}^{n} \frac{1}{2\pi\sigma_x\sigma_y\sqrt{1-\rho^2}} \exp\left\{-\frac{1}{2(1-\rho^2)} \times \left[\frac{(x_i-\mu_x)^2}{\sigma_x^2} - \frac{2\rho(x_i-\mu_x)(y_i-\mu_y)}{\sigma_x\sigma_y} + \frac{(y_i-\mu_y)^2}{\sigma_y^2}\right]\right\} \Delta x\, \Delta y \quad (15.4)$$

Given the observations $(x_i, y_i)$, $i = 1, 2, \ldots, n$, this is a function of five parameters $\mu_x$, $\mu_y$, $\sigma_x$, $\sigma_y$, and $\rho$. The factorization of this into parts depending solely on the parameters and a part which depends solely on the configuration of the sample is a relatively advanced topic and will not be derived here. The upshot of the derivation is that the probability factors into three parts:

1. The probability of the means $\bar{x}$, $\bar{y}$ given $\mu_x$, $\mu_y$, $\sigma_x^2$, $\sigma_y^2$, and $\rho$.
2. The probability of the sample variances and covariance given $\sigma_x^2$, $\sigma_y^2$, and $\rho$, i.e., of

$$s_{xx} = \sum_i (x_i - \bar{x})^2/(n-1)$$

$$s_{yy} = \sum_i (y_i - \bar{y})^2/(n-1)$$

$$s_{xy} = \sum_i (x_i - \bar{x})(y_i - \bar{y})/(n-1)$$

   (Here the divisor $(n - 1)$ is to some extent a matter of convention.)

3. The probability of the observations $(x_i, y_i)$, $i = 1, 2, \ldots, n$ given $\bar{x}$, $\bar{y}$, $s_{xx}$, $s_{yy}$, and $s_{xy}$.

In part (1) the sample means $\bar{x}$, $\bar{y}$ have a bivariate normal distribution with means $\mu_x$, $\mu_y$, variances $\sigma_x^2/n$, $\sigma_y^2/n$, and the same correlation $\rho$. So the probability that $\bar{x}$ is in $\bar{x}_0 \pm \Delta x/2$ and $\bar{y}$ is in $\bar{y}_0 \pm \Delta y/2$ is

$$\frac{n}{2\pi\sigma_x\sigma_y\sqrt{1-\rho^2}} \exp\left\{-\frac{n}{2(1-\rho^2)} \times \left[\frac{(\bar{x}_0-\mu_x)^2}{\sigma_x^2} - 2\rho\frac{(\bar{x}_0-\mu_x)(\bar{y}_0-\mu_y)}{\sigma_x\sigma_y} + \frac{(\bar{y}_0-\mu_y)^2}{\sigma_y^2}\right]\right\} \Delta x\, \Delta y \quad (15.5)$$

The part (2) which is the distribution of $s_{xx}$, $s_{yy}$, $s_{xy}$ does not depend on $\mu_x$ and $\mu_y$, but does depend on $\sigma_x^2$, $\sigma_y^2$, and $\rho$ in a complicated way which is known as the Wishart distribution. The reader may refer to an advanced text such as that of Wilks (1965) or Anderson (1958).

The general ideas about condensation of data given in Chapter 8 involve the same mathematical problems, i.e., the discovery of a minimal sufficient statistic and the determination of the probability distribution of this statistic. The deduction of the minimal sufficient statistic is obvious at a heuristic level because the likelihood is specified by $\bar{x}$, $\bar{y}$, $s_{xx}$, $s_{xy}$, and

$s_{yy}$. The determination of the distribution of the statistic is an advanced topic.

The point of maximum of the likelihood occurs, with the continuous approximation to the grouped distribution which underlies any measurement, at the values

$$\mu_x = \bar{x}$$
$$\mu_y = \bar{y}$$
$$\sigma_x^2 = \sum (x - \bar{x})^2/n$$
$$\sigma_y^2 = \sum (y - \bar{y})^2/n$$

and

$$\rho = \frac{\sum (x - \bar{x})(y - \bar{y})}{\left[\sum (x - \bar{x})^2 \sum (y - \bar{y})^2\right]^{1/2}} \tag{15.6}$$

The second-order derivatives are fairly easy to obtain, so that general asymptotic likelihood theory can be applied.

The formation of opinion about the parameters is based on parts (1) and (2). One can form an opinion about $\mu_x$, about $\sigma_x$, or about $\mu_x$ and $\sigma_x$ by the univariate procedure, and similarly for the $y$ parameters. The formation of opinion about $\rho$ is usually based on the distribution of the sample correlation coefficient

$$r = \frac{s_{xy}}{\sqrt{s_{xx}\, s_{yy}}} \tag{15.7}$$

which has a distribution depending solely on the sample size $n$ and the true correlation, $\rho$. If $n$ is "large," a mathematical result due to Fisher (the proof of which is beyond this book) is that

$$\frac{1}{2} \ln \frac{1 + r}{1 - r}$$

is distributed approximately normally with mean

$$(1/2) \ln [(1 + \rho)/(1 - \rho)]$$

and variance $1/(n - 3)$. See, for example, Kendall and Stuart (1966) or Wilks (1962).

Opinion about $(\mu_x, \mu_y)$ can be formed by a procedure due to Hotelling (1931). Let

$$\begin{pmatrix} s_{xx} & s_{xy} \\ s_{xy} & s_{yy} \end{pmatrix} \begin{pmatrix} s^{xx} & s^{xy} \\ s^{xy} & s^{yy} \end{pmatrix} = \begin{pmatrix} 1 & 0 \\ 0 & 1 \end{pmatrix}$$

Then a system of consonance regions for $\mu_x$, $\mu_y$ is given by

$$n\, s^{xx}(\bar{x} - \mu_x)^2 + 2n\, s^{xy}(\bar{x} - \mu_x)(\bar{y} - \mu_y) + n\, s^{yy}(\bar{y} - \mu_y)^2 \le (1 - b_\alpha)/b_\alpha \tag{15.8}$$

where $b_\alpha$ is the upper $\alpha$ point of the data distribution with parameters $[(1/2)(n-2), 1]$. For successive values of $\alpha$ this system gives a set of ellipses in the $(\mu_x, \mu_y)$ plane, as illustrated roughly in Figure 15.1 The formation of opinion jointly about $\sigma_x^2, \sigma_y^2$ and $\rho$ or about the totality of five parameters is beyond the level of this book.

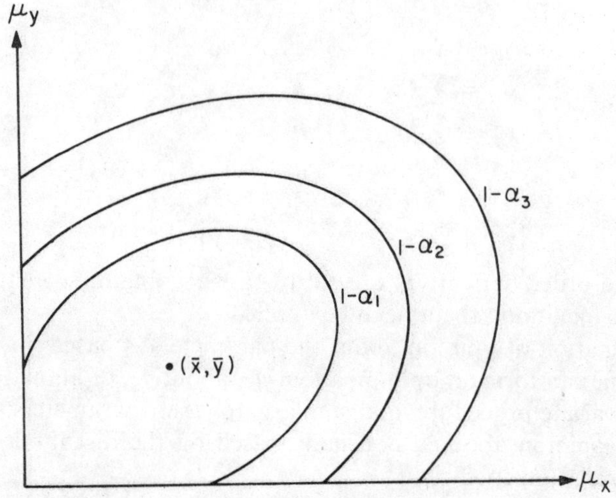


**Fig. 15.1** Consonance Regions on $\mu_x$, $\mu_y$

The formation of opinion about goodness of fit of the bivariate normal distribution will be based on the probability in part (3), though it is not clear that this matter has been adequately resolved at the present time.

There has been extensive work in recent years on Bayesian computations for this case. A full attack requires a prior distribution on $\mu_x$, $\mu_y$, $\sigma_x^2$, $\sigma_y^2$, and $\rho$. We refer to Geisser and Cornfield (1963). Certain choices related to the normalization constant of the probability density lead to easy mathematical computations, but the relevance of such arguments is obscure as in all cases of assumption of a prior distribution.

## 15.3 The Case $xRE|yRE$

To develop this case, one must make some distributional assumptions. A simple model is to suppose $x = X + e_x$ and $y = Y + e_y$, with $X \sim N(\mu, \sigma_X^2)$, $e_x \sim N(0, \sigma_{ex}^2)$, $Y \sim N(v, \sigma_Y^2)$, and $e_y \sim N(0, \sigma_{ey}^2)$ and all variables independent in all respects. A set of data will consist of $n$ pairs $(x_i, y_i)$, $i = 1, 2, \ldots, n$. The likelihood, using the continuous approximation, is

$$\frac{1}{(2\pi)^n[(\sigma_X^2 + \sigma_{ex}^2)(\sigma_Y^2 + \sigma_{ey}^2)]^{n/2}} \times$$

$$\exp\left\{-\frac{1}{2}\sum_1^n [a_{11}(x_i - \mu)^2 + a_{22}(y_i - \nu)^2]\right\} \tag{15.9}$$

where

$$1/a_{11} = \sigma_X^2 + \sigma_{ex}^2, \ 1/a_{22} = \sigma_Y^2 + \sigma_{ey}^2$$

This is a function of all the six parameters $\mu$, $\nu$, $\sigma_X^2$, $\sigma_{ex}^2$, $\sigma_Y^2$, $\sigma_{ey}^2$ but depends only on $\sum x_i^2$, $\sum y_i^2$, $\sum x_i$, $\sum y_i$, a vector of four scalar statistics. It is intuitively clear that one cannot index, in any way useful for data interpretation, the points of a six-dimensional space by the points of a four-dimensional space.

So this is a case for which no amount of data, i.e. no value of $n$, can lead to a formation of opinion or making of decisions about the parameter point in the six-dimensional space. To make any advance one must be able to incorporate more assumptions, i.e., make the model identifiable (the word here has an obvious connotation which more advanced texts make precise).

A simple way to do this, though not necessarily at all appropriate, is to impose the conditions that $\sigma_{ex}^2 = \lambda_1\sigma_X^2$, $\sigma_{ey}^2 = \lambda_2\sigma_Y^2$, where $\lambda_1$ and $\lambda_2$ are known. Any restriction which reduces the dimensionality of the parameter space to four can be used. But any such assumption may be *scientifically* invalid.

It is very important to note that with the models of the type presented in this section it is usually relevant to be interested in the variation and covariation of the underlying true random variables, $X$ and $Y$. As we have seen, if we have merely $n$ doublets $(x_i, y_i)$, we can do nothing. The idea behind the methodology of Section 15.9 below can perhaps be used. A plausible approach is to adjoin to the ideas above some techniques from Chapter 16. We give only a very succinct presentation. Suppose that we can repeat measurements of $x$ and $y$ on our random individuals. We could then have data of the following type: $(x_{ij}, y_{ij})$ with $i = 1, 2, \ldots, n$ and $j$ equal to, say, $1, 2, \ldots, r$, the $j$ subscript indicating the $j$th measurement on the $i$th individual. We might then consider the model that conditionally on $X_i$ and $Y_i$ the $x_{ij}$ are $NID(X_i, \sigma_{ex}^2)$, and the $y_{ij}$ are $NID(Y_i, \sigma_{ey}^2)$, and are independent. To this we add that $X_i$ and $Y_i$ have the bivariate normal with means $\mu_X$ and $\mu_Y$, variances $\sigma_X^2$ and $\sigma_Y^2$ and a covariance $\rho\sigma_X\sigma_Y$. From this specification, the probability of the full set of observations can be obtained and the general ideas of data reduction and statistical tests can be applied, though the development is not easy.

A very important point is that if we are concerned with the correlation of $X$ and $Y$ in a population of individuals with errors in measurement of $X$

and $Y$, the correlation of the observations can be quite misleading. With a large number of individuals, the correlation of $x$ and $y$, equal to $\text{Cov}(x, y)/\sqrt{V(x)V(y)}$, will be approximately equal with uncorrelated errors of measurement to $\text{Cov}(X, Y)/\sqrt{(\sigma_X^2 + \sigma_{ex}^2)(\sigma_Y^2 + \sigma_{ey}^2)}$, whereas the relevant correlation is often $\text{Cov}(X, Y)/(\sigma_X \sigma_Y)$. The correlation of the observed values $(x, y)$, will tend to be less than the correlation of the true values $X$ and $Y$, and often appreciably so. This phenomenon is called "attenuation," because errors of measurement result on the average in a stretching-out of the $X$ and $Y$ values, and hence lowering of regressions and correlations. It will often be critical to take into account this process and to devise some means of "adjusting for attenuation." A commonly used ad hoc procedure with estimates $s_{ex}^2$, $s_{ey}^2$ of $\sigma_{ex}^2$, $\sigma_{ey}^2$ respectively, and estimates $s_x^2$, $s_y^2$ of $\sigma_x^2$, $\sigma_y^2$ respectively is to use as an indicator of the correlation of $X$ and $Y$, the statistic $\text{Cov}(x, y)/\sqrt{(s_x^2 - s_{ex}^2)(s_y^2 - s_{ey}^2)}$. The concepts involved here are very critical in all areas in which errors of measurement cannot be neglected. In particular the ideas are essential in statistical treatment of data on quantitative inheritance and in almost all psychological data situations and interpretation of data for any type of mental test.

It is appropriate here also to refer to the nature of observations on an underlying continuous variable $X$. In Chapter 10, we took the view that a grouping of the variable $X$ is observed. An alternative idea is that the continuous variable $X$ leads to a continuous random variable $Z$, where $Z = X + e$, and $e$ follows some distribution such as the normal or uniform $(-\Delta, \Delta)$, and then that the observation is obtained by grouping on the $Z$ scale. What should be done about this very difficult problem seems to be very much a matter of empirical investigation. Observations are taken by humans, and there should be study of how observations on a "constant" actually vary. If for instance there is such an attribute as "true height" and observations are recorded "to the nearest 1/2 inch," what results will be recorded for a true height of 70.26 inches? It seems likely that it will be recorded as 70.0 with some frequency and 70.5 inches with some frequency, and perhaps other possibilities will occur. This serves to emphasize that the process of observation itself must be examined and that every observation must be interpreted in the framework of repetitions of the observation process.

## 15.4 Classical Curve Fitting: The Case $yCE|xCN$

In many respects the cases $yRE|xCN$ and $yRN|xCN$ reduce to the case $yCE|xCN$, which can also occur on its own merits. The pure case $yCE|xCN$ is somewhat idealized in that one has to imagine a population of identical specimens (or at least specimens which do not vary in regard to their true $y$ values) which have an associated errorless $x$ value and a $y$ value with measurement error. It is inherent in this description that a partially specified model is $y = f(x) + e$ in which $e$ is a measurement error. However the more common occurrence of this model is from the cases $yRE|xCN$ and $yRN|xCN$.

We have attached the name "classical curve fitting" to this situation because the great bulk of what is given in the literature on curve fitting seems to fall into this category. The situation is very much like the case of regression for a bivariate distribution as indicated in Figure 15.2.

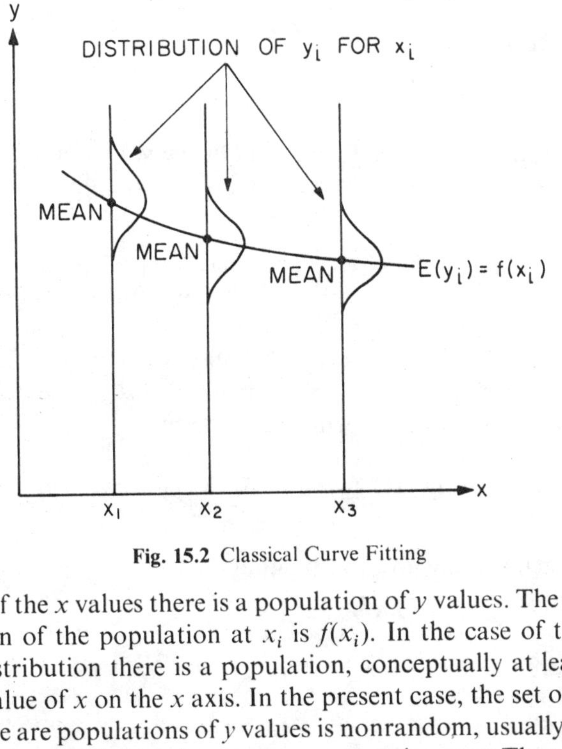


**Fig. 15.2** Classical Curve Fitting

For each of the $x$ values there is a population of $y$ values. The mean of the distribution of the population at $x_i$ is $f(x_i)$. In the case of the bivariate normal distribution there is a population, conceptually at least, at every possible value of $x$ on the $x$ axis. In the present case, the set of $x$ values at which there are populations of $y$ values is nonrandom, usually because the $x$ values are controlled by the observer or experimenter. The term "regression analysis" has come to be used in the case of the present section, although it is inappropriate. Historically, regression tells about the relation between one random variable and another random variable, and not between a random variable and a nonrandom variable.

## 15.5 The Straight Line Model

The situation we shall examine is that we have $n$ pairs of doublet observations $(x_i, y_i)$, $i = 1, 2, \ldots, n$, and we shall represent the data by the model $y_i = \alpha + \beta x_i + e_i$ in which $\alpha$, $\beta$ are unknown constants, the $x_i$ are known constants, values for the observed $y$ denoted by $y_i$, and the $e_i$ are normally and independently distributed around zero with unknown variance $\sigma^2$. In consonance with the general chapters on inference we write down the probability of the observations, assuming the usual approximation for observations from a continuous model. With grouping error $\Delta y$ on each $y_i$, this probability is essentially

$$\prod_{i=1}^{n} \frac{1}{\sqrt{2\pi}\,\sigma} \exp\left[-\frac{1}{2\sigma^2}(y_i - \alpha - \beta x_i)^2\right] \Delta y \tag{15.10}$$

The questions we ask are:

1. Does this model represent the data, or are the data consonant with this model?
2. If so, what opinions may be formed about the values of $\alpha$, $\beta$, and $\sigma^2$, and other functions of interest such as $\alpha + 5\beta$, the "true" value of $y$ at $x = 5$?

We approach the problem by examining the probability, Formula (15.10). Given the observations, this is a known function of the three parameters $\alpha$, $\beta$, and $\sigma^2$. Let us, therefore, as a first tentative step find the values of $\alpha$, $\beta$, and $\sigma^2$ at which this probability is maximized. To do the maximization by processes of elementary calculus, it helps to take logarithms first. So we have, ignoring the $\Delta y$ terms,

$$\ln P = -\frac{n}{2}\ln 2\pi - \frac{n}{2}\ln\sigma^2 - \frac{1}{2\sigma^2}\sum_i (y_i - \alpha - \beta x_i)^2 \tag{15.11}$$

This is maximized for any given $\sigma^2$ by minimizing

$$Q = \frac{1}{2}\sum_i (y_i - \alpha - \beta x_i)^2 \tag{15.12}$$

We construct the equations

$$\begin{aligned} \frac{\partial Q}{\partial \alpha} &= \sum_i (y_i - \alpha - \beta x_i)(-1) = 0 \\ \frac{\partial Q}{\partial \beta} &= \sum_i (y_i - \alpha - \beta x_i)(-x_i) = 0 \end{aligned} \tag{15.13}$$

So Equations (15.13) become, when unknowns are put on the left-hand side and known quantities on the right,

$$\begin{aligned} n\,\alpha + \beta(\textstyle\sum x) &= \textstyle\sum y \\ (\textstyle\sum x)\alpha + \beta(\textstyle\sum x^2) &= \textstyle\sum xy \end{aligned} \tag{15.14}$$

where subscripts have been suppressed for convenience.

These are called the normal equations, and the solution is called the least squares estimate of $\alpha$ and $\beta$. They are easy to solve, particularly when we make use of some elementary matrix algebra.

If we let $\begin{pmatrix} S_{11} & S_{12} \\ S_{21} & S_{22} \end{pmatrix}$ denote the matrix $\begin{pmatrix} n & \sum x \\ \sum x & \sum x^2 \end{pmatrix}$, then the normal equations are

$$\begin{pmatrix} S_{11} & S_{12} \\ S_{21} & S_{22} \end{pmatrix}\begin{pmatrix} \alpha \\ \beta \end{pmatrix} = \begin{pmatrix} \sum y \\ \sum xy \end{pmatrix} \tag{15.15}$$

Note that $S_{21} = S_{12}$.

A solution for the normal equations is

$$\begin{pmatrix} \alpha^* \\ \beta^* \end{pmatrix} = \begin{pmatrix} c_{11} & c_{12} \\ c_{21} & c_{22} \end{pmatrix}\begin{pmatrix} \sum y \\ \sum xy \end{pmatrix} \tag{15.16}$$

where we have denoted the elements of the inverse of the $S$ matrix by $c_{ij}$. Note that $c_{21}$ equals $c_{12}$. Writing $\sum y$ and $\sum xy$ in terms of the model, we have

$$\begin{aligned}\begin{pmatrix} \alpha^* \\ \beta^* \end{pmatrix} &= \begin{pmatrix} c_{11} & c_{12} \\ c_{21} & c_{22} \end{pmatrix}\begin{pmatrix} n\alpha + \sum x\beta + \sum e \\ \sum x\alpha + \sum x^2\beta + \sum xe \end{pmatrix} \\ &= \begin{pmatrix} \alpha \\ \beta \end{pmatrix} + \begin{pmatrix} c_{11} & c_{12} \\ c_{21} & c_{22} \end{pmatrix}\begin{pmatrix} \sum e \\ \sum xe \end{pmatrix}\end{aligned} \tag{15.17}$$

The statistic $(\alpha^*, \beta^*)$ is also that part of the maximum likelihood statistic which refers to the parameters $\alpha$ and $\beta$. To find the distributional properties of $\alpha^*$ and $\beta^*$, we need the distributional properties of $\sum e$ and $\sum xe$. Let $R_1 = \sum e$ and $R_2 = \sum xe$. Now

$$\begin{aligned} E(R_1) &= E(R_2) = 0 \\ V(R_1) &= n\sigma^2 = \sigma^2 S_{11} \end{aligned} \tag{15.18}$$

and

$$\begin{aligned} V(R_2) &= \sigma^2 \sum x^2 = \sigma^2 S_{22} \\ \mathrm{Cov}(R_1, R_2) &= \sigma^2 \sum x = \sigma^2 S_{12} = \sigma^2 S_{21} \end{aligned} \tag{15.19}$$

Because

$$\alpha^* = \alpha + c_{11}R_1 + c_{12}R_2$$

then

$$E(\alpha^*) = \alpha$$

and

$$\begin{aligned} V(\alpha^*) &= \sigma^2(c_{11}^2 S_{11} + c_{12}^2 S_{22} + 2c_{11}c_{12}S_{12}) \\ &= \sigma^2 c_{11}(c_{11}S_{11} + c_{12}S_{12}) + \sigma^2 c_{12}(c_{11}S_{12} + c_{12}S_{22}) \\ &= \sigma^2 c_{11} \end{aligned} \tag{15.20}$$

Similarly, $\beta^* = \beta + c_{21}R_1 + c_{22}R_2$. So

$$E(\beta^*) = \beta \tag{15.21}$$

and

$$\begin{aligned} V(\beta^*) &= \sigma^2(c_{21}^2 S_{11} + c_{22}^2 S_{22} + 2c_{21}c_{22}S_{12}) \\ &= \sigma^2 c_{21}(c_{21}S_{11} + c_{22}S_{12}) + \sigma^2 c_{22}(c_{21}S_{12} + c_{22}S_{22}) \\ &= \sigma^2 c_{22} \end{aligned} \tag{15.22}$$

Finally, $\text{Cov}(\alpha^*, \beta^*) = \sigma^2 c_{12}$. The arguments on variances and the covariance tell us why it is advantageous to compute the numbers $c_{11}$, $c_{12}$, $c_{21}$, and $c_{22}$. In fact $\alpha^*$ and $\beta^*$ have a bivariate normal distribution with means $\alpha$ and $\beta$, variances $\sigma^2 c_{11}$ and $\sigma^2 c_{22}$, and covariance $\sigma^2 c_{12}$. This is obvious because both $\alpha^*$ and $\beta^*$ are linear functions of normal variables.

What more can we say about $\alpha^*$ and $\beta^*$? It is relevant to note that we could have changed the model to read

$$y = \gamma + \delta(x - \bar{x}) + \text{error} \tag{15.23}$$

so that

$$\alpha = \gamma - \delta\bar{x}$$

and

$$\delta = \beta \tag{15.24}$$

If we had done this, the normal equations would have been

$$\begin{aligned} n\gamma &= \sum y \\ \sum (x - \bar{x})^2\delta &= \sum (x - \bar{x})y \end{aligned} \tag{15.25}$$

so that the least squares estimates of $\gamma$ and $\delta$ would have been

$$\begin{aligned} \gamma^* &= \bar{y} \\ \delta^* &= \sum (x - \bar{x})y/\sum (x - \bar{x})^2 \end{aligned} \tag{15.26}$$

The reader may verify, as would be expected, that

$$\begin{aligned} \alpha^* &= \gamma^* - \delta^*\bar{x} \\ \delta^* &= \beta^* \end{aligned} \tag{15.27}$$

Also he may verify that

$$\begin{aligned} V(\gamma^*) &= \sigma^2/n \\ V(\delta^*) &= \sigma^2/\sum (x - \bar{x})^2 \\ \text{Cov}(\gamma^*, \delta^*) &= 0 \end{aligned} \tag{15.28}$$

In the case of a single sample from a normal population, we gave a factorization of the probability into three parts: (1) the probability of $\bar{x}$, the sample mean, given $\sigma^2$; (2) the probability of $\sum (x - \bar{x})^2/(n - 1)$ given $\sigma^2$; and (3) the probability of the sample $x_1, x_2, \ldots, x_n$ given $\bar{x}$ and $\sum (x - \bar{x})^2/(n - 1)$. Part (2) did not depend on the mean $\mu$, and part (3) did not depend on $\mu$ or on $\sigma^2$. We can achieve a similar factorization in the present case. We already have what corresponds to part (1) in the distribution of $\alpha^*$ and $\beta^*$. We now want a quantity whose distribution depends only on $\sigma^2$. Consider a deviation from the fitted relationship, say

$$y_m - \alpha^* - \beta^* x_m \tag{15.29}$$

Now

$$\begin{aligned} \alpha^* &= \alpha + \text{error} \\ \beta^* &= \beta + \text{error} \end{aligned} \tag{15.30}$$

so

$$\begin{aligned} &y_m - \alpha^* - \beta^* x_m \\ &\quad = \alpha + \beta x_m + \text{error} - (\alpha + \text{error}) - (\beta + \text{error}) x_m \end{aligned} \tag{15.31}$$

and this depends solely on error, independent of the values of $\alpha$ and $\beta$. It is natural perhaps, and turns out to be appropriate by virtue of reasons we cannot give here, to consider the sum of squares of these residuals. We saw that $\alpha^* = \bar{y} - \beta^* \bar{x}$, and

$$\beta^* = \frac{\sum (x - \bar{x}) y}{\sum (x - \bar{x})^2}$$

Note in passing that the numerator of $\beta^*$ can also be written as $\sum (x - \bar{x})(y - \bar{y})$. Hence

$$\begin{aligned} y_m - \alpha^* - \beta^* x_m &= y_m - \bar{y} + \beta^* \bar{x} - \beta^* x_m \\ &= (y_m - \bar{y}) - \beta^*(x_m - \bar{x}) \end{aligned} \tag{15.32}$$

The sum of squares of these residuals is

$$\begin{aligned} &\sum [(y_m - \bar{y}) - \beta^*(x_m - \bar{x})]^2 \\ &= \sum (y_m - \bar{y})^2 - 2\beta^* \sum (y_m - \bar{y})(x_m - \bar{x}) + \beta^{*2} \sum (x_m - \bar{x})^2 \\ &= \sum (y_m - \bar{y})^2 - \frac{[\sum (y_m - \bar{y})(x_m - \bar{x})]^2}{\sum (x_m - \bar{x})^2} \end{aligned}$$

and also equals

$$\sum (y_m - \bar{y})^2 - \beta^{*2} \sum (x_m - \bar{x})^2 \tag{15.33}$$

Now taking expectations, we have

$$
\begin{aligned}
E[\textstyle\sum (y_m - \bar{y})^2] \\
&= E \textstyle\sum [\beta(x_m - \bar{x}) + (e_m - \bar{e})]^2 \\
&= \beta^2 \textstyle\sum (x_m - \bar{x})^2 + 2\beta \textstyle\sum (x_m - \bar{x}) E(e_m - \bar{e}) + E \textstyle\sum (e_m - \bar{e})^2 \\
&= \beta^2 \textstyle\sum (x_m - \bar{x})^2 + (n - 1)\sigma^2 \qquad (15.34)
\end{aligned}
$$

The reader may need to do a little algebra to verify this step. Also

$$
\begin{aligned}
E[\beta^{*2} \textstyle\sum (x_m - \bar{x})^2] &= \textstyle\sum (x_m - \bar{x})^2 E(\beta^{*2}) \\
&= \textstyle\sum (x_m - \bar{x})^2 \{[E(\beta^*)]^2 + V(\beta^*)\} \\
&= \textstyle\sum (x_m - \bar{x})^2 \left[\beta^2 + \frac{\sigma^2}{\sum (x_i - \bar{x})^2}\right] \\
&= \beta^2 \textstyle\sum (x_m - \bar{x})^2 + \sigma^2 \qquad (15.35)
\end{aligned}
$$

Hence the expectation of the sum of squares of residuals is

$$
\beta^2 \textstyle\sum (x_m - \bar{x})^2 + (n - 1)\sigma^2 - [\beta^2 \textstyle\sum (x_m - \bar{x})^2 + \sigma^2] = (n - 2)\sigma^2 \qquad (15.36)
$$

This suggests that the quantity $s^2 = \sum (\text{residual})^2/(n - 2)$ gives us an idea of the magnitude of $\sigma^2$. It can be proved that:

1. $(n - 2)\, s^2/\sigma^2$ is distributed as $\chi^2$ with $n - 2$ degrees of freedom.
2. The distribution of $s^2$ is independent of the distribution of $\alpha^*$ and $\beta^*$.
3. The probability of the sample observations in fact separates into three parts:
   a) The distribution of $\alpha^*$, $\beta^*$ given $\alpha$, $\beta$, and $\sigma^2$, which is, as stated above, the bivariate normal distribution.
   b) The distribution of $s^2$ or $(n - 2)s^2$ given $\sigma^2$.
   c) The distribution of the $y_i$, $i = 1, 2, \ldots, n$ given $\alpha^*$, $\beta^*$, and $s^2$, which is independent of $\alpha$, $\beta$, and $\sigma^2$.

Fact 3(c) tells us that $\alpha^*$, $\beta^*$, and $s^2$ comprise a minimal sufficient statistic. The process here therefore, is like that outlined in Chapter 10. These distributional facts, which we have stated and *not* proved, tell us how we may form opinions about the unknowns in the situation.

We now consider $\alpha$. We know that $\alpha^*$ is normally distributed around $\alpha$ with variance $\sigma^2 \sum x^2/[n \sum (x - \bar{x})^2]$. Let us write this in the more usual form:

$$
\sigma^2 \left[\frac{1}{n} + \frac{\bar{x}^2}{\sum (x - \bar{x})^2}\right]
$$

Also $(n - 2)s^2/\sigma^2$ is distributed as $\chi^2$ with $n - 2$ degrees of freedom, independently of $\alpha^*$. Hence

$$\frac{\alpha^* - \alpha}{\sqrt{s^2\left[\frac{1}{n} + \frac{\bar{x}^2}{\sum (x - \bar{x})^2}\right]}}$$

is distributed as $t$ with $n - 2$ degrees of freedom. It follows that values of $\alpha$ within the interval

$$\left[\alpha^* - t_{\gamma/2}s\sqrt{\frac{1}{n} + \frac{\bar{x}^2}{\sum (x - \bar{x})^2}},\ \alpha^* + t_{\gamma/2}s\sqrt{\frac{1}{n} + \frac{\bar{x}^2}{\sum (x - \bar{x})^2}}\right] \tag{15.37}$$

where $t_{\gamma/2}$ is the value of Student's $t$ with $n - 2$ degrees of freedom which is exceeded in absolute magnitude with probability $\gamma$, are consonant with the data by the two-sided $t$ test at the $\gamma$ level. Similarly $\beta^*$ is normally distributed with mean $\beta$ and variance $\sigma^2/\sum (x - \bar{x})^2$, so values of $\beta$ within the interval

$$\left[\beta^* - t_{\gamma/2}\frac{s}{\sqrt{\sum (x - \bar{x})^2}},\ \beta^* + t_{\gamma/2}\frac{s}{\sqrt{\sum (x - \bar{x})^2}}\right] \tag{15.38}$$

are consonant with the data at the $\gamma$ level. It also follows that values of $\sigma^2$ in the interval

$$\left[\frac{(n - 2)s^2}{\chi^2_{\gamma/2}},\ \frac{(n - 2)s^2}{\chi^2_{1-\gamma/2}}\right] \tag{15.39}$$

are consonant with the data by the $\chi^2$ test at the $\gamma$ level.

It can be shown, furthermore, that the set of values $\alpha$, $\beta$ such that

$$\frac{n[\alpha^* + \beta^*\bar{x} - (\alpha + \beta\bar{x})]^2 + \sum (x - \bar{x})^2(\beta^* - \beta)^2}{2s^2} \le F_{\gamma;\,2,\,n-2} \tag{15.40}$$

are consonant with the data by the $F$ test at the level $\gamma$.

We may also adjoin some other results. Consider $\alpha + \beta x_0$ which is the "true" value of $y$ when $x = x_0$. We know that $\alpha^* + \beta^* x_0$ is normally distributed with mean $\alpha + \beta x_0$ and variance

$$\sigma^2\left[\frac{1}{n} + \frac{(x_0 - \bar{x})^2}{\sum (x - \bar{x})^2}\right]$$

as the reader should check. Hence we may say that values of $\alpha + \beta x_0$ within the interval with end points

$$\alpha^* + \beta^* x_0 \pm t_{\gamma/2}s\sqrt{\frac{1}{n} + \frac{(x_0 - \bar{x})^2}{\sum (x - \bar{x})^2}} \tag{15.41}$$

are consonant with the data by the two-sided $t$ test at level $\gamma$. It is obvious from a cursory examination of this formula that the shortest interval occurs at $x_0 = \bar{x}$. This is emphasized by plotting the intervals for all $x$ as shown in Figure 15.3. Although this figure shows the limits of the $y$ population means simultaneously for all $x$ values, we are not entitled to interpret all lines lying within this region as consonant with the data at the $\gamma$ level of significance. Rather, for a given $x$ value, say $x_0$, the points between the upper and lower limits represent values for $\alpha + \beta x_0$ which are consonant with the data at the $\gamma$ level of significance. One can portray on the same graph a family of regions for different values of $x_0$.

We can construct a consonance region for the entire line by using the formula:

$$\alpha^* + \beta^* x_0 \pm s\sqrt{2F_{\gamma;\,2,\,n-2}\left[(1/n) + (x_0 - \bar{x})^2/\sum (x - \bar{x})^2\right]} \quad (15.42)$$

This region will resemble the region in Figure 15.3 but will be somewhat wider. It will have the property that all lines in the region are consonant with the data at the level of significance $\gamma$.

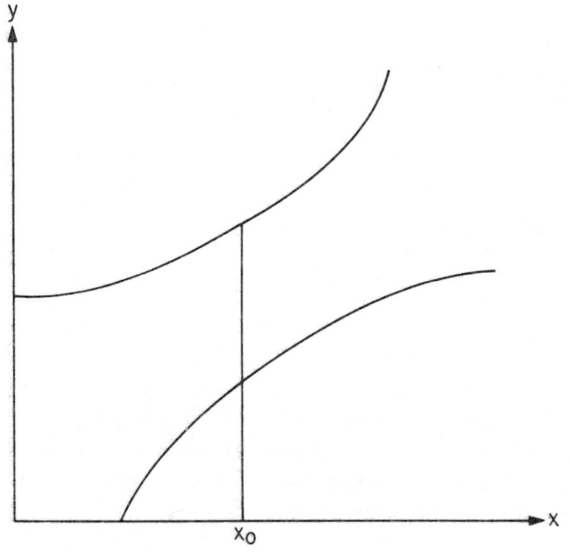


**Fig. 15.3** Consonance Intervals on $\alpha + \beta x_0$

*Example 1*

Let us now give a short numerical example illustrating the procedures we have discussed. Consider the following trivial set of data:

| $x$ | 1 | 2 | 3 | 4 | 5 |
|---|---|---|---|---|---|
| $y$ | 2 | 4 | 6 | 7 | 9 |

The normal equations are

$$\begin{pmatrix} 5 & 15 \\ 15 & 55 \end{pmatrix} \begin{pmatrix} \alpha \\ \beta \end{pmatrix} = \begin{pmatrix} 28 \\ 101 \end{pmatrix}$$

and the inverse matrix is given by

$$\begin{pmatrix} 11/10 & -3/10 \\ -3/10 & 1/10 \end{pmatrix}$$

Thus the solutions to the normal equations are given by

$$\begin{pmatrix} \alpha^* \\ \beta^* \end{pmatrix} = \begin{pmatrix} .5 \\ 1.7 \end{pmatrix}$$

with the following variances and covariance: $\mathrm{Var}(\alpha^*) = 11\sigma^2/10$, $\mathrm{Var}(\beta^*) = \sigma^2/10$, and $\mathrm{Cov}(\alpha^*, \beta^*) = -3\sigma^2/10$. To calculate the sum of squares of residuals, we can calculate each residual and then form the sum of squares as shown in Table 15.2.

**Table 15.2** Sum of Squares

| $x$ | $y$ | $y^* = \alpha^* + \beta^* x$ | $y - y^*$ | $(y - y^*)^2$ |
|---|---|---|---|---|
| 1 | 2 | 2.2 | −0.2 | .04 |
| 2 | 4 | 3.9 | 0.1 | .01 |
| 3 | 6 | 5.6 | 0.4 | .16 |
| 4 | 7 | 7.3 | −0.3 | .09 |
| 5 | 9 | 9.0 | 0.0 | .00 |
| | | | 0.0 | .30 |

From a computational viewpoint it is generally simpler to make use of the formula $\sum (y - \bar{y})^2 - [\sum (x - \bar{x})(y - \bar{y})]^2 / \sum (x - \bar{x})^2$. At any rate, $s^2$ is calculated to be $.3/3 = .1$.

## 15.6 The Problem of Goodness of Fit

The general problem of goodness of fit of a model of the type we are discussing is a very difficult problem. In fact a reasonable assessment of the present literature (1970) seems to indicate that we know very little about ways of examining the question. This is not to say that the literature contains nothing, but it does appear that there are not at present any easy procedures whose properties are simple and clear. Basic papers on the topic are those of Anscombe (1961) and Anscombe and Tukey (1963).

There is, however, one case in which a goodness of fit evaluation procedure has a simple structure—where we have more than one $y$ value for one or more $x$ values. Let us suppose that at $x_i$ we have $n_i$ $y$ values, $y_{i1}$, $y_{i2}, \ldots, y_{in_i}$. If our model is appropriate these $n_i$ values are a random sample from a normal distribution, so that

$$\sum_{j=1}^{n_i} (y_{ij} - y_{i.})^2/\sigma^2 \sim \chi^2_{n_i - 1} \tag{15.43}$$

where $y_{i.} = \sum_j y_{ij}/n_i$. Further, since the $y_{ij}$'s are independent under our model and the sum of independent $\chi^2$ variables is a $\chi^2$ variable

$$\sum_{ij} (y_{ij} - y_{i.})^2/\sigma^2 = W/\sigma^2 \sim \chi^2_{n_w} \tag{15.44}$$

where $n_w = \sum_i (n_i - 1)$.

We have already asserted that the sum of squares of residuals from the fitted line divided by $\sigma^2$ is a $\chi^2$ variable under the model. That is,

$$\sum_{ij} (y_{ij} - \alpha^* - \beta^* x_i)^2/\sigma^2 = R/\sigma^2 \sim \chi^2_{n-2} \tag{15.45}$$

where $n = \sum n_i$.

Now it can be shown that the residual sum of squares $R$ is greater than or equal to the within $x$ array sum of squares $W$, so that the sum of squares of residuals may be written as

$$R = W + (R - W) \tag{15.46}$$

It can also be shown under the assumption of the model that

$$\begin{aligned} W/\sigma^2 &\sim \chi^2_{n_w} \\ (R - W)/\sigma^2 &\sim \chi^2_{n-2-n_w} \end{aligned} \tag{15.47}$$

independently of each other. We may examine, therefore, the goodness of fit of the straight line model by forming the ratio

$$\frac{W}{n_w} \Big/ \frac{R - W}{n - 2 - n_w} \tag{15.48}$$

Under the conditions of the model this ratio is a $F$ variable with $n_w$ and $n - 2 - n_w$ degrees of freedom.

## 15.7 The Linear Calibration Problem

Suppose we have a linear calibration problem of the following type. A good measurement, $x$, of an attribute of specimens is very expensive, but we have an easy method giving $y$, which has a sizable error. A possible procedure is to use the good method and the easy method on $n$ specimens, giving results $(x_i, y_i)$, $i = 1, 2, \ldots, n$ respectively. Then we determine the dependence of $y$ on $x$ and for subsequent specimens measure $y$ and obtain the $x$ value from the fitted relationship.

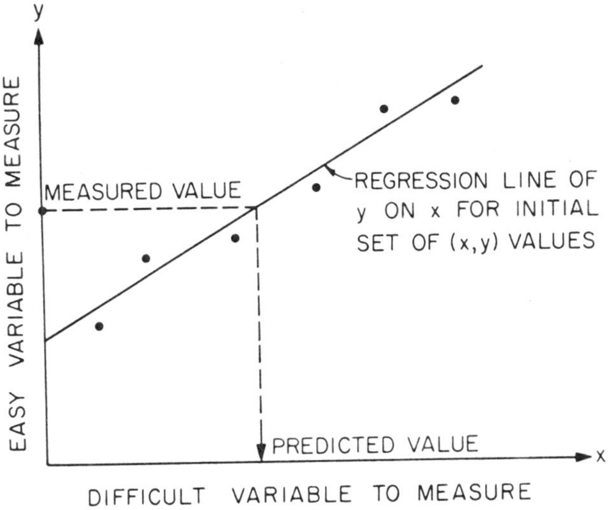


**Fig. 15.4** Linear Calibration

Let $y_0$ be the $y$ value for the new specimen and let $x_0$ be its corresponding $x$ value. Then $y_0 - \alpha^* - \beta^* x_0$ is the deviation of this $y$ value from the fitted one. This is a random variable with expectation zero and variance equal to

$$\sigma^2\left[1 + \frac{1}{n} + \frac{(x_0 - \bar{x})^2}{\sum (x - \bar{x})^2}\right]$$

and is independent of $s^2$. Hence

$$R = \frac{y_0 - \bar{y} - \beta^*(x_0 - \bar{x})}{s\left[1 + \frac{1}{n} + \frac{(x_0 - \bar{x})^2}{\sum (x - \bar{x})^2}\right]^{1/2}} \tag{15.49}$$

is a $t$ variable with $n - 2$ degrees of freedom and can be used to obtain tests of significance for hypothesized values of $x_0$ and by inversion to obtain consonance intervals for $x_0$. In other words, $R$ is a pivotal which depends only on the unknown "parameter," $x_0$. For a two-tailed $t$ test the level of significance will exceed $\gamma$ if $|R| \leq t_{\gamma/2}$. So we want to find values of $x_0$ such that $|R| \leq t_{\gamma/2}$. Let

$$\begin{aligned} S &= \sum (x - \bar{x})^2 \\ U &= y_0 - \bar{y} \\ V &= x_0 - \bar{x} \end{aligned} \tag{15.50}$$

We want to find all $x_0$ such that

$$(U - \beta^* V)^2 \leq t^2 s^2 \left(1 + \frac{1}{n} + \frac{V^2}{S}\right)$$

or

$$V^2 \left(\beta^{*2} - \frac{t^2 s^2}{S}\right) - 2V\beta^* U + U^2 - t^2 s^2 (n + 1)/n \leq 0 \quad (15.51)$$

where $t = t_{\gamma/2}$.

Then if $\beta^{*2} > t^2 s^2/S$, the values of $x_0$ consonant with the data at the level $\gamma$ are given by the interval

$$\bar{x} + \frac{\beta^* U}{\lambda} \pm \frac{ts}{\lambda}\sqrt{\frac{U^2}{S} + \frac{\lambda(n + 1)}{n}} \quad (15.52)$$

where $\lambda = \beta^{*2} - t^2 s^2/S$. If $\beta^{*2} = t^2 s^2/S$, the set of values of $x_0$ consonant with the data is a one-sided interval with boundary given by

$$- 2(x_0 - \bar{x})\beta^* U + U^2 - t^2 s^2 (n + 1)/n = 0 \quad (15.53)$$

If $\beta^{*2} < t^2 s^2/S$, the set of values consonant with the data is either the entire line or the union of two one-sided intervals.

## 15.8 The Straight Line through the Origin

One may deem it desirable on occasion to consider the case where $f(x) = \beta x$. There are no essential differences between this case and the case of $f(x) = \alpha + \beta x$ except that now $\alpha$ is forced to be zero, and only $\beta$ and $\sigma^2$ are unknown. It is easy to see that normality of error plus constant variance leads to least squares as a mode of finding the point of maximum of the likelihood and to the derivation of minimal-sufficient statistics. In fact, $\beta^*$ is given by $\sum xy/\sum x^2$. Also the minimum sum of squares is $\sum y^2 - (\sum xy)^2/\sum x^2$ and is distributed as $\chi^2\sigma^2$, independently of $\beta^*$, with $n - 1$ degrees of freedom. The ensuing developments can be provided by the reader or student as a formal exercise.

## 15.9 The Case $xCE|yCE$

We have already given an example of how this can arise. Our data will consist of $n$ pairs $(x_i, y_i)$, $i = 1, 2, \ldots, n$, and a plausible model is

$$x_i = X_i + e_i$$
$$y_i = Y_i + f_i$$

with

$$Y_i = f(X_i)$$

A simple case occurs when $Y_i = \alpha + \beta X_i$. There is a considerable literature on this. If we adjoin error assumptions, as, for instance, that the $e_i$ are $N(0, \sigma_e^2)$, the $f_i$ are $N(0, \sigma_f^2)$, and the sets $\{e_i\}$, $\{f_i\}$ are independent, the approximate likelihood can be written down immediately. It is immediately clear that we have a large list of parameters, namely, $X_1, X_2, \ldots, X_n, \alpha, \beta, \sigma_e^2, \sigma_f^2$.

This is clearly an uninformative function. The present situation brings to light a very important principle which we can place in antithesis to the likelihood principle mentioned in Chapter 11. *There are cases in which the likelihood function is totally uninformative.*

This continuized likelihood is an ill-behaved function, for we can put $x_i = X_i$ and then are open to the possibility of an indeterminate form 0/0. This topic is very deep, and we can give only very simple ideas here. To make progress one must be able either to (1) make additional assumptions, or (2) make repetitions of $x_i$ for $X_i$. The latter is really a case of the former from a certain viewpoint. If we can assume that some of the $X_i$ are equal, we can obtain some informative statistics which are based on the probability of the observations. If we can make an assumption of the magnitude of $\sigma_e^2$ in relation to the spread of the $X_i$, we can develop procedures. To exemplify, suppose for instance that the range of the $X_i$ is 20 units and that $\sigma_e$ is known (somehow or other) to be less than one. Then we can partition the $(x_i, y_i)$ into two sets $S_1$, $S_2$ and then take the line joining the points

$$[\underset{S_1}{\text{Ave}}(x), \underset{S_1}{\text{Ave}}(y)] \quad \text{and} \quad [\underset{S_2}{\text{Ave}}(x), \underset{S_2}{\text{Ave}}(y)]$$

as an estimate or indication of the "true" line.

This procedure has been examined by Wald (1940) and others from the viewpoint of consistency, i.e., behavior as $n \to \infty$, and Wald has provided approximate standard errors and statistical intervals. Consistent estimators of $\sigma_e^2$ and $\sigma_f^2$ are given by

$$\begin{aligned} \sigma_e^{*2} &= s_x^2 - s_{xy}/\beta^* \\ \sigma_f^{*2} &= s_y^2 - \beta^* s_{xy} \end{aligned} \tag{15.54}$$

where $\alpha^*$ and $\beta^*$ are obtained by fitting the line as described and

$$\begin{aligned} s_x^2 &= \sum (x - \bar{x})^2/(n - 1) \\ s_y^2 &= \sum (y - \bar{y})^2/(n - 1) \\ s_{xy} &= \sum (x - \bar{x})(y - \bar{y})/(n - 1) \end{aligned} \tag{15.55}$$

The formation of opinion about values of $\beta$ consonant with the data is possible if we assume normality for $e_i$ and $f_i$. Bartlett (1949) gives a similar method but breaks the data into three groups instead of two.

## 15.10 The Case $xRN|yRE$

We leave this case with the sole remark that it can be incorporated in the case $xRN|yRN$ described above by imposing the condition $\sigma_f^2 = 0$. This alone does not lead to identification in the sense that a very large set of data will tell us parameter values, even under the simplest normality assumptions.

## 15.11 The Use of Transformations

It will always be desired to obtain, if possible, a linear relation. To achieve this, it may be desired to transform $y$ by some monotone function of $y$. If $y$ is greater than zero, $\sqrt{y}$ or $\ln y$ or $y^2$ are possible simple candidate functions, though clearly there are many. Also one may consider making a transformation of $x$ to $g(x)$, so that the relationship of $y$ to $g(x)$ is linear. Transformations of $x$, the explanatory variable, cause little problem as long as the transformation is totally specified. Transformations to a function of $x$ with one or more unknown parameters cause problems, e.g., replacing $x$ by $\ln(x - \theta)$, where $\theta$ is to be determined. In the same way transformations of $y$ to a function of $y$ with unknown parameters cause difficulty.

Transformations to a known function of $y$ such as $\ln y$ or $y^2$ can cause other problems. The methodology described here is based on the model that the variability in $y$ values is represented by the same normal distribution for all $x$ values, actually the $N(0, \sigma^2)$ distribution. If this were actually the case, but it appeared that the use of $\ln y$ would give a straight line relationship, the variability of the quantity being explained or fitted, $\ln y$, would not be the same for $x$ values, and the methodology described here would be inappropriate.

## 15.12 Other Error Assumptions

We have seen how the assumption of a normal distribution of errors with variance independent of $x$ leads to "least squares fitting." A simple modification of this model would be to assume that the distribution of errors is normal with mean zero for each $x$ but has variance $h(x)\sigma^2$, where $h(x)$ is known or guessed. A simple transformation from $y$ to $z = y/\sqrt{h(x)}$ gives normal distributions with mean zero and the same variance $\sigma^2$. The linear model would then be

$$z = \alpha \frac{1}{\sqrt{h(x)}} + \frac{\beta x}{\sqrt{h(x)}} + \text{error} \tag{15.56}$$

and this would be fitted by least squares.

A different type of fitting would be involved with nonnormal distribution of errors. Suppose a population of errors is given by the Laplacian distribution which has the density

$$\frac{1}{2\theta}e^{-|v|/\theta}, \qquad -\infty < v < \infty \tag{15.57}$$

In this case if the function $f(x)$ is $\alpha + \beta x$, the probability of the sample is essentially

$$\frac{1}{(2\theta)^n}\exp\left(-\frac{\sum|y_i - \alpha - \beta x_i|}{\theta}\right)\prod \Delta y_i \tag{15.58}$$

The point of maximum likelihood would be given by the values for $\alpha$ and $\beta$ such that

$$\sum |y_i - \alpha - \beta x_i| \tag{15.59}$$

is a minimum. That is, we would minimize the sum of the absolute deviations. This method of fitting is rather popular with the people knowledgeable in linear programming computing techniques. It is curious however that, to our knowledge, no significance tests are available.

Another model for the distribution of errors might be the uniform distribution with density

$$\frac{1}{2\theta}, \qquad -\theta \leq v \leq \theta \tag{15.60}$$

If again $f(x)$ is $\alpha + \beta x$, this would lead to examining the function

$$\frac{1}{(2\theta)^n}\prod \phi(y_i - \alpha - \beta x_i, \theta) \tag{15.61}$$

where $\phi(a, b)$ is unity if $|a| \leq b$ and is zero if $|a| > b$. The point of maximum likelihood would be given by the value such that $\max_i |y_i - \alpha - \beta x_i|$ is a minimum, with $\theta^*$ as this value. This method has some interest, but it is unfortunate that no significance tests exist to give one an opinion of what values of $\alpha, \beta$ are consonant with data. This model does not satisfy the regularity conditions used in the likelihood theory of Chapter 10 and therefore presents considerable difficulties.

## 15.13 The General Linear Model

This is the case $yRN|xCN$ or $yRE|xCN$, in which the "$x$" is not scalar but a set of scalars. Consider, for example, the model

$$y_i = \alpha_0 + \alpha_1 x_i + \alpha_2 x_i^2 + \text{error} \tag{15.62}$$

in which $x_i$ is known without error or the case

$$y_i = \alpha_0 + \alpha_1 x_i + \alpha_2 z_i + \text{error} \tag{15.63}$$

in which $x_i$ and $z_i$ are known without error. The general class of linear models contains these two examples and many like it, the general specification being that the expectation of each $y_i$ is a *known linear* function of unknown parameters. So the total observation for the $i$th specimen in Equation (15.62) is a triplet $(y_i, x_i, x_i^2)$ or even more generally a four-tuple $(y_i, 1, x_i, x_i^2)$, a mode of statement useful in general treatment. In the case of Equation (15.63) the observation (sometimes the phrase "data set" is used) is the four-tuple $(y_i, 1, x_i, z_i)$. The character of the general linear model is given by the statement that the observation is a $(p + 1)$-tuple $(y_i, x_{i1}, x_{i2}, \ldots, x_{ip})$ in which $x_{i1}, x_{i2}, \ldots, x_{ip}$ are known for all the numbers $y_i$ to be explained by a model $y_i = x_{i1}\beta_1 + x_{i2}\beta_2 + \cdots + x_{ip}\beta_p$ + error. Note that the expectation of each $y_i$ (assuming the expectation of error is zero, as is usual) is a known linear function of unknown parameters. This is to be contrasted with a model like

$$y_i = \alpha\beta^{x_i} + \text{error} \tag{15.64}$$

in which the $x_i$ are known, but not $\alpha$ and $\beta$, or the model

$$y_i = \alpha + \beta e^{\gamma x_i} + \text{error} \tag{15.65}$$

in which the $x_i$ are known but not $\alpha$, $\beta$, and $\gamma$. Some models which might be thought to be nonlinear are, in fact, linear models. Thus for instance

$$y_i = \alpha + \beta x_i + \gamma x_i^2 + \text{error}$$

or

$$y_i = \alpha + \beta \sin x_i + \text{error} \tag{15.66}$$

are linear with regard to the unknowns, but they are not linear with regard to the explanatory variable $x$. This lack of linearity in the explanatory variable is not relevant and does not affect the methodology. From a naive point of view the point in these models is that if we know $x$, we know $x^2$, or $\sin x$.

The whole pattern which is being examined is then:

$$\begin{aligned}
y_1 &= x_{11}\beta_1 + x_{12}\beta_2 + \cdots + x_{1p}\beta_p + e_1 \\
y_2 &= x_{21}\beta_1 + x_{22}\beta_2 + \cdots + x_{2p}\beta_p + e_2 \\
&\cdots\cdots\cdots\cdots\cdots\cdots\cdots\cdots \\
y_n &= x_{n1}\beta_1 + x_{n2}\beta_2 + \cdots + x_{np}\beta_p + e_n
\end{aligned} \tag{15.67}$$

We may use matrices to give a condensed statement of these equations:

$$\begin{pmatrix} y_1 \\ y_2 \\ \cdot \\ \cdot \\ y_n \end{pmatrix} = \begin{pmatrix} x_{11} & x_{12} \cdots x_{1p} \\ x_{21} & x_{22} \cdots x_{2p} \\ \cdot & \cdot \cdots \cdot \\ \cdot & \cdot \cdots \cdot \\ x_{n1} & x_{n2} \cdots x_{np} \end{pmatrix} \begin{pmatrix} \beta_1 \\ \beta_2 \\ \cdot \\ \cdot \\ \beta_p \end{pmatrix} + \begin{pmatrix} e_1 \\ e_2 \\ \cdot \\ \cdot \\ e_n \end{pmatrix} \tag{15.68}$$

or

$$y = X\beta + e \tag{15.69}$$

where $y$ is $n \times 1$, $X$ is $n \times p$, $\beta$ is $p \times 1$, and $e$ is $n \times 1$.

*Example 2*
Suppose for instance that we have the following data:

| $i$ | 1 | 2 | 3 | 4 | 5 |
|---|---|---|---|---|---|
| $x_i$ | 2 | 3 | 4 | 8 | 12 |
| $y_i$ | 8 | 10 | 14 | 12 | 13 |

and we wish to examine the model

$$y_i = \beta_0 + \beta_1 x_i + \beta_2(1/x_i) + \beta_3 x_i^3 + \text{error}$$

These data and model may be represented by the matrix equation

$$\begin{pmatrix} 8 \\ 10 \\ 14 \\ 12 \\ 13 \end{pmatrix} = \begin{pmatrix} 1 & 2 & 1/2 & 8 \\ 1 & 3 & 1/3 & 27 \\ 1 & 4 & 1/4 & 64 \\ 1 & 8 & 1/8 & 512 \\ 1 & 12 & 1/12 & 1728 \end{pmatrix} \begin{pmatrix} \beta_0 \\ \beta_1 \\ \beta_2 \\ \beta_3 \end{pmatrix} + \begin{pmatrix} e_1 \\ e_2 \\ e_3 \\ e_4 \\ e_5 \end{pmatrix}$$

If we assume that the $e_i$ are normally and independently distributed with common variance $\sigma^2$, the probability of the $e_i$'s is

$$\left(\frac{1}{\sqrt{2\pi}\,\sigma}\right)^n \exp(-\textstyle\sum e_i^2/2\sigma^2) \prod \Delta e_i \tag{15.70}$$

But $e = y - X\beta$ and $\sum e_i^2 = e'e$ so Formula (15.70) becomes

$$\left(\frac{1}{\sqrt{2\pi}\,\sigma}\right)^n \exp[-(y - X\beta)'(y - X\beta)/2\sigma^2] \prod \Delta e_i \tag{15.71}$$

The point of maximum is obtained by differentiating it or its logarithm with respect to $\beta_1, \beta_2, \ldots, \beta_p$ and $\sigma^2$, setting the derivatives equal to zero. To obtain the derivative equations with respect to the $\beta_i$, consider

$$[y - X(\beta + \Delta\beta)]'[y - X(\beta + \Delta\beta)] - (y - X\beta)'(y - X\beta)$$
$$= 2(\Delta\beta)'X'(y - X\beta) + (\Delta\beta)'X'X(\Delta\beta) \quad (15.72)$$

If we put $(\Delta\beta)' = (\Delta\beta_1, 0, 0, \ldots, 0)$, divide by $\Delta\beta_1$, and take the limit, we get the first element of the column vector of $X'(y - X\beta)$. Putting $(\Delta\beta)' = (0, \Delta\beta_2, 0, \ldots, 0)$, we obtain the second element of $2X'(y - X\beta)$, etc. So the matrix equation obtained by setting all these derivatives equal to zero is $X'(y - X\beta) = 0$, or

$$X'X\beta = X'y \quad (15.73)$$

These are called the normal equations. Taking the derivative with respect to $\sigma^2$ and equating to zero yields the equation

$$-\frac{n}{2\sigma^2} + \frac{1}{2\sigma^4}(y - X\beta)'(y - X\beta) = 0 \quad (15.74)$$

Then the solution to this equation is

$$\sigma^{*2} = (y - X\beta^*)'(y - X\beta^*)/n \quad (15.75)$$

where $\beta^*$ is a solution to the normal equations. If we suppose that $X'X$ is nonsingular, the solution $\beta^*$ is given by

$$\beta^* = (X'X)^{-1}X'y \quad (15.76)$$

The reader may verify that $E(\beta^*) = \beta$.

Note that to be fully consistent in notation we should write the maximum likelihood statistic for $\sigma^2$ as $(\sigma^2)^*$. However, it is clear that if $\sigma^*$ is the maximum likelihood statistic for $\sigma$, then $(\sigma^2)^* = (\sigma^*)^2$. We therefore denote the maximum likelihood statistic for $\sigma^2$ by $\sigma^{*2}$.

The minimum sum of squares is

$$\begin{aligned}&(y - X\beta^*)'(y - X\beta^*)\\ &= y'y - 2\beta^{*\prime}X'y + \beta^{*\prime}X'X\beta^*\\ &= y'y - \beta^{*\prime}X'y - \beta^{*\prime}(X'y - X'X\beta^*)\\ &= y'y - \beta^{*\prime}X'y\end{aligned} \quad (15.77)$$

That is, the minimum sum of squares is given by the total sum of squares minus the sum of the products of the $\beta_i^*$'s and the corresponding right-hand sides of the normal equations.

It can be shown that $(\beta_1^*, \beta_2^*, \ldots, \beta_p^*, \sigma^{*2})$ is a minimal sufficient statistic. The joint distribution of the component variables is easily obtained. Now $\beta^*$ is a linear function of the vector of observations $y$, so we know from Chapter 4 that $\beta^*$ has a multivariate normal distribution, with mean vector

$$(X'X)^{-1}X'(X\beta) = \beta \tag{15.78}$$

and covariance matrix

$$(X'X)^{-1}X'(\sigma^2 I)X(X'X)^{-1} = \sigma^2(X'X)^{-1} \tag{15.79}$$

We use again the theorem of matrix algebra that a real symmetric matrix can be diagonalized by an orthogonal matrix. Consider the minimum sum of squares

$$\begin{aligned} y'y - \beta^{*\prime}X'y &= y'[I - X(X'X)^{-1}X']y \\ &= e'[I - P]e \end{aligned} \tag{15.80}$$

where

$$P = X(X'X)^{-1}X' \tag{15.81}$$

Note that $P = P'$ and $P^2 = P$, so $I - P$ is an idempotent matrix of rank $n - p$, and there exists an orthogonal matrix $O$ such that

$$O'[I - X(X'X)^{-1}X']O = D \tag{15.82}$$

where $D$ is a diagonal matrix with $n - p$ of the $d_i = 1$ and $p$ of the elements equal to zero. Hence with

$$\eta = O'e$$

$$y'y - \beta^{*\prime}X'y = \sum_{i=1}^{n-p} \eta_i^2 \tag{15.83}$$

Also the $\eta_i$ are independent $N(0, \sigma^2)$ variables, so

$$\frac{y'y - \beta^{*\prime}X'y}{\sigma^2} \sim \chi^2_{n-p} \tag{15.84}$$

Furthermore, putting $L' = e'(I - P) = (L_1, L_2, \ldots, L_n)$, we have,

$$\begin{aligned} e'(I - P)e &= e'(I - P)'(I - P)e \\ &= L'L \\ &= \sum L_i^2 \\ &= \sum_i [\sum_j (I - P)_{ij} e_j]^2 \end{aligned} \tag{15.85}$$

and

$$\beta_i^* = \sum_j [(X'X)^{-1}X']_{ij} y_j \tag{15.86}$$

But

$$\begin{aligned}
\text{Cov}(L_i, \beta_{i'}^*) &= \text{Cov}\{\sum_j (I - P)_{ij} e_j, \sum_j [(X'X)^{-1}X']_{i'j} e_j\} \\
&= \sigma^2 \sum_j (I - P)_{ij} [(X'X)^{-1}X']_{i'j} \\
&= \sigma^2 \sum_j [(X'X)^{-1}X']_{i'j} (I - P)_{ji} \\
&= \sigma^2 [(X'X)^{-1}X'(I - P)]_{i'i} \\
&= \sigma^2 \{(X'X)^{-1}X'[I - X(X'X)^{-1}X']\}_{i'i} \\
&= 0
\end{aligned} \tag{15.87}$$

So $\beta_{i'}^*$ is uncorrelated with $L_i$, for all $i$, $i'$ and therefore the vector random variable $\beta^*$ is independent of $\sigma^{*2}$.

From Equation 15.83 we see that if we define

$$s^2 = (y - X\beta^*)'(y - X\beta^*)/(n - p) = (y'y - \beta^{*\prime}X'y)/(n - p)$$

then $E(s^2) = \sigma^2$. If $\lambda'\beta$ is a known linear function of the parameters $\beta_1$, $\beta_2, \ldots, \beta_p$, then

$$\lambda'\beta^* \sim N[\lambda'\beta, \lambda'(X'X)^{-1}\lambda\sigma^2]$$

It can be shown that

$$(\lambda'\beta^* - \lambda'\beta)/\sqrt{\lambda'(X'X)^{-1}\lambda s^2} \sim t_{n-p}$$

a distributional fact which leads directly to intervals for the unknown $\lambda'\beta$. Consonance regions for more than one linear function of $\beta_1, \beta_2, \ldots, \beta_p$ are obtainable from the $F$ distribution. See for example, Graybill (1961) where the adjective "confidence" is used rather than "consonance."

## 15.14 Final Remarks

We should emphasize to the reader that we have given only what may be called initial ideas on this vast area of thought. We have tried to give the reader a framework which will help him think about problems for further study. The consequences of these simple results with regard to statistical tests and intervals follow fairly easily from the basic facts we have presented. However, in Chapter 16 we present results which may be transferred to the present situation with fairly obvious identification of terms.

We should also emphasize to the student and reader that we have presented only the most elementary ideas of the relationships between variables. We have discussed the relationships between a pair of variables, exemplified by data consisting of $n$ pairs $(x_i, y_i)$, $i = 1, 2, \ldots, n$. We have also discussed the case of relationship of one variable, say $y$, which is to be explained in terms of explanatory variables, say, $x_1, x_2, \ldots, x_n$. But this is only a scratching of the surface. An investigator will often meet the case in which the observation on each individual consists of an $m$-plet $(x_1, x_2, \ldots, x_m)$, with observed $m$-plets $(x_{1i}, x_{2i}, \ldots, x_{mi})$, $i = 1, 2, \ldots, n$, and for which the relationship between any one subset of the $x$'s and another disjoint subset of the $x$'s is of interest. This will lead to the ideas of partial regression, partial correlation, conditional regression, conditional variance, principal components, factor analysis, and so on which are extremely important. Also, it will lead to the necessity of considering the ideas of path analysis and path coefficients. The worker in almost any field of investigation whether in social science or physical science, has to be concerned ultimately with the idea of causation and its use. To quote a hoary example, one does not infer from the fact that there was a positive correlation of imports of bananas and divorce rate, that the way to reduce a high divorce rate is to reduce the importation of bananas. The critically important ideas of path analysis as a mode of approach and resolution are due to a single individual, Sewall Wright, a geneticist of very high stature, and basic papers are those of Wright (1921, 1934, 1954). A detailed description of the development is given by Wright (1968). See also the paper by Tukey (1954). The reader may also find Kempthorne (1957, Chapter 14) an integrated concise description of the matter. It is an unfortunate fact for the professional statistician and the user of statistics that there is now a vast literature that is relevant and much of this relevant literature is *not* in books on "mathematical statistics." The material presented in this present chapter is to "real" statistics like the "Dick, Jane, and Sally" books of our grade schools are to real literature. Both the positive and negative aspects of this analogy must be given weight. Data interpretation is, like language, a problem of the ages; there are no panaceas, but merely suggestions that have been found useful by some investigators. The reader may find papers by Kempthorne (1971a, 1971b) informative.

## PROBLEMS

1. Show that the moment generating function of the bivariate normal is

$$m(t_1, t_2) = \exp[(t_1\mu_1 + t_2\mu_2) + \frac{1}{2}(t_1^2\sigma_1^2 + 2\rho\sigma_1\sigma_2 t_1 t_2 + t_2^2\sigma_2^2)].$$

2. Show that the m.g.f. of

$$\frac{1}{1-\rho^2}\left[\frac{(x-\mu_x)^2}{\sigma_x^2} - \frac{2\rho(x-\mu_x)(y-\mu_y)}{\sigma_x\sigma_y} + \frac{(y-\mu_y)^2}{\sigma_y^2}\right]$$

is $(1 - 2t)^{-1}$, thereby showing that it is a $\chi^2$ variable with 2 degrees of freedom.

3. Verify that the point of maximum likelihood of the bivariate normal is as given by Equation (15.6).

4. Verify that

$$\sum_{i=1}^{n} (x_i - \bar{x})(y_i - \bar{y}) = \sum_{i=1}^{n} (x_i - \bar{x})y_i$$
$$= \sum_{i=1}^{n} x_i(y_i - \bar{y})$$

5. Show for the straight line model, $y_i = \alpha + \beta x_i + e_i$, with the $e_i$ uncorrelated with mean 0 and variance $\sigma^2$ that $\bar{y}$ and $\sum (x_i - \bar{x})(y_i - \bar{y})$ are uncorrelated.

6. Complete the proof of the result that the residual sum of squares for the straight line model has expectation $(n - 2)\sigma^2$.

7. Given

| $i$ | 1 | 2 | 3 | 4 | 5 |
|---|---|---|---|---|---|
| $x_i$ | 1 | 1 | 2 | 3 | 4 |
| $y_i$ | 2 | 3 | 3 | 4 | 6 |

(a) Set up the normal equations.
(b) Solve for $\alpha^*$ and $\beta^*$.
(c) Estimate the variances of $\alpha^*$ and $\beta^*$ and their covariance.
(d) Calculate the significance level for the hypothesized model $\beta = 1$ against the alternative model that $\beta \neq 1$.
(e) Calculate the significance level of the hypothesized model $\alpha = 0$ against the alternative model that $\alpha < 0$.
(f) Sketch the significance level by a two-sided $t$ test for values of the mean of the population of $y$ values at $x = 6$.
(g) Determine a consonance region for the entire regression line at the 10% level of significance.

8. Given

| $x$ | 1 | 1 | 2 | 2 | 3 | 3 | 4 | 4 |
|---|---|---|---|---|---|---|---|---|
| $y$ | 10 | 9 | 8 | 8 | 6 | 7 | 5 | 5 |

show that the least squares line is the same as that obtained from

| $x$ | 1 | 2 | 3 | 4 |
|---|---|---|---|---|
| $y$ | 9.5 | 8 | 6.5 | 5 |

9. Derive the formula for the least squares estimate of $\beta$ in the model $y_i = \beta x_i + e_i$.

10. Derive the formulas for the least squares estimates of $\beta_1$ and $\beta_2$ in the model $y_i = \beta_1 x_i + \beta_2 x_i^2 + e_i$.

11. Repeat steps (a) through (g) of Problem 7 on the following data:

| $x$ | $y$ | $x$ | $y$ |
|---|---|---|---|
| 45.2 | 1001.2 | 48.0 | 1004.0 |
| 46.1 | 1001.6 | 48.1 | 1004.1 |
| 46.7 | 1001.2 | 48.2 | 1004.7 |
| 47.0 | 1002.4 | 48.3 | 1004.8 |
| 47.2 | 1002.7 | 49.0 | 1005.1 |
| 47.2 | 1003.8 | 50.1 | 1005.0 |
| 47.3 | 1003.6 | 52.0 | 1006.2 |

Coding the data will greatly simplify the calculations.

12. An experimenter wished to set a consonance region on the entire line at the 5% level of significance, but he mistakenly used the formula

$$\alpha^* + \beta^* x \pm t_{.025} s\sqrt{\frac{1}{n} + \frac{(x - \bar{x})^2}{\sum (x - \bar{x})^2}}$$

If $n = 25$, at about what level of significance can he regard this as a consonance region on the entire line?

13. A statistics student has calculated an interval by the use of a formula but has become confused about its interpretation. He used the formula

$$\alpha^* \pm t_{.025} s\sqrt{1 + \frac{1}{n} + \frac{\bar{x}^2}{\sum (x - \bar{x})^2}}$$

What interpretations can he attach to the resulting interval?

14. Specify transformations you would make to linearize the following models:

(a) $y = \dfrac{\beta_0}{\beta_1 x}$

(b) $y = \beta_0 x^{\beta_1}$

(c) $1 = \beta_0 e^{-\beta_1 x - y}$

(d) $y = e^{\kappa_1 e^{\kappa_2 x}}$

15. Four batches of standard material for which the melting points are known are obtained from the U.S. Bureau of Standards. From each batch the melting point of two samples was determined by an analytical laboratory. The data are as follows:

| *Melting Point* U.S. Bur. Stds. | $x$ | 200 | 200 | 250 | 250 | 300 | 300 | 350 | 350 |
|---|---|---|---|---|---|---|---|---|---|
| Melting Point Laboratory | $y$ | 212 | 210 | 260 | 261 | 307 | 309 | 356 | 359 |

(a) Fit the least squares line to be used as a calibration curve.

(b) Making use of the fitted line what would be your estimate of the melting point if it were measured as 300 by the laboratory?

(c) Set a consonance interval at the 5% level upon the melting point when it is measured as 300.

16. Use Wald's method of fitting a straight line for the following data:

| $x$ | 5 | 6 | 6 | 7 | 8 | 9 | 9 | 10 | 11 | 12 | 13 |
|---|---|---|---|---|---|---|---|---|---|---|---|
| $y$ | 96 | 95 | 98 | 97 | 101 | 102 | 104 | 104 | 105 | 106 | 110 |

Can one estimate $\sigma_e^2$ and $\sigma_f^2$?

17. For the model and data of Problem 7:
(a) Use the material in Section 15.13 and exhibit the $X$ matrix and the $y$ vector.
(b) Form $X'X$ and $X'y$ and show that the normal equations are:

$$\begin{pmatrix} 5 & 11 \\ 11 & 31 \end{pmatrix} \begin{pmatrix} \alpha \\ \beta \end{pmatrix} = \begin{pmatrix} 18 \\ 47 \end{pmatrix}$$

(c) Solve these normal equations to verify that you get the same result as in Problem 7.
(d) Calculate the residuals and show that the sum of squares of these is given by using the general formula $y'y - \beta^{*\prime}X'y$.

18. Suppose a plausible model is $y_i \sim \text{NID}(\beta x_i, \sigma^2 x_i)$, where the $x_i$ are known and where $\sigma^2$ is unknown. By making the transformation $z_i = y_i/\sqrt{x_i}$ develop:
(a) The maximum likelihood estimator of $\beta$.
(b) The "obvious" estimator of $\sigma^2$.
(c) A way of forming an opinion about $\beta$.

19. The problem is the same as in Problem 18 except $y_i \sim \text{NID}(\beta x_i, \sigma^2 x_i^2)$.

20. If $y_i \sim \text{NID}(\beta x_i, g(x_i)\sigma^2)$, where $g(x_i)$ is a known function, obtain (a), (b), and (c) stated in Problem 18.

21. Suppose $x_i \sim \text{NID}(\mu, \sigma^2)$, $i = 1, 2, \ldots, n$. Obtain the linear function which is the minimum variance unbiased estimate of $\mu$.

22. Suppose $y_i$ are distributed $\text{NID}(\alpha + \beta x_i, \sigma^2)$. Obtain the minimum variance unbiased estimate of $\alpha$ and $\beta$.

# 16

# Structured Populations

## 16.1 Introduction

In the previous chapters we have been concerned primarily with a set of data which we assumed to be representable as a random sample from some distribution. Only in Chapter 15 did we consider models relating the data to explanatory variables. We now take up the case where the explanatory variables are categorical. The reason for using our title "Structural Populations" is that our original population can be partitioned into subpopulations on the basis of the categorical variables. The whole population consists of a large number of subpopulations.

It is critical to bear in mind a general picture of what one does in this type of investigative situation.

1. One decides what to observe on the basis of one's personal knowledge and opinions of associations and of factors which may prove of value for explanatory purposes.
2. With the obtained data one forms in an objective way judgments of whether potential explanatory factors influence the variable to be explained.
3. On the basis of extensive activity of the type described in (2), one arrives at a class of models indexed by parameters—mathematical variables.
4. One forms opinions of the degree of tenability of the class of models and, conditionally on a class being tenable, on the tenability of values for the parameters.

In these respects, the overall investigative problem is no different from that discussed in Chapter 8. It is just that the order of complexity is vastly greater.

## 16.2 Single Criterion of Classification: The One-Way Classification

Consider a population of individuals which can be partitioned into subpopulations by means of a single categorical variable, and suppose that the subpopulations are of equal size. Suppose also that one arithmetical variable is observed. Then the data consist of triples $(i, j, y)$ in which $i$ denotes the subpopulation, $j$ indexes the individuals in the subpopulation, and $y$ is the value of the arithmetical variable for the individual. Suppose that $i$ ranges from 1 to $I$ and that $j$ ranges from 1 to $J$. Then we can represent the data by the symbol $y_{ij}$, with $i$ ranging from 1 to $I$, and $j$ ranging from 1 to $J$. It is natural to think about the following, in the writing of which we use the standard convention of replacing a subscript by a dot to indicate that an average over the deleted subscript has been formed:

$$y_{..} = \frac{1}{IJ}\sum_i \sum_j y_{ij}, \text{ the population mean}$$

$$y_{i.} = \frac{1}{J}\sum_j y_{ij}, \text{ the mean of the } i\text{th subpopulation}$$

$$y_{i.} - y_{..} = \text{the deviation of the } i\text{th subpopulation mean from the overall mean}$$

Indeed we can write

$$y_{ij} = y_{..} + (y_{i.} - y_{..}) + (y_{ij} - y_{i.})$$

or

$$y_{ij} - y_{..} = (y_{i.} - y_{..}) + (y_{ij} - y_{i.}) \tag{16.1}$$

These are clearly identities. Consider now the variance in the overall population. It is convenient to use a divisor of population size minus one. Then

$$\sigma^2 = \frac{1}{IJ - 1}\sum_{ij} (y_{ij} - y_{..})^2 \tag{16.2}$$

We also have for $\sigma_i^2$, the variance within the $i$th subpopulation,

$$\sigma_i^2 = \frac{1}{J - 1}\sum_j (y_{ij} - y_{i.})^2 \tag{16.3}$$

Now take Equation (16.1) and square and sum. We have

$$\begin{aligned}\textstyle\sum_{ij} (y_{ij} - y_{..})^2 &= \textstyle\sum_{ij} [(y_{i.} - y_{..}) + (y_{ij} - y_{i.})]^2 \\ &= \textstyle\sum_{ij} (y_{i.} - y_{..})^2 + \sum_{ij} (y_{ij} - y_{i.})^2 + \sum_{ij} 2(y_{i.} - y_{..})(y_{ij} - y_{i.})\end{aligned} \tag{16.4}$$

But the third term on the right of Equation (16.4) is zero and hence

$$\begin{aligned}\textstyle\sum_{ij} (y_{ij} - y_{..})^2 &= \textstyle\sum_{ij} (y_{i.} - y_{..})^2 + \sum_{ij} (y_{ij} - y_{i.})^2 \\ &= J \textstyle\sum_i (y_{i.} - y_{..})^2 + \sum_{ij} (y_{ij} - y_{i.})^2 \qquad (16.5)\end{aligned}$$

The left-hand side is called, for obvious reasons, the total sum of squares. The first term in the right-hand side is called the between-group sum of squares and the second, the within-group sum of squares.

Now consider the idea that the subpopulation means $y_{i.}$ are a finite population of $I$ numbers with mean $y_{..}$. Then we can define a variance of these means in the finite population of $I$ such means. This we take to be $\sigma_b^2$, given by

$$\sigma_b^2 = \frac{1}{I-1} \sum_i (y_{i.} - y_{..})^2 \qquad (16.6)$$

We then have the identity

$$(IJ - 1)\sigma^2 = (I - 1)J\sigma_b^2 + (J - 1) \textstyle\sum_i \sigma_i^2 \qquad (16.7)$$

This can be written in the form of Table 16.1.

**Table 16.1** Population Analysis of Variance

| Source | Sum of Squares (SS) | SS in Terms of $\sigma$'s |
|---|---|---|
| Between groups | $J \sum_i (y_{i.} - y_{..})^2 = B^{\dagger}$ | $J(I - 1)\sigma_b^2$ |
| Within groups | $\sum_{ij} (y_{ij} - y_{i.})^2 = W^{\dagger}$ | $(J - 1) \sum \sigma_i^2$ |
| Total | $\sum_{ij} (y_{ij} - y_{..})^2$ | $(IJ - 1)\sigma^2$ |

This partition of the total variability as measured by sums of squares of deviations is called an analysis of variance. It is clearly informative. We can see this by considering extreme cases. If all members of any subpopulation have identical $y$'s, the within-group sum of squares is zero and the corresponding $\sigma_i^2$ is zero. If the group means for $y$ are identical, the between-group sum of squares is zero. Knowledge of $B^{\dagger}$, the between-group sum of squares, and of $W^{\dagger}$, the within-group sum of squares, gives us very quickly a moderate amount of information about the variability in the whole population. The ratio $B^{\dagger}/(B^{\dagger} + W^{\dagger})$ is a useful numerical measure of this variability. Of course it does not tell us the whole story which is given *only* by the full set of $y_{ij}$'s.

Let us now examine what happens if the subpopulations are equally variable. Then we have $\sigma_i^2 = \sigma_w^2$, say, in which the subscript $w$ denotes "within," and our identity becomes

$$(IJ - 1)\sigma^2 = (I - 1)J\sigma_b^2 + I(J - 1)\sigma_w^2 \qquad (16.8)$$

Now note that Equation (16.8) can be written

$$(IJ - 1)\sigma^2 = (IJ - 1)\sigma_w^2 + (I - 1)(J\sigma_b^2 - \sigma_w^2)$$

or

$$\sigma^2 = \sigma_w^2 + \frac{(I - 1)}{(IJ - 1)}(J\sigma_b^2 - \sigma_w^2) \tag{16.9}$$

This is as far as one can go with $I$ and $J$ arbitrary. It is reasonable next to consider the case in which $J$ is very large. In this case we get

$$\sigma^2 = \sigma_w^2 + \frac{(I - 1)}{I}\sigma_b^2 \tag{16.10}$$

Finally, we consider the case in which $I$ also is very large. We then have

$$\sigma^2 = \sigma_w^2 + \sigma_b^2 \tag{16.11}$$

in which the overall variance is expressed as the sum of two separate variances, called variance components. An interesting statistic of the population is then $\sigma_b^2/(\sigma_w^2 + \sigma_b^2)$, called the intraclass correlation coefficient. This idea is somewhat obscure, so we explain it in a little detail. Consider forming doublet observations $(u, v)$ by taking $u$ to be the $y$ of an individual and $v$ to be the $y$ of another individual in the same group or *class*. This gives us $IJ(J - 1)$ possible doublets. Now evaluate the product moment correlation of $u$ and $v$.

We have

$$\bar{u} = y_{..} = \bar{v}$$

$$\begin{aligned}\sum (u - \bar{u})^2 &= \sum_i \sum_j \sum_{\substack{j' \\ j' \neq j}} (y_{ij} - y_{..})^2 \\ &= (J - 1)\sum_{ij} (y_{ij} - y_{..})^2 \\ &= (J - 1)(IJ - 1)\sigma^2\end{aligned}$$

We get the same expression for $\sum (v - \bar{v})^2$. The sum of products around the mean is

$$\begin{aligned}&\sum_i \sum_j \sum_{\substack{j' \\ j' \neq j}} (y_{ij} - y_{..})(y_{ij'} - y_{..}) \\ &= \sum_i \sum_j (y_{ij} - y_{..})[J(y_{i.} - y_{..}) - (y_{ij} - y_{..})] \\ &= J^2 \sum_i (y_{i.} - y_{..})^2 - \sum_{ij} (y_{ij} - y_{..})^2 \\ &= J^2(I - 1)\sigma_b^2 - (IJ - 1)\sigma^2\end{aligned}$$

Hence the correlation in this artificial bivariate set of data is

$$\frac{J^2(I-1)\sigma_b^2 - (IJ-1)\sigma^2}{(J-1)(IJ-1)\sigma^2} \tag{16.12}$$

If now $I$ and $J$ are very large, Formula (16.12) becomes simply $\sigma_b^2/\sigma^2$ or $\sigma_b^2/(\sigma_w^2 + \sigma_b^2)$.

So far we have merely considered the variability as indicated by the whole population. In most practical data situations we will have available a definite amount of data, and our basis for interpreting the data will be that they have come from conceptual populations by some sort of random sampling. It is certainly possible to develop sampling theory for the various possible cases. We can imagine that we have $I$ subpopulations (groups or classes) and that we have taken all of them. We can also imagine cases in which we have only a subset of the subpopulations represented in our data. In that case we will probably have to use as a model that the subpopulations represented arose by a simple random process, as in random sampling without replacement from a finite population. In the former case our interest will probably be in the actual differences between subpopulation means. Certainly if the subpopulations are defined by variations in a stimulus (e.g., dose of a drug, type of environment, etc.) we will be interested in the means. Our interest in the variance between subpopulation means will be rather minor. This case is commonly called the "fixed model," the idea being that in the population of repetitions of the data, in which the existent data are embedded for inference purposes, the subpopulations of the existent data will always be represented.

In contrast to the fixed model we can certainly envisage a very large number of subpopulations with a subset of these represented in our data. Then for purposes of inference we may wish to take the stance that our interest in the subpopulations represented in the data will be minor, and we shall be interested in the components of variance described above. This is called the "random model."

Before taking up simple cases, it should be mentioned that there has been extensive development of elementary ideas of random sampling from finite structured populations. We refer the reader to Tukey (1950, 1956), Kempthorne (1952), Wilk and Kempthorne (1955, 1956, 1957), Cornfield and Tukey (1956). Kempthorne et al. (1961), Zyskind (1962, 1963), Zyskind et al. (1964).

We proceed to the simple cases of structured populations in which the subpopulations are normal.

## 16.3 The One-Way Fixed Model with Normality and Constant Variance

A population that can be simply partitioned into an unordered set of subpopulations is called a one-way structured population. If the variation in each subpopulation is normal, we can represent our observations by the model $y_{ij} = \mu_i + e_{ij}$, or more commonly

$$y_{ij} = \mu + a_i + e_{ij} \tag{16.13}$$

in which the $\mu_i$ are the subpopulation means, and the $e_{ij}$ are normally and independently distributed with mean zero and variance $\sigma^2$. The model Equation (16.13) is really no different from the models given in Chapter 15 represented by $y = X\beta + e$. To illustrate this, suppose $I = 2$ and there are 2 observations for each $i$. Then the explicit form is

$$\begin{pmatrix} y_{11} \\ y_{12} \\ y_{21} \\ y_{22} \end{pmatrix} = \begin{pmatrix} 1 & 1 & 0 \\ 1 & 1 & 0 \\ 1 & 0 & 1 \\ 1 & 0 & 1 \end{pmatrix} \begin{pmatrix} \mu \\ a_1 \\ a_2 \end{pmatrix} + \begin{pmatrix} e_{11} \\ e_{12} \\ e_{21} \\ e_{22} \end{pmatrix} \tag{16.14}$$

In Chapter 15 we made our development under the assumption that the matrix $X$ was such that the matrix $X'X$ is nonsingular.

In our simple case it is clear that $X_{i1} - X_{i2} - X_{i3} = 0$ for each $i$. Hence the three columns of $X$ are linearly dependent. It is easy to see that any set of two columns is linearly independent. We can also see this by noting, for instance, that if we take just the first two columns of the $X$ in Equation (16.14) as $X^*$, say, then

$$X^{*\prime}X^* = \begin{pmatrix} 4 & 2 \\ 2 & 2 \end{pmatrix}$$

which has a nonzero determinant and hence is of full rank.

Even if we had an infinite number of observations in each class with the model Equation (16.13), we would be able only to determine the quantities $\mu + a_i$. We could not determine $\mu$ and the $a_i$ separately because no linear function of the observations has expectation $\mu$ or $a_i$. In order that $\mu$ and the $a_i$ be determined explicitly, we have to place restrictions on the values, or on the parameters. A simple restriction that does the job is to require $\sum_i a_i = 0$. This restriction has the result that the number $\mu$ is what we would observe if we had $N$ observations in each subpopulation with $N$ very large and formed the average of all the data. Also $a_i$ then measures the deviation of the $i$th subpopulation mean from this overall mean.

It is reasonable for the student on first contact to ask why the field of statistics makes such wide use of models like Equation (16.13) which are said to be of "nonfull rank." The answer is that users of the models find it convenient to think of an observation as being made up of a part common to all plus a part common to all members of the $i$th subpopulation plus a part unique to the observation. In this simple case no great utility accrues, but we can readily imagine a structured population in which the structure is very complex. The writing down of a model with parameters indexed by part of the identification of the observation is found to be useful as in $y_{ijk} = \mu + r_i + c_j + e_{ijk}$, for instance. The use of nonfull rank models induces a little complexity on the one hand with a little simplicity on the other. We give a brief introduction to the matter from a general point of view in Section 16.12.

We now return to our simple model. In our introduction we discussed the partition of the sum of squares around the overall data average. Let us do this in our present case because we should be interested in seeing what can be done with weaker assumptions than the full normality ones mentioned above. Additionally, let us suppose that we do not have the same number of individuals from each subpopulation represented in our data. We will weaken the assumption to one which states that the $e_{ij}$ have expectation zero, are uncorrelated, and the variance of $e_{ij}$ is $\sigma_i^2$, not necessarily the same for all $i$. After we have done this, we can introduce the homogeneity of variance assumption and the normality assumption.

Our data are represented by the statement

$$y_{ij} = \mu + a_i + e_{ij}, \qquad i = 1, 2, \ldots, I$$
$$j = 1, 2, \ldots, n_i$$

We denote $\sum n_i$ by $N$. We can certainly do an analysis of variance. We have

$$\textstyle\sum_{ij} (y_{ij} - y_{..})^2 = \sum_i n_i(y_{i.} - y_{..})^2 + \sum_{ij} (y_{ij} - y_{i.})^2 \qquad (16.15)$$

The most elemental calculation that is informative whenever we have a model and calculate a function of the observations, i.e., a statistic, is to obtain the expected value of the statistic. In most model situations this exists, and it frequently tells us what we would be estimating by the statistic if we had an infinite amount of data. To obtain the expected values of the parts of the right-hand side of Equation (16.15), we note first that

$$\textstyle\sum_i n_i(y_{i.} - y_{..})^2 = \sum_i n_i y_{i.}^2 - N y_{..}^2$$
$$\textstyle\sum_{ij} (y_{ij} - y_{i.})^2 = \sum_{ij} y_{ij}^2 - \sum_i n_i y_{i.}^2 \qquad (16.16)$$

so we want the expected values of $Ny_{..}^2$, $\sum_i n_i y_{i.}^2$ and $\sum_{ij} y_{ij}^2$. We have $y_{ij}^2 = \mu^2 + a_i^2 + e_{ij}^2 + 2\mu a_i +$ cross-products, in which the cross-products involve the $e_{ij}$ and have zero expectations, so

$$E(y_{ij}^2) = \mu^2 + a_i^2 + 2\mu a_i + \sigma_i^2$$
$$E(\textstyle\sum_{ij} y_{ij}^2) = N\mu^2 + \sum n_i a_i^2 + 2\mu \sum n_i a_i + \sum n_i \sigma_i^2 \qquad (16.17)$$

We shall not indicate the nature of the summation when it is rather obvious. Also

$$y_{i.} = \mu + a_i + e_{i.}$$

in which

$$e_{i.} = \frac{1}{n_i}\sum_j e_{ij}$$

so

$$y_{i.}^2 = \mu^2 + a_i^2 + 2\mu a_i + e_{i.}^2 + \text{cross-products}$$

and

$$E(y_{i.}^2) = \mu^2 + a_i^2 + 2\mu a_i + \frac{\sigma_i^2}{n_i} \qquad (16.18)$$

Here we use the fact that $E(e_{i.}) = 0$ and $E(e_{i.}^2)$ equals the variance of $e_{i.}$ which is $\sigma_i^2/n_i$. Hence

$$E(\textstyle\sum n_i y_{i.}^2) = N\mu^2 + \sum n_i a_i^2 + 2\mu \sum n_i a_i + \sum \sigma_i^2 \qquad (16.19)$$

Finally

$$y_{..} = \mu + \frac{\sum n_i a_i}{N} + \frac{\sum e_{ij}}{N}$$

and

$$E(Ny_{..}^2) = N\mu^2 + \frac{(\sum n_i a_i)^2}{N} + 2\mu \sum n_i a_i + \frac{\sum n_i \sigma_i^2}{N} \qquad (16.20)$$

Hence we have

$$E\sum n_i(y_{i.} - y_{..})^2 = \sum n_i(a_i - \bar{a})^2 + \sum \sigma_i^2\left(1 - \frac{n_i}{N}\right) \qquad (16.21)$$

with $\bar{a} = (\sum n_i a_i)/N$, and

$$E\sum (y_{ij} - y_{i.})^2 = \sum (n_i - 1)\sigma_i^2 \qquad (16.22)$$

Let us now introduce homogeneity of variance so that $\sigma_i^2 = \sigma_w^2$, say. Then

$$E[\textstyle\sum (y_{ij} - y_{i.})^2] = \sigma_w^2 \sum (n_i - 1) = \sigma_w^2(N - I) \qquad (16.23)$$
$$E[\textstyle\sum n_i(y_{i.} - y_{..})^2] = (I - 1)\sigma_w^2 + \sum n_i(a_i - \bar{a})^2 \qquad (16.24)$$

From Equations (16.23) and (16.24) it is natural to consider divisors of $N - I$ and $I - 1$ respectively. These numbers are called the numbers of degrees of freedom. The basis for this terminology is that the between-group sum of squares is a weighted sum of squares of $I$ quantities $y_{i.} - y_{..}$ which are connected by one linear relation, namely

$$\sum n_i(y_{i.} - y_{..}) = 0$$

so that in a sense only $I - 1$ of them are functionally independent. Similarly the within-group sum of squares is the sum of squares of $N$ quantities $y_{ij} - y_{i.}$ among which there are $I$ dependencies

$$\sum_j (y_{ij} - y_{i.}) = 0, \qquad i = 1, 2, \ldots, I$$

It is natural to represent what has been developed so far in tabular form as in Table 16.2.

**Table 16.2** Analysis of Variance

| Source | d.f. | SS | MS | EMS |
|---|---|---|---|---|
| Between groups | $I - 1$ | $\sum n_i(y_{i.} - y_{..})^2$ | $B$ | $\sigma_w^2 + \frac{1}{I-1}\sum n_i(a_i - \bar{a})^2$ |
| Within groups | $N - I$ | $\sum (y_{ij} - y_{i.})^2$ | $W$ | $\sigma_w^2$ |
| Total | $N - 1$ | $\sum (y_{ij} - y_{..})^2$ | | |

In Table 16.2 (and in later notation) d.f. stands for number of degrees of freedom, SS for sum of squares, MS for mean square equals sum of squares divided by degrees of freedom, and finally EMS stands for expected value of mean square. It is important to realize that Table 16.2, even without the EMS column, is informative. Also note that we could have written down an EMS column without assuming $\sigma_i^2$ equals $\sigma_w^2$ for all $i$. If we had done so, we would be using only the model that the $e_{ij}$ for any one $i$ are random members from some distribution of errors with mean zero. If there are not real differences between the group means, so that $a_i - \bar{a}$ equals zero for each $i$, the numbers $B$ and $W$ should be of similar magnitude, and we would expect $B/W$ to be near unity. If on the other hand some of the $a_i - \bar{a}$ are not zero, we expect $B/W$ to be greater than unity. A real problem now, of course, is to form a better idea of what we should mean by "near unity."

Before taking up this matter, consider the $y_{i.}$. It is obvious that $y_{i.}$, being a mean of an assumed random sample, has variance $\sigma_i^2/n_i$. If we knew $\sigma_i^2$, we could use Chebyshev's inequality to obtain an idea of how close $y_{i.}$ is to $\mu + a_i$. In the absence of such knowledge we have to use $\sum_j (y_{ij} - y_{i.})^2$ to give an indicator of $\sigma_i^2$ and would normally use

$s_i^2 = \sum (y_{ij} - y_{i.})^2/(n_i - 1)$, the unbiased estimator of $\sigma_i^2$. To go farther we need a distributional assumption, and the only one that is tractable is that of normality. With this we are in the standard case of observing a random sample from a normal distribution. One aspect of interest is a comparison of means. Clearly $y_{u.} - y_{v.}$ is an indicator of $a_u - a_v$, the difference in means of the $u$th and $v$th groups. Without the homogeneity of variance assumption the variance of this is $(\sigma_u^2/n_u) + (\sigma_v^2/n_v)$. Not knowing $\sigma_u^2$ and $\sigma_v^2$, we would be forced to consider $(s_u^2/n_u) + (s_v^2/n_v)$, and we could consider a simple measure of difference

$$\frac{(y_{u.} - y_{v.})}{\left(\frac{s_u^2}{n_u} + \frac{s_v^2}{n_v}\right)^{1/2}}$$

Such a number is informative for indicating the likeness of the $u$ group and the $v$ group. If it is small there are small differences in the means relative to the spread of observations about the mean. But we do not know "how small is small," to use an informative colloquialism. One possibility is to adjoin the assumption of normality without homogeneity of variance. This, however, lands us in one of the most controversial problems of statistics, the Behrens-Fisher problem, which is extremely difficult.

We now proceed by incorporating the additional assumptions of homogeneity of variance and normality. The situation is then resolved very quickly by the following sequence of facts, each of which is essentially obvious in the light of previous chapters. We have:

1. A group mean, $y_{i.}$, is distributed $N(\mu + a_i, \sigma^2/n_i)$ and

$$\frac{(n_i - 1)s_i^2}{\sigma^2} = \frac{\sum_j (y_{ij} - y_{i.})^2}{\sigma^2}$$

   is distributed as $\chi^2$ with $n_i - 1$ degrees of freedom, and these are independent.
2. These statistics for different groups are statistically independent.
3. Because the sum of independent $\chi^2$ variables is distributed as $\chi^2$, we have that $\sum_{ij} (y_{ij} - y_{i.})^2/\sigma^2$ is distributed as $\chi^2$ with $\sum (n_i - 1)$ degrees of freedom.
4. The array of means $y_{i.}$, $i = 1, 2, \ldots, I$ has the multivariate normal distribution with mean vector $(\mu + a_1, \mu + a_2, \ldots, \mu + a_I)$ and variance-covariance matrix which is $\sigma^2 V$, say, such that $V$ is diagonal with $1/n_i$ in the $(i, i)$th position.
5. Any linear form, say, $u_1 = \lambda_{11} y_{1.} + \lambda_{12} y_{2.} + \cdots + \lambda_{1I} y_{I.}$ is normally distributed with mean $\sum \lambda_{1i}(\mu + a_i)$ and variance $\sum \lambda_{1i}^2 \sigma^2/n_i$ and independently of $\sum_{ij} (y_{ij} - y_{i.})^2$.

6. Hence the quantity

$$\frac{[u_1 - E(u_1)]}{s\sqrt{\sum \lambda_{1i}^2/n_i}}$$

is distributed as $t$ with $\sum (n_i - 1)$ degrees of freedom, where $s^2 = \sum_{ij} (y_{ij} - y_{i.})^2/\sum (n_i - 1)$. It is a pivotal quantity involving only $E(u_1)$. This enables us to set limits on $E(u_1)$ by inverting a statistical test just as in the case of a single normal mean. We have discussed in Chapter 13 the possible interpretations on such inversions. One may pick a *particular* $1 - \alpha$ and get what is called a confidence interval, but our preference is to look at the inversion for all $\alpha$ and to regard the result as indicating the tenability of every value of $E(u_1)$ or the degree of consonance of each value of $E(u_1)$ with the data.

7. Consider now two linear forms

$$u_1 = \lambda_{11}y_{1.} + \lambda_{12}y_{2.} + \cdots + \lambda_{1I}y_{I.}$$
$$u_2 = \lambda_{21}y_{1.} + \lambda_{22}y_{2.} + \cdots + \lambda_{2I}y_{I.}$$

These have a distribution independent of that of $s^2$. They are linear functions of normal variables and hence have a bivariate normal distribution with mean vector and variance-covariance matrix

$$\begin{pmatrix} E(u_1) \\ E(u_2) \end{pmatrix}, \qquad \sigma^2\begin{pmatrix} V_{11} & V_{12} \\ V_{21} & V_{22} \end{pmatrix}$$

respectively, where

$$V_{11} = \textstyle\sum_i \lambda_{1i}^2/n_i, \; V_{22} = \sum_i \lambda_{2i}^2/n_i$$
$$V_{12} = \textstyle\sum_i \lambda_{1i}\lambda_{2i}/n_i$$

Hence a pivotal quantity for finding the tenability of $E(u_1)$, $E(u_2)$ is obtained as follows: Let

$$\begin{pmatrix} c_{11} & c_{12} \\ c_{21} & c_{22} \end{pmatrix}\begin{pmatrix} V_{11} & V_{12} \\ V_{21} & V_{22} \end{pmatrix} = \begin{pmatrix} 1 & 0 \\ 0 & 1 \end{pmatrix}$$

Then

$$(1/\sigma^2)\{c_{11}[u_1 - E(u_1)]^2 + 2c_{12}[u_1 - E(u_1)][u_2 - E(u_2)] + c_{22}[u_2 - E(u_2)]^2\}$$

being essentially the exponent of a bivariate normal distribution, is distributed as $\chi^2$ with 2 degrees of freedom. It is independent of $s^2$. We now use the standard fact that if $X_1$ and $X_2$ are independent and $X_1 \sim \chi^2_{\nu_1}$, $X_2 \sim \chi^2_{\nu_2}$, then $(X_1/\nu_1)/(X_2/\nu_2)$ is distributed as $F$ with $\nu_1$ and $\nu_2$ degrees of freedom. Hence we have that

$$\frac{1}{2s^2}\{c_{11}[u_1 - E(u_1)]^2 + 2c_{12}[u_1 - E(u_1)][u_2 - E(u_2)] + c_{22}[u_2 - E(u_2)]^2\}$$

is distributed as $F$ with 2 and $\sum (n_i - 1)$ degrees of freedom. This distributional fact may now be inverted as in Chapter 13. Again our preference is to consider the whole array of inversions for the possible probability levels from one to zero. This inversion can be represented in two dimensions by a system of contours. Values of $E(u_1)$, $E(u_2)$ inside the $1 - \alpha$ contour have tenability greater than $\alpha$. From the viewpoint of accept-reject rules if we had decided beforehand to use, say a 5% reject rule so that our probability beforehand of rejecting a true value for $E(u_1)$, $E(u_2)$ were 5%, our rule would have the consequence that all values of $E(u_1)$, $E(u_2)$ *outside* a 95% contour would be rejected. This we regard as part of the intuitive basis for our opinion that values inside the 95% contour are more tenable or more consonant with the data than values outside this contour. But it is critical to emphasize that we have used a *particular* test. We could develop another test which would give us a different ordering of tenability of values for $E(u_1)$, $E(u_2)$. So the tenability is tenability as indicated by the nominated test of significance.

8. We can consider $m$ linear forms:

$$u_r = \lambda_{r1} y_{1.} + \lambda_{r2} y_{2.} + \cdots + \lambda_{rI} y_{I.}, r = 1, 2, \ldots, m$$

These have a multivariate normal distribution with mean vector and covariance matrix:

$$\begin{pmatrix} E(u_1) \\ E(u_2) \\ \cdot \\ E(u_m) \end{pmatrix}, \sigma^2 \begin{pmatrix} V_{11} & V_{12} & \cdots & V_{1m} \\ V_{21} & V_{22} & \cdots & V_{2m} \\ \cdot & \cdot & \cdots & \cdot \\ V_{m1} & V_{m2} & \cdots & V_{mm} \end{pmatrix} = \sigma^2 V$$

where $V_{rs} = \sum_i \lambda_{ri}\lambda_{si}/n_i$.

If $C = V^{-1}$, then with the straightforward generalization of the above argument for two $u$'s we have $(1/ms^2)\,[u - E(u)]'C[u - E(u)]$ which is distributed as $F$ with $m$ and $\sum (n_i - 1)$ degrees of freedom. We can then again go through all the processes of inversion.

The special case of equal numbers when each $n_i$ is $n$, is particularly easy. In this case $y_{i.}$ has variance $\sigma^2/n$. The variance of a linear form $u_r = \lambda_{r1} y_{1.} + \lambda_{r2} y_{2.} + \cdots + \lambda_{rI} y_{I.}$ is then $(\sigma^2/n) \sum_i \lambda_{ri}^2$, and the covariance of $u_r$ and $u_s$ is $(\sigma^2/n) \sum_i \lambda_{ri}\lambda_{si}$. Furthermore, if $u_r$ and $u_s$ are such that $\sum_i \lambda_{ri}\lambda_{si} = 0$, then $u_r$ and $u_s$ are said to be orthogonal and are also uncorrelated. If we have a set of $m$ such uncorrelated forms $u_1, u_2, \ldots, u_m$ which are also normalized so that $\sum_i \lambda_{ri}^2 = 1$ for each $r$, the matrix $V$ is $(1/n)I_m$ where $I_m$ is the $m \times m$ identity matrix and the matrix $C$ is $nI_m$. The result is that for any set of $m$ such orthogonal linear functions we have that $(n/ms^2) \sum [u_i - E(u_i)]^2$ is distributed as $F$ with degrees of freedom

$m$ and $\sum (n_i - 1)$ which is now $I(n - 1)$. Because we are usually interested only in comparisons among the groups, it is common to restrict the $u_i$ to be comparisons or contrasts in which we use the definition that a linear form $\sum_i \lambda_{ri} y_{i.}$ is a contrast if $\sum_i \lambda_{ri} = 0$.

Detailed examination of the type of model we discuss here is usually covered in courses on analysis of variance and the design of experiments. [See, e.g., Kempthorne (1952).] We must end our development of this model here but take up one crucial topic.

We have given one family of tests of significance, based on the $t$ test and its multidimensional analogue (in the parameter space), the $F$ test. But this is not the only test of significance which can be used. We take the case of all $n_i$ being equal to $n$. We then have that the $y_{i.}$ are normally and independently distributed with means $\mu_i$, say, and variance $\sigma^2/n$. Now we note that if random variables $X_i$, $i = 1, 2, \ldots, m$ are independently $N(0, \sigma^2)$, we can develop the distribution of the range of the $X_i$, that is, $X_{\max} - X_{\min}$. We have indicated in Chapter 6 the general form of the distribution of the range of a random sample from a given distribution. Furthermore, the distribution of what is called the studentized range $= (X_{\max} - X_{\min})/s$ has been worked out. Hence we can say that the distribution of $[\max\{y_{i.} - \mu_i\} - \min\{y_{i.} - \mu_i\}]/s$ is known. We can now use this as a test of significance on the means $y_{i.}$ and can develop a system of contours, which may be called confidence contours but which we prefer to regard as tenability or consonance contours, for the differences $\mu_i - \mu_{i'}$. This system of contours has its scope of validity analogous to the system based on the $F$ distribution. Which of these is to be preferred? We state that there is no answer. Each test of significance measures tenability or consonance in its particular direction in the parameter space. This direction is determined by the noncentrality parameter of the test of significance. To explain this, we note the following in which we keep to the equal number case for ease. Consider the variables $z_1, z_2, \ldots, z_{I-1}$ defined as follows:

$$z_1 = \frac{1}{\sqrt{2}}(y_{1.} - y_{2.})$$

$$z_2 = \frac{1}{\sqrt{6}}(y_{1.} + y_{2.} - 2y_{3.})$$

$$\cdots\cdots\cdots\cdots\cdots\cdots$$

$$z_{I-1} = \frac{1}{\sqrt{I(I-1)}}[y_{1.} + y_{2.} + \cdots + y_{(I-1).} - (I-1)y_{I.}] \quad (16.25)$$

Under our model the $z_i$ are independent normal variables with variance $\sigma^2/n$ and means $\nu_1, \nu_2, \ldots, \nu_{I-1}$, which can be written down readily.

(Note that we use in an essential way normality, homogeneity of variance, and equality of the $n_{i.}$.) Now the transformation in Equation (16.25) is orthogonal, and one can show that $\sum_1^{I-1} z_i^2$ is equal to $\sum_1^I (y_{i.} - y_{..})^2$. Hence $(n/\sigma^2)\sum_1^I (y_{i.} - y_{..})^2$ is the sum of squares of $I - 1$ independent normal variables with unit variance and means, say, $\tau_1, \tau_2, \ldots, \tau_{I-1}$. We saw in Chapter 6 that such a quantity follows the noncentral $\chi^2$ distribution. This distribution is characterized by two parameters, the degrees of freedom and the noncentrality which is $\sum_1^{I-1}\tau_i^2/2$. Our development here is very abbreviated, and we regret that we cannot give all the details. It can be seen that

$$\sum_1^{I-1}\tau_i^2 = \frac{n}{\sigma^2}\sum_1^I (\mu_i - \bar{\mu})^2$$

It follows that the $F$ test for equality of means is in this case sensitive to this function of the parameters. The $F$ test looks (if you like) at the space of the parameters, the differences of class means, in the "direction" determined by this noncentrality parameter. The range test sketched above is a test of significance with a different noncentrality parameter and looks at the parameter space in another "direction."

The reader may refer to Scheffé (1959) for discussion of his own work and that of J. W. Tukey, though from a somewhat different viewpoint than that presented here. See also Kurtz et al. (1965).

We now give an opinion of the authors. The examination of tenability of parameter values when there are two or more parameters cannot be done in a way that is uniquely appropriate and totally compelling to everyone. If we adopt an outlook of accept-reject rules and regard the matter as one of acceptance sampling, we can prove that one test is better than another, but *only* with regard to its own noncentrality parameter. Indeed, there is a theorem that the $F$ test is best in the class of tests whose sensitivity (or power) depends on the parameters only through the noncentrality parameter of the $F$ test, i.e., in our case of equal numbers on $n\sum(\mu_i - \bar{\mu})^2/\sigma^2$. This test will not be most sensitive in the class of tests whose sensitivity depends on $\max|\mu_i - \mu_j|$. A consequence of this is that the search for best tests, where tests are regarded as tests of significance aimed at judging the tenability of parameter values, is very limited right from the outset. To give a general loose statement, we can say only that a test of significance which does not use all the information available should not be used. It would be foolish, for instance, to use only part of the information available on $\sigma^2$ by disregarding some of the within-group observations. But we realize that we have not given a mathematically tight statement, and resolution of the whole matter has not yet been achieved by the statistical profession.

## 16.4 The Likelihood Function for the One-Way Fixed Model

For completeness it is appropriate to abstract part of the development in the previous section by writing down the likelihood function. We use $\mu_i$ for the group means and see that, suppressing the "errors" of grouping,

$$L = \prod_{i=1}^{I} \left\{ \frac{1}{(\sqrt{2\pi}\sigma_i)^{n_i}} \exp\left[ -\frac{1}{2\sigma_i^2} \sum_j (y_{ij} - \mu_i)^2 \right] \right\} \quad (16.26)$$

so that the logarithm of the likelihood, which is really a continuized version ignoring grouping error of measurement, is

$$\mathscr{L} = -\sum_i \frac{n_i}{2} \ln 2\pi - \sum_i \frac{n_i}{2} \ln \sigma_i^2 - \sum_{ij} \frac{1}{2\sigma_i^2} (y_{ij} - \mu_i)^2 \quad (16.27)$$

and apart from an additive constant independent of the parameters,

$$\mathscr{L} = -\sum_i \frac{n_i}{2} \ln \sigma_i^2 - \sum_i \frac{n_i(y_{i.} - \mu_i)^2}{2\sigma_i^2} - \sum_{ij} \frac{(y_{ij} - y_{i.})^2}{2\sigma_i^2} \quad (16.28)$$

The function $\mathscr{L}$ with $\sigma_i^2$ the same for all $i$ is easily written down. The log-likelihood function exhibits immediately the sufficient statistic which is a vector, as the set of means $\{y_{i.}\}$ and the set of within-group mean squares $\{s_i^2\}$. What one should do with this likelihood function is, however, quite obscure. It is easy to complete the Bayesian process with certain prior distributions, but the relevance of the answer may be questioned.

## 16.5 The One-Way Random Model with Equal Numbers

We now turn to a model closely related to that considered above. In many areas of investigation it is reasonable to consider the data in terms of an infinite population of populations, with a random sample of the populations being represented in the data. This can be formalized by the partially specified model

$$y_{ij} = \mu + \alpha_i + e_{ij}, \qquad i = 1, 2, \ldots, I; \quad j = 1, 2, \ldots, n_i \quad (16.29)$$

The case of equal $n_i$ is a specially simple one. With this model the group means $\mu + \alpha_i$ are regarded as random variables from a population and a random sample of these lead to our data. For simplicity, and without loss of generality, we can define the expectation of any $\alpha_i$ to be zero. If we now incorporate normality of all distributions, we complete the specification of the model by adjoining the additional statements:

1. $\mu$ is an unspecified constant.
2. The $\alpha_i$ are $NID(0, \sigma_\alpha^2)$.
3. The $e_{ij}$ are $NID(0, \sigma^2)$.
4. The $\alpha_i$ and $e_{ij}$ are independent. (16.30)

We are now in the position of representing our data by a model with unspecified parameters $\mu$, $\sigma_\alpha^2$, $\sigma^2$.

Before taking up the consequences of the additional assumptions, let us first consider the analysis of variance. For this we will not force the equality of the $n_i$. We can use Table 16.2 to abbreviate the development. There we took expectations with $a_i$ in place of $\alpha_i$ and regarded the $a_i$ as fixed quantities. To see what happens by merely incorporating randomness without normality, we take an additional expectation. We replace $\sigma_w^2$ by $\sigma^2$ and we note that the expectation of $\sum n_i(a_i - \bar{a})^2$, when the $a_i$ are uncorrelated with mean zero and variance $\sigma_\alpha^2$, is obtained as follows:

$$\sum n_i(a_i - \bar{a})^2 = \sum n_i a_i^2 - \frac{(\sum n_i a_i)^2}{N}$$

with $N = \sum n_i$.

$$E(\sum n_i a_i^2) = \sum n_i \sigma_\alpha^2 = N\sigma_\alpha^2$$

$$E\left[\frac{(\sum n_i a_i)^2}{N}\right] = \frac{\sum n_i^2 \sigma_\alpha^2}{N}$$

Hence

$$E\left[\sum n_i(a_i - \bar{a})^2\right] = \left(N - \frac{\sum n_i^2}{N}\right)\sigma_\alpha^2 \tag{16.31}$$

Hence under this incompletely specified random model

$$E(\text{between-group MS}) = \sigma^2 + \frac{1}{I-1}\left(N - \frac{\sum n_i^2}{N}\right)\sigma_\alpha^2$$

$$E(\text{within-group MS}) = \sigma^2 \tag{16.32}$$

These formulae are very widely used. If in addition each $n_i = n$, we have

$$N = In, \qquad \frac{\sum n_i^2}{N} = \frac{In^2}{N} = n$$

and we get

$$E(\text{between-group MS}) = \sigma^2 + n\sigma_\alpha^2$$
$$E(\text{within-group MS}) = \sigma^2 \qquad (16.33)$$

So under this model $B$, the between-group mean square, is a statistic which gives us an idea of a plausible value for $\sigma^2 + n\sigma_\alpha^2$, and $W$, the within-group mean square, gives an idea of a plausible value of $\sigma^2$. Additionally of course $(B - W)/n$ gives us an idea about $\sigma_\alpha^2$, and $(B - W)/[B + (n - 1)W]$ gives us an indication of $\sigma_\alpha^2/(\sigma^2 + \sigma_\alpha^2)$, the intraclass correlation coefficient. However, we must be careful here because it can happen with sampling fluctuations that $B - W$ is negative, so that to take $(B - W)/n$ as a *point* estimate of $\sigma_\alpha^2$ would be completely foolish because we could get a number which is negative and certainly not acceptable as an estimate of $\sigma_\alpha^2$. This problem is very deep, and the answer often used is to take $[(B - W)/n, W]$ as a point estimate of $(\sigma_\alpha^2, \sigma^2)$ if $B > W$ and to use $\{0, [(I - 1)B + (N - I)W]/(N - 1)\}$ as the point estimate of $(\sigma_\alpha^2, \sigma^2)$ if $B < W$.

To go further, it is necessary to incorporate distributional statements, and we pursue this only for the equal number case with the full assumptions stated in Equation (16.30). What we said before about the distribution of $W$ still holds, namely that $I(n - 1)W/\sigma^2$ is distributed as $\chi^2$ with $I(n - 1)$ degrees of freedom. Additionally, it can be seen that $(I - 1)B/(\sigma^2 + n\sigma_\alpha^2)$ is distributed as $\chi^2$ with $I - 1$ degrees of freedom, and $y_{..}$ is distributed normally with mean $\mu$ and variance $(1/In)(\sigma^2 + n\sigma_\alpha^2)$. To see this in an elementary but tedious argument, we have to consider a whole batch of data transformations which can be put together into a single transformation. We start off with $y_{ij}$, $i = 1, 2, \ldots, I; j = 1, 2, \ldots, n$. For each $i$ class make the following transformation, known as a Helmert type transformation:

$$z_{i1} = \frac{1}{\sqrt{n}}(y_{i1} + y_{i2} + \cdots + y_{in})$$

$$z_{i2} = \frac{1}{\sqrt{2}}(y_{i1} - y_{i2})$$

$$\cdots\cdots\cdots\cdots\cdots\cdots\cdots\cdots$$

$$z_{ik} = \frac{1}{\sqrt{k(k - 1)}}[y_{i1} + y_{i2} + \cdots + y_{i,k-1} - (k - 1)y_{ik}]$$

$$\cdots\cdots\cdots\cdots\cdots\cdots\cdots\cdots$$

$$z_{in} = \frac{1}{\sqrt{n(n - 1)}}[y_{i1} + y_{i2} + \cdots + y_{i,n-1} - (n - 1)y_{in}] \qquad (16.34)$$

Now let the $z_{ij}$ with $j > 1$ remain as they are but transform the $I$ variables $z_{i1}$, $i = 1, 2, \ldots, I$ by the same type of transformation.

$$u_1 = \frac{1}{\sqrt{I}}(z_{11} + z_{21} + \cdots + z_{I1})$$

$$u_2 = \frac{1}{\sqrt{2}}(z_{11} - z_{21})$$

. . . . . . . . . . . . . . . . . . . . . . . . . . . . . . . .

$$u_k = \frac{1}{\sqrt{k(k-1)}}[z_{11} + z_{21} + \cdots + z_{k-1,1} - (k-1)z_{k1}]$$

. . . . . . . . . . . . . . . . . . . . . . . . . . . . . . . .

$$u_I = \frac{1}{\sqrt{I(I-1)}}[z_{11} + z_{21} + \cdots + z_{I-1,1} - (I-1)z_{I1}] \qquad (16.35)$$

We have now replaced our $In$ variables $y_{ij}$ by $u_1, u_2, u_3, \ldots, u_I$ and the $z_{ij}, j = 2, 3, \ldots, n$, and $i = 1, 2, \ldots, I$. Now note that we have made a linear transformation of normally distributed variables, in that the $y_{ij}$ have a joint multivariate normal distribution. To see this, note that each $y_{ij}$ is a linear function of independent normal variables plus a constant. This distribution is of a very special form with mean vector all of whose elements are equal to $\mu$ and with variance-covariance matrix of the form exemplified by the following for the case $I = 3$.

$$V = \left(\begin{array}{c|c|c} \sigma^2 I_n + \sigma_\alpha^2 J_n & \phi & \phi \\ \hline \phi & \sigma^2 I_n + \sigma_\alpha^2 J_n & \phi \\ \hline \phi & \phi & \sigma^2 I_n + \sigma_\alpha^2 J_n \end{array}\right) \qquad (16.36)$$

It is here assumed that the observations are listed in the order $y_{11}, y_{12}, \ldots, y_{1n}; y_{21}, y_{22}, \ldots, y_{2n}$; and so on to $y_{I1}, y_{I2}, \ldots, y_{In}$. There are $I$ diagonal blocks in $V$, and each block is $n \times n$ and $J_n$ is the $n \times n$ matrix all of whose elements are equal to unity. The reader who wishes may go directly to the joint density by inverting the matrix $V$, using the fact that the inverse of a nonsingular matrix of the form $aI_n + bJ_n$ is of the form $cI_n + dJ_n$. We need not do this, though the operation is more succinct than what we have done.

We use the elementary properties of linear functions of normal variables to see that

$$u_1 \text{ is } N(\sqrt{In}\,\mu, \sigma^2 + n\sigma_\alpha^2)$$

$$u_r \text{ is } N(0, \sigma^2 + n\sigma_\alpha^2), \qquad r = 2, \ldots, I$$

$$z_{ij} \text{ is } N(0, \sigma^2), \qquad i = 1, 2, \ldots, I$$

$$j = 2, 3, \ldots, n \qquad (16.37)$$

and because of the equality of numbers all these variables are uncorrelated, and because of normality they are therefore independent. Hence we can write down the probability density of this whole set as

$$\left(\frac{1}{\sqrt{2\pi}}\right)^{In}\left(\frac{1}{\sigma^2+n\sigma_\alpha^2}\right)^{1/2}\exp\left[-\frac{(u_1-\sqrt{In}\,\mu)^2}{2(\sigma^2+n\sigma_\alpha^2)}\right]\times$$

$$\left(\frac{1}{\sigma^2+n\sigma_\alpha^2}\right)^{(I-1)/2}\exp\left[-\frac{1}{2(\sigma^2+n\sigma_\alpha^2)}\sum_{i=2}^{I}u_i^2\right]\times$$

$$\frac{1}{\sigma^{I(n-1)}}\exp\left[-\frac{1}{2\sigma^2}\sum_{i=1}^{I}\sum_{j=2}^{n}z_{ij}^2\right] \qquad (16.38)$$

with differential element $\prod du_i \prod dz_{ij}$. Now we note the following which may be verified:

$$u_1 = \sqrt{In}\,y_{..}$$

$$z_{i1} = \sqrt{n}y_{i.}$$

$$\sum_{i=2}^{I}u_i^2 = \sum_i (z_{i1} - z_{.1})^2$$

$$= n\sum_i (y_{i.} - y_{..})^2$$

$$\sum_{i=1}^{I}\sum_{j=2}^{n} z_{ij}^2 = \sum_{ij}(y_{ij} - y_{i.})^2 \qquad (16.39)$$

The basis for these calculations is that a Helmert transformation is orthogonal so that, for instance

$$\sum_{i=1}^{I} u_i^2 = \sum_{i=1}^{I} z_{i1}^2$$

and

$$u_1 = \sqrt{I}z_{.1} \qquad (16.40)$$

The upshot of this is that the transformed probability density is

$$\frac{1}{(2\pi)^{(In/2)}}\frac{\sqrt{In}}{\psi}\exp\left[-\frac{In}{2\psi^2}(y_{..}-\mu)^2\right]\frac{1}{\psi^{I-1}}\times$$

$$\exp\left[-\frac{(I-1)B}{2\psi^2}\right]\frac{1}{\sigma^{I(n-1)}}\exp\left[-\frac{I(n-1)W}{2\sigma^2}\right] \qquad (16.41)$$

where $\psi^2 = \sigma^2 + n\sigma_\alpha^2$, $B$ is the between-group mean square, and $W$ is the within-group mean square. This shows that $y_{..}$, $B$, and $W$ comprise a sufficient statistic, and a little thought shows that this is a minimal sufficient statistic. By analogy with the univariate normal distribution, we see that

$$y_{..} \text{ is } N\left(\mu, \frac{\sigma^2 + n\sigma_\alpha^2}{In}\right)$$

$$\frac{(I-1)B}{\sigma^2 + n\sigma_\alpha^2} \text{ is } \chi^2 \text{ with } I - 1 \text{ d.f.}$$

$$\frac{I(n-1)W}{\sigma^2} \text{ is } \chi^2 \text{ with } I(n-1) \text{ d.f.} \qquad (16.42)$$

and these quantities are statistically independent.

The development of inferential procedures for this situation, even though it is very elementary and basic, has not reached any answers which can be presented as a consensus of professional statisticians. There are approximate confidence interval procedures which are superficially reasonable but have deep logical difficulties. The authors' preference is to make significance tests on $\mu$ by using $y_{..}$ and $B$ as one would with a simple random sample from a normal population. Here we treat $I(n-1)W/\sigma^2$ as a pivotal quantity telling us about $\sigma^2$. To make judgments about $\sigma_\alpha^2/\sigma^2$, we would use the $F$ distribution, and calculate the tenability of a value $\theta_o$, say, on the basis of a probability goodness of fit test. To make a judgment on $\sigma_\alpha^2$ is most difficult. To make a judgment on $(\sigma_\alpha^2, \sigma^2)$, we would use the probability goodness of fit procedure with the joint distribution of $B$ and $W$, though this presents considerable computational difficulties. But we emphasize that *there is no simple answer* which survives all reasonable criteria of logic.

## 16.6 Crossed and Nested Classification Models

We now consider data analysis for a set of individuals which have been classified by two or more factors of classification. First, we define a factor of classification to be a partition of our individuals into disjoint sets. We may label the disjoint subsets by a variable $u$ which is one of the numbers 1, 2, . . . , $K$, where $K$ is the number of disjoint sets. A second factor of classification will partition the individuals into disjoint sets, and we may label these by $v$ which is one of the numbers 1, 2, . . . , $L$, where $L$ is the number of sets in this partition. A simple example is where we have data on people obtained in a survey in which they are classified by sex, male or female, and by location, urban or rural. Of course, strict operational definitions must be used in making the classifications. Then our individuals lie in one of four possible cells. We realize that both males and females occur in both urban and rural locations. For this reason the classification is called a cross-classification.

To obtain a contrast to this case, consider a geophysical exploration study which involved soil samples from four geographical regions. Suppose also that each of the four regions was partitioned into three areas and soil samples were taken from each area. Each soil sample can be recognized as belonging to a particular region, so region is a factor with four levels. Each soil sample also can be recognized as belonging to a particular area. There are 12 areas, so area is a factor with 12 levels. However, the soil samples which belong to a particular area must all belong to a particular region because area is a partition within region.

For the case of the people in the survey, we say that the two factors of classification, sex and location, are crossed. For the case of the soil samples, we say that the factor area is nested in the factor region because all samples which belong to the same area must necessarily belong to the same region. We say that region is a nestor factor which has area as its nested factor. If *all* individuals that have any particular level $j$, say, of a factor $B$ have a definite level $i$, say, of factor $A$, we say that factor $B$ is nested in factor $A$.

In the case of the cross-classification with factor $A$ with levels $1, 2, \ldots, r$, and factor $B$ with levels $1, 2, \ldots, s$, it is natural and useful to identify an observation by the ordered subscripts $i, j, k$ in which $i$ denotes the level of $A$, $j$ denotes the level of $B$, and $k$ serves to label the observations within the class of observations which have level $i$ of factor $A$ and level $j$ of factor $B$.

In the case of a nested classification there are two possible ways of indexing the observations. We can use a subscript $i$ to indicate the level of the nestor factor and use a subscript $j$ to indicate the level of the nested factor. We would then note that all observations with a particular value for $j$, say $j = 5$, for the nested factor would have the same level of the nestor factor, say $i = 2$. If we use this mode of indexing, we would have the possibility of a list of classes like the following: (1, 1), (1, 2), (2, 3), (2, 4), (2, 5), (3, 6), (3, 7), .... This mode of indexing has the merit of indicating exactly in which class any observation falls; but it has the demerit that the range for the second subscript may be very large, and the level of the second factor does not automatically tell the level of the first factor.

It will be seen later that this would be very clumsy for many purposes. A simpler way to index the observations is by another doublet $(i, j)$ in which $i$ denotes the level of the nestor factor, and $(i, j)$ is a composite label for the level of the nested factor. The subscript $j$ will range over the positive integers $1, 2, \ldots$ with a maximum value of $n_i$, where there are $n_i$ classes according to the nested factor within the $i$th level of the nestor factor. If we use this mode of indexing, we say that the label "$j$", which appears to refer to the nested factor, is meaningless when considered alone. In this case "$j$" is *not* a level of the nested factor. It would be meaningless with observations labeled as $y_{ij}$, for instance, to perform a sum or average over the values of "$j$" except within a level $i$ of the nested factor. Observations with a particular value for "$j$", say $j = 2$, do not all lie in one class of a partition of the observations.

From another viewpoint, if we use this mode of labeling and have two observations $y_{12}$ and $y_{32}$, there is no relation between these two observations in terms of the classification structure. If the whole set of observations had come to us in a different order, the one labeled $y_{12}$ might have been labeled $y_{13}$ and the one labeled as $y_{32}$ might instead have been labeled $y_{31}$.

The second mode of labeling, which is that in which the two factors are regarded as crossed for labeling purposes, is the one used most exclusively.

Let us now consider the balanced case in which we have observations indexed by $i, j, k$, in which $i$ goes from 1 to $I$ and $j$ goes from 1 to $J$, with $k$ as the label of a particular observation in the set of observations with particular values for $i$ and $j$. We then have $IJ$ classes of observation in all, and we can display the means of these classes in tabular form as in Table 16.3. We say that we have two factors $A$ and $B$.

**Table 16.3** Class Means for a Two-Factor Classification

| $i$ \ $j$ | 1 | 2 | ... | $J$ | Mean |
|---|---|---|---|---|---|
| 1 | $\mu_{11}$ | $\mu_{12}$ | ... | $\mu_{1J}$ | $\mu_{1.}$ |
| 2 | $\mu_{21}$ | $\mu_{22}$ | ... | $\mu_{2J}$ | $\mu_{2.}$ |
| . | . | . | ... | . | . |
| . | . | . | ... | . | . |
| . | . | . | ... | . | . |
| $I$ | $\mu_{I1}$ | $\mu_{I2}$ | ... | $\mu_{IJ}$ | $\mu_{I.}$ |
| Mean | $\mu_{.1}$ | $\mu_{.2}$ | ... | $\mu_{.J}$ | $\mu_{..}$ |

If the two factors $A$ and $B$ are crossed, observations and means with the same value for $i$ belong to a class of a factor with $I$ levels, and observations with the same value for $j$ belong to a class of a factor with $J$ levels. We may then write the identity

$$\mu_{ij} = \mu_{..} + (\mu_{i.} - \mu_{..}) + (\mu_{.j} - \mu_{..}) + (\mu_{ij} - \mu_{i.} - \mu_{.j} + \mu_{..}) \qquad (16.43)$$

in which

$$\mu_{i.} = \frac{1}{J} \sum_j \mu_{ij}$$

$$\mu_{.j} = \frac{1}{I} \sum_i \mu_{ij}$$

and

$$\mu_{..} = \frac{1}{IJ} \sum_{ij} \mu_{ij}$$

Each of these terms has a meaning and is given a name. The first term $\mu_{..}$ is just the overall mean. The deviation of the $A_i$ category mean from the overall mean is called the effect of category $A_i$. Similarly, the quantity $\mu_{.j} - \mu_{..}$ is called the effect of category $B_j$. The last term is called the interaction of category $A_i$ with category $B_j$ and measures the amount by which $\mu_{ij}$ fails to be the sum of the overall mean and the effects of $A_i$ and $B_j$. In this case we see that the means $\mu_{ij}$, $\mu_{i.}$, $\mu_{.j}$, and $\mu_{..}$ have utility. They are said to be admissible.

In contrast suppose that factor $B$ is nested within factor $A$. It is still convenient to represent the populations and their means by Table 16.3. However, the index $j$ alone does not represent the level of a factor. It represents the level of a factor within a level of a factor $A$. The set of subpopulations with, say $j = 3$ does not lie in a class formed by a partition. We say that the subscript $j$ is not meaningful *across* levels of factor $A$. The nested factor has in fact $IJ$ levels. For this case an obvious identity is

$$\mu_{ij} = \mu_{..} + (\mu_{i.} - \mu_{..}) + (\mu_{ij} - \mu_{i.}) \tag{16.44}$$

The mean in cell $(i, j)$ is given by the overall mean $\mu_{..}$ plus the effect of level $i$ of $A$ plus the effect of level $j$ of $B$ *within* level $i$ of $A$. We say that the admissible means are $\mu_{ij}$, $\mu_{i.}$, and $\mu_{..}$. Also, we say that the mean $\mu_{.j}$ is not admissible.

If we labeled our classes by $i, j$ with the factor indexed by $i$ nested in the factor indexed by $j$, we would write instead the identity

$$\mu_{ij} = \mu_{..} + (\mu_{.j} - \mu_{..}) + (\mu_{ij} - \mu_{.j}) \tag{16.45}$$

However, it is conventional and useful to write subscripts indicating level of a nested factor after the subscript indicating level of the nestor factor.

We may note also that there is a relation between the effects in a nested model which has been treated as a crossed model, namely,

$\mu_{ij} - \mu_{i.} = (\mu_{.j} - \mu_{..}) + (\mu_{ij} - \mu_{i.} - \mu_{.j} + \mu_{..})$. This says that the effect of $B_j$ in the $A_i$ class equals the $B_j$ effect plus the $A_iB_j$ interaction.

It is important to note the role of the interaction term in Equation (16.43). Suppose that we have a cross-classification with two factors, which we call a two-way classification, and there is no interaction, i.e.,

$$\mu_{ij} - \mu_{i.} - \mu_{.j} + \mu_{..} = 0 \tag{16.46}$$

Then we can write

$$\mu_{ij} = \mu_{..} + (\mu_{i.} - \mu_{..}) + (\mu_{.j} - \mu_{..}) \tag{16.47}$$

or

$$\mu_{ij} = \mu + a_i + b_j \tag{16.48}$$

with an obvious correspondence of symbols. In this case the whole array of $IJ$ means is describable in terms of $1 + I + J$ numbers

$$\mu_{..}, \mu_{i.} - \mu_{..}, \mu_{.j} - \mu_{..}$$

or $\mu$, $a_i$, $b_j$. We see also that we have the conditions on $a_i$ and $b_j$:

$$\textstyle\sum_i a_i = \sum_i (\mu_{i.} - \mu_{..}) = 0$$

and

$$\textstyle\sum_j b_j = \sum_j (\mu_{.j} - \mu_{..}) = 0$$

Note also that the condition of zero interaction can also be written as

$$\mu_{ij} - \mu_{i'j} - \mu_{ij'} + \mu_{i'j'} = 0 \text{ for all } i, i', j, j'$$

Alternatively, we can write

$$\mu_{ij} = \mu_{11} + a_i' + b_j'$$

in which

$$a_i' = a_i - a_1, \qquad b_j' = b_j - b_1 \tag{16.49}$$

It is very fortunate when this happens, because we can talk about the effect of factor $A$, as given by the numbers $(a_i - a_1)$, $i = 2, \ldots, I$ and the effect of factor $B$, as given by the numbers $(b_j - b_1)$, $j = 2, \ldots, J$. If we had interaction, the effect of changing a level of $A$ from, say, $i$ to $i'$ depends on the level of $B$ at which the change takes place. The role of absence of interaction (or of small interaction, measured in some way) is tremendous in that we assume very often that interaction does not exist. Consider, for example, males or females, drug $A$ or drug $B$. In the absence of interaction we can say that the difference between drug $B$ and drug $A$ is the same for both males and females. We believe that "aspirins" or "no aspirins" does not interact with sex with respect to some measure of health (e.g., temperature). We would not think the same about some potential hormonal treatments. The general paradigm that humanity follows, and is forced to follow, is that it assumes interactions do not exist in a situation until data give some evidence that they do exist. As always, by "data" we mean all the relevant information that is available and not merely the data of the particular study.

## 16.7 Nested Models: Random Effects

There is one situation in which the nested (hierarchal) model seems far more reasonable than the crossed model. This occurs when subsampling is employed. A random sample is taken of categories of $A$. Within each category of $A$ a random sample is taken of categories of $B$. Within $B$ a random sample of observations is taken. An example of this is the case of an industrial engineering study of worker productivity which involved selecting ten factories at random from a population of factories, within each factory selecting some departments at random, and from each department selecting a random sample of workers.

As in all cases of structured data we represent an observation by the mean of the class in which it lies plus a deviation from the mean, so that if we have $N_{ij}$ observations in the $(i, j)$ cell we can write:

$$\begin{aligned} y_{ijk} &= \mu_{ij} + e_{ijk}, \qquad k = 1, 2, \ldots, N_{ij} \\ &= \mu_{..} + (\mu_{i.} - \mu_{..}) + (\mu_{ij} - \mu_{i.}) + e_{ijk} \\ &= \mu + a_i + b_{ij} + e_{ijk} \end{aligned}$$

where

$$\begin{aligned} \mu &= \mu_{..} \\ a_i &= \mu_{i.} - \mu_{..} \\ b_{ij} &= \mu_{ij} - \mu_{i.} \\ i &= 1, 2, \ldots, r \\ j &= 1, 2, \ldots, s_i \end{aligned} \tag{16.50}$$

Because the factories in our data were chosen at random and the departments were chosen at random also, both the $a_i$ and the $b_{ij}$ are random variables; and we find their distributional properties by the theory of sampling from finite populations. However, it is more customary and easiest to incorporate some mathematical distribution into the model. The simplest assumptions are

$$a_i \sim \text{NID}(0, \sigma_a^2),\ b_{ij} \sim \text{NID}(0, \sigma_b^2),\ e_{ijk} \sim \text{NID}(0, \sigma^2) \tag{16.51}$$

and that all the random variables are independent.

Note here that there are $r$ "$a$" classes, $s_i$ "$b$" classes within the $i$th "$a$" class, and $N_{ij}$ individuals in the $(i, j)$th $ab$ class. We shall not carry the computations through for this case of "general numbers," but shall confine ourselves to the case where $s_1 = \cdots = s_r = s$ and $N_{ij} = n$ for all $i$ and $j$. The data layout showing means and totals would be as in Table 16.4.

**Table 16.4** Data Layout—Nested Model

| | $A$ | | | | | | | | | | | | |
|---|---|---|---|---|---|---|---|---|---|---|---|---|---|
| | 1 | | | | 2 | | | | . . . | $r$ | | | |
| | 1 | 2 | . . . | $s$ | 1 | 2 | . . . | $s$ | . . . | 1 | 2 | . . . | $s$ |
| | $y_{111}$ | $y_{121}$ | . . . | $y_{1s1}$ | $y_{211}$ | $y_{221}$ | . . . | $y_{2s1}$ | . . . | $y_{r11}$ | $y_{r21}$ | . . . | $y_{rs1}$ |
| | $y_{112}$ | $y_{122}$ | . . . | $y_{1s2}$ | $y_{212}$ | $y_{222}$ | . . . | $y_{2s2}$ | . . . | $y_{r12}$ | $y_{r22}$ | . . . | $y_{rs2}$ |
| $B$ | . | . | . . . | . | . | . | . . . | . | . . . | . | . | . . . | . |
| | . | . | . . . | . | . | . | . . . | . | . . . | . | . | . . . | . |
| | $y_{11n}$ | $y_{12n}$ | . . . | $y_{1sn}$ | $y_{21n}$ | $y_{22n}$ | . . . | $y_{2sn}$ | . . . | $y_{r1n}$ | $y_{r2n}$ | . . . | $y_{rsn}$ |
| $B$ totals | $Y_{11.}$ | $Y_{12.}$ | . . . | $Y_{1s.}$ | $Y_{21.}$ | $Y_{22.}$ | . . . | $Y_{2s.}$ | . . . | $Y_{r1.}$ | $Y_{r2.}$ | . . . | $Y_{rs.}$ |
| $B$ means | $y_{11.}$ | $y_{12.}$ | . . . | $y_{1s.}$ | $y_{21.}$ | $y_{22.}$ | . . . | $y_{2s.}$ | . . . | $y_{r1.}$ | $y_{r2.}$ | . . . | $y_{rs.}$ |
| $A$ totals | $Y_{1..}$ | | | | $Y_{2..}$ | | | | . . . | $Y_{r..}$ | | | |
| $A$ means | $y_{1..}$ | | | | $y_{2..}$ | | | | . . . | $y_{r..}$ | | | |

Next we describe the computation of the quantities appearing in the analysis of variance:

$$\begin{aligned}
\text{Total SS} &= \textstyle\sum_{ijk} y_{ijk}^2 - Y_{...}^2/rsn \\
A \text{ SS} &= \textstyle\sum_i Y_{i..}^2/sn - Y_{...}^2/rsn \\
B \text{ in } A \text{ SS} &= \textstyle\sum_i \left(\sum_j Y_{ij.}^2/n - Y_{i..}^2/sn\right) \\
&= \textstyle\sum_{ij} Y_{ij.}^2/n - \sum_i Y_{i..}^2/sn \\
\text{Error SS} &= \textstyle\sum_{ij} \left(\sum_k y_{ijk}^2 - Y_{ij.}^2/n\right) \\
&= \textstyle\sum_{ijk} y_{ijk}^2 - \sum_{ij} Y_{ij.}^2/n
\end{aligned} \tag{16.52}$$

We use the convention here that replacement of a subscript by a dot and capitalization of the letter $y$ indicates summation over the suppressed subscript.

The error sum of squares may be obtained by subtraction. That is, error SS = total SS − $A$ SS − $B$ in $A$ SS. The analysis of variance, including the expected mean squares, is given in Table 16.5. We state without proof that the following distributional properties hold:

**Table 16.5** Hierarchal Analysis of Variance

| Source | d.f. | SS | MS | EMS |
|---|---|---|---|---|
| $A$ | $r - 1$ | | MS($A$) | $\sigma^2 + n\sigma_b^2 + sn\sigma_a^2$ |
| $B$ in $A$ | $r(s - 1)$ | as given | MS($B$) | $\sigma^2 + n\sigma_b^2$ |
| Error | $rs(n - 1)$ | in text | MS($E$) | $\sigma^2$ |
| Total | $rsn - 1$ | | | |

$$y_{...} \text{ is } N\left(\mu, \frac{\sigma^2 + n\sigma_b^2 + ns\sigma_a^2}{rsn}\right)$$

$$\frac{(r-1)\text{MS}(A)}{(\sigma^2 + n\sigma_b^2 + ns\sigma_a^2)} \text{ is } \chi^2 \text{ with } r - 1 \text{ d.f.}$$

$$\frac{r(s-1)\text{MS}(B)}{(\sigma^2 + n\sigma_b^2)} \text{ is } \chi^2 \text{ with } r(s - 1) \text{ d.f.}$$

$$\frac{rs(n-1)\text{MS}(E)}{\sigma^2} \text{ is } \chi^2 \text{ with } rs(n - 1) \text{ d.f.} \tag{16.53}$$

and these variables are independently distributed. We can also see that $y_{...}$, MS($A$), MS($B$), and MS($E$) comprise a minimal sufficient statistic. The whole analysis of variance is more difficult, and the distributional properties with normality are very complex with unequal members.

The mean squares MS($A$), MS($B$), and MS($E$) give us some indication of plausible values for $\sigma^2$, $\sigma_b^2$, and $\sigma_a^2$. Just how we use these statistics is difficult to decide and really quite obscure in the present state of knowledge of statistics. It is common to abstract what are called unbiased estimates of $\sigma^2$, $\sigma_a^2$, $\sigma_b^2$ as

$$\begin{aligned} \hat{\sigma}^2 &= \text{MS}(E) \\ \hat{\sigma}_b^2 &= [\text{MS}(B) - \text{MS}(E)]/n \\ \hat{\sigma}_a^2 &= [\text{MS}(A) - \text{MS}(B)]/ns \end{aligned} \tag{16.54}$$

but the terminology is very poor because these formulas can give negative values for $\hat{\sigma}_b^2$ and $\hat{\sigma}_a^2$, values which do not lie in the parameter space. There is no doubt, however, that these are *informative* statistics, and approximate standard errors for them can be computed.

The ratio MS($B$)/MS($E$) gives us an indication of $\sigma_b^2/\sigma^2$ and thus can lead to a test of significance of $\sigma_b^2$, with sensitivity depending on $n\sigma_b^2/\sigma^2$. Similarly, MS($A$)/MS($B$) gives an indication of $\sigma_a^2/(\sigma^2 + n\sigma_b^2)$ and can lead to a test of significance of $\sigma_a^2$, with sensitivity depending on $ns\sigma_a^2/(\sigma^2 + n\sigma_b^2)$.

**Table 16.6** Percent of HCl in Samples

| Batch | 1 | | 2 | | 3 | |
|---|---|---|---|---|---|---|
| Sample | 1 | 2 | 1 | 2 | 1 | 2 |
| Analysis 1 | 20 | 19 | 22 | 21 | 18 | 20 |
| Analysis 2 | 21 | 22 | 24 | 22 | 14 | 18 |
| Analysis 3 | 18 | 23 | 20 | 26 | 16 | 18 |
| Sample total | 59 | 64 | 66 | 69 | 48 | 56 |
| Sample mean | 19.67 | 21.33 | 22.00 | 23.00 | 16.00 | 18.67 |
| Batch total | 123 | | 135 | | 104 | |
| Batch mean | 20.50 | | 22.50 | | 17.33 | |
| Grand total | 362 | | | | | |
| Grand mean | 20.11 | | | | | |

*Example 1*
A chemical process used large quantities of an etch. From time to time the percent of HC1 was checked. The variation in percent of HC1 was alarming, and a study was conducted to determine the major causes. Three batches were selected at random from the incoming stock; two samples were chosen from each batch, and three analyses were made on each sample. The data are given in Table 16.6.

The computations for the analysis of variance are as follows:

1. Total SS $= \sum_{ijk} y_{ijk}^2 - Y_{...}^2/rsn$
$= 143.78.$
2. Batch SS $= \sum_i Y_{i..}^2/sn - Y_{...}^2/rsn$
$= 81.45.$
3. Sample in batch SS $= \sum_{ij} Y_{ij.}^2/n - \sum_i Y_{i..}^2/sn$
$= 17.66.$
4. Error SS = total SS − batch SS − sample in batch SS
$= 44.67.$

The analysis of variance is given in Table 16.7.

To evaluate the model that $\sigma_s^2 = 0$, we form $F = 5.89/3.72 = 1.58$. Since $F_{.05;3,12} = 3.49$ and $F_{.50;3,12}$ is near unity, we do not have strong evidence that there is variation between samples within batches. However, the data and consideration of limits will tell us that our experiment is very imprecise on this question.

**Table 16.7** Analysis of Variance for Data of Table 16.6

| Source | d.f. | SS | MS | EMS |
|---|---|---|---|---|
| Batches | 2 | 81.45 | 40.72 | $\sigma^2 + 3\sigma_s^2 + 6\sigma_b^2$ |
| Sample in batches | 3 | 17.66 | 5.89 | $\sigma^2 + 3\sigma_s^2$ |
| Error | 12 | 44.67 | 3.72 | $\sigma^2$ |
| Total | 17 | 143.78 | | |

The naive estimates of components of variance are

$$\hat{\sigma}^2 = 3.72$$
$$\hat{\sigma}_s^2 = (5.89 - 3.72)/3 = 0.72$$
$$\hat{\sigma}_b^2 = (40.72 - 5.89)/6 = 5.80$$

Thus we estimate that batches account for about 56% of the variation, samples for about 7% of the variation, and analyses for about 36% of the variation.

## 16.8 Cross-Classification: Fixed Effects

We have already described in some detail the cross-classification model for the means of populations. Suppose that we have a random sample of $n$ observations from each of $rs$ subpopulations which fit a cross-classification.

A possible model for this situation is

$$y_{ijk} = \mu + a_i + b_j + (ab)_{ij} + e_{ijk}$$
$$i = 1, 2, \ldots, r$$
$$j = 1, 2, \ldots, s$$
$$k = 1, 2, \ldots, n$$

where

$$\textstyle\sum_i a_i = \sum_j b_j = \sum_i (ab)_{ij} = \sum_j (ab)_{ij} = 0$$

and

$$e_{ijk} \sim \text{NID}(0, \sigma^2) \qquad (16.55)$$

Actually, this is a situation of considerable complexity. We simplify it in many ways and give what can only be termed an elementary, simplified naive presentation. The data layout, including totals and means, is shown in Table 16.8.

**Table 16.8** Data Layout: Two-Way Classification

| Level of $B$ / Level of $A$ | 1 | 2 | . . . | $s$ | Combining $B$ Levels: Total | Combining $B$ Levels: Mean |
|---|---|---|---|---|---|---|
| 1 | $y_{11k}$ | $y_{12k}$ | . . . | $y_{1sk}$ | $Y_{1..}$ | $y_{1..}$ |
| Cell totals | $Y_{11.}$ | $Y_{12.}$ | . . . | $Y_{1s.}$ | | |
| Cell means | $y_{11.}$ | $y_{12.}$ | . . . | $y_{1s.}$ | | |
| 2 | $y_{21k}$ | $y_{22k}$ | . . . | $y_{2sk}$ | $Y_{2..}$ | $y_{2..}$ |
| Cell totals | $Y_{21.}$ | $Y_{22.}$ | . . . | $Y_{2s.}$ | | |
| Cell means | $y_{21.}$ | $y_{22.}$ | . . . | $y_{2s.}$ | | |
| . | . | . | . . . | . | . | . |
| $r$ | $y_{r1k}$ | $y_{r2k}$ | . . . | $y_{rsk}$ | $Y_{r..}$ | $y_{r..}$ |
| Cell totals | $Y_{r1.}$ | $Y_{r2.}$ | . . . | $Y_{rs.}$ | | |
| Cell means | $y_{r1.}$ | $y_{r2.}$ | . . . | $y_{rs.}$ | | |
| Combining $A$ levels | | | | | | |
| Total | $Y_{.1.}$ | $Y_{.2.}$ | . . . | $Y_{.s.}$ | $Y_{...}$ | |
| Mean | $y_{.1.}$ | $y_{.2.}$ | . . . | $y_{.s.}$ | | $y_{...}$ |

In each cell $k$ runs from 1 to $n$.

**Table 16.9** Analysis of Variance of Cross-Classification of Table 16.8

| Source | d.f. | SS | MS | EMS |
|---|---|---|---|---|
| $A$ | $r - 1$ | as given by Equation 16.56 | MS($A$) | $\sigma^2 + sn \sum_i a_i^2/(r - 1)$ |
| $B$ | $s - 1$ | | MS($B$) | $\sigma^2 + rn \sum_j b_j^2/(s - 1)$ |
| $AB$ | $(r - 1)(s - 1)$ | | MS($AB$) | $\sigma^2 + n \sum_{ij} (ab)_{ij}^2/(r - 1)(s - 1)$ |
| Error | $rs(n - 1)$ | | MS($E$) | $\sigma^2$ |
| Total | $rsn - 1$ | | | |

Computation for the analysis of variance proceeds as follows:

$$\begin{aligned}
\text{Total SS} &= \textstyle\sum_{ijk} y_{ijk}^2 - Y_{...}^2/rsn \\
A \text{ SS} &= \textstyle\sum_i Y_{i..}^2/sn - Y_{...}^2/rsn \\
B \text{ SS} &= \textstyle\sum_j Y_{.j.}^2/rn - Y_{...}^2/rsn \\
AB \text{ SS} &= \textstyle\sum_{ij} Y_{ij.}^2/n - Y_{...}^2/rsn - A \text{ SS} - B \text{ SS} \\
\text{Error SS} &= \text{total SS} - A \text{ SS} - B \text{ SS} - AB \text{ SS}
\end{aligned} \qquad (16.56)$$

The analysis of variance is given in Table 16.9. The evaluation of models is simple. The model that $a_1 = a_2 = \cdots = a_r$ is examined by forming $F = \text{MS}(A)/\text{MS}(E)$ and comparing with the appropriate $F$ distribution. The $F$ ratio for evaluating $b_1 = b_2 = \cdots = b_s$ is given by $\text{MS}(B)/\text{MS}(E)$. The $F$ ratio for evaluating the model with $(ab)_{ij} = 0$ for all $i$ and $j$ is given by $\text{MS}(AB)/\text{MS}(E)$. In all cases the degrees of freedom for $F$ ratios are those in Table 16.9.

Fitting the model is also simple. An initial condensation of the data is given by the following statistics, with $i = 1, 2, \ldots, r$, and $j = 1, 2, \ldots, s$:

$$\begin{aligned} a_i^* &= y_{i..} - y_{...} \\ b_j^* &= y_{.j.} - y_{...} \\ (ab)_{ij}^* &= y_{ij.} - y_{i..} - y_{.j.} + y_{...} \\ \mu^* &= y_{...} \\ \hat{\sigma}^2 &= \text{error MS} \end{aligned} \tag{16.57}$$

The quantities in Equation (16.57) comprise a sufficient statistic which becomes minimal if some are deleted to remove the arithmetical dependencies which exist in the set.

## 16.9 Cross-Classification: Random Effects

In the event that the $r$ categories or levels of $A$ and the $s$ categories of $B$ were chosen at random from a population of categories, certain modifications can be made in our outlook. Since the levels used in our sample have been chosen at random, we would not be interested in them individually. Rather, we are interested in the population from whence they came. Accordingly, certain changes are made in the model. The computations for the analysis of variance proceed in the same manner, but the problems of inference are different. The arithmetical partition of the total sum of squares is the same as that given in Section 16.8, i.e., by Equation (16.56).

A reasonable model is now the following:

$$\begin{aligned} y_{ijk} &= \mu + a_i + b_j + (ab)_{ij} + e_{ijk} \\ i &= 1, 2, \ldots, r \\ j &= 1, 2, \ldots, s \\ k &= 1, 2, \ldots, n \end{aligned} \tag{16.58}$$

in which all terms but $\mu$ on the right-hand side of Equation (16.58) are random variables.

It is possible to develop properties of the sums of squares and hence mean squares with the theory of finite sampling based on the idea that we have observed a random set of $r$ levels from a population of $R$ levels of $A$, a random set of $s$ levels from a population of $S$ levels of $B$, and random sampling of $n$ individuals from a population of $N$ individuals in each of the selected $rs$ subpopulations. We shall examine a completely specified model using assumptions of normality. We assume

$$\begin{aligned} a_i &\sim \text{NID}(0, \sigma_a^2) \\ b_j &\sim \text{NID}(0, \sigma_b^2) \\ (ab)_{ij} &\sim \text{NID}(0, \sigma_{ab}^2) \\ e_{ijk} &\sim \text{NID}(0, \sigma^2) \end{aligned} \tag{16.59}$$

and these sets of random variables are independent.

The nature of the analysis of variance is given in Table 16.10. The quantities MS($A$), MS($B$), MS($AB$), and MS($E$) together with $y_{...}$ comprise a minimal sufficient statistic.

**Table 16.10** Analysis of Variance for Balanced Two-Way Random Model

| Source | d.f. | MS | EMS |
|---|---|---|---|
| $A$ | $r - 1$ | MS($A$) | $\sigma^2 + n\sigma_{ab}^2 + sn\sigma_a^2$ |
| $B$ | $s - 1$ | MS($B$) | $\sigma^2 + n\sigma_{ab}^2 + rn\sigma_b^2$ |
| $AB$ | $(r - 1)(s - 1)$ | MS($AB$) | $\sigma^2 + n\sigma_{ab}^2$ |
| Error | $rs(n - 1)$ | MS($E$) | $\sigma^2$ |
| Total | $rsn - 1$ | | |

In the previous section, $F$ ratios were formed by dividing the appropriate mean square by the error mean square. Here the situation is more complicated. To evaluate the hypothesis that $\sigma_{ab}^2 = 0$, we form $F = \text{MS}(AB)/\text{MS}(E)$ and refer to the $F$ distribution with $(r - 1)(s - 1)$ and $rs(n - 1)$ degrees of freedom. To evaluate the hypothesis that $\sigma_b^2 = 0$, we form $F = \text{MS}(B)/\text{MS}(AB)$ and refer to the $F$ distribution with $s - 1$ and $(r - 1)(s - 1)$ degrees of freedom. Similarly, for the hypothesis $\sigma_a^2 = 0$ we form $F = \text{MS}(A)/\text{MS}(AB)$ and use $r - 1$ and $(r - 1)(s - 1)$ degrees of freedom.

Exact consonance regions exist for $\sigma^2$, but not for the other variance components. In regard to $\sigma^2$, it is fairly easy to see that

$$\frac{\text{error SS}}{\chi_{\alpha/2}^2} \leq \sigma^2 \leq \frac{\text{error SS}}{\chi_{1-\alpha/2}^2} \tag{16.60}$$

gives a $(1 - \alpha)$ consonance interval on $\sigma^2$. In regard to the other variance components, there is no simple prescription. Texts on statistical methods give suggestions, but the problems are very difficult. Research is still being done but is inconclusive so far. Again Bayesian computations can be performed if a prior distribution can be specified. We believe any such prior should be based on historical evidence, and not merely on considerations of ease of integration.

## 16.10 Cross-Classification: Mixed Model

Consider a case for which there are two factors of classification, $A$ and $B$, and where factor $A$ is fixed, which is to say identifiable and repeated in the data and repeatable in the future. On the other hand, factor $B$ is random in the sense that there is a population of possible levels of which we have a random sample. An example of $A$ could be methods of processing. Factor $B$ could be batches of material as they arrive at the delivery point. We suppose that it is reasonable to assume as a working basis that we have a random sample of an infinity of possible batches. Each batch is divided into $r$ parts, so that there is one part for each method of processing. The observation is, say, the percentage conversion to a desired end product. We could take either of two points of view:

1. That we wish to form an opinion about what happened with the particular batches we had.
2. That we wish to form an opinion about what would happen or what we would find if we could subject all batches to all the methods of processing.

If our interest is in point (1), we use the fixed model; but if our interest is in point (2), we wish to take account in our model of the fact that we have a random sample of the population of batches which is of interest.

A possible model for this situation with the second point in mind is the following:

$$\begin{aligned} y_{ijk} &= \mu + a_i + b_j + (ab)_{ij} + e_{ijk}, & i &= 1, 2, \ldots, r \\ \sum a_i &= 0 & j &= 1, 2, \ldots, s \\ b_j &\sim \text{NID}(0, \sigma_b^2) & k &= 1, 2, \ldots, n \\ (ab)_{ij} &\sim \text{NID}(0, \sigma_{ab}^2) \\ e_{ijk} &\sim \text{NID}(0, \sigma^2) \end{aligned} \tag{16.61}$$

with independence of the sets of random variables. We shall now be rather "sloppy" and say that it is reasonable to require that the terms $(ab)_{ij}$ are such as to add to zero over $i$ for each and every $j$. If one thinks about the conceptual two-way table containing the methods of processing and all the possible batches, this seems reasonable. So we incorporate the additional model property: $\sum_i (ab)_{ij} = 0$, for every $j$. This has the consequence that the independence assumption with regard to the $(ab)_{ij}$ which is given in Equation (16.61) must be modified. A partial description of the nature of the resulting analysis of variance is given in Table 16.11. Note that the term $\sigma_{ab}^2$ does not appear in the expected mean

**Table 16.11** Analysis of Variance for Balanced Two-Way Mixed Model

| Source | d.f. | MS | EMS |
|---|---|---|---|
| $A$ | $r - 1$ | MS($A$) | $\sigma^2 + n\sigma_{ab}^2 + sn \sum a_i^2/(r - 1)$ |
| $B$ | $s - 1$ | MS($B$) | $\sigma^2 + rn\sigma_b^2$ |
| $AB$ | $(r - 1)(s - 1)$ | MS($AB$) | $\sigma^2 + n\sigma_{ab}^2$ |
| Error | $rs(n - 1)$ | MS($E$) | $\sigma^2$ |
| Total | $rsn - 1$ | | |

square for factor $B$ but does for factor $A$. This situation would be reversed if $A$ were random and $B$ fixed. As far as estimation is concerned, the estimation of variance components proceeds in standard fashion. Since the levels of $A$ are fixed, we would be interested in estimating such quantities as $a_i - a_j$. In regard to testing, the main effect $B$ and interaction mean squares are compared with the error mean square. The main effect $A$ mean square, however, is compared with the interaction mean square. The values for expected mean squares indicate which mean squares should be compared with which in order to lead to the formation of some opinion.

## 16.11 The Diagrammatic Representation of Classification Structures

A very useful device for conveying a population structure,* which one is going to use as a basis for interpreting a set of data, is given by diagrams that originated in pure mathematics and are called Hasse diagrams. We give two examples in Figure 16.1.

The basic rule is that if a factor $Y$ is nested by a factor $X$, then $Y$ is connected to $X$ by links going upward. Every o in the picture corresponds to a factor of classification. The top one is in a certain sense

*The first use of these diagrams is apparently in a Ph.D. thesis in statistics by N. Throckmorton done at Iowa State University.

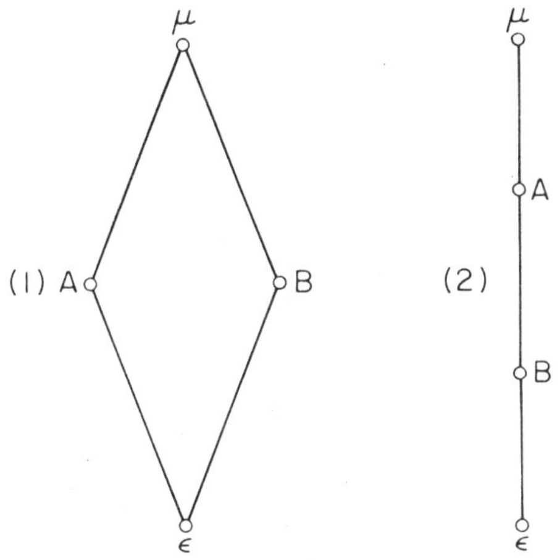


**Fig. 16.1** The Two Possible Structures with Two Factors of Classification

unnecessary but is useful in indicating that the diagram is complete in an upward direction. We attach a label $\mu$ to it to indicate that all observations are in the same body of data to be interpreted. Diagram (1) suggests that an individual can be in an $A$ class, say the $i$th, and in a $B$ class, say the $j$th, independently. Diagram (2) suggests that once an observation is in a certain $B$ class, it is in an $A$ class. The bottom o in the diagram is useful in indicating that one has concluded the diagram. It is labeled $\varepsilon$ to suggest that every observation is unique in one respect at least, and failing all else it is unique with regard to measurement error.

Diagram (1) then tells us that we have a two-way classification with factors $A$ and $B$. It also suggests that a model for an observation can be written as $\mu + A_i + B_j + (AB)_{ij} +$ error. Diagram (2) tells us that we have a nested structure with factor $B$ nested in factor $A$. It also suggests, after we have been motivated, that a model for an observation is $\mu + A_i + B_{ij} +$ error.

We think it will be of interest to the reader to illustrate these diagrams for more complex models. Consider, then, the following diagrams with five circles given in Figure 16.2. With the suggestions and motivations given, the reader can probably interpret these diagrams. Let us explain them.

1. Here we have three factors of classification $A$, $B$ in $A$, and $C$ in $B$. An example is the identification of houses $C$ in counties $B$ in states $A$ in the United States $\mu$.
2. Here again we have three factors of classification $A$, $B$, and $C$: $B$ and $C$ are nested in $A$, but $B$ and $C$ are crossed. To continue our analogy

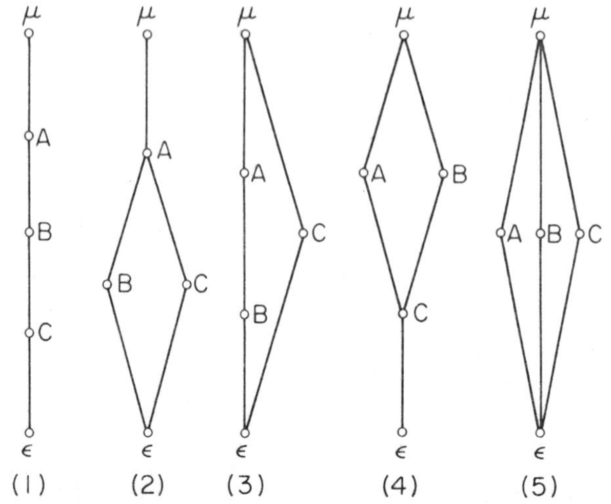


**Fig. 16.2** The Five Possible Structures with Three Factors of Classification

with houses: $\mu$ could be the United States, $A$ the county, $B$ the type of house (wood or brick), and $C$ the number of rooms 1, 2, 3, . . . , while $\varepsilon$ denotes the individual house.

3. Again we have three factors $A$, $B$, and $C$. Let $\mu$ denote the United States, $A$ the state, $B$ the county, and $C$ the location (rural or urban), with $\varepsilon$ again denoting the individual house.

4. A possible naming of the three factors and of $\mu$ and $\varepsilon$ could be:
   $\mu$: the United States
   $A$: rural or urban
   $B$: owned or rented
   $C$: individual house within $AB$ classes
   $\varepsilon$: the individual measurement (e.g., amount of electricity consumed in the year 1966)

5. A possible naming of the three factors $A$, $B$, and $C$ and of $\mu$ and $\varepsilon$ could be:
   $\mu$: the United States
   $A$: rural or urban
   $B$: owned or rented
   $C$: with or without a telephone
   $\varepsilon$: the individual house

It is an elementary fact that these are the only possible structures with three "real" factors of classification.

## 16.12 The General Linear Fixed Model

In Chapter 15 we discussed the model

$$y = X\beta + e \tag{16.62}$$

in which $y$ is a vector of $n$ observations, $\beta$ is a vector of $p$ unspecified constants or parameters, $e$ is a vector of $n$ errors, and $X$ is a known $n \times p$ matrix. There we used the assumption that $X'X$, which is a $p \times p$ matrix, is nonsingular, i.e., of rank $p$. It is desirable to present an introduction to the case in which the rank is $r$, which is less than $p$, because classification linear models have this attribute. First, we define a classification linear model as one in which observations are labeled by subscripts $i, j, \ldots$, so that $i$ is the level of one factor of classification of the individuals, $j$ is the level of another, etc., and furthermore, the observation is expressed as a *sum* of a contribution for error and contributions which depend on how the individual occurs in the classification structure. Examples are:

$$y_{ijk} = \mu + r_i + c_j + (rc)_{ij} + t_k + e_{ijk}$$

or

$$y_{ijk} = \mu + r_i + c_j + e_{ijk}$$

with $i = 1, 2, \ldots, I$; $j = 1, 2, \ldots, J$; and $k = 1, 2, \ldots, n$. We call such a linear model a classification linear model. Each parameter enters the model with coefficient zero or one, so that each element of $X$ is zero or one. For example with $y_{ijk} = \mu + r_i + c_j + e_{ijk}$, $I = 2$, $J = 3$, $n = 1$, we have

$$\begin{pmatrix} y_{111} \\ y_{121} \\ y_{131} \\ y_{211} \\ y_{221} \\ y_{231} \end{pmatrix} = \begin{pmatrix} 1 & 1 & 0 & 1 & 0 & 0 \\ 1 & 1 & 0 & 0 & 1 & 0 \\ 1 & 1 & 0 & 0 & 0 & 1 \\ 1 & 0 & 1 & 1 & 0 & 0 \\ 1 & 0 & 1 & 0 & 1 & 0 \\ 1 & 0 & 1 & 0 & 0 & 1 \end{pmatrix} \begin{pmatrix} \mu \\ r_1 \\ r_2 \\ c_1 \\ c_2 \\ c_3 \end{pmatrix} + \begin{pmatrix} e_{111} \\ e_{121} \\ e_{131} \\ e_{211} \\ e_{221} \\ e_{231} \end{pmatrix}$$

It follows for classification linear models in general that the normal equations can be written down merely by inspection of the data and the model. There will be dependencies in the columns of $X$, e.g., the addition of columns for $r_1, r_2, \ldots, r_I$ term by term gives the column for $\mu$ in $X$. The singularity causes no trouble provided it is recognized and one takes appropriate steps. One can modify the columns of $X$ by imposing conditions like $r_1 + r_2 + \cdots + r_I = 0$. This enables one to express $r_I$

in terms of the other $r_i$ in the original model. This leads to modification of the original $X$ matrix by first subtracting the column associated with $r_I$ from the other $r_i$ columns and then deleting the $r_I$ column. This really amounts to a reparametrization of the model (a representation of the data-model situation with different parameters), and in any particular case many reparametrizations to full rank are possible.

For mathematical purposes it is convenient to consider the matter without doing this. A very basic idea is that of estimability. A linear function $\lambda'\beta$ is said to be linearly estimable in $X$ if there exists a vector $a$ such that $a'X = \lambda'$ or $X'a = \lambda$. Now consider the normal equations

$$X'X\beta = X'y \tag{16.63}$$

These are consistent, using the elementary rank condition for consistency of a set of linear equations (see any text on linear equations or linear algebra). This is the critical nonelementary fact. Hence the equations have one solution or an infinity of solutions. Suppose $\beta_1$ and $\beta_2$ are two different solutions. Then

$$\begin{aligned} & X'X\beta_1 = X'X\beta_2 \\ & \rightarrow X'X(\beta_1 - \beta_2) = \phi \rightarrow (\beta_1 - \beta_2)'X'X(\beta_1 - \beta_2) = 0 \\ & \rightarrow (X\beta_1 - X\beta_2)'(X\beta_1 - X\beta_2) = 0 \\ & \rightarrow X\beta_1 = X\beta_2 \end{aligned} \tag{16.64}$$

So we get the same answer for the fitted values of the $y_i$, whatever solution we take. Now take $y$ to be successively the columns of $I_n$. Then the above implies that there is a solution to the matrix equation

$$X'XB = X' \tag{16.65}$$

and that $XB$ is the same for all solutions. Also, transposing, we get

$$B'X'X = X \tag{16.66}$$

Now note that from Equation (16.65) $B'X'XB = B'X'$, and from Equation (16.66) $B'X'XB = XB$, so $XB = B'X'$. Denote this by $P$. Then we also have $P' = P$ and $P^2 = P$. Consider now the possible solution $\hat{\beta} = By$. Substituting in Equation (16.63), we have $X'X\hat{\beta} = X'XBy = X'y$. Hence $\hat{\beta} = By$ satisfies the normal equations, and $X\hat{\beta} = XBy = Py$ is a unique transform of $y$. The above is the basis for most distribution theory of the linear fixed model, but space limitations prevent our pursuing this.

Consider the approximate likelihood under the assumption of normal independent errors (ignoring the grouping errors),

$$\frac{1}{(2\pi)^{n/2}}\frac{1}{\sigma^n}\exp\left[-\frac{1}{2\sigma^2}(y - X\beta)'(y - X\beta)\right] \tag{16.67}$$

The logarithm of the likelihood is, apart from an additive constant,

$$-\frac{n}{2}\ln\sigma^2 - \frac{1}{2\sigma^2}(y - X\beta)'(y - X\beta) \tag{16.68}$$

Maximization with respect to $\beta$ leads to the normal equations

$$X'X\beta = X'y$$

and to

$$\sigma^2 = \frac{1}{n}\min_{\beta}(y - X\beta)'(y - X\beta) \tag{16.69}$$

Hence the maximum likelihood statistic for $X\beta$ is $Py$, and the maximum likelihood statistic for $\sigma^2$ is

$$\sigma^{*2} = \frac{1}{n}\min_{\beta}(y - X\beta)'(y - X\beta) \tag{16.70}$$

We see that the maximum likelihood statistic for $X\beta$ is $X\hat{\beta}$. We shall use this expression rather than the more clumsy $(X\beta)^*$.

Now the minimum here is

$$y'y - \hat{\beta}'X'y - y'X\hat{\beta} + \hat{\beta}'X'X\hat{\beta} \tag{16.71}$$

and for all solutions of the normal equations this gives

$$y'y - \hat{\beta}'X'y = y'(I - P)y$$

Hence the maximum of the likelihood is equal to

$$\frac{1}{(2\pi)^{n/2}}\frac{1}{(\sigma^{*2})^{n/2}}\exp\left(\frac{-n}{2}\right) \tag{16.72}$$

The tests of significance on aspects of $\beta$ given earlier can be deduced from this beginning by the likelihood ratio method.

Consider the testing of

$$\lambda_i'\beta = c_i, \qquad i = 1, 2, \ldots, k \tag{16.73}$$

We suppose that each of these $\lambda_i'\beta$ is estimable because no test of significance can have sensitivity with regard to a nonestimable function. Also we suppose that the set of functions $\lambda_i'\beta$ are linearly independent. We write these in matrix form as

$$\Lambda'\beta = c \tag{16.74}$$

The requirement that these be estimable is that there exists a matrix $A$, such that

$$E(A'y) = \Lambda'\beta \text{ or } A'X = \Lambda' \tag{16.75}$$

Now consider the maximum of the likelihood under the condition that Equation (16.74) holds. We have to minimize

$$\frac{n}{2}\ln\sigma^2 + \frac{1}{2\sigma^2}(y - X\beta)'(y - X\beta) \tag{16.76}$$

subject to $\Lambda'\beta = c$. We introduce Lagrangian multiplers $\rho_1/\sigma^2, \rho_2/\sigma^2, \ldots, \rho_k/\sigma^2$ and differentiate

$$\frac{n}{2}\ln\sigma^2 + \frac{1}{2\sigma^2}[(y - X\beta)'(y - X\beta) - 2\rho'(\Lambda'\beta - c)] \tag{16.77}$$

where $\rho' = (\rho_1, \rho_2, \ldots, \rho_k)$ with respect to $\beta_1, \beta_2, \ldots, \beta_p$.
This gives equations for $\beta$:

$$X'X\beta - X'y - \Lambda\rho = \phi$$

or

$$X'X\beta = X'y + \Lambda\rho \tag{16.78}$$

or because

$$\Lambda = X'A$$
$$X'X\beta = X'y + X'A\rho = X'(y + A\rho) \tag{16.79}$$

These are just like the original normal equations, and a solution is given by

$$\tilde{\beta} = B(y + A\rho) \tag{16.80}$$

and

$$X\tilde{\beta} = XB(y + A\rho)$$

or

$$X\tilde{\beta} = Py + PA\rho \tag{16.81}$$

We now solve for $\rho$ by substituting:

$$\Lambda'\tilde{\beta} = A'X\tilde{\beta} = A'Py + A'PA\rho = c \tag{16.82}$$

so that

$$\rho = (A'PA)^{-1}(c - A'Py) \tag{16.83}$$

We use here without proof that $A'PA$ is nonsingular under our conditions. But

$$A'Py = A'X\hat{\beta} = \widehat{\Lambda'\beta} = \Lambda'\hat{\beta}$$

so

$$\rho = (A'PA)^{-1}(c - \widehat{\Lambda'\beta}) \tag{16.84}$$

Now we are in a position to write down the maximum of the likelihood subject to the constraints. We have

$$\tilde{\sigma}^2 = \frac{1}{n}(y - X\tilde{\beta})'(y - X\tilde{\beta}) \tag{16.85}$$

But

$$\begin{aligned}(y - X\tilde{\beta})'(y - X\tilde{\beta}) &\\ &= [y - P(y + A\rho)]'[y - P(y + A\rho)] \\ &= [(I - P)y - PA\rho]'[(I - P)y - PA\rho] \\ &= y'(I - P)^2 y - 2\rho'A'P'(I - P)y + \rho'A'P'PA\rho \end{aligned} \tag{16.86}$$

Using

$$(I - P)^2 = I - P, \qquad P' = P, \qquad P = P^2, \qquad P'(I - P) = \phi$$

we have

$$\begin{aligned} n\tilde{\sigma}^2 &= y'(I - P)y + \rho'A'PA\rho \\ &= n\sigma^{*2} + (c - \widehat{\Lambda'\beta})'(A'PA)^{-1}(c - \widehat{\Lambda'\beta}) \end{aligned} \tag{16.87}$$

Hence the maximum of the likelihood subject to the constraints is

$$\frac{1}{(2\pi)^{n/2}} \frac{1}{(\tilde{\sigma}^2)^{n/2}} \exp\left(-\frac{n}{2}\right) \tag{16.88}$$

Therefore, the likelihood ratio is

$$\left(\frac{\tilde{\sigma}^2}{\sigma^{*2}}\right)^{n/2} \tag{16.89}$$

which is

$$\left\{1 + \frac{[(c - \widehat{\Lambda'\beta})'(A'PA)^{-1}(c - \widehat{\Lambda'\beta})]}{n\sigma^{*2}}\right\}^{n/2} \tag{16.90}$$

Large values of this are given by large values of

$$\frac{(c - \widehat{\Lambda'\beta})'(A'PA)^{-1}(c - \widehat{\Lambda'\beta})}{\sigma^{*2}}$$

so we can use this as the criterion.

To relate this to the commonly presented form of the test criterion it is necessary to use the fact, which we shall not prove, that $n\sigma^{*2}/\sigma^2$ is distributed as $\chi^2_{n-r}$ independently of $X\hat{\beta}$, where $r$ is the rank of matrix $X$. If then we write

$$(n - r)s^2 = (y - X\hat{\beta})'(y - X\hat{\beta})$$

we have that $s^2$ is the error mean square after fitting $y = X\beta + e$, and

$$(n - r)s^2 = n\sigma^{*2}$$

Hence the likelihood ratio method of developing a test of significance is equivalent to using as a test criterion the function

$$\frac{(c - \widehat{\Lambda'\beta})'(A'PA)^{-1}(c - \widehat{\Lambda'\beta})}{ks^2} \tag{16.91}$$

Finally, because $\Lambda'\beta = A'X\beta$ and $\widehat{\Lambda'\beta} = A'X\hat{\beta} = A'Py$, the variance-covariance matrix of $\widehat{\Lambda'\beta}$ is $\sigma^2 A'PA$. Hence the likelihood ratio method for testing significance is equivalent to the use of

$$\frac{(c - \widehat{\Lambda'\beta})'\left[\frac{1}{\sigma^2}\operatorname{Var}(\widehat{\Lambda'\beta})\right]^{-1}(c - \widehat{\Lambda'\beta})}{ks^2} \tag{16.92}$$

which is the statistic presented earlier.

We state without proof that the statistic (16.92) is distributed according to the $F$ distribution with $k$ and $n - r$ degrees of freedom. This fact tells us therefore that the form (16.92) with the elements of the vector $c$ regarded as unknown parameters is a pivotal for these unknowns. This tells us also that values of the elements of $c$ that are consonant with the data by the $F$ test at the $\alpha$ level of significance are given by

$$\frac{(c - \widehat{\Lambda'\beta})'\left[\frac{1}{\sigma^2}\operatorname{Var}(\widehat{\Lambda'\beta})\right]^{-1}(c - \widehat{\Lambda'\beta})}{ks^2} \le F_{\alpha;k,n-r}$$

For each possible value of $\alpha$, this gives us a $k$-dimensional ellipsoid, and clearly ellipsoids for different values of $\alpha$ have the desired containment property. These facts, rather than a derivation by the likelihood ratio method, seem to us to be the basis for the appeal of the test procedure.

Finally, we note the following which is very useful in actual data analysis:

$$(y - X\hat{\beta})'(y - X\hat{\beta}) = y'y - \hat{\beta}'X'y \tag{16.93}$$

We call $\hat{\beta}'X'y$ the sum of squares due to fitting the model $y = X\beta + e$.

Hence if we have two linear models:

$$\begin{aligned} M_0 &: y = X\beta + e \\ M_1 &: y = Z\gamma + e \end{aligned} \tag{16.94}$$

which are such that $M_1$ is a restriction of $M_0$ obtained by imposing conditions on functions estimable under $M_0$, then the test statistic is given by

$$\frac{[\min \mathrm{SS}(M_1) - \min \mathrm{SS}(M_0)]}{r - r_1}\Bigg/\frac{\min \mathrm{SS}(M_0)}{n - r} \tag{16.95}$$

**Table 16.12** Analysis of Variance for Aspects of a Linear Model

| Source | d.f. | SS | MS |
|---|---|---|---|
| Fitting $M_1$ | $r_1$ | $\hat{\gamma}'Z'y$ | |
| Difference | $r - r_1$ | * | $s_1^2$ |
| Fitting $M_0$ | $r$ | $\hat{\beta}'X'y$ | |
| Difference | $n - r$ | * | $s^2$ |
| Total | $n$ | $y'y$ | |

where $r$ is the rank of $M_0$ and $r_1$ is the rank of $M_1$. The computations are represented nicely in an analysis of variance as in Table 16.12.

The items in Table 16.12 marked * are obtained by subtraction. Sometimes the "sources" are called "reduction sums of squares." The test of significance of the restrictions of $M_0$ which give $M_1$ is to evaluate $s_1^2/s^2$ by the $F$ distribution with $r - r_1$ and $n - r$ degrees of freedom. The utility of this approach for classification models is that interesting restrictions on a linear model are often given by merely eliminating terms, so that for instance, $y_{ijk} = \mu + r_i + c_j + e_{ijk}$ is a restriction of the model $y_{ijk} = \mu + r_i + c_j + t_k + e_{ijk}$, obtained by imposing the conditions that $t_k - t_{k'} = 0$ for all $k \neq k'$.

## 16.13 The "Analysis of Covariance" Model

We now give a very short description of linear models which involve both classification and functional terms. We can write this as

$$y = X\beta + Z\delta + e \tag{16.96}$$

in which the elements of $Z$ are arithmetic variables. We may as well assume that if $\delta$ is $q \times 1$, then the rank of $Z$ is $q$. This gives normal equations

$$X'X\beta + X'Z\delta = X'y \tag{16.97}$$

$$Z'X\beta + Z'Z\delta = Z'y \tag{16.98}$$

Now solving Equation (16.97), we have $X\beta = P(y - Z\delta)$, so, substituting, Equation (16.98) becomes $Z'(I - P)Z\delta = Z'(I - P)y$, and we have equations for $\delta$ which are like normal equations and are called, in fact, "reduced normal equations." The general topic can be pursued with moderate ease from this initial point.

## 16.14 Final Remarks

The extensions of the material of this chapter are manifold. The case of given fixed linear models is straightforward, though there are very difficult problems in formulating strategies of model search. The case of random linear models leading to estimation of components of variance is very difficult, and we have only ad hoc techniques of little logical basis. The case of mixed linear models is being explored almost continuously because the problems are extremely challenging, and also because scientific investigators *have* to work with such models.

Another class of extensions is to the case in which the variables to be explained are multivariate. Some elementary aspects of this are given by Anderson (1958) for example. See also Tukey (1946b).

We also mention that there are deep problems of testing goodness of fit of linear models. The particular case of linear models with structured data is discussed by Anscombe (1961), Anscombe and Tukey (1963), and Tukey (1949a).

We mention that we have given only a very brief introduction to linear models. We refer the reader to Kempthorne (1952), Graybill (1961), Rao (1965), Plackett (1960), Draper and Smith (1966), and Daniel and Wood (1971) for treatments from different points of view and with different aims.

### PROBLEMS

1. Perform an analysis of variance on the following data:

| Groups | | |
|---|---|---|
| 1 | 2 | 3 |
| 11.2 | 8.7 | 12.1 |
| 10.3 | 10.1 | 11.7 |
| 11.8 | 9.8 | 10.9 |
| 9.4 | 9.2 | 12.0 |

Obtain the within-group SS by subtraction *and* by pooling the sum of squares of deviations within groups.

2. A research and development laboratory of a battery manufacturer has been conducting research with battery additives. With optimism the project leader presents the following results on the average efficiency obtained:

| Standard | Additive $A$ | Additive $B$ |
|---|---|---|
| 35 | 40 | 45 |

The laboratory head is pessimistic and asks to see the entire set of data. These data are as follows:

| Standard | Additive $A$ | Additive $B$ |
|---|---|---|
| 36 | 46 | 45 |
| 32 | 37 | 46 |
| 41 | 42 | 38 |
| 39 | 38 | 46 |
| 27 | 37 | 50 |

(a) Use the analysis of variance and $F$ test to evaluate the significance level for the model where the three group means are equal.

(b) Use the two-tailed $t$ test to calculate the significance level for the model that the two additives have the same average effect.

(c) Use any other significance tests you choose and summarize your opinion about the effect of additives.

3. Derive the expected mean squares in the analysis of variance for the random model in Sect. 16.5.

4. A city engineer is disturbed about complaints received because of voltage fluctuation in the city power supply. Accordingly, he conducts a study to determine the extent and nature of the fluctuation. Voltage readings are taken at random times each day for one week. The data are as follows:

| Mon. | Tues. | Wed. | Thur. | Fri. | Sat. |
|---|---|---|---|---|---|
| 109 | 111 | 115 | 104 | 110 | 110 |
| 107 | 105 | 116 | 112 | 112 | 112 |
| 110 | 112 | 110 | 110 | 114 | 108 |
| 112 | | 109 | 108 | | 106 |
| 115 | | 110 | | | 110 |

Assuming that these days represent a random selection of days, use the material in Sect. 16.5 to determine the significance level for the model where the day-to-day variance $\sigma_a^2$ is zero.

5. Given the following data:

| | $A_1$ | $A_2$ | $A_3$ |
|---|---|---|---|
| $B_1$ | 4 | 5 | 6 |
| $B_2$ | 3 | 5 | 4 |
| $B_3$ | 1 | 4 | 2 |
| $B_4$ | 2 | 3 | 2 |

(a) Perform the analysis of variance, giving sums of squares for total, $A$, and $B$ in $A$.
(b) Perform the analysis of variance, giving sums of squares for total, $B$, and $A$ in $B$.
(c) Perform the analysis of variance, giving sums of squares for total, $A$, $B$, and $AB$.
(d) Verify from the preceding that

$$A \text{ in } B \text{ SS} = A \text{ SS} + AB \text{ SS}$$

$$B \text{ in } A \text{ SS} = B \text{ SS} + AB \text{ SS}$$

6. Verify part (d) of Problem 5 algebraically for an $r \times s$ two-way classification.

7. Given the following data:

| | $A_1$ | $A_2$ | $A_3$ |
|---|---|---|---|
| $B_1$ | 2<br>1 | 3<br>4 | 4<br>5 |
| $B_2$ | 4<br>3 | 6<br>6 | 8<br>7 |
| $B_3$ | 2<br>2 | 4<br>2 | 6<br>5 |
| $B_4$ | 4<br>2 | 6<br>5 | 8<br>7 |

(a) Consider this data as 12 groups of 2 observations each. Perform an analysis of variance giving total SS, between-group SS, and within-group SS.
(b) Perform the analysis of variance given in Section 16.8 giving total SS, $A$ SS, $B$ SS, $AB$ SS, and error SS.
(c) Verify from the preceding that between-group SS $= A$ SS $+ B$ SS $+ AB$ SS.

8. Do part (c) of Problem 7 algebraically for an $r \times s$ two-way classification with $n$ observations per cell.

9. Derive the expected mean squares in the analysis of variance for the random model given in Section 16.7.

10. Derive the expected mean squares in the analysis of variance for the fixed effects model given in Section 16.8.

11. Derive the expected mean squares for the random effects model in Section 16.9.

12. An experiment was conducted to study the effect of four lubricants upon drilling speed of metal drills. The drilling rates obtained while using the four additives are as follows:

| | | Additive | | | |
|---|---|---|---|---|---|
| | | 1 | 2 | 3 | 4 |
| Drilling Speed | 1 | 20 | 22 | 19 | 25 |
| | 2 | 21 | 23 | 20 | 23 |
| | 3 | 18 | 16 | 14 | 19 |
| | 4 | 19 | 20 | 18 | 22 |

Assuming that the four drills were selected at random, use the material in Section 16.10 to determine the significance level for the model, where there is no difference in additives, and to "estimate" the drill component of variance.

13. An experiment was conducted to study the filtration of gases. Five different brands of filters were available. These were used with three flow rates, and the efficiency in terms of percent of known impurities removed was recorded.

| | | Brand | | | | |
|---|---|---|---|---|---|---|
| | | 1 | 2 | 3 | 4 | 5 |
| Flow Rate | 1 | 30 | 50 | 30 | 23 | 65 |
| | | 40 | 60 | 36 | 19 | 59 |
| | 2 | 33 | 40 | 31 | 26 | 33 |
| | | 32 | 44 | 33 | 24 | 32 |
| | 3 | 71 | 35 | 31 | 27 | 22 |
| | | 68 | 26 | 32 | 26 | 24 |

Use the material of Section 16.8 to calculate the significance level for the model where there is no interaction.

14. If the two-way table of means $\mu_{ij}$, $i = 1, 2, \ldots, r, j = 1, 2, \ldots, s$ are in fact given by a function of $i$ and $j$, e.g., $\mu_{ij} = i^2 + j^2$, give three examples of functions with interaction and three examples of functions without interaction.

15. Derive the expected mean squares in the analysis of variance for the mixed model described in Section 16.10.

# 17
# Epilogue

## 17.1 Introduction

In the foregoing chapters we have developed the basic ideas of probability models and of the fitting of these models to data. We must emphasize that we have given only an introduction. We hope the reader will be able to apply the ideas to investigative situations he meets. We also hope that he will pursue the area of thought in the directions that interest him and are relevant to his own needs. We have presented a first course in statistics which leads into other courses in mathematical statistics and applications. We suggest that the student consider following the present material by courses in mathematical probability, mathematical statistics, theory of sampling, design of experiments, theory of games, decision theory, econometrics, psychometrics, etc.

At this point we feel it useful to discuss in some detail the meaning of probability. Everyone uses ideas of probability, and it is appropriate that we present briefly our understanding of the nature of this usage.

## 17.2 The Probability Calculus

Probability calculus is strictly a process by which we calculate a number, called a probability to associate with a complex event, from numbers called probabilities associated with elementary events. What we define as elementary events is a matter of intellectual choice. They are usually simple aspects of a whole concatenation of occurrences leading to the complex event. For instance, we may ask for the probability that 100 tosses of a penny will give a proportion of heads that is between 0.48 and 0.52. To do this, we have to specify a probability model, the simple one being that of independence of successive tosses and the same probability of head on each toss. Without some probability model we are unable to answer our question, and the result of a calculation is entirely dependent on our assumed probability model. The process of calculating the probability from the assumed elementary probabilities may be very difficult, for instance, in the case of long or infinite sequences of elementary trials. However, the logic of the process is clear and unambiguous. Given a probability model that is completely specified, there is a definite answer for the probability of a specified complex event, and we may call this "the probability" if the model is understood to be agreed on. So we may state the probability that one player in bridge will be dealt the 13 cards of one suit if we have agreed that all hands are equally likely.

## 17.3 Probabilities in the Real World

While many mathematicians are interested in the probability calculus solely as an interesting branch of mathematics, most of humanity is interested in probability as a way of quantifying uncertainty about unknown facts of the real world. Usually the facts are unknown because they are in the future. Will there be rain next January 1 in Ames, Iowa? Will the value of shares of IBM be higher on next July 1 than it is now? Will we survive to age seventy at least? These are all questions about the future. However, many present and past facts are also unknown. Did Shakespeare really write the plays usually attributed to him? Is there oil under county $Z$ in Wyoming? These are questions to which there is a definite answer of Yes or No which we do not know. Definite answers for some of the questions may result from investigation. If we drill in county $Z$ in Wyoming and find oil, the answer to that question is Yes. If we drill and do not find oil, the provisional answer is No, but we may not have drilled in the right place and to the right depth. The question of authorship of the Shakespeare plays, however, is one for which there never will be a categorical answer. All we can do is search the past for evidence, and the evidence will make the hypothesis that Shakespeare actually wrote the plays more or less tenable, with some definition of tenability. Also, the hypothesis that Bacon wrote the plays may become more or less tenable.

The questions we mentioned above are questions of fact, but we are unable in our present state of knowledge to reach a definite answer. For all such questions the individual has to express *in some way* the uncertainty of his opinion. The probability scale is a natural candidate for expressing such uncertainty. For instance, we may say that our probability is one if we are sure it will rain in Ames next January 1; and if we say our probability is zero, we are sure it will not rain. What do we mean by saying that our probability is 0.25? We mean the following. We are familiar with the concept of a perfect coin; i.e., a coin which will give head $H$ with relative frequency very close to 0.5 in a long series of careful trials designed so that the coin does not wear and the flipping is haphazard. We know that the probability of $HH$ on two tosses is 0.25 for this ideal system. We say that our uncertainty of opinion about whether it will rain on next January 1 in Ames, Iowa, is like our uncertainty on whether two flips of the ideal coin will give $HH$. We use conceptual idealized probability mechanisms to give us a probability scale. It is therefore essential in our view to note that a statement of probability is a quantification of personal opinion based on a given amount of data.

We can form a probability assessment by reviewing our own recollections of past years. Perhaps we can remember ten years and we recall rain on January 1 for one of these. We shall then give our probability as 0.10. However, suppose we consult meteorological records which exist, say, for the past 85 years, and from those records we see that there was rain in 19 of those years. We would then give our probability as 19/85. We make a judgment on the basis of a simple probability model applied to the data we have. If we can get more data, our assessment will change. If we find that a historical record shows that rain on January 1 in one year is followed in the record by rain on the next January 1 in 50% of the cases and we have the information that it rained on January 1, 1970, we shall say that our probability of rain on January 1, 1971, is 0.50 (actually, no such correlations are observed).

To make crystal clear the role of the data, we consider the following: Suppose a coin which has been found on investigation to be reasonably fair is shaken inside a closed square box and the shaking is such that we can be highly sure that the coin will come to rest in a flat position on the bottom of the box. After the shaking, the coin shows either head or tail. There is no uncertainty about the "state of nature." We shall say, however, our probability that the coin shows head is 1/2. Another person who has been permitted to raise the lid of the box and look will make one of the two statements, "The probability of head is 1," or "the probability of tail is 1." Thus our personal probability of 1/2 is a quantification of our uncertainty based on our knowledge, uncertain though it is.

It is clear from these examples that when we state our probability of an uncertain result $H$, say, is 0.3, we are giving a summary of a very complicated personal mental process. We contemplate the "future trial." We search our memory for instances that are like the future trial in as many respects as possible. We then say that in our historical record of instances similar to the future trial the relative frequency of $H$ was 0.3. The consequences of this are as follows:

1. A probability statement about a future outcome in the real world is a result of specification of the future trial and a selection from the whole

historical record of a section of cases that are like the future trial in some respects.

2. A probability statement requires the examination of the historical record for a probability model. The simplest probability model commonly used is that of independent Bernoulli trials; and if we wish to use this, we must examine the data and convince ourselves that the model is reasonable for the historical record available to us.
3. Without some historical record we judge to be relevant, we are unable to make a probability assessment in which we have any confidence.

This of course begs the question of what we mean by "relevant." From a certain viewpoint each one of us has a historical record that can be applied to any situation. At times it requires considerable restraint for any of us to say that his personal historical record is so poor that it is not worthwhile trying to make a probability assessment. Most people can be induced to express a probability statement on anything. Suppose, for instance, you are asked if the word *zahdu* in Urdu means "goat." You have never been exposed to Urdu, we presume. However, we offer you a sequence of wagers in the following way. We use a number of coins that have been checked out to be close to fair. We ask which payoff you would prefer: $1 if you say Yes with regard to *zahdu* and are correct or $1 if you say head and a toss of coin gives head. If you prefer the latter, your probability as to whether *zahdu* means "goat" is less than 1/2.

Now we take two coins and the alternative proposition as: $1 if you say "two heads" and a toss of two coins gives two heads. If you prefer the payoff with regard to the meaning of *zahdu*, your personal probability that it means "goat" is greater than 1/4. Even this type of analysis requires an assumption, namely, that the basis of information does not change. You would find in fact that the more you thought about the matter the more you could recollect what might be relevant, so "probability" may change under this examination.

The upshot of the above discussion is that a probability about an uncertain fact of the real world, whether past or future, is based on the application of introspection and statistical methods to the selection of a portion of the available historical record judged to be relevant to the uncertain fact. A person will examine his historical record, select a portion of it which appears to be relevant, and follow a model based on mathematical probability.

Can we expect agreement between two individuals on their probability assessments? It appears not. Individual *A* may use a portion of the historical record available to him which he judges to be relevant to the unknown fact. Individual *B* does likewise. The only way individuals *A* and *B* can reach agreement is to agree on a common historical record that is relevant, *and* to agree on how this record will be represented by a probability model, *and* to agree on how unspecified constants in the probability model shall be estimated from the data. The possibility of agreement is obviously quite remote in most situations.

## 17.4 Comments on the Processes

The following comments on the processes outlined above seem appropriate:

1. A probability about a future trial determined from the historical record has no direct relation to an infinite set of future trials. We can, if we wish, give the probability a quasi-operational meaning by saying that if our probability is $P$ and we could have an infinite number of future trials like the one under consideration, we are of the opinion that in $P$ of them the outcome would be such and such. Instead, a probability is a look backward into history starting from our present position.
2. A determined probability $P$ has no implications that we should be prepared to wager on the outcome of the future trial with odds $P$ to $1 - P$. However, if we are forced to wager, we would use $P$. Also, we can turn the matter around and say that in making plans for ourselves, we have to make wagers continually. For instance, we have to make a choice of whether or not to take out flight insurance on the next plane trip that we take. We may "decide" this by not even thinking about the matter and not taking insurance. Or we may decide after reflecting on the question. As another example, it may happen that a completely innocuous-looking walk in the country will result in being shot by a hunter, struck by lightning, bitten by a poisonous spider, etc. So whatever we do, we are exposed to danger to a greater or lesser degree. We may and do choose to ignore this most of the time but in general, are forced to make assessments of what the future contains. Therefore we make judgments which are like assessments of probability.

The argument requiring a choice of alternative wagers given above appears to have complete force under some assumptions. It also has the property of forcing a relation between a person's assessment of probability and probability as a relative frequency. In general, however, individuals use "probability," "probable," and "likely" in ordinary discourse; and we have no way of knowing just what they mean except by extensive interrogation. Whether they mean a probability as a relative frequency in a "similar series of trials" or as a "logical" probability of, say, 1/4 arrived at because there are four possible outcomes which seem to them equally probable, we can discover only by psychological examination. We become involved in the whole question of the meaning of language as used by the individual. The pursuit of this topic as an aspect of developing an understanding—a model—of human discourse is very important and is the basic aspect of communicable human knowledge. A critical role of education is the development of an agreed meaning of words, and over the centuries word usage becomes more and more precise in some sense. It is hoped that the teaching of probability and statistics, to an increasing extent, will have this effect in the use of the words associated with probability.

The fact that an individual can be forced to express a probability does not mean that he has any confidence in his figure. If it is based on a probability model that is highly tenable because it is based on an extensive historical record and if the individual is highly confident that this record is relevant, he will have high confidence. On the other hand, when he is pushed to the wall on a question for which he has little relevant data, he will say, "If you insist on my giving a figure, I will say 1/100, but I have no faith in this figure other than that it is the best I can do." The relevance, therefore, of forcing each individual to determine his probability with regard to questions such as whether *zahdu* in Urdu means "goat" in English seems highly tangential.

We have described in Chapter 11 how a prior distribution can be combined with data to give a posterior distribution. The above implies that any thinking person can develop a prior for any situation. But whenever the prior has been forced or chosen by quasi-logical rules, the utility of the posterior distribution seems very questionable. Additionally, there is a deep question underlying any use of prior distributions. Even under the best circumstances, the prior is supposed to represent past data apart from the present data. The question of agreement of past and present data must then be raised, and if the agreement is poor in some sense, the problem of interpreting the present data stands on its own. *In the last resort we must use a model that is consonant with all the data, both past and present.*

## 17.5 Publicly Agreed Probabilities

Every individual can express his own probability about an outcome, so there is no such thing as "the probability of outcome $X$." We must recognize, however, that there are many matters on which there are generally accepted probabilities. If we hear that James Smith and his wife are going to have a baby, we will, in the absence of any special information about Smith and his wife, accept as 1/2 (or very near this) the probability that the baby will be male. This example is interesting because if we have the additional information that Smith and his wife already have five children all of whom are female, most of us would not agree that our probability is 1/2. We would say that if the commonly used model of sex determination holds, the probability is 1/2, but we do not know whether there exist subpopulations of circumstances in which the probability is different from 1/2. The authors would not be prepared to make an even bet in this case.

Similarly, we have many probabilities based on historical frequencies of survival of humans to various ages. We are quite prepared to accept these as reasonable for an unspecified individual, John Doe. But if we have the information that a particular individual's parents both died in their forties and a large proportion of near ancestors died at early ages, we would not be prepared to accept the figures obtained from the whole historical record. If we are asked, "What is your probability that John Doe will survive to seventy?" we have one answer based on the model that John Doe is like those on whom the history is available. If we are told in addition that John Doe is six feet tall, blonde, wears spectacles, was raised in New York City, and had appendicitis at age eight, we may well change our probability if we have information about a subclass who are like John Doe in these respects. If we continue our investigation of John Doe, we will find that he is quite unlike any individual occurring in the record. We may then prefer to say that we cannot make a probability assessment in which we have any confidence. If, however, we are forced to make an assessment by the wagering process described above, we will adopt a probability model which seems the most tenable to us and calculate a probability on the basis of this.

We must then differentiate between models based on extensive historical records and shown tenable in some sense and models we are forced to use because we have very little relevant record.

## 17.6 Probability in the Accumulation of Empirical Knowledge

Because the probability scale extends from zero meaning "never" to "one" meaning "always," there is a tendency to think that all empirical thought is based on probability of hypotheses or models. However, this does not appear to be the case. It seems clear that knowledge in all fields of science consists of the development of models which enable us to make good predictions. In making a prediction, we will use the model which is most tenable and most supported in some senses by the data, but we never take a model as being the complete definite representation of reality. It is a model which works as far as we can see at present. We feel that it will be found to be defective as we search more and more deeply for verification of its consequences. Each model is a step along the way to improvement of understanding and prediction. Each model provides us with a basis for asking questions, and without a model we cannot attempt to answer them. A consequence is that whether a model is right or wrong is not the appropriate question. The appropriate questions are, How good are the predictions it makes?, and What questions does it suggest which will lead to a better model? Furthermore in asking How good are the predictions it makes?, we are not asking how good a *single* prediction is. We have to make many predictions and consider in some sense how good they all are. The path to knowledge is strewn with models which have been found inappropriate or improvable. The question of truth of a model appears to be irrelevant because one can be sure that any model will be found to be erroneous in some respects.

It seems then that a use of "probability" as in "the probability that the theory of relativity is correct" does not really enter at all into the building up of knowledge. But, on the other hand, the statement, "The theory of relativity gives us the model with best tenability among competing models," is used widely and is informative. Indeed, the theory of relativity appears to provide us at present with the only model that is tenable in the light of data accumulated to the present. But this is quite different from any statement such as "The probability that the theory of relativity gives a true model is 1.0." We see no need for statements of this type. Even in the case of the theory of relativity, we note in the scientific journals serious questioning of some aspects.

We note that if tenability of a probability model is judged by any of the types of significance test we have described in this book and if we are indeed obtaining data by random sampling from a specified distribution $F$ indexed by a parameter $\theta$, as the amount of data increases, the tenability of the "true" value $\theta_T$ will remain high and the tenability of any other value of $\theta$ will tend to zero with increasing amount of data. Thus the supposed advantage of a Bayesian argument that any prior distribution is increasingly overcome by an increasing amount of data is not unique.

It must be recognized, however, that our decisions on what to do next in an investigative process are based on some sort of personal probability. This does not mean that this is at all a complete summarization of our knowledge or lack thereof. A fundamental idea of science is the reporting of data in such a way that every individual can form his own personal probability if he wishes, or can form in some way a judgment of the extent to which various models are consonant with the data. The most basic need of science is, therefore, the development of generally acceptable modes of reducing data.

## 17.7 The Use of Random Devices in Forming Opinions

We have seen throughout that data analysis is based on the search for a class of probability models that adequately represents the data, followed by application of inferential procedures to the class that is obtained. The role of probability in our model is that it gives an explanation of aspects of the data that we cannot explain in other ways. The search for a probability model can never lead to a sure result. We can only say that the model we obtain is a reasonable representation of all reported relevant data insofar as we can judge.

We do not in this way obtain probabilities that have properties of relative frequency in a large population of possible repetitions. If we wish our procedures to have such properties, we have to introduce probability as a relative frequency into our data collection process. To do this, we have to have a real process which is an accurate reflection of probability as relative frequency in our probability models. We assume that such processes exist, e.g., dice and a tossing machine that have been very carefully designed and machined and have been checked to give results like our conceptual probability models.

We can make the application of probability models to data more valid by using such a device in the collection of data on some simple questions. Suppose our job is to set up an accept-reject rule for boxes containing 100 switches. Suppose we wish to reject boxes that contain more than 5% of defective switches. We can use our random device to draw a random sample of ten switches and can use our mathematics to obtain an accept-reject rule such as "Reject if there are more than 3 defective switches in our sample of 10." The chain of inference, to the point of obtaining probabilities which are observationally verifiable frequencies, of rejecting boxes of different qualities and the formulation of a rule is now much more nearly complete than if we had merely the rule: Take ten switches from the box as *you* see fit. This is the case because, to evaluate this latter rule, we would have to make studies of how the switches are packed in the box and of how *you* make a selection of ten.

The same type of reasoning is applied very widely in experimentation. Consider the question of whether supplementation of the diet of eight-year-old children with a particular daily vitamin pill results in better growth. Suppose that a set of 100 eight-year-old children are available for the study. A procedure used very widely is to choose 50 of the children who are to receive the vitamin pill by the use of a random device. The consequences of following this rule of obtaining data are many, and we merely state what we think, is the crucial one. If the rule for the experiment is used, there is a definite probability—not a subjective one based on search for a model, etc.—that significance of a statistical test will achieve any particular level in the absence of effect of the treatment. This is closely related to the nonparametric tests described in Chapter 12. The point is that the use of the random device to determine what observations will be taken induces a certain probability structure for the observations obtained. As a consequence the application of probability models is considerably less subjective. The residual subjective element is the acceptance of the tossing mechanism as a process of producing mathematical randomness.

The introduction of randomization into the collection of data is, in the opinion of the great bulk of statisticians, one of the landmarks in the development of scientific method. Furthermore, it has been accepted as an essential part of scientific method by almost all investigators in "noisy" sciences. The great difficulty of development of social science lies, it is clear, in the incontrovertible fact that randomized experiments are not possible. Lack of association between "treatments" and "units" must be assumed, rather than built in by the data-collecting procedure. This landmark of method is due to R. A. Fisher. The definitive work is Fisher (1935a), though journal papers have appeared from 1926 or earlier. The paper of Neyman et al. (1935) merits study. More recent descriptions of the ideas are given by Kempthorne (1952, 1955, 1966) and numerous writings from the Statistical Laboratory of Iowa State University.

## 17.8 Final Remarks

To trace the whole history of ideas on what inference is and what are suitable rules of inference is a huge task which would require one to evaluate the whole history of human thought processes. Traditionally, it appears that two types of inference have been admitted, deduction and induction. One exception to this exhaustive dichotomy is the outlook of C. S. Pierce. The authors have found the description by Gallie (1966) informative. From this we learn that Pierce distinguished three types of inference—deduction, induction, and hypothesis. The third Pierce preferred to call "abduction." Pierce said, it appears, that induction is a method of testing, rather than of developing, knowledge. The method of abduction or hypothesis seems to be the development of ideas from examination of observational data. Gallie (1966) summarizes the method by the sequence: (1) A surprising fact $C$ is observed; (2) if a model $A$ were true, fact $C$ would not be at all surprising; (3) hence there is reason to suspect that model $A$ may be true. It seems clear that this process is used, and it is also relevant that this process seems to be closely related to the ideas advanced in earlier chapters of this book with regard to data examination and choice of a model. The nature of abduction to Pierce was, it appears, that it is the one form of inference that initiates new ideas and leads to the advancement of knowledge. This activity was according to Gallie's interpretation of Pierce, "the essential function of the cognitive mind." It is important to note that item (3) above says "to suspect that model $A$ may be true." It does not say "to believe model $A$ is true." The main conception of Pierce was one of "habits of inference," and perhaps we are correct in stating that any habit must be validated by consideration of how it works out in idealized cases, that is, by its operating characteristics. The general philosophy of pragmatism as put forward by Pierce seems to lie at the root of statistical practice. We recommend that the reader study this outlook. It is interesting that in Pierce's outlook absolute certainty is beyond us, but this does not imply that the only possible attitude is one of complete skepticism. The conception of truth enters and is essential as an ideal for science. It is relevant, we think, to mention that Pierce talked about "fair" samples and "fair" judgments on the basis of these. The meaning of "fair," which is used by other philosophers also (e.g., Reichenbach, 1949), is obscure. In consonance with earlier parts of the present book, we would say that a "fair" sample is one taken by a process the operating characteristics of which have been studied and judged to have the properties of "random sampling" in the frequency sense. It is interesting to note that the widespread appeal of random sampling defined in this way is somewhat embarrassing to some of the recent Bayesians, in that it does not appear to have any status in their outlook. One meets the phrase "Draw a sample at haphazard," as though this is a definitive (or even partially definitive) prescription. The drift of Pierce's writing that knowledge is public and not personal seems to be a reasonable counter to the recent Bayesian view that the critical aspect is the opinion of the individual. We close with the opinion that the role and relevance of statistical methodology is a very deep philosophical question which cannot be dismissed with a few simple and seemingly plausible statements, whatever system of interpreting data is being presented.

## 17.9 A Partial Guide to Texts on Statistics

We have attempted to present in this book the basic ideas of probability theory, statistics, and data analysis. Our material is strictly introductory but will serve as a background for the broad spectrum of all aspects of statistics. We hope that we shall have stimulated the reader to pursue some aspects in depth. The total literature is possibly bewildering, so we have included a classification of some texts which will aid the student who wishes to explore a particular aspect of the overall subject. This list of books serves also to indicate many which have contributed to our outlook. We consider it essential to warn the student that while there is no basic controversy on the mathematics of statistics, deep controversies exhibited by the various books on foundations have effects with regard to other areas, so that many of the cited books have an outlook which is dominated by one viewpoint not accepted by many workers as having totally compelling force. The student should therefore sample the offerings.

Our list reflects our personal experience and tastes and should not be regarded as definitive. There are many other excellent texts and monographs. In addition, there are many papers in the journals which have strong bearing on the topics which form the main theme of our book but which we are unable to cover.

We wish to say also that much of the philosophical literature is relevant background. We feel it desirable to warn the reader that almost all writings except some recent ones on the philosophy of knowledge become involved in probability, but then give it only very superficial discussion and "hand-waving" arguments. Additionally, very few philosophers have been personally involved in developing knowledge of the real world. In statistics as in almost all aspects of life, arts, and sciences, it is performance and not pure theory that must be the ultimate yardstick. We suggest particularly the writings of Aristotle, Bacon, Hume, Poincaré, and Popper as relevant and informative. See also Wittgenstein (1958).

We have indicated our estimate of the degree of difficulty of cited material by the code, * intermediate, ** somewhat advanced and *** advanced or esoteric. We judge unmarked references to be moderately accessible to the group at which we are aiming.

1. PROBABILITY: ***Chung (1968), **Cramér (1946), Cramér (1955), ***Doob (1953), *Feller (1950), ***Feller (1966), **Gnedenko (1967), Hodges and Lehmann (1965), Jeffreys (1961), **Kolmogorov (1956), **Lamperti (1966), ***Lévy (1937), ***Loéve (1963), ***Neveu (1965), Mosteller, Rourke, and Thomas (1961), Parzen (1960), **Tucker (1967), **Uspensky (1937).

II. FOUNDATIONS OF PROBABILITY AND STATISTICS: Boole (1854, 1953), Brown (1957), Campbell (1957), Carnap (1962), Cox (1961), Fisher (1956), **Fraser (1968), Good (1950), Hacking (1965), Jeffreys (1961), Kyburg and Smokler (1964), Lindley (1965), Ogden and Richards (1956), Reichenbach (1949), Savage, L. J. (1954), Smart (1958), Venn (1866), von Mises (1935).

III. STOCHASTIC PROCESSES: Bailey (1964), **Bartlett (1960, 1966), *Bharucha-Reid (1960), ***Billingsley (1961), ***Chung (1967), *Cox and Lewis (1965), Cramér and Leadbetter (1967), *Karlin (1966), Kemeny and Snell (1960), **Moran (1962), *Parzen (1962, *Takács (1960).

IV. MATHEMATICAL STATISTICS: Anderson and Bancroft (1952), Birnbaum, Z. W. (1962), Brunk (1965), *Cramér (1946), *Ferguson (1967), **Fisher (1950a), *Fisz (1963), Freeman (1963), Freund (1960), Hodges and Lehmann (1964), Hoel (1962), Hogg and Craig (1970), Huntsberger (1961), **Kendall and Stuart (1966), Larson (1969), **Lehmann (1959), Lindgren (1968), Meyer (1965), Mood and Graybill (1963), Neyman (1950), **Neyman (1967), **Neyman and Pearson (1967), **Pearson (1966), **Rao (1965), von Mises (1964), **Wilks (1962).

V. SPECIAL TOPICS:

*Bayesian Statistics:* Jeffreys (1961), Lindley (1965), Raiffa and Schlaifer (1961), Savage, I. R. (1968), Savage, L. J. (1954), Schlaifer (1959).

*Biology and Genetics:* Bliss (1967, 1970), Finney (1964), Kempthorne (1957).

*Combinatorics:* David and Barton (1962), Riordan (1958).

*Computation:* Elderton (1938), Hastings (1955).

*Decision Theory:* *Blackwell and Girshick (1954), Chernoff and Moses (1959), *De Groot (1970), *Luce and Raiffa (1957), ***Wald (1950), *Weiss (1961).

*Design of Experiments:* Cochran and Cox (1957), Cox (1958), Davis (1954), Federer (1955), Finney (1960), Fisher (1935a), Kempthorne (1952), Mann (1949), Winer (1962), Yates (1937).

*Econometrics:* Johnston (1963), Tintner (1952).

*History:* Todhunter (1949), Walker (1931).

*Information and Communication:* Khinchin (1957), *Kullback (1959), Middleton (1960).

*Analysis of Variance:* Acton (1959), Draper and Smith (1966), *Graybill (1961), *Plackett (1960), **Rao (1965), **Scheffé (1959), Williams (1959).

*Multivariate Analysis:* *Anderson (1958), Lawley and Maxwell (1963), *Roy (1957).

*Non-Parametric Statistics:* Bradley (1968), **Fraser (1957), Kendall (1955), Siegel (1956), **Walsh (1962, 1968).

*Quality Control:* Burr (1953), Duncan (1965), Grant (1952).

*Reliability:* Barlow and Proschan (1965).

*Sequential Analysis:* **Wald (1947), Wetherill (1966).

*Survey Sampling:* Cochran (1963), Deming (1950), Hansen, Hurwitz, and Madow (1953), Sampford (1962), Sukhatme and Sukhatme (1970), Yates (1953).

*Tables:* Fisher and Yates (1963), Hald (1952a), Owen (1962), Pearson and Hartley (1954), Rand Corporation (1955).

*Time Series:* **Blackman and Tukey (1959), Box and Jenkins (to be published), Cox and Lewis (1966), **Grenander and Rosenblatt (1957), *Hannan (1960), Jenkins and Watts (1968), Tintner (1952), *Wold (1954).

VI. General Applications: Bancroft (1968), Bennett and Franklin (1954), Bliss (1967, 1970), Bowker and Liebermann (1959), Brownlee (1965), Dixon and Massey (1969), Fisher (1958), Fryer (1966), Goulden (1952), Hald (1952a), Johnson and Leone (1964), Li (1965), Natrella (1963), Ostle (1963), *Rao (1952), Snedecor and Cochran (1967), Steel and Torrie (1960), Wallis and Roberts (1956), Youden (1951).

# Appendix

**Table A.1** Random Sampling Numbers

| | | | | | | | | | | | | | | | | | | | | | | | | |
|---|---|---|---|---|---|---|---|---|---|---|---|---|---|---|---|---|---|---|---|---|---|---|---|---|
| 97 | 58 | 55 | 23 | 12 | 87 | 39 | 84 | 32 | 23 | 26 | 91 | 01 | 11 | 26 | 01 | 24 | 06 | 58 | 20 | 33 | 46 | 38 | 86 | 23 |
| 84 | 95 | 87 | 34 | 95 | 31 | 23 | 12 | 64 | 75 | 89 | 28 | 38 | 15 | 91 | 81 | 89 | 08 | 86 | 08 | 88 | 20 | 02 | 11 | 67 |
| 11 | 52 | 38 | 09 | 94 | 32 | 47 | 35 | 42 | 67 | 39 | 33 | 89 | 97 | 16 | 28 | 94 | 86 | 93 | 86 | 96 | 13 | 43 | 85 | 99 |
| 38 | 69 | 94 | 97 | 10 | 44 | 42 | 85 | 46 | 88 | 56 | 56 | 63 | 58 | 22 | 89 | 19 | 26 | 82 | 25 | 94 | 15 | 54 | 65 | 62 |
| 23 | 99 | 36 | 33 | 41 | 99 | 76 | 22 | 29 | 19 | 92 | 53 | 92 | 15 | 71 | 47 | 57 | 74 | 69 | 03 | 65 | 57 | 90 | 53 | 17 |
| 09 | 15 | 95 | 74 | 87 | 09 | 63 | 82 | 63 | 29 | 84 | 57 | 45 | 80 | 07 | 13 | 57 | 40 | 58 | 34 | 21 | 93 | 90 | 39 | 21 |
| 55 | 75 | 91 | 36 | 57 | 38 | 30 | 89 | 64 | 42 | 01 | 84 | 83 | 12 | 79 | 32 | 09 | 56 | 03 | 81 | 90 | 88 | 00 | 71 | 02 |
| 84 | 62 | 29 | 92 | 42 | 03 | 92 | 37 | 46 | 19 | 90 | 75 | 68 | 84 | 49 | 53 | 80 | 62 | 19 | 20 | 31 | 14 | 42 | 11 | 17 |
| 79 | 25 | 70 | 07 | 80 | 85 | 32 | 53 | 87 | 11 | 33 | 79 | 14 | 20 | 04 | 12 | 40 | 31 | 74 | 39 | 80 | 21 | 37 | 65 | 20 |
| 40 | 10 | 91 | 52 | 27 | 21 | 18 | 64 | 61 | 04 | 85 | 55 | 16 | 90 | 71 | 31 | 95 | 15 | 86 | 74 | 87 | 80 | 75 | 71 | 27 |
| 93 | 18 | 86 | 63 | 72 | 22 | 53 | 44 | 23 | 89 | 38 | 06 | 46 | 04 | 79 | 67 | 77 | 33 | 21 | 75 | 40 | 51 | 74 | 60 | 53 |
| 63 | 71 | 69 | 30 | 23 | 12 | 85 | 90 | 05 | 07 | 67 | 33 | 56 | 52 | 60 | 21 | 50 | 72 | 26 | 28 | 48 | 67 | 31 | 87 | 61 |
| 05 | 29 | 95 | 78 | 06 | 10 | 41 | 62 | 18 | 37 | 42 | 91 | 98 | 43 | 33 | 20 | 58 | 62 | 80 | 65 | 19 | 90 | 07 | 84 | 49 |
| 30 | 04 | 29 | 90 | 89 | 64 | 25 | 66 | 36 | 41 | 99 | 59 | 15 | 43 | 86 | 34 | 10 | 05 | 99 | 83 | 08 | 02 | 18 | 01 | 22 |
| 75 | 50 | 83 | 42 | 46 | 80 | 76 | 77 | 34 | 16 | 04 | 05 | 06 | 28 | 86 | 60 | 70 | 04 | 13 | 28 | 98 | 76 | 78 | 43 | 69 |
| 68 | 82 | 44 | 11 | 33 | 11 | 20 | 42 | 00 | 22 | 40 | 03 | 06 | 12 | 45 | 06 | 32 | 34 | 44 | 18 | 01 | 26 | 36 | 78 | 42 |
| 51 | 38 | 78 | 69 | 65 | 25 | 98 | 73 | 40 | 31 | 12 | 04 | 99 | 51 | 09 | 49 | 04 | 32 | 68 | 68 | 54 | 64 | 15 | 25 | 68 |
| 98 | 41 | 81 | 63 | 70 | 58 | 43 | 39 | 93 | 18 | 54 | 46 | 98 | 33 | 01 | 47 | 85 | 39 | 81 | 11 | 48 | 84 | 07 | 64 | 76 |
| 08 | 44 | 37 | 01 | 53 | 59 | 67 | 11 | 11 | 53 | 16 | 98 | 16 | 52 | 52 | 39 | 32 | 22 | 18 | 22 | 04 | 03 | 06 | 77 | 17 |
| 17 | 30 | 92 | 82 | 09 | 42 | 37 | 88 | 43 | 35 | 11 | 54 | 89 | 05 | 61 | 10 | 46 | 27 | 43 | 33 | 88 | 92 | 72 | 62 | 01 |
| 74 | 87 | 89 | 10 | 02 | 19 | 45 | 29 | 65 | 70 | 77 | 81 | 98 | 78 | 67 | 05 | 62 | 57 | 08 | 79 | 30 | 32 | 62 | 91 | 87 |
| 61 | 81 | 52 | 99 | 80 | 11 | 55 | 21 | 98 | 02 | 08 | 26 | 01 | 20 | 16 | 07 | 42 | 88 | 56 | 51 | 31 | 96 | 14 | 85 | 49 |
| 55 | 08 | 43 | 08 | 22 | 50 | 28 | 03 | 18 | 00 | 80 | 79 | 60 | 18 | 33 | 92 | 36 | 13 | 50 | 41 | 43 | 59 | 82 | 16 | 65 |
| 44 | 38 | 47 | 15 | 16 | 96 | 03 | 51 | 42 | 15 | 35 | 96 | 40 | 87 | 91 | 56 | 91 | 13 | 58 | 85 | 40 | 06 | 36 | 04 | 30 |
| 12 | 45 | 97 | 68 | 57 | 62 | 36 | 61 | 03 | 29 | 46 | 60 | 79 | 85 | 99 | 91 | 13 | 99 | 95 | 58 | 75 | 14 | 74 | 88 | 12 |
| 19 | 95 | 23 | 05 | 45 | 01 | 87 | 81 | 18 | 92 | 36 | 94 | 07 | 14 | 08 | 90 | 32 | 51 | 29 | 61 | 50 | 60 | 34 | 92 | 25 |
| 71 | 55 | 86 | 72 | 94 | 77 | 08 | 55 | 65 | 50 | 33 | 53 | 94 | 81 | 52 | 36 | 31 | 53 | 12 | 74 | 88 | 59 | 99 | 35 | 95 |
| 07 | 32 | 94 | 03 | 20 | 66 | 29 | 98 | 75 | 65 | 70 | 30 | 56 | 59 | 08 | 24 | 51 | 75 | 48 | 73 | 11 | 29 | 77 | 08 | 36 |
| 10 | 35 | 58 | 59 | 25 | 89 | 62 | 60 | 77 | 71 | 24 | 13 | 38 | 20 | 83 | 02 | 48 | 11 | 67 | 95 | 38 | 97 | 15 | 58 | 18 |
| 62 | 99 | 34 | 08 | 06 | 81 | 46 | 09 | 16 | 82 | 95 | 17 | 13 | 46 | 36 | 51 | 36 | 87 | 56 | 10 | 80 | 79 | 40 | 48 | 82 |
| 19 | 44 | 35 | 31 | 20 | 16 | 05 | 25 | 26 | 38 | 98 | 94 | 18 | 38 | 88 | 10 | 90 | 29 | 01 | 12 | 48 | 85 | 52 | 97 | 22 |
| 77 | 76 | 94 | 64 | 49 | 45 | 39 | 58 | 07 | 88 | 32 | 11 | 43 | 09 | 51 | 32 | 69 | 31 | 63 | 02 | 33 | 47 | 08 | 94 | 85 |
| 97 | 43 | 81 | 59 | 46 | 59 | 26 | 04 | 63 | 86 | 87 | 31 | 55 | 50 | 66 | 11 | 37 | 04 | 68 | 14 | 57 | 17 | 08 | 82 | 48 |
| 09 | 77 | 93 | 46 | 95 | 36 | 98 | 08 | 77 | 39 | 71 | 44 | 48 | 10 | 19 | 54 | 80 | 24 | 83 | 47 | 06 | 79 | 01 | 78 | 43 |
| 71 | 09 | 43 | 23 | 16 | 33 | 93 | 21 | 87 | 89 | 16 | 53 | 05 | 53 | 16 | 98 | 96 | 30 | 89 | 49 | 83 | 32 | 23 | 13 | 32 |
| 25 | 19 | 47 | 70 | 48 | 16 | 91 | 39 | 59 | 80 | 66 | 77 | 96 | 02 | 08 | 59 | 58 | 48 | 91 | 81 | 04 | 31 | 64 | 65 | 15 |
| 43 | 23 | 23 | 81 | 42 | 61 | 42 | 37 | 17 | 76 | 75 | 40 | 18 | 81 | 33 | 51 | 68 | 04 | 41 | 00 | 72 | 82 | 28 | 68 | 03 |
| 50 | 57 | 81 | 53 | 79 | 98 | 04 | 75 | 77 | 30 | 49 | 18 | 17 | 01 | 70 | 06 | 01 | 53 | 04 | 76 | 49 | 93 | 39 | 68 | 00 |
| 81 | 04 | 78 | 50 | 20 | 33 | 21 | 64 | 10 | 00 | 49 | 43 | 08 | 86 | 53 | 25 | 50 | 24 | 70 | 63 | 01 | 08 | 52 | 66 | 67 |
| 19 | 62 | 59 | 60 | 23 | 26 | 11 | 30 | 12 | 63 | 26 | 60 | 61 | 15 | 83 | 27 | 41 | 02 | 61 | 80 | 72 | 19 | 91 | 56 | 53 |
| 32 | 52 | 48 | 94 | 61 | 60 | 43 | 08 | 29 | 67 | 86 | 20 | 90 | 03 | 18 | 48 | 22 | 42 | 82 | 59 | 84 | 31 | 00 | 92 | 15 |
| 79 | 73 | 88 | 64 | 27 | 89 | 92 | 95 | 64 | 78 | 40 | 06 | 16 | 28 | 66 | 54 | 93 | 14 | 19 | 00 | 39 | 11 | 13 | 27 | 55 |
| 05 | 12 | 93 | 24 | 38 | 18 | 25 | 64 | 65 | 51 | 81 | 15 | 80 | 43 | 36 | 94 | 49 | 89 | 58 | 80 | 80 | 76 | 25 | 65 | 69 |
| 59 | 72 | 45 | 18 | 64 | 49 | 67 | 78 | 83 | 66 | 72 | 92 | 63 | 42 | 78 | 21 | 14 | 35 | 00 | 16 | 05 | 92 | 74 | 20 | 31 |
| 22 | 75 | 30 | 52 | 34 | 00 | 43 | 50 | 50 | 91 | 10 | 64 | 18 | 60 | 30 | 48 | 99 | 84 | 23 | 37 | 20 | 03 | 50 | 50 | 05 |
| 86 | 21 | 48 | 23 | 45 | 01 | 80 | 49 | 33 | 99 | 57 | 92 | 46 | 06 | 55 | 60 | 98 | 81 | 40 | 20 | 72 | 45 | 67 | 83 | 67 |
| 47 | 02 | 27 | 40 | 96 | 41 | 44 | 06 | 54 | 76 | 83 | 52 | 32 | 56 | 15 | 09 | 45 | 22 | 54 | 07 | 49 | 70 | 54 | 48 | 84 |
| 36 | 76 | 21 | 72 | 44 | 85 | 55 | 63 | 87 | 29 | 62 | 84 | 18 | 48 | 29 | 23 | 75 | 29 | 90 | 68 | 02 | 56 | 04 | 32 | 34 |
| 43 | 84 | 04 | 45 | 20 | 18 | 42 | 25 | 25 | 95 | 70 | 15 | 92 | 80 | 82 | 47 | 10 | 21 | 18 | 57 | 83 | 54 | 02 | 09 | 53 |
| 88 | 82 | 00 | 84 | 16 | 82 | 67 | 66 | 77 | 89 | 78 | 31 | 98 | 11 | 56 | 27 | 07 | 76 | 59 | 71 | 87 | 56 | 99 | 27 | 28 |

## Table A.1 (*continued*)

| | | | | |
|---|---|---|---|---|
| 90 78 82 54 47 | 20 83 80 10 41 | 35 22 23 03 98 | 79 74 41 35 05 | 78 73 95 47 83 |
| 78 58 68 87 41 | 11 08 81 29 89 | 71 23 10 01 79 | 25 06 00 45 80 | 64 70 95 34 29 |
| 51 42 21 03 88 | 20 05 35 93 00 | 68 12 09 55 09 | 36 54 95 22 82 | 48 30 09 56 87 |
| 93 15 07 60 86 | 67 37 94 24 35 | 82 44 19 92 96 | 21 84 29 04 29 | 83 32 05 10 48 |
| 27 12 31 66 62 | 09 54 17 31 23 | 27 30 37 36 79 | 75 50 39 57 12 | 67 23 22 09 33 |
| 79 44 83 55 47 | 96 50 93 56 82 | 58 16 35 18 87 | 64 08 22 47 93 | 86 43 43 30 17 |
| 89 73 43 91 03 | 57 91 35 40 64 | 13 61 94 37 16 | 09 93 96 25 87 | 30 23 42 54 31 |
| 29 30 90 00 58 | 15 99 93 33 67 | 80 08 59 21 66 | 13 54 56 85 25 | 05 32 03 52 52 |
| 97 33 17 26 25 | 04 73 18 10 05 | 34 40 32 65 07 | 28 68 29 31 97 | 89 57 95 55 16 |
| 07 15 44 92 47 | 28 50 93 03 53 | 37 70 19 68 59 | 95 39 87 90 46 | 98 64 46 24 71 |
| 82 50 35 50 80 | 23 67 81 25 02 | 83 08 12 70 00 | 25 31 33 80 06 | 19 86 14 59 27 |
| 59 21 86 16 30 | 27 85 16 26 34 | 50 15 87 22 69 | 71 36 95 90 76 | 90 99 79 63 21 |
| 04 19 60 33 05 | 29 02 33 74 56 | 38 84 21 07 35 | 93 54 70 18 47 | 14 62 75 45 02 |
| 96 91 44 09 94 | 06 89 50 88 83 | 82 50 11 82 51 | 30 68 91 06 28 | 86 65 17 45 20 |
| 31 71 03 53 38 | 94 02 52 72 15 | 44 49 53 42 43 | 00 36 97 67 64 | 12 27 46 00 18 |
| 03 70 22 67 59 | 98 10 64 68 08 | 79 06 89 48 41 | 85 72 10 87 24 | 96 04 20 68 00 |
| 08 45 79 46 89 | 74 73 67 60 15 | 70 37 61 44 07 | 67 89 81 54 26 | 57 17 63 27 74 |
| 37 80 05 75 64 | 48 51 68 68 27 | 71 75 45 32 27 | 76 35 26 58 88 | 67 74 48 90 94 |
| 90 63 56 69 37 | 19 74 48 63 31 | 52 36 84 40 66 | 72 66 03 41 87 | 65 29 12 36 64 |
| 22 69 38 02 88 | 89 71 43 01 87 | 41 79 42 99 29 | 41 08 47 32 19 | 45 29 59 69 90 |
| 05 79 69 67 64 | 36 14 82 65 26 | 40 51 63 42 48 | 85 48 34 12 04 | 33 26 52 26 52 |
| 48 91 53 03 82 | 64 24 06 31 03 | 97 44 82 24 89 | 88 48 66 54 10 | 41 27 09 11 61 |
| 94 64 97 27 25 | 62 23 94 40 54 | 56 32 97 78 90 | 58 86 41 75 19 | 42 90 85 36 68 |
| 15 85 82 52 08 | 52 96 26 92 88 | 93 11 03 23 52 | 78 23 57 85 43 | 53 90 42 22 22 |
| 09 81 37 66 56 | 99 08 59 19 48 | 29 69 21 64 95 | 12 08 15 24 45 | 59 25 22 76 96 |
| 43 83 99 02 76 | 12 16 45 52 66 | 35 70 93 09 52 | 75 40 34 35 62 | 65 42 27 20 59 |
| 31 98 09 80 62 | 75 26 64 57 26 | 46 41 47 90 97 | 99 46 10 51 42 | 73 28 98 89 91 |
| 81 35 42 62 84 | 37 02 59 78 16 | 17 96 05 71 39 | 88 05 34 05 92 | 22 43 89 66 89 |
| 97 95 56 39 75 | 65 47 61 86 33 | 14 88 55 33 69 | 70 87 79 94 46 | 17 61 72 27 01 |
| 37 63 35 93 23 | 17 30 14 51 51 | 17 28 21 74 67 | 12 11 57 19 27 | 38 70 73 82 92 |
| 39 22 96 00 48 | 52 49 62 09 40 | 08 30 27 54 70 | 96 06 52 12 80 | 36 12 38 68 05 |
| 61 29 84 34 51 | 60 19 77 82 16 | 64 45 02 27 04 | 65 55 90 95 04 | 20 39 29 96 28 |
| 38 84 18 10 29 | 19 09 66 06 78 | 37 09 60 50 21 | 52 72 01 52 70 | 29 65 05 37 16 |
| 64 29 48 04 08 | 55 72 25 25 77 | 54 26 27 24 39 | 66 67 06 40 00 | 99 35 70 69 58 |
| 64 02 32 99 63 | 62 42 89 32 20 | 81 14 08 40 45 | 22 15 37 49 38 | 96 51 19 08 27 |
| 13 83 39 51 30 | 31 49 94 83 66 | 02 50 95 18 98 | 58 84 90 58 81 | 00 40 91 12 46 |
| 83 30 90 09 35 | 41 12 87 93 66 | 85 96 20 65 34 | 23 13 05 41 01 | 91 48 95 59 45 |
| 46 63 53 97 63 | 18 86 37 56 20 | 35 62 66 11 37 | 30 91 89 97 51 | 64 78 06 95 65 |
| 54 43 40 02 41 | 55 70 52 96 87 | 02 82 61 21 88 | 60 65 98 42 09 | 03 61 20 83 01 |
| 27 18 65 62 01 | 97 45 79 51 37 | 74 47 20 11 48 | 97 93 73 86 50 | 46 61 95 01 24 |
| 45 42 16 13 20 | 34 51 08 71 52 | 39 17 71 39 84 | 97 27 72 49 42 | 81 62 32 87 22 |
| 35 92 97 02 34 | 93 32 95 81 13 | 92 05 40 70 95 | 71 66 61 24 08 | 77 32 73 66 79 |
| 60 55 35 57 24 | 52 95 84 90 64 | 38 39 72 70 17 | 98 42 85 96 67 | 41 11 83 17 78 |
| 43 17 21 09 60 | 58 86 12 31 11 | 66 61 43 96 00 | 93 97 00 15 20 | 37 96 73 56 63 |
| 07 85 74 58 28 | 38 74 68 32 61 | 87 14 71 83 47 | 90 11 96 70 08 | 67 04 34 46 08 |
| 33 00 29 08 87 | 42 59 40 24 97 | 44 99 13 56 87 | 95 02 47 97 89 | 23 51 45 37 83 |
| 97 14 00 42 23 | 72 03 19 02 41 | 11 23 36 98 32 | 19 91 42 03 58 | 62 23 74 45 06 |
| 68 58 32 80 82 | 40 49 71 83 37 | 93 49 99 60 72 | 88 14 26 88 95 | 48 69 35 40 63 |
| 39 87 38 16 06 | 82 92 62 32 75 | 67 64 50 49 39 | 29 55 53 92 97 | 04 48 60 53 90 |
| 37 73 01 84 87 | 42 88 30 93 75 | 01 18 34 73 30 | 28 44 28 18 01 | 00 38 26 38 57 |

Source: A. Hald, *Statistical Tables and Formulas*. John Wiley and Sons, Inc., New York, 1952, pp. 94–95.

## Table A.2 Binomial Distribution

A table of $\binom{n}{x} p^x(1-p)^{n-x}$ for $n = 2, 3, \ldots, 10$ and

$p = 0.01, 0.05(0.05)0.30, \frac{1}{3}, 0.35(0.05)0.50$, and $p = 0.49$

| $n$ | $x$ | $p$ .01 | .05 | .10 | .15 | .20 | .25 | .30 | $\frac{1}{3}$ | .35 | .40 | .45 | .49 | .50 |
|---|---|---|---|---|---|---|---|---|---|---|---|---|---|---|
| 2 | 0 | .9801 | .9025 | .8100 | .7225 | .6400 | .5625 | .4900 | .4444 | .4225 | .3600 | .3025 | .2601 | .2500 |
| | 1 | .0198 | .0950 | .1800 | .2550 | .3200 | .3750 | .4200 | .4444 | .4550 | .4800 | .4950 | .4998 | .5000 |
| | 2 | .0001 | .0025 | .0100 | .0225 | .0400 | .0625 | .0900 | .1111 | .1225 | .1600 | .2025 | .2401 | .2500 |
| 3 | 0 | .9703 | .8574 | .7290 | .6141 | .5120 | .4219 | .3430 | .2963 | .2746 | .2160 | .1664 | .1327 | .1250 |
| | 1 | .0294 | .1354 | .2430 | .3251 | .3840 | .4219 | .4410 | .4444 | .4436 | .4320 | .4084 | .3823 | .3750 |
| | 2 | .0003 | .0071 | .0270 | .0574 | .0960 | .1406 | .1890 | .2222 | .2389 | .2880 | .3341 | .3674 | .3750 |
| | 3 | .0000 | .0001 | .0010 | .0034 | .0080 | .0156 | .0270 | .0370 | .0429 | .0640 | .0911 | .1176 | .1250 |
| 4 | 0 | .9606 | .8145 | .6561 | .5220 | .4096 | .3164 | .2401 | .1975 | .1785 | .1296 | .0915 | .0677 | .0625 |
| | 1 | .0388 | .1715 | .2916 | .3685 | .4096 | .4219 | .4116 | .3951 | .3845 | .3456 | .2995 | .2600 | .2500 |
| | 2 | .0006 | .0135 | .0486 | .0975 | .1536 | .2109 | .2646 | .2963 | .3105 | .3456 | .3675 | .3747 | .3750 |
| | 3 | .0000 | .0005 | .0036 | .0115 | .0256 | .0469 | .0756 | .0988 | .1115 | .1536 | .2005 | .2400 | .2500 |
| | 4 | .0000 | .0000 | .0001 | .0005 | .0016 | .0039 | .0081 | .0123 | .0150 | .0256 | .0410 | .0576 | .0625 |
| 5 | 0 | .9510 | .7738 | .5905 | .4437 | .3277 | .2373 | .1681 | .1317 | .1160 | .0778 | .0503 | .0345 | .0312 |
| | 1 | .0480 | .2036 | .3280 | .3915 | .4096 | .3955 | .3602 | .3292 | .3124 | .2592 | .2059 | .1657 | .1562 |
| | 2 | .0010 | .0214 | .0729 | .1382 | .2048 | .2637 | .3087 | .3292 | .3364 | .3456 | .3369 | .3185 | .3125 |
| | 3 | .0000 | .0011 | .0081 | .0244 | .0512 | .0879 | .1323 | .1646 | .1811 | .2304 | .2757 | .3060 | .3125 |
| | 4 | .0000 | .0000 | .0004 | .0022 | .0064 | .0146 | .0284 | .0412 | .0488 | .0768 | .1128 | .1470 | .1562 |
| | 5 | .0000 | .0000 | .0000 | .0001 | .0003 | .0010 | .0024 | .0041 | .0053 | .0102 | .0185 | .0283 | .0312 |
| 6 | 0 | .9415 | .7351 | .5314 | .3771 | .2621 | .1780 | .1176 | .0878 | .0754 | .0467 | .0277 | .0176 | .0156 |
| | 1 | .0571 | .2321 | .3543 | .3993 | .3932 | .3560 | .3025 | .2634 | .2437 | .1866 | .1359 | .1014 | .0938 |
| | 2 | .0014 | .0305 | .0984 | .1762 | .2458 | .2966 | .3241 | .3292 | .3280 | .3110 | .2780 | .2437 | .2344 |
| | 3 | .0000 | .0021 | .0146 | .0415 | .0819 | .1318 | .1852 | .2195 | .2355 | .2765 | .3032 | .3121 | .3125 |
| | 4 | .0000 | .0001 | .0012 | .0055 | .0154 | .0330 | .0595 | .0823 | .0951 | .1382 | .1861 | .2249 | .2344 |
| | 5 | .0000 | .0000 | .0001 | .0004 | .0015 | .0044 | .0102 | .0165 | .0205 | .0369 | .0609 | .0864 | .0938 |
| | 6 | .0000 | .0000 | .0000 | .0000 | .0001 | .0002 | .0007 | .0014 | .0018 | .0041 | .0083 | .0139 | .0156 |
| 7 | 0 | .9321 | .6983 | .4783 | .3206 | .2097 | .1335 | .0824 | .0585 | .0490 | .0280 | .0152 | .0090 | .0078 |
| | 1 | .0659 | .2573 | .3720 | .3960 | .3670 | .3115 | .2471 | .2048 | .1848 | .1306 | .0872 | .0603 | .0547 |
| | 2 | .0020 | .0406 | .1240 | .2097 | .2753 | .3115 | .3177 | .3073 | .2985 | .2613 | .2140 | .1740 | .1641 |
| | 3 | .0000 | .0036 | .0230 | .0617 | .1147 | .1730 | .2269 | .2561 | .2679 | .2903 | .2918 | .2786 | .2734 |
| | 4 | .0000 | .0002 | .0026 | .0109 | .0287 | .0577 | .0972 | .1280 | .1442 | .1935 | .2388 | .2676 | .2734 |
| | 5 | .0000 | .0000 | .0002 | .0012 | .0043 | .0115 | .0250 | .0384 | .0466 | .0774 | .1172 | .1543 | .1641 |
| | 6 | .0000 | .0000 | .0000 | .0001 | .0004 | .0013 | .0036 | .0064 | .0084 | .0172 | .0320 | .0494 | .0547 |
| | 7 | .0000 | .0000 | .0000 | .0000 | .0000 | .0001 | .0002 | .0005 | .0006 | .0016 | .0037 | .0068 | .0078 |
| 8 | 0 | .9227 | .6634 | .4305 | .2725 | .1678 | .1001 | .0576 | .0390 | .0319 | .0168 | .0084 | .0046 | .0039 |
| | 1 | .0746 | .2793 | .3826 | .3847 | .3355 | .2670 | .1977 | .1561 | .1373 | .0896 | .0548 | .0352 | .0312 |
| | 2 | .0026 | .0515 | .1488 | .2376 | .2936 | .3115 | .2965 | .2731 | .2587 | .2090 | .1569 | .1183 | .1094 |
| | 3 | .0001 | .0054 | .0331 | .0839 | .1468 | .2076 | .2541 | .2731 | .2786 | .2787 | .2568 | .2273 | .2188 |
| | 4 | .0000 | .0004 | .0046 | .0185 | .0459 | .0865 | .1361 | .1707 | .1875 | .2322 | .2627 | .2730 | .2734 |
| | 5 | .0000 | .0000 | .0004 | .0026 | .0092 | .0231 | .0467 | .0683 | .0808 | .1239 | .1719 | .2098 | .2188 |
| | 6 | .0000 | .0000 | .0000 | .0002 | .0011 | .0038 | .0100 | .0171 | .0217 | .0413 | .0703 | .1008 | .1094 |
| | 7 | .0000 | .0000 | .0000 | .0000 | .0001 | .0004 | .0012 | .0024 | .0033 | .0079 | .0164 | .0277 | .0312 |
| | 8 | .0000 | .0000 | .0000 | .0000 | .0000 | .0000 | .0001 | .0002 | .0002 | .0007 | .0017 | .0033 | .0039 |

### Table A.2 (*continued*)

| n | x | p .01 | .05 | .10 | .15 | .20 | .25 | .30 | $\frac{1}{3}$ | .35 | .40 | .45 | .49 | .50 |
|---|---|---|---|---|---|---|---|---|---|---|---|---|---|---|
| 9 | 0 | .9135 | .6302 | .3874 | .2316 | .1342 | .0751 | .0404 | .0260 | .0207 | .0101 | .0046 | .0023 | .0020 |
| | 1 | .0830 | .2985 | .3874 | .3679 | .3020 | .2253 | .1556 | .1171 | .1004 | .0605 | .0339 | .0202 | .0176 |
| | 2 | .0034 | .0629 | .1722 | .2597 | .3020 | .3003 | .2668 | .2341 | .2162 | .1612 | .1110 | .0776 | .0703 |
| | 3 | .0001 | .0077 | .0446 | .1069 | .1762 | .2336 | .2668 | .2731 | .2716 | .2508 | .2119 | .1739 | .1641 |
| | 4 | .0000 | .0006 | .0074 | .0283 | .0661 | .1168 | .1715 | .2048 | .2194 | .2508 | .2600 | .2506 | .2461 |
| | 5 | .0000 | .0000 | .0008 | .0050 | .0165 | .0389 | .0735 | .1024 | .1181 | .1672 | .2128 | .2408 | .2461 |
| | 6 | .0000 | .0000 | .0001 | .0006 | .0028 | .0087 | .0210 | .0341 | .0424 | .0743 | .1160 | .1542 | .1641 |
| | 7 | .0000 | .0000 | .0000 | .0000 | .0003 | .0012 | .0039 | .0073 | .0098 | .0212 | .0407 | .0635 | .0703 |
| | 8 | .0000 | .0000 | .0000 | .0000 | .0000 | .0001 | .0004 | .0009 | .0013 | .0035 | .0083 | .0153 | .0176 |
| | 9 | .0000 | .0000 | .0000 | .0000 | .0000 | .0000 | .0000 | .0001 | .0001 | .0003 | .0008 | .0016 | .0020 |
| 10 | 0 | .9044 | .5987 | .3487 | .1969 | .1074 | .0563 | .0282 | .0173 | .0135 | .0060 | .0025 | .0012 | .0010 |
| | 1 | .0914 | .3151 | .3874 | .3474 | .2684 | .1877 | .1211 | .0867 | .0725 | .0403 | .0207 | .0114 | .0098 |
| | 2 | .0042 | .0746 | .1937 | .2759 | .3020 | .2816 | .2335 | .1951 | .1757 | .1209 | .0763 | .0495 | .0439 |
| | 3 | .0001 | .0105 | .0574 | .1298 | .2013 | .2503 | .2668 | .2601 | .2522 | .2150 | .1665 | .1267 | .1172 |
| | 4 | .0000 | .0010 | .0112 | .0401 | .0881 | .1460 | .2001 | .2276 | .2377 | .2508 | .2384 | .2130 | .2051 |
| | 5 | .0000 | .0001 | .0015 | .0085 | .0264 | .0584 | .1029 | .1366 | .1536 | .2007 | .2340 | .2456 | .2461 |
| | 6 | .0000 | .0000 | .0001 | .0012 | .0055 | .0162 | .0368 | .0569 | .0689 | .1115 | .1596 | .1966 | .2051 |
| | 7 | .0000 | .0000 | .0000 | .0001 | .0008 | .0031 | .0090 | .0163 | .0212 | .0425 | .0746 | .1080 | .1172 |
| | 8 | .0000 | .0000 | .0000 | .0000 | .0001 | .0004 | .0014 | .0030 | .0043 | .0106 | .0229 | .0389 | .0439 |
| | 9 | .0000 | .0000 | .0000 | .0000 | .0000 | .0000 | .0001 | .0003 | .0005 | .0016 | .0042 | .0083 | .0098 |
| | 10 | .0000 | .0000 | .0000 | .0000 | .0000 | .0000 | .0000 | .0000 | .0000 | .0001 | .0003 | .0008 | .0010 |

Source: E. Parzen, *Modern Probability Theory and Its Applications*. John Wiley and Sons, Inc., New York, 1960, pp. 442–43.

### Table A.3 Cumulative Poisson Distribution

$$\text{Values of } \sum_{x=X}^{\infty} \frac{e^{-\theta}\theta^{x}}{x!} = 1 - F(x)_{x=X-1}$$

| X | θ 0.1 | 0.2 | 0.3 | 0.4 | 0.5 | 0.6 | 0.7 | 0.8 | 0.9 | 1.0 |
|---|---|---|---|---|---|---|---|---|---|---|
| 0 | 1.0000 | 1.0000 | 1.0000 | 1.0000 | 1.0000 | 1.0000 | 1.0000 | 1.0000 | 1.0000 | 1.0000 |
| 1 | 0.0952 | 0.1813 | 0.2592 | 0.3297 | 0.3935 | 0.4512 | 0.5034 | 0.5507 | 0.5934 | 0.6321 |
| 2 | 0.0047 | 0.0175 | 0.0369 | 0.0616 | 0.0902 | 0.1219 | 0.1558 | 0.1912 | 0.2275 | 0.2642 |
| 3 | 0.0002 | 0.0011 | 0.0036 | 0.0079 | 0.0144 | 0.0231 | 0.0341 | 0.0474 | 0.0629 | 0.0803 |
| 4 | 0.0000 | 0.0001 | 0.0003 | 0.0008 | 0.0018 | 0.0034 | 0.0058 | 0.0091 | 0.0135 | 0.0190 |
| 5 | 0.0000 | 0.0000 | 0.0000 | 0.0001 | 0.0002 | 0.0004 | 0.0008 | 0.0014 | 0.0023 | 0.0037 |
| 6 | 0.0000 | 0.0000 | 0.0000 | 0.0000 | 0.0000 | 0.0000 | 0.0001 | 0.0002 | 0.0003 | 0.0006 |
| 7 | 0.0000 | 0.0000 | 0.0000 | 0.0000 | 0.0000 | 0.0000 | 0.0000 | 0.0000 | 0.0000 | 0.0001 |

| X | θ 1.1 | 1.2 | 1.3 | 1.4 | 1.5 | 1.6 | 1.7 | 1.8 | 1.9 | 2.0 |
|---|---|---|---|---|---|---|---|---|---|---|
| 0 | 1.0000 | 1.0000 | 1.0000 | 1.0000 | 1.0000 | 1.0000 | 1.0000 | 1.0000 | 1.0000 | 1.0000 |
| 1 | 0.6671 | 0.6988 | 0.7275 | 0.7534 | 0.7769 | 0.7981 | 0.8173 | 0.8347 | 0.8504 | 0.8647 |
| 2 | 0.3010 | 0.3374 | 0.3732 | 0.4082 | 0.4422 | 0.4751 | 0.5068 | 0.5372 | 0.5663 | 0.5940 |
| 3 | 0.0996 | 0.1205 | 0.1429 | 0.1665 | 0.1912 | 0.2166 | 0.2428 | 0.2694 | 0.2963 | 0.3233 |
| 4 | 0.0257 | 0.0338 | 0.0431 | 0.0537 | 0.0656 | 0.0788 | 0.0932 | 0.1087 | 0.1253 | 0.1429 |
| 5 | 0.0054 | 0.0077 | 0.0107 | 0.0143 | 0.0186 | 0.0237 | 0.0296 | 0.0364 | 0.0441 | 0.0527 |
| 6 | 0.0010 | 0.0015 | 0.0022 | 0.0032 | 0.0045 | 0.0060 | 0.0080 | 0.0104 | 0.0132 | 0.0166 |
| 7 | 0.0001 | 0.0003 | 0.0004 | 0.0006 | 0.0009 | 0.0013 | 0.0019 | 0.0026 | 0.0034 | 0.0045 |
| 8 | 0.0000 | 0.0000 | 0.0001 | 0.0001 | 0.0002 | 0.0003 | 0.0004 | 0.0006 | 0.0008 | 0.0011 |
| 9 | 0.0000 | 0.0000 | 0.0000 | 0.0000 | 0.0000 | 0.0000 | 0.0001 | 0.0001 | 0.0002 | 0.0002 |

**Table A.3** (*continued*)

| | $\theta$ | | | | | | | | | |
|---|---|---|---|---|---|---|---|---|---|---|
| $X$ | 2.1 | 2.2 | 2.3 | 2.4 | 2.5 | 2.6 | 2.7 | 2.8 | 2.9 | 3.0 |
| 0 | 1.0000 | 1.0000 | 1.0000 | 1.0000 | 1.0000 | 1.0000 | 1.0000 | 1.0000 | 1.0000 | 1.0000 |
| 1 | 0.8775 | 0.8892 | 0.8997 | 0.9093 | 0.9179 | 0.9257 | 0.9328 | 0.9392 | 0.9450 | 0.9502 |
| 2 | 0.6204 | 0.6454 | 0.6691 | 0.6916 | 0.7127 | 0.7326 | 0.7513 | 0.7689 | 0.7854 | 0.8009 |
| 3 | 0.3504 | 0.3773 | 0.4040 | 0.4303 | 0.4562 | 0.4816 | 0.5064 | 0.5305 | 0.5540 | 0.5768 |
| 4 | 0.1614 | 0.1806 | 0.2007 | 0.2213 | 0.2424 | 0.2640 | 0.2859 | 0.3081 | 0.3304 | 0.3528 |
| 5 | 0.0621 | 0.0725 | 0.0838 | 0.0959 | 0.1088 | 0.1226 | 0.1371 | 0.1523 | 0.1682 | 0.1847 |
| 6 | 0.0204 | 0.0249 | 0.0300 | 0.0357 | 0.0420 | 0.0490 | 0.0567 | 0.0651 | 0.0742 | 0.0839 |
| 7 | 0.0059 | 0.0075 | 0.0094 | 0.0116 | 0.0142 | 0.0172 | 0.0206 | 0.0244 | 0.0287 | 0.0335 |
| 8 | 0.0015 | 0.0020 | 0.0026 | 0.0033 | 0.0042 | 0.0053 | 0.0066 | 0.0081 | 0.0099 | 0.0119 |
| 9 | 0.0003 | 0.0005 | 0.0006 | 0.0009 | 0.0011 | 0.0015 | 0.0019 | 0.0024 | 0.0031 | 0.0038 |
| 10 | 0.0001 | 0.0001 | 0.0001 | 0.0002 | 0.0003 | 0.0004 | 0.0005 | 0.0007 | 0.0009 | 0.0011 |
| 11 | 0.0000 | 0.0000 | 0.0000 | 0.0000 | 0.0001 | 0.0001 | 0.0001 | 0.0002 | 0.0002 | 0.0003 |
| 12 | 0.0000 | 0.0000 | 0.0000 | 0.0000 | 0.0000 | 0.0000 | 0.0000 | 0.0000 | 0.0001 | 0.0001 |

| | $\theta$ | | | | | | | | | |
|---|---|---|---|---|---|---|---|---|---|---|
| $X$ | 3.1 | 3.2 | 3.3 | 3.4 | 3.5 | 3.6 | 3.7 | 3.8 | 3.9 | 4.0 |
| 0 | 1.0000 | 1.0000 | 1.0000 | 1.0000 | 1.0000 | 1.0000 | 1.0000 | 1.0000 | 1.0000 | 1.0000 |
| 1 | 0.9550 | 0.9592 | 0.9631 | 0.9666 | 0.9698 | 0.9727 | 0.9753 | 0.9776 | 0.9798 | 0.9817 |
| 2 | 0.8153 | 0.8288 | 0.8414 | 0.8532 | 0.8641 | 0.8743 | 0.8838 | 0.8926 | 0.9008 | 0.9084 |
| 3 | 0.5988 | 0.6201 | 0.6406 | 0.6603 | 0.6792 | 0.6973 | 0.7146 | 0.7311 | 0.7469 | 0.7619 |
| 4 | 0.3752 | 0.3975 | 0.4197 | 0.4416 | 0.4634 | 0.4848 | 0.5058 | 0.5265 | 0.5468 | 0.5665 |
| 5 | 0.2018 | 0.2194 | 0.2374 | 0.2558 | 0.2746 | 0.2936 | 0.3128 | 0.3322 | 0.3516 | 0.3712 |
| 6 | 0.0943 | 0.1054 | 0.1171 | 0.1295 | 0.1424 | 0.1559 | 0.1699 | 0.1844 | 0.1994 | 0.2149 |
| 7 | 0.0388 | 0.0446 | 0.0510 | 0.0579 | 0.0653 | 0.0733 | 0.0818 | 0.0909 | 0.1005 | 0.1107 |
| 8 | 0.0142 | 0.0168 | 0.0198 | 0.0231 | 0.0267 | 0.0308 | 0.0352 | 0.0401 | 0.0454 | 0.0511 |
| 9 | 0.0047 | 0.0057 | 0.0069 | 0.0083 | 0.0099 | 0.0117 | 0.0137 | 0.0160 | 0.0185 | 0.0214 |
| 10 | 0.0014 | 0.0018 | 0.0022 | 0.0027 | 0.0033 | 0.0040 | 0.0048 | 0.0058 | 0.0069 | 0.0081 |
| 11 | 0.0004 | 0.0005 | 0.0006 | 0.0008 | 0.0010 | 0.0013 | 0.0016 | 0.0019 | 0.0023 | 0.0028 |
| 12 | 0.0001 | 0.0001 | 0.0002 | 0.0002 | 0.0003 | 0.0004 | 0.0005 | 0.0006 | 0.0007 | 0.0009 |
| 13 | 0.0000 | 0.0000 | 0.0000 | 0.0001 | 0.0001 | 0.0001 | 0.0001 | 0.0002 | 0.0002 | 0.0003 |
| 14 | 0.0000 | 0.0000 | 0.0000 | 0.0000 | 0.0000 | 0.0000 | 0.0000 | 0.0000 | 0.0001 | 0.0001 |

| | $\theta$ | | | | | | | | | |
|---|---|---|---|---|---|---|---|---|---|---|
| $X$ | 4.1 | 4.2 | 4.3 | 4.4 | 4.5 | 4.6 | 4.7 | 4.8 | 4.9 | 5.0 |
| 0 | 1.0000 | 1.0000 | 1.0000 | 1.0000 | 1.0000 | 1.0000 | 1.0000 | 1.0000 | 1.0000 | 1.0000 |
| 1 | 0.9834 | 0.9850 | 0.9864 | 0.9877 | 0.9889 | 0.9899 | 0.9909 | 0.9918 | 0.9926 | 0.9933 |
| 2 | 0.9155 | 0.9220 | 0.9281 | 0.9337 | 0.9389 | 0.9437 | 0.9482 | 0.9523 | 0.9561 | 0.9596 |
| 3 | 0.7762 | 0.7898 | 0.8026 | 0.8149 | 0.8264 | 0.8374 | 0.8477 | 0.8575 | 0.8667 | 0.8753 |
| 4 | 0.5858 | 0.6046 | 0.6228 | 0.6406 | 0.6577 | 0.6743 | 0.6903 | 0.7058 | 0.7207 | 0.7350 |
| 5 | 0.3907 | 0.4102 | 0.4296 | 0.4488 | 0.4679 | 0.4868 | 0.5054 | 0.5237 | 0.5418 | 0.5595 |
| 6 | 0.2307 | 0.2469 | 0.2633 | 0.2801 | 0.2971 | 0.3142 | 0.3316 | 0.3490 | 0.3665 | 0.3840 |
| 7 | 0.1214 | 0.1325 | 0.1442 | 0.1564 | 0.1689 | 0.1820 | 0.1954 | 0.2092 | 0.2233 | 0.2378 |
| 8 | 0.0573 | 0.0639 | 0.0710 | 0.0786 | 0.0866 | 0.0951 | 0.1040 | 0.1133 | 0.1231 | 0.1334 |
| 9 | 0.0245 | 0.0279 | 0.0317 | 0.0358 | 0.0403 | 0.0451 | 0.0503 | 0.0558 | 0.0618 | 0.0681 |
| 10 | 0.0095 | 0.0111 | 0.0129 | 0.0149 | 0.0171 | 0.0195 | 0.0222 | 0.0251 | 0.0283 | 0.0318 |
| 11 | 0.0034 | 0.0041 | 0.0048 | 0.0057 | 0.0067 | 0.0078 | 0.0090 | 0.0104 | 0.0120 | 0.0137 |
| 12 | 0.0011 | 0.0014 | 0.0017 | 0.0020 | 0.0024 | 0.0029 | 0.0034 | 0.0040 | 0.0047 | 0.0055 |
| 13 | 0.0003 | 0.0004 | 0.0005 | 0.0007 | 0.0008 | 0.0010 | 0.0012 | 0.0014 | 0.0017 | 0.0020 |
| 14 | 0.0001 | 0.0001 | 0.0002 | 0.0002 | 0.0003 | 0.0003 | 0.0004 | 0.0005 | 0.0006 | 0.0007 |
| 15 | 0.0000 | 0.0000 | 0.0000 | 0.0001 | 0.0001 | 0.0001 | 0.0001 | 0.0001 | 0.0002 | 0.0002 |
| 16 | 0.0000 | 0.0000 | 0.0000 | 0.0000 | 0.0000 | 0.0000 | 0.0000 | 0.0000 | 0.0001 | 0.0001 |

**Table A.3** (*continued*)

| X | θ = 5.1 | 5.2 | 5.3 | 5.4 | 5.5 | 5.6 | 5.7 | 5.8 | 5.9 | 6.0 |
|---|---|---|---|---|---|---|---|---|---|---|
| 0 | 1.0000 | 1.0000 | 1.0000 | 1.0000 | 1.0000 | 1.0000 | 1.0000 | 1.0000 | 1.0000 | 1.0000 |
| 1 | 0.9939 | 0.9945 | 0.9950 | 0.9955 | 0.9959 | 0.9963 | 0.9967 | 0.9970 | 0.9973 | 0.9975 |
| 2 | 0.9628 | 0.9658 | 0.9686 | 0.9711 | 0.9734 | 0.9756 | 0.9776 | 0.9794 | 0.9811 | 0.9826 |
| 3 | 0.8835 | 0.8912 | 0.8984 | 0.9052 | 0.9116 | 0.9176 | 0.9232 | 0.9285 | 0.9334 | 0.9380 |
| 4 | 0.7487 | 0.7619 | 0.7746 | 0.7867 | 0.7983 | 0.8094 | 0.8200 | 0.8300 | 0.8396 | 0.8488 |
| 5 | 0.5769 | 0.5939 | 0.6105 | 0.6267 | 0.6425 | 0.6579 | 0.6728 | 0.6873 | 0.7013 | 0.7149 |
| 6 | 0.4016 | 0.4191 | 0.4365 | 0.4539 | 0.4711 | 0.4881 | 0.5050 | 0.5217 | 0.5381 | 0.5543 |
| 7 | 0.2526 | 0.2676 | 0.2829 | 0.2983 | 0.3140 | 0.3297 | 0.3456 | 0.3616 | 0.3776 | 0.3937 |
| 8 | 0.1440 | 0.1551 | 0.1665 | 0.1783 | 0.1905 | 0.2030 | 0.2159 | 0.2290 | 0.2424 | 0.2560 |
| 9 | 0.0748 | 0.0819 | 0.0894 | 0.0974 | 0.1056 | 0.1143 | 0.1234 | 0.1328 | 0.1426 | 0.1528 |

| X | θ = 5.1 | 5.2 | 5.3 | 5.4 | 5.5 | 5.6 | 5.7 | 5.8 | 5.9 | 6.0 |
|---|---|---|---|---|---|---|---|---|---|---|
| 10 | 0.0356 | 0.0397 | 0.0441 | 0.0488 | 0.0538 | 0.0591 | 0.0648 | 0.0708 | 0.0772 | 0.0839 |
| 11 | 0.0156 | 0.0177 | 0.0200 | 0.0225 | 0.0253 | 0.0282 | 0.0314 | 0.0349 | 0.0386 | 0.0426 |
| 12 | 0.0063 | 0.0073 | 0.0084 | 0.0096 | 0.0110 | 0.0125 | 0.0141 | 0.0160 | 0.0179 | 0.0201 |
| 13 | 0.0024 | 0.0028 | 0.0033 | 0.0038 | 0.0045 | 0.0051 | 0.0059 | 0.0068 | 0.0078 | 0.0088 |
| 14 | 0.0008 | 0.0010 | 0.0012 | 0.0014 | 0.0017 | 0.0030 | 0.0023 | 0.0027 | 0.0031 | 0.0036 |
| 15 | 0.0003 | 0.0003 | 0.0004 | 0.0005 | 0.0006 | 0.0007 | 0.0009 | 0.0010 | 0.0012 | 0.0014 |
| 16 | 0.0001 | 0.0001 | 0.0001 | 0.0002 | 0.0002 | 0.0002 | 0.0003 | 0.0004 | 0.0004 | 0.0005 |
| 17 | 0.0000 | 0.0000 | 0.0000 | 0.0001 | 0.0001 | 0.0001 | 0.0001 | 0.0001 | 0.0001 | 0.0002 |
| 18 | 0.0000 | 0.0000 | 0.0000 | 0.0000 | 0.0000 | 0.0000 | 0.0000 | 0.0000 | 0.0000 | 0.0001 |

| X | θ = 6.1 | 6.2 | 6.3 | 6.4 | 6.5 | 6.6 | 6.7 | 6.8 | 6.9 | 7.0 |
|---|---|---|---|---|---|---|---|---|---|---|
| 0 | 1.0000 | 1.0000 | 1.0000 | 1.0000 | 1.0000 | 1.0000 | 1.0000 | 1.0000 | 1.0000 | 1.0000 |
| 1 | 0.9978 | 0.9980 | 0.9982 | 0.9983 | 0.9985 | 0.9986 | 0.9988 | 0.9989 | 0.9990 | 0.9991 |
| 2 | 0.9841 | 0.9854 | 0.9866 | 0.9877 | 0.9887 | 0.9897 | 0.9905 | 0.9913 | 0.9920 | 0.9927 |
| 3 | 0.9423 | 0.9464 | 0.9502 | 0.9537 | 0.9570 | 0.9600 | 0.9629 | 0.9656 | 0.9680 | 0.9704 |
| 4 | 0.8575 | 0.8658 | 0.8736 | 0.8811 | 0.8882 | 0.8948 | 0.9012 | 0.9072 | 0.9129 | 0.9182 |
| 5 | 0.7281 | 0.7408 | 0.7531 | 0.7649 | 0.7763 | 0.7873 | 0.7978 | 0.8080 | 0.8177 | 0.8270 |
| 6 | 0.5702 | 0.5859 | 0.6012 | 0.6163 | 0.6310 | 0.6453 | 0.6594 | 0.6730 | 0.6863 | 0.6993 |
| 7 | 0.4098 | 0.4258 | 0.4418 | 0.4577 | 0.4735 | 0.4892 | 0.5047 | 0.5201 | 0.5353 | 0.5503 |
| 8 | 0.2699 | 0.2840 | 0.2983 | 0.3127 | 0.3272 | 0.3419 | 0.3567 | 0.3715 | 0.3864 | 0.4013 |
| 9 | 0.1633 | 0.1741 | 0.1852 | 0.1967 | 0.2084 | 0.2204 | 0.2327 | 0.2452 | 0.2580 | 0.2709 |
| 10 | 0.0910 | 0.0984 | 0.1061 | 0.1142 | 0.1226 | 0.1314 | 0.1404 | 0.1498 | 0.1505 | 0.1695 |
| 11 | 0.0469 | 0.0514 | 0.0563 | 0.0614 | 0.0688 | 0.0726 | 0.0786 | 0.0849 | 0.0916 | 0.0985 |
| 12 | 0.0224 | 0.0250 | 0.0277 | 0.0307 | 0.0339 | 0.0373 | 0.0409 | 0.0448 | 0.0490 | 0.0534 |
| 13 | 0.0100 | 0.0113 | 0.0127 | 0.0143 | 0.0160 | 0.0179 | 0.0199 | 0.0221 | 0.0245 | 0.0270 |
| 14 | 0.0042 | 0.0048 | 0.0055 | 0.0063 | 0.0071 | 0.0080 | 0.0091 | 0.0102 | 0.0115 | 0.0128 |
| 15 | 0.0016 | 0.0019 | 0.0022 | 0.0026 | 0.0030 | 0.0034 | 0.0039 | 0.0044 | 0.0050 | 0.0057 |
| 16 | 0.0006 | 0.0007 | 0.0008 | 0.0010 | 0.0012 | 0.0014 | 0.0016 | 0.0018 | 0.0021 | 0.0024 |
| 17 | 0.0002 | 0.0003 | 0.0003 | 0.0004 | 0.0004 | 0.0005 | 0.0006 | 0.0007 | 0.0008 | 0.0010 |
| 18 | 0.0001 | 0.0001 | 0.0001 | 0.0001 | 0.0002 | 0.0002 | 0.0002 | 0.0003 | 0.0003 | 0.0004 |
| 19 | 0.0000 | 0.0000 | 0.0000 | 0.0000 | 0.0001 | 0.0001 | 0.0001 | 0.0001 | 0.0001 | 0.0001 |

**Table A.3** (*continued*)

| | θ | | | | | | | | | |
|---|---|---|---|---|---|---|---|---|---|---|
| *X* | 7.1 | 7.2 | 7.3 | 7.4 | 7.5 | 7.6 | 7.7 | 7.8 | 7.9 | 8.0 |
| 0 | 1.0000 | 1.0000 | 1.0000 | 1.0000 | 1.0000 | 1.0000 | 1.0000 | 1.0000 | 1.0000 | 1.0000 |
| 1 | 0.9992 | 0.9993 | 0.9993 | 0.9994 | 0.9994 | 0.9995 | 0.9995 | 0.9996 | 0.9996 | 0.9997 |
| 2 | 0.9933 | 0.9939 | 0.9944 | 0.9949 | 0.9953 | 0.9957 | 0.9961 | 0.9964 | 0.9967 | 0.9970 |
| 3 | 0.9725 | 0.9745 | 0.9764 | 0.9781 | 0.9797 | 0.9812 | 0.9826 | 0.9839 | 0.9851 | 0.9862 |
| 4 | 0.9233 | 0.9281 | 0.9326 | 0.9368 | 0.9409 | 0.9446 | 0.9482 | 0.9515 | 0.9547 | 0.9576 |
| 5 | 0.8359 | 0.8445 | 0.8527 | 0.8605 | 0.8679 | 0.8751 | 0.8819 | 0.8883 | 0.8945 | 0.9004 |
| 6 | 0.7119 | 0.7241 | 0.7360 | 0.7474 | 0.7586 | 0.7693 | 0.7797 | 0.7897 | 0.7994 | 0.8088 |
| 7 | 0.5651 | 0.5796 | 0.5940 | 0.6080 | 0.6218 | 0.6354 | 0.6486 | 0.6616 | 0.6743 | 0.6866 |
| 8 | 0.4162 | 0.4311 | 0.4459 | 0.4607 | 0.4754 | 0.4900 | 0.5044 | 0.5188 | 0.5330 | 0.5470 |
| 9 | 0.2840 | 0.2973 | 0.3108 | 0.3243 | 0.3380 | 0.3518 | 0.3657 | 0.3796 | 0.3935 | 0.4075 |
| 10 | 0.1798 | 0.1904 | 0.2012 | 0.2123 | 0.2236 | 0.2351 | 0.2469 | 0.2589 | 0.2710 | 0.2834 |
| 11 | 0.1058 | 0.1133 | 0.1212 | 0.1293 | 0.1378 | 0.1465 | 0.1555 | 0.1648 | 0.1743 | 0.1841 |
| 12 | 0.0580 | 0.0629 | 0.0681 | 0.0735 | 0.0792 | 0.0852 | 0.0915 | 0.0980 | 0.1048 | 0.1119 |
| 13 | 0.0297 | 0.0327 | 0.0358 | 0.0391 | 0.0427 | 0.0464 | 0.0504 | 0.0546 | 0.0591 | 0.0638 |
| 14 | 0.0143 | 0.0159 | 0.0176 | 0.0195 | 0.0216 | 0.0238 | 0.0261 | 0.0286 | 0.0313 | 0.0342 |
| 15 | 0.0065 | 0.0073 | 0.0082 | 0.0092 | 0.0103 | 0.0114 | 0.0127 | 0.0141 | 0.0156 | 0.0173 |
| 16 | 0.0028 | 0.0031 | 0.0036 | 0.0041 | 0.0046 | 0.0052 | 0.0059 | 0.0066 | 0.0074 | 0.0082 |
| 17 | 0.0011 | 0.0013 | 0.0015 | 0.0017 | 0.0020 | 0.0022 | 0.0026 | 0.0029 | 0.0033 | 0.0037 |
| 18 | 0.0004 | 0.0005 | 0.0006 | 0.0007 | 0.0008 | 0.0009 | 0.0011 | 0.0012 | 0.0014 | 0.0016 |
| 19 | 0.0002 | 0.0002 | 0.0002 | 0.0003 | 0.0003 | 0.0004 | 0.0004 | 0.0005 | 0.0006 | 0.0006 |
| 20 | 0.0001 | 0.0001 | 0.0001 | 0.0001 | 0.0001 | 0.0001 | 0.0002 | 0.0002 | 0.0002 | 0.0003 |
| 21 | 0.0000 | 0.0000 | 0.0000 | 0.0000 | 0.0000 | 0.0000 | 0.0001 | 0.0001 | 0.0001 | 0.0001 |

| | θ | | | | | | | | | |
|---|---|---|---|---|---|---|---|---|---|---|
| *X* | 8.1 | 8.2 | 8.3 | 8.4 | 8.5 | 8.6 | 8.7 | 8.8 | 8.9 | 9.0 |
| 0 | 1.0000 | 1.0000 | 1.0000 | 1.0000 | 1.0000 | 1.0000 | 1.0000 | 1.0000 | 1.0000 | 1.0000 |
| 1 | 0.9997 | 0.9997 | 0.9998 | 0.9998 | 0.9998 | 0.9998 | 0.9998 | 0.9998 | 0.9999 | 0.9999 |
| 2 | 0.9972 | 0.9975 | 0.9977 | 0.9979 | 0.9981 | 0.9982 | 0.9984 | 0.9985 | 0.9987 | 0.9988 |
| 3 | 0.9873 | 0.9882 | 0.9891 | 0.9900 | 0.9907 | 0.9914 | 0.9921 | 0.9927 | 0.9932 | 0.9938 |
| 4 | 0.9604 | 0.9630 | 0.9654 | 0.9677 | 0.9699 | 0.9719 | 0.9738 | 0.9756 | 0.9772 | 0.9788 |
| 5 | 0.9060 | 0.9113 | 0.9163 | 0.9211 | 0.9256 | 0.9299 | 0.9340 | 0.9379 | 0.9416 | 0.9450 |
| 6 | 0.8178 | 0.8264 | 0.8347 | 0.8427 | 0.8504 | 0.8578 | 0.8648 | 0.8716 | 0.8781 | 0.8843 |
| 7 | 0.6987 | 0.7104 | 0.7219 | 0.7330 | 0.7438 | 0.7543 | 0.7645 | 0.7744 | 0.7840 | 0.7932 |
| 8 | 0.5609 | 0.5746 | 0.5881 | 0.6013 | 0.6144 | 0.6272 | 0.6398 | 0.6522 | 0.6643 | 0.6761 |
| 9 | 0.4214 | 0.4353 | 0.4493 | 0.4631 | 0.4769 | 0.4906 | 0.5042 | 0.5177 | 0.5311 | 0.5443 |
| 10 | 0.2959 | 0.3085 | 0.3212 | 0.3341 | 0.3470 | 0.3600 | 0.3731 | 0.3863 | 0.3994 | 0.4126 |
| 11 | 0.1942 | 0.2045 | 0.2150 | 0.2257 | 0.2366 | 0.2478 | 0.2591 | 0.2706 | 0.2822 | 0.2940 |
| 12 | 0.1193 | 0.1269 | 0.1348 | 0.1429 | 0.1513 | 0.1600 | 0.1689 | 0.1780 | 0.1874 | 0.1970 |
| 13 | 0.0687 | 0.0739 | 0.0793 | 0.0850 | 0.0909 | 0.0971 | 0.1035 | 0.1102 | 0.1171 | 0.1242 |
| 14 | 0.0372 | 0.0405 | 0.0439 | 0.0476 | 0.0514 | 0.0555 | 0.0597 | 0.0642 | 0.0689 | 0.0739 |
| 15 | 0.0190 | 0.0209 | 0.0229 | 0.0251 | 0.0274 | 0.0299 | 0.0325 | 0.0353 | 0.0383 | 0.0415 |
| 16 | 0.0092 | 0.0102 | 0.0113 | 0.0125 | 0.0138 | 0.0152 | 0.0168 | 0.0184 | 0.0202 | 0.0220 |
| 17 | 0.0042 | 0.0047 | 0.0053 | 0.0059 | 0.0066 | 0.0074 | 0.0082 | 0.0091 | 0.0101 | 0.0111 |
| 18 | 0.0018 | 0.0021 | 0.0023 | 0.0027 | 0.0030 | 0.0034 | 0.0038 | 0.0043 | 0.0048 | 0.0053 |
| 19 | 0.0008 | 0.0009 | 0.0010 | 0.0011 | 0.0013 | 0.0015 | 0.0017 | 0.0019 | 0.0022 | 0.0024 |
| 20 | 0.0003 | 0.0003 | 0.0004 | 0.0005 | 0.0005 | 0.0006 | 0.0007 | 0.0008 | 0.0009 | 0.0011 |
| 21 | 0.0001 | 0.0001 | 0.0002 | 0.0002 | 0.0002 | 0.0002 | 0.0003 | 0.0003 | 0.0004 | 0.0004 |
| 22 | 0.0000 | 0.0000 | 0.0001 | 0.0001 | 0.0001 | 0.0001 | 0.0001 | 0.0001 | 0.0002 | 0.0002 |
| 23 | 0.0000 | 0.0000 | 0.0000 | 0.0000 | 0.0000 | 0.0000 | 0.0000 | 0.0000 | 0.0001 | 0.0001 |

**Table A.3** (*continued*)

| X | θ 9.1 | 9.2 | 9.3 | 9.4 | 9.5 | 9.6 | 9.7 | 9.8 | 9.9 | 10 |
|---|---|---|---|---|---|---|---|---|---|---|
| 0 | 1.0000 | 1.0000 | 1.0000 | 1.0000 | 1.0000 | 1.0000 | 1.0000 | 1.0000 | 1.0000 | 1.0000 |
| 1 | 0.9999 | 0.9999 | 0.9999 | 0.9999 | 0.9999 | 0.9999 | 0.9999 | 0.9999 | 1.0000 | 1.0000 |
| 2 | 0.9989 | 0.9990 | 0.9991 | 0.9991 | 0.9992 | 0.9993 | 0.9993 | 0.9994 | 0.9995 | 0.9995 |
| 3 | 0.9942 | 0.9947 | 0.9951 | 0.9955 | 0.9958 | 0.9962 | 0.9965 | 0.9967 | 0.9970 | 0.9972 |
| 4 | 0.9802 | 0.9816 | 0.9828 | 0.9840 | 0.9851 | 0.9862 | 0.9871 | 0.9880 | 0.9889 | 0.9897 |
| 5 | 0.9483 | 0.9514 | 0.9544 | 0.9571 | 0.9597 | 0.9622 | 0.9645 | 0.9667 | 0.9688 | 0.9707 |
| 6 | 0.8902 | 0.8959 | 0.9014 | 0.9065 | 0.9115 | 0.9162 | 0.9207 | 0.9250 | 0.9290 | 0.9329 |
| 7 | 0.8022 | 0.8108 | 0.8192 | 0.8273 | 0.8351 | 0.8426 | 0.8498 | 0.8567 | 0.8634 | 0.8699 |
| 8 | 0.6877 | 0.6990 | 0.7101 | 0.7208 | 0.7313 | 0.7416 | 0.7515 | 0.7612 | 0.7706 | 0.7798 |
| 9 | 0.5574 | 0.5704 | 0.5832 | 0.5958 | 0.6082 | 0.6204 | 0.6324 | 0.6442 | 0.6558 | 0.6672 |
| 10 | 0.4258 | 0.4389 | 0.4521 | 0.4651 | 0.4782 | 0.4911 | 0.5040 | 0.5168 | 0.5295 | 0.5421 |
| 11 | 0.3059 | 0.3180 | 0.3301 | 0.3424 | 0.3547 | 0.3671 | 0.3795 | 0.3920 | 0.4045 | 0.4170 |
| 12 | 0.2068 | 0.2168 | 0.2270 | 0.2374 | 0.2480 | 0.2588 | 0.2697 | 0.2807 | 0.2919 | 0.3032 |
| 13 | 0.1316 | 0.1393 | 0.1471 | 0.1552 | 0.1636 | 0.1721 | 0.1809 | 0.1899 | 0.1991 | 0.2084 |
| 14 | 0.0790 | 0.0844 | 0.0900 | 0.0958 | 0.1019 | 0.1081 | 0.1147 | 0.1214 | 0.1284 | 0.1355 |
| 15 | 0.0448 | 0.0483 | 0.0529 | 0.0559 | 0.0600 | 0.0643 | 0.0688 | 0.0735 | 0.0784 | 0.0835 |
| 16 | 0.0240 | 0.0262 | 0.0285 | 0.0309 | 0.0335 | 0.0362 | 0.0391 | 0.0421 | 0.0454 | 0.0487 |
| 17 | 0.0122 | 0.0135 | 0.0148 | 0.0162 | 0.0177 | 0.0194 | 0.0211 | 0.0230 | 0.0249 | 0.0270 |
| 18 | 0.0059 | 0.0066 | 0.0073 | 0.0081 | 0.0089 | 0.0098 | 0.0108 | 0.0119 | 0.0130 | 0.0143 |
| 19 | 0.0027 | 0.0031 | 0.0034 | 0.0038 | 0.0043 | 0.0048 | 0.0053 | 0.0059 | 0.0065 | 0.0072 |
| 20 | 0.0012 | 0.0014 | 0.0015 | 0.0017 | 0.0020 | 0.0022 | 0.0025 | 0.0028 | 0.0031 | 0.0035 |
| 21 | 0.0005 | 0.0006 | 0.0007 | 0.0008 | 0.0009 | 0.0010 | 0.0011 | 0.0013 | 0.0014 | 0.0016 |
| 22 | 0.0002 | 0.0002 | 0.0003 | 0.0003 | 0.0004 | 0.0004 | 0.0005 | 0.0005 | 0.0006 | 0.0007 |
| 23 | 0.0001 | 0.0001 | 0.0001 | 0.0001 | 0.0001 | 0.0002 | 0.0002 | 0.0002 | 0.0003 | 0.0003 |
| 24 | 0.0000 | 0.0000 | 0.0000 | 0.0000 | 0.0001 | 0.0001 | 0.0001 | 0.0001 | 0.0001 | 0.0001 |

| X | θ 11 | 12 | 13 | 14 | 15 | 16 | 17 | 18 | 19 | 20 |
|---|---|---|---|---|---|---|---|---|---|---|
| 0 | 1.0000 | 1.0000 | 1.0000 | 1.0000 | 1.0000 | 1.0000 | 1.0000 | 1.0000 | 1.0000 | 1.0000 |
| 1 | 1.0000 | 1.0000 | 1.0000 | 1.0000 | 1.0000 | 1.0000 | 1.0000 | 1.0000 | 1.0000 | 1.0000 |
| 2 | 0.9998 | 0.9999 | 1.0000 | 1.0000 | 1.0000 | 1.0000 | 1.0000 | 1.0000 | 1.0000 | 1.0000 |
| 3 | 0.9988 | 0.9995 | 0.9998 | 0.9999 | 1.0000 | 1.0000 | 1.0000 | 1.0000 | 1.0000 | 1.0000 |
| 4 | 0.9951 | 0.9977 | 0.9990 | 0.9995 | 0.9998 | 0.9999 | 1.0000 | 1.0000 | 1.0000 | 1.0000 |
| 5 | 0.9849 | 0.9924 | 0.9963 | 0.9982 | 0.9991 | 0.9996 | 0.9998 | 0.9999 | 1.0000 | 1.0000 |
| 6 | 0.9625 | 0.9797 | 0.9893 | 0.9945 | 0.9972 | 0.9986 | 0.9993 | 0.9997 | 0.9998 | 0.9999 |
| 7 | 0.9214 | 0.9542 | 0.9741 | 0.9858 | 0.9924 | 0.9960 | 0.9979 | 0.9990 | 0.9995 | 0.9997 |
| 8 | 0.8568 | 0.9105 | 0.9460 | 0.9684 | 0.9820 | 0.9900 | 0.9946 | 0.9971 | 0.9985 | 0.9992 |
| 9 | 0.7680 | 0.8450 | 0.9002 | 0.9379 | 0.9626 | 0.9780 | 0.9874 | 0.9929 | 0.9961 | 0.9979 |
| 10 | 0.6595 | 0.7576 | 0.8342 | 0.8906 | 0.9301 | 0.9567 | 0.9739 | 0.9846 | 0.9911 | 0.9950 |
| 11 | 0.5401 | 0.6528 | 0.7483 | 0.8243 | 0.8815 | 0.9226 | 0.9509 | 0.9696 | 0.9817 | 0.9892 |
| 12 | 0.4207 | 0.5384 | 0.6468 | 0.7400 | 0.8152 | 0.8730 | 0.9153 | 0.9451 | 0.9653 | 0.9786 |
| 13 | 0.3113 | 0.4240 | 0.5369 | 0.6415 | 0.7324 | 0.8069 | 0.8650 | 0.9083 | 0.9394 | 0.9610 |
| 14 | 0.2187 | 0.3185 | 0.4270 | 0.5356 | 0.6368 | 0.7255 | 0.7991 | 0.8574 | 0.9016 | 0.9339 |
| 15 | 0.1460 | 0.2280 | 0.3249 | 0.4296 | 0.5343 | 0.6325 | 0.7192 | 0.7919 | 0.8503 | 0.8951 |
| 16 | 0.0926 | 0.1556 | 0.2364 | 0.3306 | 0.4319 | 0.5333 | 0.6285 | 0.7133 | 0.7852 | 0.8435 |
| 17 | 0.0559 | 0.1013 | 0.1645 | 0.2441 | 0.3359 | 0.4340 | 0.5323 | 0.6250 | 0.7080 | 0.7789 |
| 18 | 0.0332 | 0.0630 | 0.1095 | 0.1728 | 0.2511 | 0.3407 | 0.4360 | 0.5314 | 0.6216 | 0.7030 |
| 19 | 0.0177 | 0.0374 | 0.0698 | 0.1174 | 0.1805 | 0.2577 | 0.3450 | 0.4378 | 0.5305 | 0.6186 |

Table A.3 (*continued*)

| X | θ 11 | 12 | 13 | 14 | 15 | 16 | 17 | 18 | 19 | 20 |
|---|---|---|---|---|---|---|---|---|---|---|
| 20 | 0.0093 | 0.0213 | 0.0427 | 0.0765 | 0.1248 | 0.1878 | 0.2637 | 0.3491 | 0.4394 | 0.5297 |
| 21 | 0.0047 | 0.0116 | 0.0250 | 0.0479 | 0.0830 | 0.1318 | 0.1945 | 0.2693 | 0.3528 | 0.4409 |
| 22 | 0.0023 | 0.0061 | 0.0141 | 0.0288 | 0.0531 | 0.0892 | 0.1385 | 0.2009 | 0.2745 | 0.3563 |
| 23 | 0.0010 | 0.0030 | 0.0076 | 0.0167 | 0.0327 | 0.0582 | 0.0953 | 0.1449 | 0.2069 | 0.2794 |
| 24 | 0.0005 | 0.0015 | 0.0040 | 0.0093 | 0.0195 | 0.0367 | 0.0633 | 0.1011 | 0.1510 | 0.2125 |
| 25 | 0.0002 | 0.0007 | 0.0020 | 0.0050 | 0.0112 | 0.0223 | 0.0406 | 0.0683 | 0.1067 | 0.1568 |
| 26 | 0.0001 | 0.0003 | 0.0010 | 0.0026 | 0.0062 | 0.0131 | 0.0252 | 0.0446 | 0.0731 | 0.1122 |
| 27 | 0.0000 | 0.0001 | 0.0005 | 0.0013 | 0.0033 | 0.0075 | 0.0152 | 0.0282 | 0.0486 | 0.0779 |
| 28 | 0.0000 | 0.0001 | 0.0002 | 0.0006 | 0.0017 | 0.0041 | 0.0088 | 0.0173 | 0.0313 | 0.0525 |
| 29 | 0.0000 | 0.0000 | 0.0001 | 0.0003 | 0.0009 | 0.0022 | 0.0050 | 0.0103 | 0.0195 | 0.0343 |
| 30 | 0.0000 | 0.0000 | 0.0000 | 0.0001 | 0.0004 | 0.0011 | 0.0027 | 0.0059 | 0.0118 | 0.0218 |
| 31 | 0.0000 | 0.0000 | 0.0000 | 0.0001 | 0.0002 | 0.0006 | 0.0014 | 0.0033 | 0.0070 | 0.0135 |
| 32 | 0.0000 | 0.0000 | 0.0000 | 0.0000 | 0.0001 | 0.0003 | 0.0007 | 0.0018 | 0.0040 | 0.0081 |
| 33 | 0.0000 | 0.0000 | 0.0000 | 0.0000 | 0.0000 | 0.0001 | 0.0004 | 0.0010 | 0.0022 | 0.0047 |
| 34 | 0.0000 | 0.0000 | 0.0000 | 0.0000 | 0.0000 | 0.0001 | 0.0002 | 0.0005 | 0.0012 | 0.0027 |
| 35 | 0.0000 | 0.0000 | 0.0000 | 0.0000 | 0.0000 | 0.0000 | 0.0001 | 0.0002 | 0.0006 | 0.0015 |
| 36 | 0.0000 | 0.0000 | 0.0000 | 0.0000 | 0.0000 | 0.0000 | 0.0000 | 0.0001 | 0.0003 | 0.0008 |
| 37 | 0.0000 | 0.0000 | 0.0000 | 0.0000 | 0.0000 | 0.0000 | 0.0000 | 0.0001 | 0.0002 | 0.0004 |
| 38 | 0.0000 | 0.0000 | 0.0000 | 0.0000 | 0.0000 | 0.0000 | 0.0000 | 0.0000 | 0.0001 | 0.0002 |
| 39 | 0.0000 | 0.0000 | 0.0000 | 0.0000 | 0.0000 | 0.0000 | 0.0000 | 0.0000 | 0.0000 | 0.0001 |
| 40 | 0.0000 | 0.0000 | 0.0000 | 0.0000 | 0.0000 | 0.0000 | 0.0000 | 0.0000 | 0.0000 | 0.0001 |

Source: N. L. Johnson and F. C. Leone, *Statistics and Experimental Design in Engineering and the Physical Sciences*, I. John Wiley and Sons, Inc., New York, 1964, pp. 455–59. Originally from *Poisson's Exponential Limit* by E. C. Molina, Copyright © 1942 by Litton Educational Publishing, Inc., by permission of D. Van Nostrand Company, Inc.

## Table A.4 Cumulative Normal Distribution

$$F(x) = \int_{-\infty}^{x} \frac{1}{\sqrt{2\pi}} e^{-t^2/2} \, dt$$

| $x$ | .00 | .01 | .02 | .03 | .04 | .05 | .06 | .07 | .08 | .09 |
|---|---|---|---|---|---|---|---|---|---|---|
| .0 | .5000 | .5040 | .5080 | .5120 | .5160 | .5199 | .5239 | .5279 | .5319 | .5359 |
| .1 | .5398 | .5438 | .5478 | .5517 | .5557 | .5596 | .5636 | .5675 | .5714 | .5753 |
| .2 | .5793 | .5832 | .5871 | .5910 | .5948 | .5987 | .6026 | .6064 | .6103 | .6141 |
| .3 | .6179 | .6217 | .6255 | .6293 | .6331 | .6368 | .6406 | .6443 | .6480 | .6517 |
| .4 | .6554 | .6591 | .6628 | .6664 | .6700 | .6736 | .6772 | .6808 | .6844 | .6879 |
| .5 | .6915 | .6950 | .6985 | .7019 | .7054 | .7088 | .7123 | .7157 | .7190 | .7224 |
| .6 | .7257 | .7291 | .7324 | .7357 | .7389 | .7422 | .7454 | .7486 | .7517 | .7549 |
| .7 | .7580 | .7611 | .7642 | .7673 | .7704 | .7734 | .7764 | .7794 | .7823 | .7852 |
| .8 | .7881 | .7910 | .7939 | .7967 | .7995 | .8023 | .8051 | .8078 | .8106 | .8133 |
| .9 | .8159 | .8186 | .8212 | .8238 | .8264 | .8289 | .8315 | .8340 | .8365 | .8389 |
| 1.0 | .8413 | .8438 | .8461 | .8485 | .8508 | .8531 | .8554 | .8577 | .8599 | .8621 |
| 1.1 | .8643 | .8665 | .8686 | .8708 | .8729 | .8749 | .8770 | .8790 | .8810 | .8830 |
| 1.2 | .8849 | .8869 | .8888 | .8907 | .8925 | .8944 | .8962 | .8980 | .8997 | .9015 |
| 1.3 | .9032 | .9049 | .9066 | .9082 | .9099 | .9115 | .9131 | .9147 | .9162 | .9177 |
| 1.4 | .9192 | .9207 | .9222 | .9236 | .9251 | .9265 | .9279 | .9292 | .9306 | .9319 |
| 1.5 | .9332 | .9345 | .9357 | .9370 | .9382 | .9394 | .9406 | .9418 | .9429 | .9441 |
| 1.6 | .9452 | .9463 | .9474 | .9484 | .9495 | .9505 | .9515 | .9525 | .9535 | .9545 |
| 1.7 | .9554 | .9564 | .9573 | .9582 | .9591 | .9599 | .9608 | .9616 | .9625 | .9633 |
| 1.8 | .9641 | .9649 | .9656 | .9664 | .9671 | .9678 | .9686 | .9693 | .9699 | .9706 |
| 1.9 | .9713 | .9719 | .9726 | .9732 | .9738 | .9744 | .9750 | .9756 | .9761 | .9767 |
| 2.0 | .9772 | .9778 | .9783 | .9788 | .9793 | .9798 | .9803 | .9808 | .9812 | .9817 |
| 2.1 | .9821 | .9826 | .9830 | .9834 | .9838 | .9842 | .9846 | .9850 | .9854 | .9857 |
| 2.2 | .9861 | .9864 | .9868 | .9871 | .9875 | .9878 | .9881 | .9884 | .9887 | .9890 |
| 2.3 | .9893 | .9896 | .9898 | .9901 | .9904 | .9906 | .9909 | .9911 | .9913 | .9916 |
| 2.4 | .9918 | .9920 | .9922 | .9925 | .9927 | .9929 | .9931 | .9932 | .9934 | .9936 |
| 2.5 | .9938 | .9940 | .9941 | .9943 | .9945 | .9946 | .9948 | .9949 | .9951 | .9952 |
| 2.6 | .9953 | .9955 | .9956 | .9957 | .9959 | .9960 | .9961 | .9962 | .9963 | .9964 |
| 2.7 | .9965 | .9966 | .9967 | .9968 | .9969 | .9970 | .9971 | .9972 | .9973 | .9974 |
| 2.8 | .9974 | .9975 | .9976 | .9977 | .9977 | .9978 | .9979 | .9979 | .9980 | .9981 |
| 2.9 | .9981 | .9982 | .9982 | .9983 | .9984 | .9984 | .9985 | .9985 | .9986 | .9986 |
| 3.0 | .9987 | .9987 | .9987 | .9988 | .9988 | .9989 | .9989 | .9989 | .9990 | .9990 |
| 3.1 | .9990 | .9991 | .9991 | .9991 | .9992 | .9992 | .9992 | .9992 | .9993 | .9993 |
| 3.2 | .9993 | .9993 | .9994 | .9994 | .9994 | .9994 | .9994 | .9995 | .9995 | .9995 |
| 3.3 | .9995 | .9995 | .9995 | .9996 | .9996 | .9996 | .9996 | .9996 | .9996 | .9997 |
| 3.4 | .9997 | .9997 | .9997 | .9997 | .9997 | .9997 | .9997 | .9997 | .9997 | .9998 |

| $x$ | 1.282 | 1.645 | 1.960 | 2.326 | 2.576 | 3.090 | 3.291 | 3.891 | 4.417 |
|---|---|---|---|---|---|---|---|---|---|
| $F(x)$ | .90 | .95 | .975 | .99 | .995 | .999 | .9995 | .99995 | .999995 |
| $2[1 - F(x)]$ | .20 | .10 | .05 | .02 | .01 | .002 | .001 | .0001 | .00001 |

Source: A. M. Mood and F. A. Graybill, *Introduction to the Theory of Statistics*. McGraw-Hill, New York, 1963, p. 431.

**Table A.5** Cumulative Chi-Square Distribution

$$F(u) = \int_0^u \frac{x^{(n-2)/2} e^{-x/2}\, dx}{2^{n/2}[(n-2)/2]!}$$

| n \ F | .005 | .010 | .025 | .050 | .100 | .250 | .500 | .750 | .900 | .950 | .975 | .990 | .995 |
|---|---|---|---|---|---|---|---|---|---|---|---|---|---|
| 1 | $.0^4393$ | $.0^3157$ | $.0^3982$ | $.0^2393$ | .0158 | .102 | .455 | 1.32 | 2.71 | 3.84 | 5.02 | 6.63 | 7.88 |
| 2 | .0100 | .0201 | .0506 | .103 | .211 | .575 | 1.39 | 2.77 | 4.61 | 5.99 | 7.38 | 9.21 | 10.6 |
| 3 | .0717 | .115 | .216 | .352 | .584 | 1.21 | 2.37 | 4.11 | 6.25 | 7.81 | 9.35 | 11.3 | 12.8 |
| 4 | .207 | .297 | .484 | .711 | 1.06 | 1.92 | 3.36 | 5.39 | 7.78 | 9.49 | 11.1 | 13.3 | 14.9 |
| 5 | .412 | .554 | .831 | 1.15 | 1.61 | 2.67 | 4.35 | 6.63 | 9.24 | 11.1 | 12.8 | 15.1 | 16.7 |
| 6 | .676 | .872 | 1.24 | 1.64 | 2.20 | 3.45 | 5.35 | 7.84 | 10.6 | 12.6 | 14.4 | 16.8 | 18.5 |
| 7 | .989 | 1.24 | 1.69 | 2.17 | 2.83 | 4.25 | 6.35 | 9.04 | 12.0 | 14.1 | 16.0 | 18.5 | 20.3 |
| 8 | 1.34 | 1.65 | 2.18 | 2.73 | 3.49 | 5.07 | 7.34 | 10.2 | 13.4 | 15.5 | 17.5 | 20.1 | 22.0 |
| 9 | 1.73 | 2.09 | 2.70 | 3.33 | 4.17 | 5.90 | 8.34 | 11.4 | 14.7 | 16.9 | 19.0 | 21.7 | 23.6 |
| 10 | 2.16 | 2.56 | 3.25 | 3.94 | 4.87 | 6.74 | 9.34 | 12.5 | 16.0 | 18.3 | 20.5 | 23.2 | 25.2 |
| 11 | 2.60 | 3.05 | 3.82 | 4.57 | 5.58 | 7.58 | 10.3 | 13.7 | 17.3 | 19.7 | 21.9 | 24.7 | 26.8 |
| 12 | 3.07 | 3.57 | 4.40 | 5.23 | 6.30 | 8.44 | 11.3 | 14.8 | 18.5 | 21.0 | 23.3 | 26.2 | 28.3 |
| 13 | 3.57 | 4.11 | 5.01 | 5.89 | 7.04 | 9.30 | 12.3 | 16.0 | 19.8 | 22.4 | 24.7 | 27.7 | 29.8 |
| 14 | 4.07 | 4.66 | 5.63 | 6.57 | 7.79 | 10.2 | 13.3 | 17.1 | 21.1 | 23.7 | 26.1 | 29.1 | 31.3 |
| 15 | 4.60 | 5.23 | 6.26 | 7.26 | 8.55 | 11.0 | 14.3 | 18.2 | 22.3 | 25.0 | 27.5 | 30.6 | 32.8 |
| 16 | 5.14 | 5.81 | 6.91 | 7.96 | 9.31 | 11.9 | 15.3 | 19.4 | 23.5 | 26.3 | 28.8 | 32.0 | 34.3 |
| 17 | 5.70 | 6.41 | 7.56 | 8.67 | 10.1 | 12.8 | 16.3 | 20.5 | 24.8 | 27.6 | 30.2 | 33.4 | 35.7 |
| 18 | 6.26 | 7.01 | 8.23 | 9.39 | 10.9 | 13.7 | 17.3 | 21.6 | 26.0 | 28.9 | 31.5 | 34.8 | 37.2 |
| 19 | 6.84 | 7.63 | 8.91 | 10.1 | 11.7 | 14.6 | 18.3 | 22.7 | 27.2 | 30.1 | 32.9 | 36.2 | 38.6 |
| 20 | 7.43 | 8.26 | 9.59 | 10.9 | 12.4 | 15.5 | 19.3 | 23.8 | 28.4 | 31.4 | 34.2 | 37.6 | 40.0 |
| 21 | 8.03 | 8.90 | 10.3 | 11.6 | 13.2 | 16.3 | 20.3 | 24.9 | 29.6 | 32.7 | 35.5 | 38.9 | 41.4 |
| 22 | 8.64 | 9.54 | 11.0 | 12.3 | 14.0 | 17.2 | 21.3 | 26.0 | 30.8 | 33.9 | 36.8 | 40.3 | 42.8 |
| 23 | 9.26 | 10.2 | 11.7 | 13.1 | 14.8 | 18.1 | 22.3 | 27.1 | 32.0 | 35.2 | 38.1 | 41.6 | 44.2 |
| 24 | 9.89 | 10.9 | 12.4 | 13.8 | 15.7 | 19.0 | 23.3 | 28.2 | 33.2 | 36.4 | 39.4 | 43.0 | 45.6 |
| 25 | 10.5 | 11.5 | 13.1 | 14.6 | 16.5 | 19.9 | 24.3 | 29.3 | 34.4 | 37.7 | 40.6 | 44.3 | 46.9 |
| 26 | 11.2 | 12.2 | 13.8 | 15.4 | 17.3 | 20.8 | 25.3 | 30.4 | 35.6 | 38.9 | 41.9 | 45.6 | 48.3 |
| 27 | 11.8 | 12.9 | 14.6 | 16.2 | 18.1 | 21.7 | 26.3 | 31.5 | 36.7 | 40.1 | 43.2 | 47.0 | 49.6 |
| 28 | 12.5 | 13.6 | 15.3 | 16.9 | 18.9 | 22.7 | 27.3 | 32.6 | 37.9 | 41.3 | 44.5 | 48.3 | 51.0 |
| 29 | 13.1 | 14.3 | 16.0 | 17.7 | 19.8 | 23.6 | 28.3 | 33.7 | 39.1 | 42.6 | 45.7 | 49.6 | 52.3 |
| 30 | 13.8 | 15.0 | 16.8 | 18.5 | 20.6 | 24.5 | 29.3 | 34.8 | 40.3 | 43.8 | 47.0 | 50.9 | 53.7 |

Source: A. M. Mood and F. A. Graybill, *Introduction to the Theory of Statistics*. McGraw-Hill, New York, 1963, p. 432. By Permission of the publishers of *Biometrika*. (See Vol. 32, 1941.)

**Table A.6** Cumulative Student's $t$ Distribution

$$F(t) = \int_{-\infty}^{t} \frac{\left(\frac{n-1}{2}\right)!}{\left(\frac{n-2}{2}\right)! \sqrt{\pi n}\left(1 + \frac{x^2}{n}\right)^{(n+1)/2}} dx$$

| $n$ \ $F$ | .75 | .90 | .95 | .975 | .99 | .995 | .9995 |
|---|---|---|---|---|---|---|---|
| 1 | 1.000 | 3.078 | 6.314 | 12.706 | 31.821 | 63.657 | 636.619 |
| 2 | .816 | 1.886 | 2.920 | 4.303 | 6.965 | 9.925 | 31.598 |
| 3 | .765 | 1.638 | 2.353 | 3.182 | 4.541 | 5.841 | 12.941 |
| 4 | .741 | 1.533 | 2.132 | 2.776 | 3.747 | 4.604 | 8.610 |
| 5 | .727 | 1.476 | 2.015 | 2.571 | 3.365 | 4.032 | 6.859 |
| 6 | .718 | 1.440 | 1.943 | 2.447 | 3.143 | 3.707 | 5.959 |
| 7 | .711 | 1.415 | 1.895 | 2.365 | 2.998 | 3.499 | 5.405 |
| 8 | .706 | 1.397 | 1.860 | 2.306 | 2.896 | 3.355 | 5.041 |
| 9 | .703 | 1.383 | 1.833 | 2.262 | 2.821 | 3.250 | 4.781 |
| 10 | .700 | 1.372 | 1.812 | 2.228 | 2.764 | 3.169 | 4.587 |
| 11 | .697 | 1.363 | 1.796 | 2.201 | 2.718 | 3.106 | 4.437 |
| 12 | .695 | 1.356 | 1.782 | 2.179 | 2.681 | 3.055 | 4.318 |
| 13 | .694 | 1.350 | 1.771 | 2.160 | 2.650 | 3.012 | 4.221 |
| 14 | .692 | 1.345 | 1.761 | 2.145 | 2.624 | 2.977 | 4.140 |
| 15 | .691 | 1.341 | 1.753 | 2.131 | 2.602 | 2.947 | 4.073 |
| 16 | .690 | 1.337 | 1.746 | 2.120 | 2.583 | 2.921 | 4.015 |
| 17 | .689 | 1.333 | 1.740 | 2.110 | 2.567 | 2.898 | 3.965 |
| 18 | .688 | 1.330 | 1.734 | 2.101 | 2.552 | 2.878 | 3.992 |
| 19 | .688 | 1.328 | 1.729 | 2.093 | 2.539 | 2.861 | 3.883 |
| 20 | .687 | 1.325 | 1.725 | 2.086 | 2.528 | 2.845 | 3.850 |
| 21 | .686 | 1.323 | 1.721 | 2.080 | 2.518 | 2.831 | 3.819 |
| 22 | .686 | 1.321 | 1.717 | 2.074 | 2.508 | 2.819 | 3.792 |
| 23 | .685 | 1.319 | 1.714 | 2.069 | 2.500 | 2.807 | 3.767 |
| 24 | .685 | 1.318 | 1.711 | 2.064 | 2.492 | 2.797 | 3.745 |
| 25 | .684 | 1.316 | 1.708 | 2.060 | 2.485 | 2.787 | 3.725 |
| 26 | .684 | 1.315 | 1.706 | 2.056 | 2.479 | 2.779 | 3.707 |
| 27 | .684 | 1.314 | 1.703 | 2.052 | 2.473 | 2.771 | 3.690 |
| 28 | .683 | 1.313 | 1.701 | 2.048 | 2.467 | 2.763 | 3.674 |
| 29 | .683 | 1.311 | 1.699 | 2.045 | 2.462 | 2.756 | 3.659 |
| 30 | .683 | 1.310 | 1.697 | 2.042 | 2.457 | 2.750 | 3.646 |
| 40 | .681 | 1.303 | 1.684 | 2.021 | 2.423 | 2.704 | 3.551 |
| 60 | .679 | 1.296 | 1.671 | 2.000 | 2.390 | 2.660 | 3.460 |
| 120 | .677 | 1.289 | 1.658 | 1.980 | 2.358 | 2.617 | 3.373 |
| $\infty$ | .674 | 1.282 | 1.645 | 1.960 | 2.326 | 2.576 | 3.291 |

Source: Table III of Fisher and Yates: *Statistical Tables for Biological, Agricultural and Medical Research*, published by Oliver and Boyd Ltd., Edinburgh, and by permission of the authors and publishers. (See also Mood and Graybill, *Introduction to the Theory of Statistics*, p. 433.)

**Table A.7** Cumulative $F$ Distribution

$m$ degrees of freedom in numerator; $n$ in denominator

$$G(F) = \int_0^F \frac{[(m+n-2)/2]!\,m^{m/2} n^{n/2} x^{(m-2)/2} (n+mx)^{-(m+n)/2}}{[(m-2)/2]!\,[(n-2)/2]!}\,dx$$

| G | n \| m | 1 | 2 | 3 | 4 | 5 | 6 | 7 | 8 | 9 | 10 | 12 | 15 | 20 | 30 | 60 | 120 | ∞ |
|---|---|---|---|---|---|---|---|---|---|---|---|---|---|---|---|---|---|---|
| .90 | | 39.9 | 49.5 | 53.6 | 55.8 | 57.2 | 58.2 | 58.9 | 59.4 | 59.9 | 60.2 | 60.7 | 61.2 | 61.7 | 62.3 | 62.8 | 63.1 | 63.3 |
| .95 | | 161 | 200 | 216 | 225 | 230 | 234 | 237 | 239 | 241 | 242 | 244 | 246 | 248 | 250 | 252 | 253 | 254 |
| .975 | 1 | 648 | 800 | 864 | 900 | 922 | 937 | 948 | 957 | 963 | 969 | 977 | 985 | 993 | 1000 | 1010 | 1010 | 1020 |
| .99 | | 4,050 | 5,000 | 5,400 | 5,620 | 5,760 | 5,860 | 5,930 | 5,980 | 6,020 | 6,060 | 6,110 | 6,160 | 6,210 | 6,260 | 6,310 | 6,340 | 6,370 |
| .995 | | 16,200 | 20,000 | 21,600 | 22,500 | 23,100 | 23,400 | 23,700 | 23,900 | 24,100 | 24,200 | 24,400 | 24,600 | 24,800 | 25,000 | 25,200 | 25,400 | 25,500 |
| .90 | | 8.53 | 9.00 | 9.16 | 9.24 | 9.29 | 9.33 | 9.35 | 9.37 | 9.38 | 9.39 | 9.41 | 9.42 | 9.44 | 9.46 | 9.47 | 9.48 | 9.49 |
| .95 | | 18.5 | 19.0 | 19.2 | 19.2 | 19.3 | 19.3 | 19.4 | 19.4 | 19.4 | 19.4 | 19.4 | 19.4 | 19.5 | 19.5 | 19.5 | 19.5 | 19.5 |
| .975 | 2 | 38.5 | 39.0 | 39.2 | 39.2 | 39.3 | 39.3 | 39.4 | 39.4 | 39.4 | 39.4 | 39.4 | 39.4 | 39.4 | 39.5 | 39.5 | 39.5 | 39.5 |
| .99 | | 98.5 | 99.0 | 99.2 | 99.2 | 99.3 | 99.3 | 99.4 | 99.4 | 99.4 | 99.4 | 99.4 | 99.4 | 99.4 | 99.5 | 99.5 | 99.5 | 99.5 |
| .995 | | 199 | 199 | 199 | 199 | 199 | 199 | 199 | 199 | 199 | 199 | 199 | 199 | 199 | 199 | 199 | 199 | 199 |
| .90 | | 5.54 | 5.46 | 5.39 | 5.34 | 5.31 | 5.28 | 5.27 | 5.25 | 5.24 | 5.23 | 5.22 | 5.20 | 5.18 | 5.17 | 5.15 | 5.14 | 5.13 |
| .95 | | 10.1 | 9.55 | 9.28 | 9.12 | 9.01 | 8.94 | 8.89 | 8.85 | 8.81 | 8.79 | 8.74 | 8.70 | 8.66 | 8.62 | 8.57 | 8.55 | 8.53 |
| .975 | 3 | 17.4 | 16.0 | 15.4 | 15.1 | 14.9 | 14.7 | 14.6 | 14.5 | 14.5 | 14.4 | 14.3 | 14.3 | 14.2 | 14.1 | 14.0 | 13.9 | 13.9 |
| .99 | | 34.1 | 30.8 | 29.5 | 28.7 | 28.2 | 27.9 | 27.7 | 27.5 | 27.3 | 27.2 | 27.1 | 26.9 | 26.7 | 26.5 | 26.3 | 26.2 | 26.1 |
| .995 | | 55.6 | 49.8 | 47.5 | 46.2 | 45.4 | 44.8 | 44.4 | 44.1 | 43.9 | 43.7 | 43.4 | 43.1 | 42.8 | 42.5 | 42.1 | 42.0 | 41.8 |
| .90 | | 4.54 | 4.32 | 4.19 | 4.11 | 4.05 | 4.01 | 3.98 | 3.95 | 3.93 | 3.92 | 3.90 | 3.87 | 3.84 | 3.82 | 3.79 | 3.78 | 3.76 |
| .95 | | 7.71 | 6.94 | 6.59 | 6.39 | 6.26 | 6.16 | 6.09 | 6.04 | 6.00 | 5.96 | 5.91 | 5.86 | 5.80 | 5.75 | 5.69 | 5.66 | 5.63 |
| .975 | 4 | 12.2 | 10.6 | 9.98 | 9.60 | 9.36 | 9.20 | 9.07 | 8.98 | 8.90 | 8.84 | 8.75 | 8.66 | 8.56 | 8.46 | 8.36 | 8.31 | 8.26 |
| .99 | | 21.2 | 18.0 | 16.7 | 16.0 | 15.5 | 15.2 | 15.0 | 14.8 | 14.7 | 14.5 | 14.4 | 14.2 | 14.0 | 13.8 | 13.7 | 13.6 | 13.5 |
| .995 | | 31.3 | 26.3 | 24.3 | 23.2 | 22.5 | 22.0 | 21.6 | 21.4 | 21.1 | 21.0 | 20.7 | 20.4 | 20.2 | 19.9 | 19.6 | 19.5 | 19.3 |
| .90 | | 4.06 | 3.78 | 3.62 | 3.52 | 3.45 | 3.40 | 3.37 | 3.34 | 3.32 | 3.30 | 3.27 | 3.24 | 3.21 | 3.17 | 3.14 | 3.12 | 3.11 |
| .95 | | 6.61 | 5.79 | 5.41 | 5.19 | 5.05 | 4.95 | 4.88 | 4.82 | 4.77 | 4.74 | 4.68 | 4.62 | 4.56 | 4.50 | 4.43 | 4.40 | 4.37 |
| .975 | 5 | 10.0 | 8.43 | 7.76 | 7.39 | 7.15 | 6.98 | 6.85 | 6.76 | 6.68 | 6.62 | 6.52 | 6.43 | 6.33 | 6.23 | 6.12 | 6.07 | 6.02 |
| .99 | | 16.3 | 13.3 | 12.1 | 11.4 | 11.0 | 10.7 | 10.5 | 10.3 | 10.2 | 10.1 | 9.89 | 9.72 | 9.55 | 9.38 | 9.20 | 9.11 | 9.02 |
| .995 | | 22.8 | 18.3 | 16.5 | 15.6 | 14.9 | 14.5 | 14.2 | 14.0 | 13.8 | 13.6 | 13.4 | 13.1 | 12.9 | 12.7 | 12.4 | 12.3 | 12.1 |

**Table A.7** *(continued)*

| $G$ | $n$ | $m$: 1 | 2 | 3 | 4 | 5 | 6 | 7 | 8 | 9 | 10 | 12 | 15 | 20 | 30 | 60 | 120 | ∞ |
|---|---|---|---|---|---|---|---|---|---|---|---|---|---|---|---|---|---|---|
| .90 | | 3.78 | 3.46 | 3.29 | 3.18 | 3.11 | 3.05 | 3.01 | 2.98 | 2.96 | 2.94 | 2.90 | 2.87 | 2.84 | 2.80 | 2.76 | 2.74 | 2.72 |
| .95 | | 5.99 | 5.14 | 4.76 | 4.53 | 4.39 | 4.28 | 4.21 | 4.15 | 4.10 | 4.06 | 4.00 | 3.94 | 3.87 | 3.81 | 3.74 | 3.70 | 3.67 |
| .975 | 6 | 8.81 | 7.26 | 6.60 | 6.23 | 5.99 | 5.82 | 5.70 | 5.60 | 5.52 | 5.46 | 5.37 | 5.27 | 5.17 | 5.07 | 4.96 | 4.90 | 4.85 |
| .99 | | 13.7 | 10.9 | 9.78 | 9.15 | 8.75 | 8.47 | 8.26 | 8.10 | 7.98 | 7.87 | 7.72 | 7.56 | 7.40 | 7.23 | 7.06 | 6.97 | 6.88 |
| .995 | | 18.6 | 14.5 | 12.9 | 12.0 | 11.5 | 11.1 | 10.8 | 10.6 | 10.4 | 10.2 | 10.0 | 9.81 | 9.59 | 9.36 | 9.12 | 9.00 | 8.88 |
| .90 | | 3.59 | 3.26 | 3.07 | 2.96 | 2.88 | 2.83 | 2.78 | 2.75 | 2.72 | 2.70 | 2.67 | 2.63 | 2.59 | 2.56 | 2.51 | 2.49 | 2.47 |
| .95 | | 5.59 | 4.74 | 4.35 | 4.12 | 3.97 | 3.87 | 3.79 | 3.73 | 3.68 | 3.64 | 3.57 | 3.51 | 3.44 | 3.38 | 3.30 | 3.27 | 3.23 |
| .975 | 7 | 8.07 | 6.54 | 5.89 | 5.52 | 5.29 | 5.12 | 4.99 | 4.90 | 4.82 | 4.76 | 4.67 | 4.57 | 4.47 | 4.36 | 4.25 | 4.20 | 4.14 |
| .99 | | 12.2 | 9.55 | 8.45 | 7.85 | 7.46 | 7.19 | 6.99 | 6.84 | 6.72 | 6.62 | 6.47 | 6.31 | 6.16 | 5.99 | 5.82 | 5.74 | 5.65 |
| .995 | | 16.2 | 12.4 | 10.9 | 10.1 | 9.52 | 9.16 | 8.89 | 8.68 | 8.51 | 8.38 | 8.18 | 7.97 | 7.75 | 7.53 | 7.31 | 7.19 | 7.08 |
| .90 | | 3.46 | 3.11 | 2.92 | 2.81 | 2.73 | 2.67 | 2.62 | 2.59 | 2.56 | 2.54 | 2.50 | 2.46 | 2.42 | 2.38 | 2.34 | 2.31 | 2.29 |
| .95 | | 5.32 | 4.46 | 4.07 | 3.84 | 3.69 | 3.58 | 3.50 | 3.44 | 3.39 | 3.35 | 3.28 | 3.22 | 3.15 | 3.08 | 3.01 | 2.97 | 2.93 |
| .975 | 8 | 7.57 | 6.06 | 5.42 | 5.05 | 4.82 | 4.65 | 4.53 | 4.43 | 4.36 | 4.30 | 4.20 | 4.10 | 4.00 | 3.89 | 3.78 | 3.73 | 3.67 |
| .99 | | 11.3 | 8.65 | 7.59 | 7.01 | 6.63 | 6.37 | 6.18 | 6.03 | 5.91 | 5.81 | 5.67 | 5.52 | 5.36 | 5.20 | 5.03 | 4.95 | 4.86 |
| .995 | | 14.7 | 11.0 | 9.60 | 8.81 | 8.30 | 7.95 | 7.69 | 7.50 | 7.34 | 7.21 | 7.01 | 6.81 | 6.61 | 6.40 | 6.18 | 6.06 | 5.95 |
| .90 | | 3.36 | 3.01 | 2.81 | 2.69 | 2.61 | 2.55 | 2.51 | 2.47 | 2.44 | 2.42 | 2.38 | 2.34 | 2.30 | 2.25 | 2.21 | 2.18 | 2.16 |
| .95 | | 5.12 | 4.26 | 3.86 | 3.63 | 3.48 | 3.37 | 3.29 | 3.23 | 3.18 | 3.14 | 3.07 | 3.01 | 2.94 | 2.86 | 2.79 | 2.75 | 2.71 |
| .975 | 9 | 7.21 | 5.71 | 5.08 | 4.72 | 4.48 | 4.32 | 4.20 | 4.10 | 4.03 | 3.96 | 3.87 | 3.77 | 3.67 | 3.56 | 3.45 | 3.39 | 3.33 |
| .99 | | 10.6 | 8.02 | 6.99 | 6.42 | 6.06 | 5.80 | 5.61 | 5.47 | 5.35 | 5.26 | 5.11 | 4.96 | 4.81 | 4.65 | 4.48 | 4.40 | 4.31 |
| .995 | | 13.6 | 10.1 | 8.72 | 7.96 | 7.47 | 7.13 | 6.88 | 6.69 | 6.54 | 6.42 | 6.23 | 6.03 | 5.83 | 5.62 | 5.41 | 5.30 | 5.19 |
| .90 | | 3.29 | 2.92 | 2.73 | 2.61 | 2.52 | 2.46 | 2.41 | 2.38 | 2.35 | 2.32 | 2.28 | 2.24 | 2.20 | 2.15 | 2.11 | 2.08 | 2.06 |
| .95 | | 4.96 | 4.10 | 3.71 | 3.48 | 3.33 | 3.22 | 3.14 | 3.07 | 3.02 | 2.98 | 2.91 | 2.84 | 2.77 | 2.70 | 2.62 | 2.58 | 2.54 |
| .975 | 10 | 6.94 | 5.46 | 4.83 | 4.47 | 4.24 | 4.07 | 3.95 | 3.85 | 3.78 | 3.72 | 3.62 | 3.52 | 3.42 | 3.31 | 3.20 | 3.14 | 3.08 |
| .99 | | 10.0 | 7.56 | 6.55 | 5.99 | 5.64 | 5.39 | 5.20 | 5.06 | 4.94 | 4.85 | 4.71 | 4.56 | 4.41 | 4.25 | 4.08 | 4.00 | 3.91 |
| .995 | | 12.8 | 9.43 | 8.08 | 7.34 | 6.87 | 6.54 | 6.30 | 6.12 | 5.97 | 5.85 | 5.66 | 5.47 | 5.27 | 5.07 | 4.86 | 4.75 | 4.64 |
| .90 | | 3.18 | 2.81 | 2.61 | 2.48 | 2.39 | 2.33 | 2.28 | 2.24 | 2.21 | 2.19 | 2.15 | 2.10 | 2.06 | 2.01 | 1.96 | 1.93 | 1.90 |
| .95 | | 4.75 | 3.89 | 3.49 | 3.26 | 3.11 | 3.00 | 2.91 | 2.85 | 2.80 | 2.75 | 2.69 | 2.62 | 2.54 | 2.47 | 2.38 | 2.34 | 2.30 |
| .975 | 12 | 6.55 | 5.10 | 4.47 | 4.12 | 3.89 | 3.73 | 3.61 | 3.51 | 3.44 | 3.37 | 3.28 | 3.18 | 3.07 | 2.96 | 2.85 | 2.79 | 2.72 |
| .99 | | 9.33 | 6.93 | 5.95 | 5.41 | 5.06 | 4.82 | 4.64 | 4.50 | 4.39 | 4.30 | 4.16 | 4.01 | 3.86 | 3.70 | 3.54 | 3.45 | 3.36 |
| .995 | | 11.8 | 8.51 | 7.23 | 6.52 | 6.07 | 5.76 | 5.52 | 5.35 | 5.20 | 5.09 | 4.91 | 4.72 | 4.53 | 4.33 | 4.12 | 4.01 | 3.90 |

## Table A.7 (continued)

| G | n \ m | 1 | 2 | 3 | 4 | 5 | 6 | 7 | 8 | 9 | 10 | 12 | 15 | 20 | 30 | 60 | 120 | ∞ |
|---|---|---|---|---|---|---|---|---|---|---|---|---|---|---|---|---|---|---|
| .90 | | 3.07 | 2.70 | 2.49 | 2.36 | 2.27 | 2.21 | 2.16 | 2.12 | 2.09 | 2.06 | 2.02 | 1.97 | 1.92 | 1.87 | 1.82 | 1.79 | 1.76 |
| .95 | | 4.54 | 3.68 | 3.29 | 3.06 | 2.90 | 2.79 | 2.71 | 2.64 | 2.59 | 2.54 | 2.48 | 2.40 | 2.33 | 2.25 | 2.16 | 2.11 | 2.07 |
| .975 | 15 | 6.20 | 4.77 | 4.15 | 3.80 | 3.58 | 3.41 | 3.29 | 3.20 | 3.12 | 3.06 | 2.96 | 2.86 | 2.76 | 2.64 | 2.52 | 2.46 | 2.40 |
| .99 | | 8.68 | 6.36 | 5.42 | 4.89 | 4.56 | 4.32 | 4.14 | 4.00 | 3.89 | 3.80 | 3.67 | 3.52 | 3.37 | 3.21 | 3.05 | 2.96 | 2.87 |
| .995 | | 10.8 | 7.70 | 6.48 | 5.80 | 5.37 | 5.07 | 4.85 | 4.67 | 4.54 | 4.42 | 4.25 | 4.07 | 3.88 | 3.69 | 3.48 | 3.37 | 3.26 |
| .90 | | 2.97 | 2.59 | 2.38 | 2.25 | 2.16 | 2.09 | 2.04 | 2.00 | 1.96 | 1.94 | 1.89 | 1.84 | 1.79 | 1.74 | 1.68 | 1.64 | 1.61 |
| .95 | | 4.35 | 3.49 | 3.10 | 2.87 | 2.71 | 2.60 | 2.51 | 2.45 | 2.39 | 2.35 | 2.28 | 2.20 | 2.12 | 2.04 | 1.95 | 1.90 | 1.84 |
| .975 | 20 | 5.87 | 4.46 | 3.86 | 3.51 | 3.29 | 3.13 | 3.01 | 2.91 | 2.84 | 2.77 | 2.68 | 2.57 | 2.46 | 2.35 | 2.22 | 2.16 | 2.09 |
| .99 | | 8.10 | 5.85 | 4.94 | 4.43 | 4.10 | 3.87 | 3.70 | 3.56 | 3.46 | 3.37 | 3.23 | 3.09 | 2.94 | 2.78 | 2.61 | 2.52 | 2.42 |
| .995 | | 9.94 | 6.99 | 5.82 | 5.17 | 4.76 | 4.47 | 4.26 | 4.09 | 3.96 | 3.85 | 3.68 | 3.50 | 3.32 | 3.12 | 2.92 | 2.81 | 2.69 |
| .90 | | 2.88 | 2.49 | 2.28 | 2.14 | 2.05 | 1.98 | 1.93 | 1.88 | 1.85 | 1.82 | 1.77 | 1.72 | 1.67 | 1.61 | 1.54 | 1.50 | 1.46 |
| .95 | | 4.17 | 3.32 | 2.92 | 2.69 | 2.53 | 2.42 | 2.33 | 2.27 | 2.21 | 2.16 | 2.09 | 2.01 | 1.93 | 1.84 | 1.74 | 1.68 | 1.62 |
| .975 | 30 | 5.57 | 4.18 | 3.59 | 3.25 | 3.03 | 2.87 | 2.75 | 2.65 | 2.57 | 2.51 | 2.41 | 2.31 | 2.20 | 2.07 | 1.94 | 1.87 | 1.79 |
| .99 | | 7.56 | 5.39 | 4.51 | 4.02 | 3.70 | 3.47 | 3.30 | 3.17 | 3.07 | 2.98 | 2.84 | 2.70 | 2.55 | 2.39 | 2.21 | 2.11 | 2.01 |
| .995 | | 9.18 | 6.35 | 5.24 | 4.62 | 4.23 | 3.95 | 3.74 | 3.58 | 3.45 | 3.34 | 3.18 | 3.01 | 2.82 | 2.63 | 2.42 | 2.30 | 2.18 |
| .90 | | 2.79 | 2.39 | 2.18 | 2.04 | 1.95 | 1.87 | 1.82 | 1.77 | 1.74 | 1.71 | 1.66 | 1.60 | 1.54 | 1.48 | 1.40 | 1.35 | 1.29 |
| .95 | | 4.00 | 3.15 | 2.76 | 2.53 | 2.37 | 2.25 | 2.17 | 2.10 | 2.04 | 1.99 | 1.92 | 1.84 | 1.75 | 1.65 | 1.53 | 1.47 | 1.39 |
| .975 | 60 | 5.29 | 3.93 | 3.34 | 3.01 | 2.79 | 2.63 | 2.51 | 2.41 | 2.33 | 2.27 | 2.17 | 2.06 | 1.94 | 1.82 | 1.67 | 1.58 | 1.48 |
| .99 | | 7.08 | 4.98 | 4.13 | 3.65 | 3.34 | 3.12 | 2.95 | 2.82 | 2.72 | 2.63 | 2.50 | 2.35 | 2.20 | 2.03 | 1.84 | 1.73 | 1.60 |
| .995 | | 8.49 | 5.80 | 4.73 | 4.14 | 3.76 | 3.49 | 3.29 | 3.13 | 3.01 | 2.90 | 2.74 | 2.57 | 2.39 | 2.19 | 1.96 | 1.83 | 1.69 |
| .90 | | 2.75 | 2.35 | 2.13 | 1.99 | 1.90 | 1.82 | 1.77 | 1.72 | 1.68 | 1.65 | 1.60 | 1.54 | 1.48 | 1.41 | 1.32 | 1.26 | 1.19 |
| .95 | | 3.92 | 3.07 | 2.68 | 2.45 | 2.29 | 2.18 | 2.09 | 2.02 | 1.96 | 1.91 | 1.83 | 1.75 | 1.66 | 1.55 | 1.43 | 1.35 | 1.25 |
| .975 | 120 | 5.15 | 3.80 | 3.23 | 2.89 | 2.67 | 2.52 | 2.39 | 2.30 | 2.22 | 2.16 | 2.05 | 1.94 | 1.82 | 1.69 | 1.53 | 1.43 | 1.31 |
| .99 | | 6.85 | 4.79 | 3.95 | 3.48 | 3.17 | 2.96 | 2.79 | 2.66 | 2.56 | 2.47 | 2.34 | 2.19 | 2.03 | 1.86 | 1.66 | 1.53 | 1.38 |
| .995 | | 8.18 | 5.54 | 4.50 | 3.92 | 3.55 | 3.28 | 3.09 | 2.93 | 2.81 | 2.71 | 2.54 | 2.37 | 2.19 | 1.98 | 1.75 | 1.61 | 1.43 |
| .90 | | 2.71 | 2.30 | 2.08 | 1.94 | 1.85 | 1.77 | 1.72 | 1.67 | 1.63 | 1.60 | 1.55 | 1.49 | 1.42 | 1.34 | 1.24 | 1.17 | 1.00 |
| .95 | | 3.84 | 3.00 | 2.60 | 2.37 | 2.21 | 2.10 | 2.01 | 1.94 | 1.88 | 1.83 | 1.75 | 1.67 | 1.57 | 1.46 | 1.32 | 1.22 | 1.00 |
| .975 | ∞ | 5.02 | 3.69 | 3.12 | 2.79 | 2.57 | 2.41 | 2.29 | 2.19 | 2.11 | 2.05 | 1.94 | 1.83 | 1.71 | 1.57 | 1.39 | 1.27 | 1.00 |
| .99 | | 6.63 | 4.61 | 3.78 | 3.32 | 3.02 | 2.80 | 2.64 | 2.51 | 2.41 | 2.32 | 2.18 | 2.04 | 1.88 | 1.70 | 1.47 | 1.32 | 1.00 |
| .995 | | 7.88 | 5.30 | 4.28 | 3.72 | 3.35 | 3.09 | 2.90 | 2.74 | 2.62 | 2.52 | 2.36 | 2.19 | 2.00 | 1.79 | 1.53 | 1.36 | 1.00 |

Source: A. M. Mood and F. A. Graybill, *Introduction to the Theory of Statistics*. McGraw-Hill, New York, 1963, pp. 434–35. By permission of the publishers of *Biometrika*. (See Vol. 33, 1943.)

**Table A.8** Expected Order Statistics for Standard Normal Distribution

| $n$ | $i$ | $E(x_{(i)}, n)$ | $n$ | $i$ | $E(x_{(i)}, n)$ | $n$ | $i$ | $E(x_{(i)}, n)$ |
|---|---|---|---|---|---|---|---|---|
| 2 | 1 | .56418 95835 | 12 | 5 | .31224 88787 | 17 | 5 | .61945 76511 |
| 3 | 1 | .84628 43753 | 12 | 6 | .10258 96798 | 17 | 6 | .45133 34467 |
| 4 | 1 | 1.02937 53730 | 13 | 1 | 1.66799 01770 | 17 | 7 | .29518 64872 |
| 4 | 2 | .29701 13823 | 13 | 2 | 1.16407 71937 | 17 | 8 | .14598 74231 |
| 5 | 1 | 1.16296 44736 | 13 | 3 | .84983 46324 | 18 | 1 | 1.82003 18790 |
| 5 | 2 | .49501 89705 | 13 | 4 | .60285 00882 | 18 | 2 | 1.35041 37134 |
| 6 | 1 | 1.26720 63606 | 13 | 5 | .38832 71210 | 18 | 3 | 1.06572 81829 |
| 6 | 2 | .64175 50388 | 13 | 6 | .19052 36911 | 18 | 4 | .84812 50190 |
| 6 | 3 | .20154 68338 | 14 | 1 | 1.70338 15541 | 18 | 5 | .66479 46127 |
| 7 | 1 | 1.35217 83756 | 14 | 2 | 1.20790 22754 | 18 | 6 | .50158 15510 |
| 7 | 2 | .75737 42706 | 14 | 3 | .90112 67039 | 18 | 7 | .35083 72382 |
| 7 | 3 | .35270 69592 | 14 | 4 | .66176 37035 | 18 | 8 | .20773 53071 |
| 8 | 1 | 1.42360 03060 | 14 | 5 | .45556 60500 | 18 | 9 | .06880 25682 |
| 8 | 2 | .85222 48625 | 14 | 6 | .26729 70489 | 19 | 1 | 1.84448 15116 |
| 8 | 3 | .47282 24949 | 14 | 7 | .08815 92141 | 19 | 2 | 1.37993 84915 |
| 8 | 4 | .15251 43995 | 15 | 1 | 1.73591 34449 | 19 | 3 | 1.09945 30994 |
| 9 | 1 | 1.48501 31622 | 15 | 2 | 1.24793 50823 | 19 | 4 | .88586 19615 |
| 9 | 2 | .93229 74567 | 15 | 3 | .94768 90303 | 19 | 5 | .70661 14847 |
| 9 | 3 | .57197 07829 | 15 | 4 | .71487 73983 | 19 | 6 | .54770 73710 |
| 9 | 4 | .27452 59191 | 15 | 5 | .51570 10430 | 19 | 7 | .40164 22742 |
| 10 | 1 | 1.53875 27308 | 15 | 6 | .33529 60639 | 19 | 8 | .26374 28909 |
| 10 | 2 | 1.00135 70446 | 15 | 7 | .16529 85263 | 19 | 9 | .13072 48795 |
| 10 | 3 | .65605 91057 | 16 | 1 | 1.76599 13931 | 20 | 1 | 1.86747 50598 |
| 10 | 4 | .37576 46970 | 16 | 2 | 1.28474 42232 | 20 | 2 | 1.40760 40959 |
| 10 | 5 | .12266 77523 | 16 | 3 | .99027 10960 | 20 | 3 | 1.13094 80522 |
| 11 | 1 | 1.58643 63519 | 16 | 4 | .76316 67458 | 20 | 4 | .92098 17004 |
| 11 | 2 | 1.06191 65201 | 16 | 5 | .57000 93557 | 20 | 5 | .74538 30058 |
| 11 | 3 | .72883 94047 | 16 | 6 | .39622 27551 | 20 | 6 | .59029 69215 |
| 11 | 4 | .46197 83072 | 16 | 7 | .23375 15785 | 20 | 7 | .44833 17532 |
| 11 | 5 | .22489 08792 | 16 | 8 | .07728 74593 | 20 | 8 | .31493 32416 |
| 12 | 1 | 1.62922 76399 | 17 | 1 | 1.79394 19809 | 20 | 9 | .18695 73647 |
| 12 | 2 | 1.11573 21843 | 17 | 2 | 1.31878 19878 | 20 | 10 | .06199 62865 |
| 12 | 3 | .79283 81991 | 17 | 3 | 1.02946 09889 | | | |
| 12 | 4 | .53684 30214 | 17 | 4 | .80738 49287 | | | |

Source: A. E. Sarhan and B. G. Greenberg, *Contributions to Order Statistics*, John Wiley and Sons, Inc., New York, 1961, p. 193.

Since $x_{(1)} \leq x_{(2)} \cdots \leq x_{(n)}$, all expected values shown in this table are negative. Their counterparts on the other side are positive since $E(x_{(i)}) = -E(x_{(n-i+1)})$.

This table is reproduced from D. Teichroew, "Tables of expected values of order statistics and products of order statistics from samples of size 20 and less from the normal distribution," *Ann. Math. Statist.*, Vol. 27 (1956), pp. 410–426, with permission of W. Kruskal, editor of the *Annals of Mathematical Statistics* and the author.

**Table A.9** Percentage Points of the Kolmogorov-Smirnov Statistic

| Sample size ($N$) | Level of significance for $D$ = maximum $\lvert F_0(X) - S_F(X)\rvert$ | | | | |
|---|---|---|---|---|---|
| | .20 | .15 | .10 | .05 | .01 |
| 1 | .900 | .925 | .950 | .975 | .995 |
| 2 | .684 | .726 | .776 | .842 | .929 |
| 3 | .565 | .597 | .642 | .708 | .828 |
| 4 | .494 | .525 | .564 | .624 | .733 |
| 5 | .446 | .474 | .510 | .565 | .669 |
| 6 | .410 | .436 | .470 | .521 | .618 |
| 7 | .381 | .405 | .438 | .486 | .577 |
| 8 | .358 | .381 | .411 | .457 | .543 |
| 9 | .339 | .360 | .388 | .432 | .514 |
| 10 | .322 | .342 | .368 | .410 | .490 |
| 11 | .307 | .326 | .352 | .391 | .468 |
| 12 | .295 | .313 | .338 | .375 | .450 |
| 13 | .284 | .302 | .325 | .361 | .433 |
| 14 | .274 | .292 | .314 | .349 | .418 |
| 15 | .266 | .283 | .304 | .338 | .404 |
| 16 | .258 | .274 | .295 | .328 | .392 |
| 17 | .250 | .266 | .286 | .318 | .381 |
| 18 | .244 | .259 | .278 | .309 | .371 |
| 19 | .237 | .252 | .272 | .301 | .363 |
| 20 | .231 | .246 | .264 | .294 | .356 |
| 25 | .21 | .22 | .24 | .27 | .32 |
| 30 | .19 | .20 | .22 | .24 | .29 |
| 35 | .18 | .19 | .21 | .23 | .27 |
| Over 35 | $\frac{1.07}{\sqrt{N}}$ | $\frac{1.14}{\sqrt{N}}$ | $\frac{1.22}{\sqrt{N}}$ | $\frac{1.36}{\sqrt{N}}$ | $\frac{1.63}{\sqrt{N}}$ |

Source: S. Siegel, *Non-Parametric Statistics*. By permission of the American Statistical Association, and Frank J. Massey, Jr. (See *J. Am. Stat. Assoc*. 46: 70, 1951.)

# Bibliography

*Abbreviations for Journals*

*Ann. Math. Stat.*—Annals of Mathematical Statistics
*Phil. Trans. Roy. Soc.*—Philosophical Transactions of the Royal Society of London
*J. Roy. Stat. Soc.*—Journal of the Royal Statistical Society
*J. Am. Stat. Assoc.*—Journal of the American Statistical Association
*Bull. Am. Math. Soc.*—Bulletin of the American Mathematical Society
*Ann. Math.*—Annals of Mathematics
*Phil. Mag.*—Philosophical Magazine
*Ann. Eug.*—Annals of Eugenics

*Note:*
We have attached a system of asterisks: * denoting nonelementary, ** somewhat advanced, and *** advanced, to indicate degree of mathematical difficulty of material. Pages on which material is cited are given in [ ].


Acton, Forman S. 1959. *Analysis of straight-line data.* New York: Wiley. [510]

Aitken, A. C. 1935. On least squares and linear combinations of observations. *Proc. Roy. Soc. Edin.* 55: 42–48.

Anderson, R. L., and T. A. Bancroft. 1952. *Statistical theory in research.* New York: McGraw-Hill. [510]

*Anderson, T. W. 1959. *An introduction to multivariate statistical analysis.* New York: Wiley. [422, 494, 510]

Anscombe, F. J. 1948. The validity of comparative experiments. *J. Roy. Stat. Soc.* (A)111: 181–211.

———. 1961. Examination of residuals. *Proc. 4th Berkeley Symp. Math. Stat. Prob.* 1: 1–36. [435, 494]

Anscombe, F. J., and J. W. Tukey. 1963. The examination and analysis of residuals. *Technometrics* 5: 141–60. [435, 494]

Arbuthnott, J. 1710. An argument for divine providence taken from the constant regularity of the births of both sexes. *Phil. Trans. Roy. Soc.* 27: 186–90. [313]

Bahadur, R. R. 1954. Sufficiency and statistical decision functions. *Ann. Math. Stat.* 25: 423–62.

———. 1960. Stochastic comparison of tests. *Ann. Math. Stat.* 31: 276–95.

Bailey, Norman T. J. 1964. *The elements of stochastic processes*. New York: Wiley. [510]

Bancroft, T. A. 1968. *Topics in intermediate statistical methods*. Ames: Iowa State Univ. Press. [511]

**Barankin, E. W., and M. Katz, Jr. 1959. Sufficient statistics of minimal dimension. *Sankhyā* 21: 217–46. [262]

Barlow, Richard E., and Frank Proschan. 1965. *Mathematical theory of reliability*. New York: Wiley. [510]

Barnard, G. A. 1947. The meaning of a significance level. *Biometrika* 34: 179–82. [315]

**———. 1962. Comments on Stein's "A remark on the likelihood principle." *J. Roy. Stat. Soc.* (A)125: 569–73.

Barnard, G. A., and D. A. Sprott. 1970. A note on Basu's examples of anomalous ancillary statistics. Presented at Waterloo Symposium on Inference. [316]

Barnard, G. A., G. M. Jenkins, and C. B. Winsten. 1962. Likelihood inference and time series. *J. Roy. Stat. Soc.* (A)125: 321–72. [147, 295]

Bartlett, M. S. 1949. Fitting a straight line when both variables are subject to error. *Biometrics* 5: 207–12 [439]

———. 1960. *Stochastic population models in ecology and epidemiology*. London: Methuen. [510]

**———. 1966. *An introduction to stochastic processes*. Cambridge: Cambridge Univ. Press. [510]

Basu, D. 1955. On statistics independent of a complete sufficient statistic. *Sankhyā* 15: 377–80.

**———. 1959. The family of ancillary statistics. *Sankhyā* 21: 247–56. [316]

**———. 1964. Recovery of ancillary information. *Sankhyā* 26: 3–16. [316]

Batschelet, E. 1965. Statistical methods for the analysis of problems in animal orientation and certain biological rhythms. Washington: American Institute of Biological Sciences. [20]

Bayes, T. 1763. An essay towards solving a problem in the doctrine of chances. *Phil. Trans. Roy. Soc.* 53: 370–418. [67]

Bennett, Carl A., and Norman L. Franklin. 1954. *Statistical analysis in chemistry and the chemical industry*. New York: Wiley. [511]

*Bharucha-Reid, A. T. 1960. *Elements of the theory of Markov processes and their applications*. New York: McGraw-Hill. [510]

***Billingsley, P. 1961. *Statistical inference for Markov processes*. Chicago: Univ. Chicago Press. [188, 510]

**Birch, M. W. 1964. A new proof of the Fisher-Pearson theorem. *Ann. Math. Stat.* 35: 817–24. [268, 283]

Birnbaum, A. 1962. On the foundations of statistical inference (with discussion). *J. Am. Stat. Assoc.* 57: 269–326. [295]

———. 1969. Concepts of statistical evidence. In S. Morganbesser et al., eds., *Philosophy, science and method: Essays in honor of E. Nagel*. New York: St. Martins Press. [295, 296]

Birnbaum, Z. W. 1962. *Introduction to probability and statistics*. New York: Harper. [510]

**Blackman, R. B., and J. W. Tukey. 1959. *The measurement of power spectra*. New York: Dover. [207, 511]

**Blackwell, D. 1947. Conditional expectation and unbiased sequential estimation. *Ann. Math. Stat.* 18: 105–10. [401]

*Blackwell, David, and M. A. Girshick. 1954. *Theory of games and statistical decisions*. New York: Wiley. [411, 510]

Bliss, C. I. 1967, 1970. *Statistics in biology*, 2 vols. New York: McGraw-Hill. [510, 511]
Boole, G. 1854. *An investigation of the laws of thought*. New York: Dover. [510]
———. 1953 (reprint). *Studies in logic and probability*. London: Watts. [510]
Bowker, Albert H., and Gerald J. Lieberman. 1959. *Engineering statistics*. Englewood Cliffs, N.J.: Prentice-Hall. [382, 511]
Box, G. E. P., and N. R. Draper. 1969. *Evolutionary operation; a statistical method for process improvement*. New York: Wiley.
Box, G. E. P., and S. M. Jenkins. (In press.) *Statistical methods for forecasting and control*. San Francisco: Holden-Day. [208, 511]
Bradley, James V. 1968. *Distribution-free statistical tests*. Englewood Cliffs, N.J.: Prentice-Hall. [510]
Bromwich, T. J. I'A. 1965 (orig. ed., 1908). *An introduction to the theory of infinite series*. London: Macmillan. [262]
Brown, G. S. 1957. *Probability and scientific inference*. London: Longmans. [510]
Brownlee, K. A. 1965. *Statistical theory and methodology in science and engineering*, 2nd ed. New York: Wiley. [511]
Brunk, H. D. 1965. *Mathematical statistics*, 2nd ed. New York: Blaisdell. [510]
*Buehler, Robert J. 1959. Some validity criteria for statistical inferences. *Ann. Math. Stat.* 30: 845–63. [364, 376]
*Buehler, R. J., and A. P. Feddersen. 1963. Note on a conditional property of Student's t. *Ann. Math. Stat.* 34: 1098–1100. [376]
***Burkholder, D. L. 1961. Sufficiency in the undominated case. *Ann. Math. Stat.* 32: 1191–1200.
Burr, Irving W. 1953. *Engineering statistics and quality control*. New York: McGraw-Hill. [510]
Cajori, F. 1919. *A history of mathematics*. New York: Macmillan. [79]
Campbell, N. R. 1957. *Foundations of science: The philosophy of theory and experiment*. New York: Dover. [259, 510]
Carnap, Rudolf. 1962. *Logical foundations of probability*, 2nd ed. Chicago: Univ. Chicago Press. [510]
Chandrasekhar, S. 1960. *Principles of stellar dynamics*. New York: Dover. [214]
***Chernoff, H. 1952. A measure of asymptotic efficiency for tests of a hypothesis based on the sum of observations. *Ann. Math. Stat.* 23: 493–507. [326]
———. 1956. Large sample theory: Parametric case. *Ann. Math. Stat.* 27: 1–22. [133]
Chernoff, Herman, and Lincoln E. Moses. 1959. *Elementary decision theory*. New York: Wiley. [510]
***Chung, K. L. 1967. *Markov chains with stationary transition probabilities*. 2nd ed. Berlin: Springer-Verlag. [510]
***———. 1968. *A course in probability theory*. New York: Harcourt, Brace & World. [509]
***Church, A. 1940. On the concept of a random sequence. *Bull. Am. Math. Soc.* 46: 130–35.
Clopper, C. J., and E. S. Pearson. 1934. The use of confidence or fiducial limits in the case of the binomial. *Biometrika* 26: 404–13. [368]
Cochran, W. G. 1934. The distribution of quadratic forms in a normal system, with applications to the analysis of covariance. *Proc. Camb. Phil. Soc.* 30: 178–91.
———. 1963. *Sampling techniques*, 2nd ed. New York: Wiley. [511]
Cochran, William G., and Gertrude M. Cox. 1957. *Experimental designs*, 2nd ed. New York: Wiley. [510]

Copas, J. B. 1969. Compound decisions and empirical Bayes (with discussion). *J. Roy. Stat. Soc.* (A)31: 397–425. [414]

Cornfield, J. 1969. The Bayesian outlook and its application (with discussion). *Biometrics* 25: 617–57. [308]

Cornfield, J., and J. W. Tukey. 1956. Average values of mean squares in factorials. *Ann. Math. Stat.* 27: 907–49. [455]

Cox, D. R. 1958. *Planning of experiments*. New York: Wiley. [510]

*Cox, D. R., and H. D. Lewis. 1965. *The theory of stochastic processes*. New York: Wiley. [188, 510]

Cox, D. R., and P. A. W. Lewis. 1966. *The statistical analysis of series of events*. London: Methuen. [511]

Cox, Richard T. 1961. *The algebra of probable inference*. Baltimore: Johns Hopkins Univ. Press. [510]

Craig, A. T. 1943. Note on the independence of certain quadratic forms. *Ann. Math. Stat.* 14: 195–97.

*Cramér, Harald. 1946. *Mathematical methods of statistics*. Princeton: Princeton Univ. Press. [99, 401, 509, 510]

———. 1955. *The elements of probability theory and some of its applications*. New York: Wiley. [509]

***Cramér, Harald, and M. A. Leadbetter. 1967. *Stationary and related stochastic processes*. New York: Wiley. [510]

**Curtis, J. H. 1942. A note on the theory of moment generating functions. *Ann. Math. Stat.* 13: 430–33. [99]

Daniel, C. 1959. Use of half-normal plots in interpreting factorial two-level experiments. *Technometrics* 1: 311–41.

Daniel, C., and F. S. Wood. 1971. *Fitting equations to data*. New York: Wiley.

David, F. N., and D. E. Barton. 1962. *Combinatorial chance*. London: Griffin. [510]

David, H. A. 1954. The distribution of range in certain non-normal populations. *Biometrika* 41: 463–68.

Davis, D. J. 1952. An analysis of some failure data. *J. Am. Stat. Assoc.* 47: 113–50.

Davies, O. L. (ed). 1954. *The design and analysis of industrial experiments*. New York: Hafner. [510]

*De Groot, Morris H. 1970. *Optimal statistical decisions*. New York: McGraw-Hill. [510]

Deming, W. E. 1950. *Some theory of sampling*. New York: Wiley. [511]

Dempster, A. P. 1964. On the difficulties inherent in Fisher's fiducial argument. *J. Am. Stat. Assoc.* 59: 56–66. [358]

Dixon, W. J. 1950. Analysis of extreme values. *Ann. Math. Stat.* 21: 488–506.

Dixon, W. J., and F. J. Massey. 1969. *Introduction to statistical analysis*, 3rd ed. New York: McGraw-Hill. [368, 511]

***Doob, J. L. 1953. *Stochastic processes*. New York: Wiley. [509]

Draper, N. R., and H. Smith. 1966. *Applied regression analysis*. New York: Wiley. [510]

Duncan, Acheson J. 1965. *Quality control and industrial statistics*, 3rd ed. Homewood, Ill.: Irwin. [510]

Edwards, Ward, H. Lindman, and L. J. Savage. 1963. Bayesian statistical inference for psychological research. *Psych. Rev.* 70: 193–242.

Eisenhart, C. 1947. The assumptions underlying the analysis of variance. *Biometrics* 3: 1–21.

Elderton, W. P. 1938. *Frequency curves and correlation*. Cambridge: Cambridge Univ. Press. [244, 510]

Epstein, B., and M. Sobel. 1954. Some theorems relevant to life testing for an exponential distribution. *Ann. Math. Stat.* 25: 373–81.

Federer, Walter, T. 1955. *Experimental design: Theory and application.* New York: Macmillan. [510]

Feller, W. 1936. Note on regions similar to the sample space. *Stat. Res. Mem.* 2: 117–25.

———. 1957. *An introduction to probability theory and its applications,* vol. I. 2nd ed. New York: Wiley. [125, 509]

***———. 1966. *An introduction to probability theory and its applications,* vol. II. New York: Wiley. [509]

*Ferguson, Thomas S. 1967. *Mathematical statistics.* New York: Academic Press. [510]

Finney, D. J. 1960. *An introduction to the theory of experimental design.* Chicago: Univ. Chicago Press. [510]

———. 1964. *Statistical method in biological assay,* 2nd ed. London: Griffin. [510]

Fisher, R. A. 1922. On the mathematical foundations of theoretical statistics. *Phil. Trans. Roy. Soc.* (A)222: 309–68.

———. 1924. The conditions under which chi square measures the discrepancy between observations and hypothesis. *J. Roy. Stat. Soc.* 137: 442–50. [283]

———. 1925. Theory of statistical estimation. *Proc. Camb. Phil. Soc.* 22: 700–725. [278, 286]

———. 1926. The arrangement of field experiments. *J. Min. Agr. England* 33: 503–13. [507]

———. 1934a. Randomization, and an old enigma of card play. *Mathematical Gazette.* 18: 294–97. [411]

———. 1934b. Two new properties of mathematical likelihood. *Proc. Roy. Soc.* (A)144: 285–307. [281]

———. 1935a (rev. ed., 1960). *The design of experiments.* Edinburgh: Oliver and Boyd. [79, 342, 507, 510]

———. 1935b. The fiducial argument in statistical inference. *Ann. Eug.* 6: 391–98. [346, 378]

**———. 1950a. *Contributions to mathematical statistics.* New York: Wiley. [510]

———. 1950b. The significance of deviations from expectations in a Poisson series *Biometrics* 6: 17–24. [248]

———. 1956. *Statistical methods and scientific inference.* London: Oliver and Boyd. [295, 296, 310, 314, 315, 380, 510]

———. 1958. *Statistical methods for research workers,* 13th ed. Edinburgh: Oliver and Boyd. [79, 511]

Fisher, R. A., and F. Yates. 1963. *Statistical tables for biological, agricultural, and medical research,* 6th ed. Edinburgh: Oliver and Boyd. [178, 511]

*Fisz, Marek. 1963. *Probability theory and mathematical statistics,* 3rd ed. New York: Wiley. [510]

Fix, E., and J. L. Hodges. 1955. Significance probabilities of the Wilcoxon test. *Ann. Math. Stat.* 26: 301–12. [343]

Folks, J. L., and C. E. Antle. 1967. Straight line confidence regions for linear models. *J. Am. Stat. Assoc.* 62: 1365–74.

Folks, J. L., and J. H. Blankenship. 1967. A note on probability plotting. *Ind. Qual. Control* 23: 495–96.

Folks, J. L., D. A. Pierce, and C. Stewart. 1965. Estimating the fraction of acceptable product. *Technometrics* 7: 43–50.

**Fraser, D. A. S. 1957. *Nonparametric methods in statistics*. New York: Wiley. [510]

***———. 1961. The fiducial method and invariance. *Biometrika* 48: 261–80.

**———. 1968. *The structure of inference*. New York: Wiley. [380, 510]

Freeman, Harold. 1963. *Introduction to statistical inference*. Reading, Mass.: Addison-Wesley. [510]

Freund, John E. 1960. *Modern elementary statistics*, 2nd ed. Englewood Cliffs, N.J.: Prentice-Hall. [510]

Fryer, H. C. 1966. *Concepts and methods of experimental statistics*. Boston: Allyn and Bacon. [511]

Gabbe, J. D., M. B. Wilk, and W. L. Brown. 1967. Statistical analysis and modeling of the high-energy proton from the Telstar I satellite. *Bell System Tech. J.* 46: 1301–1450.

Gallie, W. B. 1966 (Penguin, 1952). *Pierce and pragmatism*. New York: Dover. [507, 508]

**Gani, J. 1955. Some theorems and sufficiency conditions for the maximum likelihood estimator of an unknown parameter in a single Markov chain. *Biometrika* 42: 342–59. [188]

**Geisser, S. 1964. Estimation in the uniform covariance case. *J. Roy. Stat. Soc.* (B)26: 477–83.

**———. 1965. Bayesian estimation in multivariate analysis. *Ann. Math. Stat.* 36: 150–59. [309, 310]

**Geisser, S., and J. Cornfield. 1963. Posterior distribution for multivariate normal parameters. *J. Roy. Stat. Soc.* (B)25: 368–76. [380, 424]

Gnanadesikan, R., and M. B. Wilk. 1969. Data analytic methods in multivariate statistical analysis. In P. R. Krishnaiah, ed., *Multivariate Analysis II*. New York: Academic Press.

**Gnedenko, B. V. 1967. *The theory of probability*, 4th ed. New York: Chelsea. [509]

Good, I. J. 1950. *Probability and the weighing of evidence*. London: Griffin. [79, 308, 510]

Goulden, Cyril H. 1952. *Methods of statistical analysis*, 2nd ed. New York: Wiley, [511]

Grant, Eugene L. 1952. *Statistical quality control*. New York: McGraw-Hill. [510]

*Graybill, Franklin A. 1961. *Introduction to linear statistical models*, vol I. New York: McGraw-Hill. [446, 494, 510]

**Grenander, Ulf, and Murray Rosenblatt. 1957. *Statistical analysis of stationary time series*. New York: Wiley. [511]

Hacking, I. 1965. *Logic of statistical inference*. Cambridge: Cambridge Univ. Press. [295, 380, 510]

———. 1967. Slightly more realistic personal probability. *Philosophy of Science* 34: 311–25.

Hald, A. 1952a. *Statistical tables and formulas*. New York: Wiley. [178, 511]

———. 1952b. *Statistical theory with engineering applications*. New York: Wiley. [511]

Halmos, P. R., and L. J. Savage. 1949. Application of the Radon-Nikodym theorem to the theory of sufficient statistics. *Ann. Math. Stat.* 20: 225–41.

*Hannan, E. J. 1960. *Time series analysis*. London: Methuen. [511]

Hansen, Morris H., William N. Hurwitz, and William G. Madow. 1953. *Sample survey methods and theory*, 2 vols. New York: Wiley. [510]

Hartley, H. O. 1938. Studentization and large sample theory. *J. Roy. Stat. Soc.* (suppl.) 5: 80–88.

———. 1942. The range in random samples. *Biometrika* 32: 334–48.

———. 1944. Studentization. *Biometrika* 33: 173–80.

Hartley, H. O., and H. A. David. 1954. Universal bounds for mean range and extreme observation. *Ann. Math. Stat.* 25: 85–99.

Hastings, Cecil, Jr. 1955. *Approximations for digital computers.* Princeton: Princeton Univ. Press. [97, 230, 510]

Herdan, G. 1960. *Small particle statistics*, 2nd ed. London: Butterworth's. [25]

Hodges, J. L., Jr., and E. L. Lehmann. 1950. Some problems in minimax point estimation. *Ann. Math. Stat.* 21: 182–98. [398]

———. 1964. *Basic concepts of probability and statistics.* San Francisco: Holden-Day. [510]

———. 1965. *Elements of finite probability.* San Francisco: Holden-Day. [509]

***Hoeffding, W. 1965. Asymptotically optimal tests for multinomial distributions (with discussion). *Ann. Math. Stat.* 36: 369–408. [327]

Hoel, Paul G. 1962. *Introduction to mathematical statistics,* 3rd ed. New York: Wiley. [510]

Hogben, L. 1957. *Statistical theory,* London: George Allen and Unwin.

Hogg, Robert V., and Allen T. Craig. 1970. *Introduction to mathematical statistics,* 3rd ed. New York: Macmillan. [510]

Hotelling, H. 1931. The generalization of "Student's" ratio. *Ann. Math. Stat.* 2: 360–78 [423]

Hudson, D. 1968. Interval estimation from the likelihood function. Unpubl. tech. rept., Bell Telephone Laboratories, Inc., New Jersey. [297]

Huntsberger, David V. 1961. *Elements of statistical inference.* Boston: Allyn and Bacon. [510]

IBM Manual C20-8011. Random number generation and testing. [180]

Jeffreys, H. 1961. *Theory of probability,* 3rd ed. Oxford: Clarendon Press. [79, 308, 309, 509, 510]

Jenkins, G. M., and D. G. Watts. 1968. *Spectral analysis and its applications.* San Francisco: Holden-Day. [207, 511]

Johnson, Norman L., and Fred C. Leone. 1964. *Statistics and experimental design in engineering and the physical sciences,* vols. I, II. New York: Wiley. [511]

Johnston, J. 1963. *Econometric methods.* New York: McGraw-Hill. [510]

*Karlin, Samuel. 1966. *A first course in stochastic processes.* New York: Academic Press. [191, 510]

Kemeny, John G., and J. Laurie Snell. 1960. *Finite Markov chains.* Princeton: Van Nostrand. [510]

Kempthorne, Oscar. 1952. *The design and analysis of experiments.* New York: Wiley. [455, 463, 494, 507, 510]

———. 1955. The randomization theory of experimental inference. *J. Am. Stat. Assoc.* 50: 946–67. [507]

———. 1957 (reprint 1969, Iowa State Univ. Press). *An introduction to genetic statistics.* New York: Wiley. [447, 510]

———. 1966. Some aspects of experimental inference. *J. Am. Stat. Assoc.* 61: 11–34 [507]

———. 1971a (in press). Theories of inference and data analysis. In T. A. Bancroft, ed., *Statistical papers in honor of George W. Snedecor.* Ames: Iowa State Univ. Press. [447]

———. 1971b (in press). Probability, statistics, and the knowledge business. In Proceedings of Waterloo International Symposium on Statistical Inference. [447]

Kempthorne, O., and T. E. Doerfler. 1969. The behavior of some significance tests under experiment randomization. *Biometrika* 56: 231–48. [345]

Kempthorne, O., G. Zyskind, S. Addelman, T. N. Throckmorton, and R. F. White. 1961. Analysis of variance procedures. Aeronautical Research Laboratory Tech. Rept. 149, Wright-Patterson Air Force Base, Ohio. [455]

Kendall, M. G. 1955. *Rank correlation methods*. 2nd ed. New York: Hafner. [510]

**Kendall, M. G., and Alan Stuart. 1966. *The advanced theory of statistics*. New York: Hafner. [34, 207, 272, 423, 510]

Keynes, J. M. 1921 (reprint, 1962). *A treatise on probabilities*. New York: Harper and Row. [79]

*Khinchin, A. I. 1957. *Mathematical foundations of information theory*. New York: Dover. [27, 34, 510]

**Kolmogorov, A. N. 1956. *Foundations of the theory of probability*, 2nd Engl. ed. New York: Chelsea. [509]

Koopman, B. O. 1936. On distributions admitting a sufficient statistic. *Trans. Am. Math. Soc.* 39: 399–409. [278]

***———. 1940. The axioms and algebra of intuitive probability. *Ann. Math.* 41: 269–92. [80]

Kruskal, J. B., and J. D. Carroll. 1969. Geometric models and badness of fit functions. In P. R. Krishnaiah, ed., *Multivariate analysis* II. New York: Academic Press.

Krutchkoff, R. G. 1970. Empirical Bayes estimation for the applied statistician. Proceedings of Army Conference on Design of Experiments. (To be published.) [413]

*Kullback, Solomon. 1959. *Information theory and statistics*. New York: Wiley. [386, 510]

Kurtz, T. E., B. F. Link, J. W. Tukey, and D. L. Wallace. 1965. Short-cut multiple comparisons for balanced single and double classifications, part I. *Technometrics* 7: 95–162. [464]

Kyburg, H. E., Jr., and H. E. Smokler. 1964. *Studies in subjective probability*. New York: Wiley. [80, 510]

**Lamperti, J. 1966. *Probability*. New York: Benjamin. [509]

Laplace, Pierre Simon de. 1814 (reprint, 1952, Dover). *A philosophical essay on probabilities*. New York: Dover. [79]

Larson, H. J. 1969. *Introduction to probability theory and statistical inference*. New York: Wiley. [510]

Lawley, D. N., and A. E. Maxwell. 1963. *Factor analysis as a statistical method*. London: Butterworth's. [510]

**Lehmann, E. L. 1959. *Testing statistical hypotheses*. New York: Wiley. [328, 329, 330, 332, 354, 373, 374, 510]

***Lehmann, E. L., and H. Scheffé. 1950. Completeness, similar regions, and unbiased estimates, part I. *Sankhyā* 10: 305–40. [255, 402, 404]

***Lévy, P. 1937. *Théorie de l'addition des variables aléatoires*. Paris: Gauthier-Villars. [99, 510]

Li, J. C. R. 1964. *Statistical inference*. Ann Arbor: Edwards. [511]

Lindgren, B. W. 1968. *Statistical theory*, 2nd ed. New York: Macmillan. [510]

Lindgren, B. W., and G. W. McElrath. 1969. *Introduction to probability and statistics*, 3rd ed. New York: Macmillan.

Lindley, D. V. 1958. Fiducial distributions and Bayes' theorem. *J. Roy. Stat. Soc.* (B)20: 102–7. [380]

———. 1961. The use of prior probability distributions in statistical inference and decision. *Proc. 4th Berkeley Symp. Math. Stat. Prob.* 1: 453–68.

———. 1965. *Introduction to probability and statistics from a Bayesian viewpoint*. Cambridge: Cambridge Univ. Press. [79, 308, 510]

———. 1969. Review of *The structure of inference* by D. A. S. Fraser. *Biometrika* 56: 453–56.

***Loève, Michel. 1963. *Probability theory*, 3rd ed. Princeton: Van Nostrand. [510]

Littell, R. C., and J. L. Folks. 1970. On the relation between Bahadur efficiency and power. *Ann. Math. Stat.* 41: 2135–36.

*Luce, R. D., and H. Raiffa. 1957. *Games and decisions*. New York: Wiley. [411, 510]

*Lukacs, E. 1960. *Characteristic functions*. London: Griffin. [24]

*Lukacs, E., and R. G. Laha. 1964. *Applications of characteristic functions*. London: Griffin. [24]

*Mann, H. B. 1949. *Analysis and design of experiments*. New York: Dover. [510]

Mann, H. B., and A. Wald. 1943. On stochastic limits and order relationships. *Ann. Math. Stat.* 14: 217–26. [133]

Meyer, Paul L. 1965. *Introductory probability and statistical applications*. Reading, Mass.: Addison-Wesley. [510]

Middleton, D. 1960. *An introduction to statistical communication theory*. New York: McGraw-Hill. [510]

Mood, Alexander M. 1940. The distribution theory of runs. *Ann. Math. Stat.* 11: 367–92. [233]

Mood, Alexander M., and Franklin A. Graybill. 1963. *Introduction to the theory of statistics*, 2nd ed. New York: McGraw-Hill. [510]

**Moran, P. A. P. 1962. *The statistical processes of evolutionary theory*. Oxford: Clarendon Press. [510]

Mosteller, R., R. E. K. Rourke, and G. B. Thomas, Jr. 1961. *Probability and statistics*. Reading, Mass: Addison-Wesley. [510]

Natrella, Mary Gibbons. 1963. *Experimental statistics*. National Bureau of Standards Handbook 91, USDC. [511]

***Neveu, Jacques. 1965. *Mathematical foundations of the calculus of probability*. San Francisco: Holden-Day. [510]

Neyman, J. 1934. On the two different aspects of the representative method. *J. Roy. Stat. Soc.* 97: 558–625.

———. 1935. On the problem of confidence intervals. *Ann. Math. Stat.* 6: 111–16. [374, 376]

———. 1941. Fiducial argument and the theory of confidence intervals. *Biometrika* 32: 128–50. [310, 380]

**———. 1949. Contribution to the theory of the $\chi^2$ test. *Proc. Berkeley Symp. Math. Stat. Prob.* pp. 239–73. [278, 407]

———. 1950. *First course in probability and statistics*. New York: Holt. [510]

**———. 1967. *A selection of early statistical papers*. Berkeley: Univ. Calif. Press. [510]

Neyman, J., K. Iwaskiewicz, and St. Kolodziejczyk. 1935. Statistical problems in agricultural experimentation. *J. Roy. Stat. Soc.* (suppl.) 2: 107–54. [507]

**Neyman, J., and E. S. Pearson. 1933. On the problem of the most efficient tests of statistical hypotheses. *Phil. Trans. Roy. Soc.* (A)231: 289–337. [333]

**———. 1967. *Joint statistical papers*. Berkeley: Univ. Calif. Press. [510]

Ogden, C. K., and I. A. Richards. 1956. *The meaning of meaning*, 8th ed. New York: Harcourt, Brace. [510]

Ostle, Bernard. 1963. *Statistics in research*, 2nd ed. Ames: Iowa State Univ. Press. [511]

Owen, D. B. 1962. *Handbook of statistical tables*. Reading, Mass.: Addison-Wesley, [511]

Parzen, Emanuel. 1960. *Modern probability theory and its applications*. New York: Wiley. [510]

*———. 1962. *Stochastic processes*. San Francisco: Holden-Day.
**Pearson, E. S. 1966. *Selected papers*. Berkeley: Univ. Calif. Press. [510]
Pearson, E. S., and H. O. Hartley. 1954. *Biometrika tables for statisticians*. Cambridge: Cambridge Univ. Press. [511]
Pearson, K. 1900. On a criterion that a given set of deviations from the probable in the case of a correlated system of variables is such that it can be reasonably supposed to have arisen in random sampling. *Phil. Mag.* 50: 157–76. [313]
PSSC (Physical Science Study Committee). 1967. *Physics*, 2nd ed. Boston: Heath. [212]
Pierce, D. A., and J. L. Folks. 1969. Sensitivity of Bayes procedures to the prior distribution. *Operations Research* 17: 344–50.
***Pitcher, T. S. 1957. Sets of measures not admitting necessary and sufficient statistics or subfields. *Ann. Math. Stat.* 28: 267–68. [262, 287]
*Plackett, R. L. 1960. *Principles of regression analysis*. Oxford: Clarendon Press. [494, 510]
Poincaré, H. 1907. *Science and hypothesis*. New York: C. Scribner.
———. 1914. *Science and method*. New York: T. Nelson and Sons.
Popper, K. R. 1945. *The open society and its enemies*. London: Routledge.
———. 1959. *The logic of scientific discovery*. New York: Basic Books.
**Pratt, J. W. 1959. On a general concept of "in probability." *Ann. Math. Stat.* 30: 549–58. [133]
———. 1965. Bayesian interpretation of standard inference statements (with discussion). *J. Roy. Stat. Soc.* (B)27: 169–203. [310]
Pratt, J. W., H. Raiffa, and R. Schlaifer. 1964. The foundations of decision under uncertainty: An elementary exposition. *J. Am. Stat. Assoc.* 59: 353–75.
*Pyke, R. 1965. Spacings (with discussion). *J. Roy. Stat. Soc.* (B)13: 395–449. [232]
Raiffa, Howard, and Robert Schlaifer. 1961. *Applied statistical decision theory*. Boston: Div. Res., Harvard Univ. Graduate School of Bus. Admin. [309, 510]
Ramsey, F. P. 1926. Truth and probability. In R. B. Braithwaite, ed., *The foundations of mathematics and other logical essays*. London: Kegan, Paul. [79]
Rand Corporation. 1955. *A million random digits with 100,000 normal deviates*. Glencoe, Ill.: Free Press. [179, 511]
Rao, C. R. 1945. Information and accuracy attainable in estimation of statistical parameters. *Bull. Calcutta Math. Soc.* 37: 81–91. [401]
*———. 1952. *Advanced statistical methods in biometric research*. New York: Wiley. [511]
———. 1965. *Linear statistical inference and its applications* New York: Wiley. [101, 268, 399, 494, 510]
Reichenbach, Hans. 1949. *The theory of probability*, 2nd ed. Berkeley: Univ. Calif. Press. [79, 510]
Riordan, John. 1958. *An introduction to combinatorial analysis*. New York: Wiley. [510]
Robbins, H. 1955. An empirical Bayes approach to statistics. *Proc. 3rd Berkeley Symp. Math. Stat. Prob.* 1: 157–63. [413]
———. 1963. The empirical Bayes approach to testing statistical hypotheses. *Rev. Intern. Stat. Inst.* 31: 195–208.
———. 1964. The empirical Bayes approach to statistical decision problems. *Ann. Math. Stat.* 35: 1–20.
*Roy, S. N. 1957. *Some aspects of multivariate analysis*. New York: Wiley. [510]
Sampford, M. R. 1962. *An introduction to sampling theory*. Edinburgh: Oliver and Boyd. [511]

Sarhan, A. E., and B. G. Greenberg. 1962. *Contributions to order statistics.* New York: Wiley.

Savage, I. R. 1968. *Statistics: Uncertainty and behavior.* Boston: Houghton Mifflin. [510]

**Savage, Leonard J. 1954. *The foundations of statistics.* New York: Wiley. [79, 308, 510]

———. 1961. The foundations of statistics reconsidered. *Proc. 4th Berkeley Symp. Math. Stat. Prob.* 1: 575–86. [308]

———. 1962. *The foundations of statistical inference.* London: Methuen. [300, 308]

———. 1967a. Difficulties in the theory of personal probability. *Philosophy of Science.* 34: 305–10. [308]

———. 1967b. Implications of personal probability for induction. *J. Philosophy* 64: 593–97. [308]

Scheffé, H. 1947. A useful convergence theorem for probability distributions. *Ann. Math. Stat.* 18: 434–38.

**———. 1959. *The analysis of variance.* New York: Wiley. [464, 510]

Schlaifer, R. 1959. *Probability and statistics for business decisions.* New York: McGraw-Hill. [308, 510]

Shapiro, S. S., and M. B. Wilk. 1965. An analysis of variance test for normality (complete samples). *Biometrika* 52: 591–611. [286]

Shimony, A. 1967. Amplifying personal probability: Comments on L. J. Savage's "Difficulties in the theory of personal probability." *Philosophy of Science.* 34: 326–32.

Siegel, Sidney. 1956. *Nonparametric statistics for the behavioral sciences.* New York: McGraw-Hill. [510]

Smart, W. M. 1958. *Combination of observations.* Cambridge: Cambridge Univ. Press. [510]

Smirnov, N. 1939. On the estimation of the discrepancy between empirical curves of distribution for two independent samples. *Bull. Math. Univ. Moscow* 2: 3–14.

Snedecor George W., and William G. Cochran. 1967. *Statistical methods*, 6th ed. Ames: Iowa State Univ. Press. [511]

**Steck, G. P. 1957. Limit theorems for conditional distributions. *Univ. Calif. Pub. Stat.* 2: 237–84. [229, 283]

Steel, Robert G. D., and James H. Torrie. 1960. *Principles and procedures of statistics.* New York: McGraw-Hill. [511]

"Student" (W. S. Gossett). 1908. The probable error of a mean. *Biometrika* 6: 1–25. [313]

Sukhatme, B. V., and P. V. Sukhatme. 1970. *Sampling theory of surveys with applications,* 2nd ed. Ames: Iowa State Univ. Press. [511]

*Takács, L. 1960. *Stochastic processes.* London: Methuen. [510]

Teicher, H. 1954. On the multivariate Poisson distribution. *Skand. Aktuarietidskrift* 37: 1–9. [157]

*Tintner, G. 1952. *Econometrics.* New York: Wiley. [510, 511]

Todhunter, I. 1949. *A history of the mathematical theory of probability.* London: Chelsea. [79, 510]

**Tucker, Howard. 1967. *A graduate course in probability.* New York: Academic Press. [510]

Tukey, J. W. 1949a. One degree of freedom for non-additivity. *Biometrics* 5: 232–42. [494]

———. 1949b. Dyadic anova, an analysis of variance of vectors. *Human Biology* 21: 65–110. [494]

———. 1950. Some sampling simplified. *J. Am. Stat. Assoc.* 45: 501–19. [455]

———. 1954 (reprint, 1964). Correlation, regression and path analysis. In O. Kempthorne et al., eds., *Statistics and mathematics in biology.* New York: Hafner. [447]

———. 1956. Variances of variance components. I. Balanced designs. *Ann. Math. Stat.* 27: 722–36. [455]

———. 1960. Conclusions vs. decisions. *Technometrics* 2: 423–34. [390]

———. 1962. The future of data analysis. *Ann. Math. Stat.* 33: 1–67.

**Uspensky, J. V. 1937. *Introduction to mathematical probability.* New York: McGraw-Hill. [510]

Venn, J. A. 1866 (reprint, 1962). *The logic of chance.* London: Chelsea. [79, 309, 510]

Von Mises, R. 1935. *Probability, statistics, and truth.* London: Hodge. [79, 510]

———. 1964. *Mathematical theory of probability and statistics*, edited and complemented by H. Geiringer. New York: Academic Press. [79, 510]

*Von Neumann, J., and O. Morgenstern. 1947. *Theory of games and economic behavior*, 2nd ed. Princeton: Princeton Univ. Press. [411]

Wald, Abraham. 1940. The fitting of straight lines if both variables are subject to error. *Ann. Math. Stat.* 11: 284–300. [439]

**———. 1947. *Sequential analysis.* New York: Wiley. [289, 290, 316, 355, 510]

***———. 1950. *Statistical decision functions.* New York: Wiley. [390, 510]

Walker, H. M. 1931. *Studies in the history of statistical method.* Baltimore: Williams and Wilkins. [79, 510]

Wallis, W. A., and H. V. Roberts. 1956. *Statistics: A new approach.* Glencoe: Free Press. [511]

**Walsh, J. E. 1962, 1968. *Handbook of nonparametric statistics.* New York: Van Nostrand. [510]

Watson, G. S. 1970. Orientation statistics in the earth sciences. *Bull. Geol. Inst. Univ. Uppsala* (N.S.2) 9: 73–90. [20]

*Weiss, Lionel. 1961. *Statistical decision theory.* New York: McGraw-Hill. [510]

Welch, B. L. 1939. On confidence limits and sufficiency with particular reference to parameters and location. *Ann. Math. Stat.* 10: 58–69. [377]

Wetherill, G. B. 1966. *Sequential methods in statistics.* London: Methuen. [510]

Wilcoxon, F. 1947. Probability tables for individual comparisons by ranking methods. *Biometrics* 3: 119–22. [340]

Wilk, M. B., and R. Gnanadesikan. 1968. Probability plotting methods for the analysis of data. *Biometrika* 55: 1–17. [219]

Wilk, M. B., and O. Kempthorne. 1955. Fixed, mixed, and random models. *J. Am. Stat. Assoc.* 50: 1144–67. [455]

———. 1956. Some aspects of the analysis of factorial experiments in a completely randomized design. *Ann. Math. Stat.* 27: 950–85. [455]

———. 1957. Non-additivities in a Latin square design. *J. Am. Stat. Assoc.* 52: 218–36. [455]

Wilks, S. S. 1948. Order statistics. *Bull. Am. Math. Soc.* 54: 5–50.

**———. 1962. *Mathematical statistics.* New York: Wiley. [422, 423, 510]

Williams, E. J. 1959. *Regression analysis.* New York: Wiley. [510]

Winer, B. J. 1962. *Statistical principles in experimental design.* New York: McGraw-Hill. [510]

Wise, M. E. 1963. Multinomial probabilities and the $\chi^2$ and $X^2$ distributions. *Biometrika* 50: 145–54. [326]

Wittgenstein, L. 1958 (reprint, Harper Torch Books, 1965). *The blue and brown books.* New York: Harper and Row. [509]

*Wold, Herman. 1954. *A study in the analysis of stationary time series*. Stockholm: Almquist and Wiksell. [511]

Wright, S. 1921. Correlation and causation. *J. Agr. Res.* 20: 557–85. [447]

———. 1934. The method of path coefficients. *Ann. Math. Stat.* 5: 161–215. [447]

———. 1954 (reprint, 1964). The interpretation of multivariate systems. In O. Kempthorne, et al., eds., *Statistics and mathematics in biology*. New York: Hafner. [447]

———. 1968. *Evolution and the genetics of populations*, vol. I. *Genetic and biometric foundations*. Chicago: Univ. Chicago Press. [447]

Yates, F. 1937. The design and analysis of factorial experiments. Harpenden, England: Imperial Bureau of Soil Science. [510]

———. 1953. *Sampling methods for censuses and surveys*, 2nd ed. London: Griffin. [511]

Youden, W. J. 1951. *Statistical methods for chemists*. New York: Wiley. [511]

Yule, G. U. 1900. On the association of attributes in statistics. *Phil. Trans. Roy. Soc.* (A)194: 257–319. [33]

Zyskind, G. 1962. On structure, relation, Σ, and expectation of mean squares. *Sankhyā* 24: 115–48. [455]

———. 1963. Some consequences of randomization in a generalization of the balanced incomplete block design. *Ann. Math. Stat.* 34: 1569–81. [455]

Zyskind, G., O. Kempthorne, R. F. White, E. E. Dayhoff, and T. E. Doerfler. 1964. Research on analysis of variance and related topics. Aeronautical Research Laboratory Tech. Rept. 64–193, Wright-Patterson Air Force Base, Ohio. [455]

# Index